The Chemistry of Contrast Agents in Medical Magnetic Resonance Imaging

The Chemistry of Contrast Agents in Medical Magnetic Resonance Imaging

Edited by

André E. Merbach
Éva Tóth
University of Lausanne, Switzerland

JOHN WILEY & SONS, LTD
Chichester · New York · Weinheim · Brisbane · Singapore · Toronto

National 01243 779777
International (+44) 1243 779777

e-mail (for orders and customer service enquiries): cs-books@wiley.co.uk
Visit our Home Page on http://www.wiley.co.uk or http://www.wiley.com

Other Wiley Editorial Offices

John Wiley & Sons, Inc., 605 Third Avenue,
New York, NY 10158-0012, USA

Wiley-VCH Verlag GmbH
Pappelallee 3, D-69469 Weinheim, Germany

Jacaranda Wiley Ltd, 33 Park Road, Milton,
Queensland 4064, Australia

John Wiley & Sons (Asia) Pte Ltd, 2 Clementi Loop #02–01,
Jin Xing Distripark, Singapore 129809

John Wiley & Sons (Canada) Ltd, 22 Worcester Road,
Rexdale, Ontario, M9W 1L1, Canada

Library of Congress Cataloging-in-Publication Data

The chemistry of contrast agents in medical magnetic resonance imaging / edited by
André Merbach, Éva Tóth.
p. cm.
Includes bibliographical references and index.
ISBN 0-471-60778-9 (alk. paper)
1. Contrast media. 2. Magnetic resonance imaging. I. Merbach, André E. II. Tóth, Éva.
RC78.7.C65 C48 2001
616.07'548—dc21 00-043387

British Library Cataloguing in Publication Data

A catalogue record for this book is available from the British Library

ISBN 0-471-60778-9

Typeset in 10/12pt Times by Kolam Information Services Pvt Ltd, Pondicherry, India.
Printed and bound in Great Britain by Biddles Ltd, Guildford and King's Lynn.
This book is printed on acid-free paper responsibly manufactured from sustainable forestry in which at least two trees are planted for each one used for paper production.

CONTENTS

CONTRIBUTORS

Silvio Aime
Dipartimento di Chimica Inorganica Chimica Fisica e Chimica dei Materiali, Università degli Studi di Torino, Via P. Giuria, 7, 10125 Torino, Italy

Pier Lucio Anelli
Milano Research Center, Bracco s.p.a, Via E. Folli, 50, 20134 Milano, Italy

R. Linn Belford
Department of Chemistry, University of Illinois at Urbana/Champaign, 47 Roger Adams Laboratory, MC 712, 600 s'Mathews, Urbana, IL 61801, USA

Atle Bjørnerud
Nycomed-Amersham Nycoveien 1–2, PO Box 4220, Torshov, 0401 Oslo, Norway

Mauro Botta
Dipartimento di Scienze e Tecnologie Avanzate, Università del Piemonte Orientale 'Amedeo Avogadro', Palazzo Borsalino, Via Cavour 84, 15100 Alessandria, Italy

Ernö Brücher
Department of Inorganic and Analytical Chemistry, Lajos Kossuth University, PO Box 37, 4010 Debrecen, Hungary

James I. Bruce
Department of Chemistry, University of Durham, South Road, Durham DH1 3LE, UK

Robert B. Clarkson
Veterinary Biosciences, Medical Information Science and Bioengineering, University of Illinois, 257 LAC, 1008 W. Hazelwood Drive, Urbana, IL 61801, USA

Jean-Marie Colet
Departement de Chimie Organique, Laboratoire RMN, Université du Mons-Hainaut, Av du Champ de Mars 24, 7000 Mons, Belgium

Daniele M. Corsi
Department of Applied Organic Chemistry and Catalysis, Delft University of Technology, Julianalaan 136, 2628 BL Delft, The Netherlands

Jean-François Desreux
Chimie de Coordination et Radiochimie, Université de Liège, Sart Tilman B16, B-4000 Liège, Belgium

Mauro Fasano
Dipartimento di Chimica Inorganica Chimica Fisica e Chimica dei Materiali, Università degli Studi di Torino, Via P. Giuria, 7, 10125 Torino, Italy

Carlos F. G. C. Geraldes
Department of Biochemistry, Faculty of Science and Technology and Center of NeuroSciences, Universidade of Coimbra, PO BOX 3126, 3000 Coimbra, Portugal

Pierre Gillis
Departement de Chimie Organique, Laboratoire de Biologie Physique, Université du Mons-Hainaut, Av du Champ de Mars 24, 7000 Mons, Belgium

Lothar Helm
Institut de Chimie Minérale et Analytique, BCH, Faculté des Sciences, Université de Lausanne, 1015 Lausanne, Switzerland

Vincent Jacques
Chimie de Coordination et Radiochimie, Université de Liège, Sart Tilman B16, B–4000 Liège, Belgium

Luciano Lattuada
Milano Research Center, Bracco s.p.a, Via E. Folli, 50, 20134 Milano, Italy

Mark P. Lowe
Department of Chemistry, University of Durham, South Road, Durham DH1 3LE, UK

Sven Månsson
Department of Experimental Research, Malmö University Hospital, Södra Förstadsgatan 101, 205 02 Malmö, Sweden

André E. Merbach
Institut de Chimie Minérale et Analytique, BCH, Faculté des Sciences, Université de Lausanne, 1015 Lausanne, Switzerland

Robert N. Müller
Departement de Chimie Organique, Laboratoire RMN, Université du Mons-Hainaut, Av du Champ de Mars 24, 7000 Mons, Belgium

Assisa Ouakssim
Departement de Chimie Organique, Laboratoire RMN, Université du Mons-Hainaut, Av du Champ de Mars 24, 7000 Mons, Belgium

David Parker
Department of Chemistry, University of Durham, South Road, Durham DH1 3LE, UK

Joop A. Peters
Department of Applied Organic Chemistry and Catalysis, Delft University of Technology, Julianalaan 136, 2628 BL Delft, The Netherlands

Johannes Platzek
Research Laboratories of Schering AG, Müllerstrasse 178, 13342 Berlin, Germany

Bernd Radüchel
Research Laboratories of Schering AG, Müllerstrasse 178, 13342 Berlin, Germany

Alain Roch
Departement de Chimie Organique, Laboratoire RMN, Université du Mons-Hainaut, Av du Champ de Mars 24, 7000 Mons, Belgium

Heribert Schmitt-Willich
Research Laboratories of Schering AG, Müllerstrasse 178, 13342 Berlin, Germany

A. Dean Sherry
Department of Chemistry, University of Texas at Dallas, PO Box 830688, Richardson, TX 75083-0688, USA

Alexej I. Smirnov
Department of Medical Information Science, University of Illinois, 61a MSB, MC 714, 506 s'Mathews, 257 LAC, 1008 W. Hazelwood Drive, Urbana, IL 61801, USA

Tatyana Ivanovna Smirnova
Department of Veterinary Clinical Medicine, University of Illinois, 367 Noyes Laboratory, MC 712, 600 s'Mathews, 257 LAC, 1008 W. Hazelwood Drive, Urbana, IL 61801, USA

Detlev Sülzle
Research Laboratories of Schering AG, Müllerstrasse 178, 13342 Berlin, Germany

Enzo Terreno
Dipartimento di Chimica Inorganica Chimica Fisica e Chimica dei Materiali, Università degli Studi di Torino, Via P. Giuria, 7, 10125 Torino, Italy

Éva Tóth
Institut de Chimie Minérale et Analytique, BCH, Faculté des Sciences, Université de Lausanne, 1015 Lausanne, Switzerland

Emrin Zitha-Bovens
Department of Applied Organic Chemistry and Catalysis, Delft University of Technology, Julianalaan 136, 2628 BL Delft, The Netherlands

Preface

Over the last decade, Magnetic Resonance Imaging (MRI) has evolved into one of the most powerful techniques in diagnostic clinical medicine and biomedical research. MRI is primarily used to produce anatomical images, but it also gives information on the physico-chemical state of tissues, flow diffusion and motion. The strong expansion of medical MRI has prompted the development of a new class of pharmacological products, called contrast agents, which are designed for administration to patients in order either to enhance the contrast between normal and diseased tissue or to indicate organ function or blood flow. Nowadays, around 30 % of all MRI investigations use a contrast medium, the number of which will certainly further increase with the development of new agents and applications.

The chemistry and design of new contrast agents is a very active domain of research, both in universities and industry. This interest gave birth to the idea of preparing a monograph–the first in the field–that would cover the whole spectrum of chemical aspects of MRI contrast agents, from the basic science through to more specialized topics. With this perspective, this present book intends to offer the basic principles of the functioning of contrast agents, as well as 'state-of-the-art' reviews on recent developments. Moreover, the reader will also find a compilation of relevant physico-chemical data on MRI contrast agents.

The first chapter discusses the physical principles of medical magnetic resonance imaging by describing the nuclear magnetic resonance NMR phenomenon and its relaxation processes, as well as the imaging hardware and image generation. This is followed by the theory and mechanism of proton relaxivity for gadolinium-based contrast agents, which, in addition to one approved manganese chelate material and a few paramagnetic particulate agents, represent the dominant part of the contrast agent market. The design of more powerful contrast agents requires not only the understanding of nuclear relaxation principles in the presence of unpaired electrons but also the exploration of the relationship between molecular structure and each of the factors which determine relaxivity. Certainly, contrast agent development is not possible without efficient synthetic methods. Two of the contributions deal with ligand synthesis, representing the two main streams (linear and macrocyclic ligands) in the chelation of gadolinium. An entire chapter is devoted to protein-bound chelates, which is a rapidly evolving and promising field of contrast agent research. Besides the goal of pursuing the highest relaxivities, the targeting procedures realized through non-covalent metal chelate–protein interactions may lead to a number of applications in different areas of clinical diagnosis by designing systems whose relaxation enhancement is associated with the recognition of a specific protein. The basic requirement of safety in the case of gadolinium chelates is inherently related to their thermodynamic and kinetic stabilities. These aspects are also addressed in detail. Computational studies on Gd(III)-based contrast agents are often neglected from the panoply of possible methods used to assess their physico-chemical properties. We wanted to fill this gap by devoting a chapter to this topic. Each of the parameters governing relaxivity, and thus the efficacy of contrast agents, is related to their solution structure and/or dynamics. These aspects, indispensable for further developments in the field, are also discussed in detail. Probably the least understood part of the relaxation theory applying to Gd(III) chelates is electron spin relaxation. This provided a good reason to devote an entire chapter to the new possibilities in electron paramagnetic resonance (EPR), such as the technique of multifrequency and high-frequency EPR, which can provide invaluable information in this respect. One further contribution deals with the classification, synthesis and characterization of particulate MRI contrast agents. Although only indirectly related to MRI contrast agents, the last chapter is devoted to photophysical aspects of lanthanide(III) complexes. The luminescent ions, Eu^{3+} and Tb^{3+}, flank Gd^{3+} in the lanthanide series and thus studies of their photophysical properties can throw light on the solution behaviour of the corresponding Gd species.

The initiative to prepare this monograph was launched in January 1999. In a short period of time, a distinguished team of experts, from both the academic and industrial sectors, had accepted the invitation to contribute to this project. Most of them have already been actively collaborating in the frame of the successive COST (European Co-operation in the Field of Scientific and Technical Research) D1, D8, and more recently, D18 actions. These collaborations, as well as the annual COST meetings, have contributed to a large extent to the development of our knowledge in the field of MRI contrast agents. We should also mention the fruitful co-operations between various academic and industrial groups, which is nicely manifested in the participation of several industrial contributors to the realization of this book.

André E. Merbach
Éva Tóth
Lausanne, April 2000

1 Physical Principles of Medical Imaging by Nuclear Magnetic Resonance

SVEN MÅNSSON
Malmö University Hospital, Malmö, Sweden

and

ATLE BJØRNERUD
Nycomed-Amersham, Oslo, Norway

1 THE NMR PHENOMENON

1.1 NUCLEAR SPIN AND MAGNETIC MOMENT

It has been known for more than 50 years that atomic nuclei, which possess a spin angular momentum, will interact with magnetic fields. The discovery of this interaction has led to the development of Magnetic Resonance Imaging (MRI), one of the most powerful diagnostic imaging techniques available today.

Many atomic nuclei have a spin angular momentum. The spin of the nucleus is composed of the spins of the individual protons and neutrons in the nucleus. Nuclear spin is, in a strict sense, a purely quantum mechanical quantity, but in terms of classical physics it may be visualized as an angular momentum due to the rotation of the nucleus around an axis through its center. The nuclear spin is a vector which is orientated parallel to the 'axis of rotation', with its magnitude being given by the following:

$$|\mathbf{I}| = \hbar\sqrt{I(I+1)}, \ \hbar = h/2\pi \tag{1.1}$$

where h is the Planck constant and I is the spin quantum number[1]. Nuclei with an even mass number (the sum of the number of protons and neutrons) will have their spin quantum numbers as even multiples of $1/2$ ($I = 0, 1, \cdots$), while nuclei with an odd mass number will have spin quantum numbers as

[1] Vectors and matrices will be denoted by a bold typeface, while vector magnitudes and scalars will be denoted by a regular typeface.

The Chemistry of Contrast Agents in Medical Magnetic Resonance Imaging
Edited by A. E. Merbach and É. Tóth.

odd multiples of 1/2 ($I = 1/2, 3/2, \ldots$). Nuclei with a spin quantum number of $I = 0$ can not be studied by using NMR (these nuclei are insensitive to magnetic fields). To this latter group belong all even–even nuclei (nuclei with an even number of protons and an even number of neutrons) and most of the odd–odd nuclei. The most commonly studied nuclei in NMR experiments are those nuclei with a spin quantum number $I = 1/2$, e.g. ^{1}H, ^{19}F, ^{13}C or ^{31}P.

When placed in an external magnetic field, nuclei with a spin angular momentum can be pictured as microscopic compass needles, orientating themselves along the external magnetic field. As distinguished from a compass needle, the nuclei can not, for quantum mechanical reasons, be orientated in an arbitrary direction relative to the magnetic field. Only discrete directions are allowed and each direction corresponds to an energy level of the nucleus. If a nucleus with spin quantum number I is placed in a static magnetic field $\boldsymbol{B}_0$, the projection of the spin vector on the $\boldsymbol{B}_0$-vector will be quantified, so that the length of the projection only can be $m_I\hbar$, with the magnetic quantum number $m_I = -I, -I+1, \ldots, I$ (i.e. a total of $2I + 1$ possible orientations, see Figure 1.1).

Again following a classical argument, the positively charged protons in a nucleus with spin angular momentum constitute a ring current with negligible dimension, which in turn gives rise to a dipolar magnetic moment $\boldsymbol{\mu}$. This magnetic moment $\boldsymbol{\mu}$ is, like the nuclear spin $\boldsymbol{I}$, a vector aligned parallel to the 'rotation axis' of the nucleus and thus parallel to $\boldsymbol{I}$. The magnitude of the magnetic moment does not, in contrast to the nuclear spin, follow directly from quantum mechanics. However, the spin of any isotope is proportional to its magnetic moment, according to the following relationship:

$$\boldsymbol{\mu} = \gamma \boldsymbol{I} \tag{1.2}$$

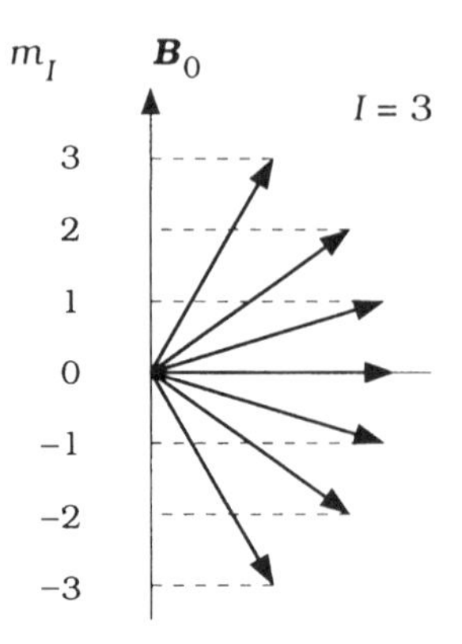

Figure 1.1 The possible orientations for the spin vector of a nucleus with spin quantum number $I = 3$, in the presence of an external magnetic field $\boldsymbol{B}_0$.

where the constant γ is the gyromagnetic ratio, characteristic for each spinning isotope. The $\boldsymbol{B}_0$-field is by convention orientated along the z-axis in an orthogonal coordinate system. Due to the relationship between the nuclear spin and the magnetic moment (Equation (1.2)), the z-component of the magnetic moment (i.e. the projection of $\boldsymbol{\mu}$ on the $\boldsymbol{B}_0$-field), μ_z, is restricted to the discrete values $\mu_z = \gamma m_I \hbar$, with $m_I = -I, -I+1, \ldots, I$.

A magnetic moment $\boldsymbol{\mu}$, placed in a magnetic field $\boldsymbol{B}_0$, represents a magnetic energy E, given by the following:

$$E = -\boldsymbol{\mu} \cdot \boldsymbol{B}_0 \tag{1.3}$$

The energy states of the nucleus are accordingly given by m_I as follows:

$$E_{mI} = -m_I \gamma \hbar B_0 \tag{1.4}$$

which can be exemplified with the energy-term scheme for a nucleus with a spin quantum number of $I = 1/2$ and $I = 1$, respectively (see Figure 1.2).

The energy difference, ΔE, between two neighboring m-states is, for a given nucleus, always given as follows:

$$\Delta E = \gamma \hbar B_0 \tag{1.5}$$

and is thus proportional to the magnetic field strength B_0. A nucleus is allowed to change its energy state, but the transition may only occur to a neighboring state, since the magnetic quantum number m_I can only be changed by $+1$ or -1. In order to move to a higher energy level ('one m-step up'), the nucleus has to be supplied with the necessary energy difference ΔE in form of electromagnetic radiation. Vice versa, in order to move to a lower energy state, the nucleus has to emit the energy difference ΔE, also in the form of electromagnetic radiation. Both of these energy transitions must obey the following relationship:

$$\Delta E = h\nu_0 \tag{1.6}$$

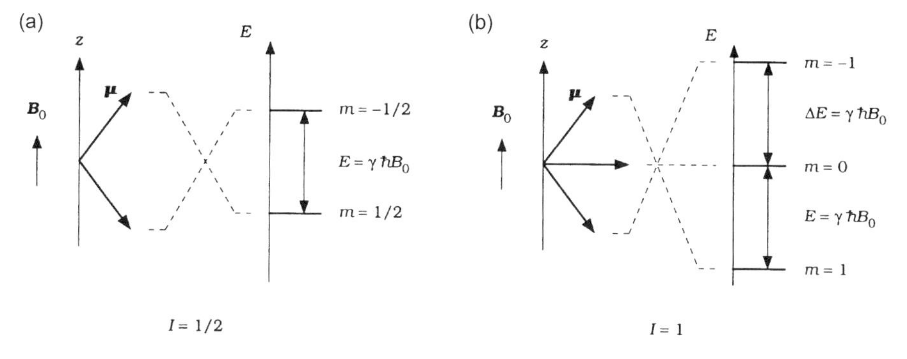

Figure 1.2 Energy-term schemes for nuclei with spin quantum numbers of (a) $I = 1/2$ and (b) $I = 1$.

for the frequency ν_0 of the electromagnetic radiation. Combination of Equations (1.5) and (1.6) gives the relationship

$$h\nu_0 = \gamma\hbar B_0 \tag{1.7}$$

and thus the angular frequency, $\omega_0 = 2\pi\nu_0$, for the electromagnetic radiation at an energy transition:

$$\omega_0 = \gamma B_0 \tag{1.8}$$

Due to the coupling between the magnetic moment $\boldsymbol{\mu}$ and the nuclear spin $\boldsymbol{I}$, $\boldsymbol{\mu}$ will always be orientated at an angle to $\boldsymbol{B}_0$, and will thus experience a torque trying to turn $\boldsymbol{\mu}$ parallel to $\boldsymbol{B}_0$. Due to this coupling, the nuclear spin $\boldsymbol{I}$ will also experience the same torque. The mechanical torque $\boldsymbol{F}$ acting on $\boldsymbol{I}$ is given by the following:

$$\boldsymbol{F} = \boldsymbol{\mu} \times \boldsymbol{B}_0 \tag{1.9}$$

According to the laws of motion for an angular momentum vector, we have:

$$\frac{d\boldsymbol{I}}{dt} = \boldsymbol{F} \tag{1.10}$$

If Equations (1.2), (1.9) and (1.10) are combined, the motion of the magnetic moment $\boldsymbol{\mu}$ can be described by the equation:

$$\frac{d\boldsymbol{\mu}}{dt} = \gamma\boldsymbol{\mu} \times \boldsymbol{B}_0 \tag{1.11}$$

showing that the vector $\boldsymbol{\mu}$ moves perpendicularly to both $\boldsymbol{B}_0$ and $\boldsymbol{\mu}$, i.e. $\boldsymbol{\mu}$ revolves around $\boldsymbol{B}_0$. It can easily be verified that the angular velocity, $\boldsymbol{\omega}_L$, of its motion will be given by:

$$\boldsymbol{\omega}_L = -\gamma\boldsymbol{B}_0 \tag{1.12}$$

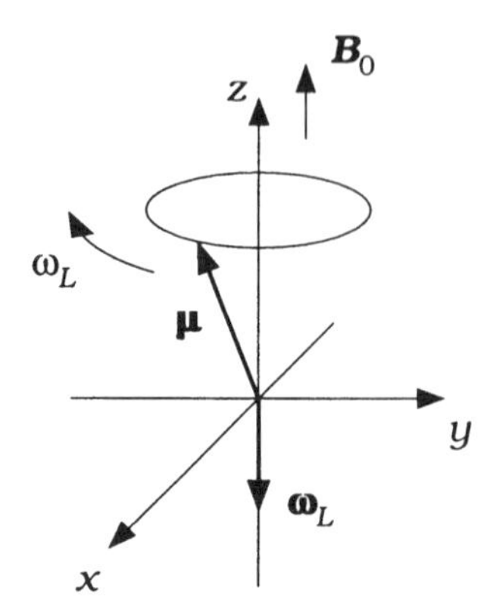

Figure 1.3 The magnetic moment $\boldsymbol{\mu}$ rotates around the $\boldsymbol{B}_0$-field rotates with Larmor frequency $\boldsymbol{\omega}_L$.

as shown in Figure 1.3. The angular velocity ω_L is called the 'Larmor frequency'. We now see that the Larmor frequency is identical with the angular velocity ω_0 required to obtain a transition between two energy states of the nucleus. For this reason, ω_0 is thus also called 'Larmor frequency'.

1.2 THE MACROSCOPIC MAGNETIZATION

After this excursion into quantum mechanics and a microscopic study of the behavior of a single nucleus, we will now consider NMR from a macroscopic point of view. To be able to detect the electromagnetic radiation emitted at the transition of the nucleus from a higher to a lower energy state, of course more than one single nucleus is necessary. Within the limits of technically achievable magnetic field strengths (these limits are about 3 T for medical NMR systems and up to 19 T for high-resolution NMR spectrometers), the NMR signal is weak when compared to many other spectroscopic methods. However, due to the fact that the number of atoms is very large in a normal NMR sample ($\approx 10^{22}/\text{cm}^3$), it is still possible to obtain a detectable signal. For a population of nuclei with a spin quantum number I, and thus $2I+1$ energy states, the distribution of the spins in each state is, in thermal equilibrium, governed by the Boltzmann law, as follows:

$$N_m = N_0 \frac{\exp\left(-E_m/(k_B T)\right)}{\sum_{n=-I}^{n=I} \exp\left(-E_n/(k_B T)\right)} \tag{1.13}$$

where N_m is the number of spins in the state m, N_0 is the total number of spins, E_m is the energy of the state m, T is the absolute temperature and k_B is the Boltzmann constant.

The number of spins in each energy state decreases with increasing energy. In a macroscopic sample, containing a very large number of spins, a macroscopic magnetization, $\boldsymbol{M}$, will be obtained as the vector sum of all of the individual microscopic magnetic moments $\boldsymbol{\mu}$:

$$\boldsymbol{M} = \sum \boldsymbol{\mu} \tag{1.14}$$

In an unexcited sample, the precession of an individual magnetic moment around the $\boldsymbol{B}_0$-field will not be coherent with the other spins. At a given moment, the components in the xy-plane of the individual spins $\boldsymbol{\mu}$ will thus be randomly distributed, so causing the x- and y-components of $\boldsymbol{M}$ to equal zero, i.e. $M_x = 0$, and $M_y = 0$; M_z, however, will not be zero. As we have seen before, the z-component of an individual spin, μ_z, is restricted to the discrete values $\mu_z = \gamma m_I \hbar$, with $m_I = -I, -I+1, \ldots I$. If we calculate the number of spins in each m-state with the help of Equations (1.4) and (1.13) and then sum the contributions to M_z from each m-state, we will have the following expression for M_z:

$$M_z = \frac{N_0\gamma^2\hbar^2 I(I+1)}{3k_B T}B_0 \tag{1.15}$$

where the approximation $e^x \approx 1 + x$ for small values of x has been used. Equation (1.15) is known as Curie's Law, and shows that the macroscopic magnetization is directly proportional to the magnetic field strength B_0. Purely from this point of view, it is thus desirable to create as strong values of the magnetic fields as is technically possible. The relationship also shows why the proton (^1H) is the most observable of naturally occurring nuclei, as it has the highest γ and is present in high concentrations in most tissues.

For nuclei with a spin quantum number $I = 1/2$, the spins are distributed in only two states, namely spin 'up' (low energy) and spin 'down' (high energy). The two populations, N_{up} and N_{down}, are according to Equations (1.4) and (1.13) related by the following:

$$\frac{N_{up}}{N_{down}} = \exp\left(\gamma\hbar B_0/(k_B T)\right) \tag{1.16}$$

For example, for a magnetic field strength of $B_0 = 2.35$ T, at room temperature (300 K), the ratio N_{up}/N_{down}, will for protons ($\gamma = 2.68 \times 10^8\ \mathrm{s^{-1}\,T^{-1}}$) be $1 + 1.6 \times 10^{-5}$, showing that the difference in population size, which give rise to the macroscopic magnetization $\boldsymbol{M}$, is very small compared to the total number of nuclei.

1.3 MACROSCOPIC DESCRIPTION OF THE EXCITATION PROCESS

Assume that a coil is placed close to the sample with its long axis parallel to the x-axis. The coil is fed with a sinusoidal current with angular frequency Ω. The current through the coil generates an oscillating magnetic field $\boldsymbol{B}_1$ along the x-axis (see Figure 1.4).

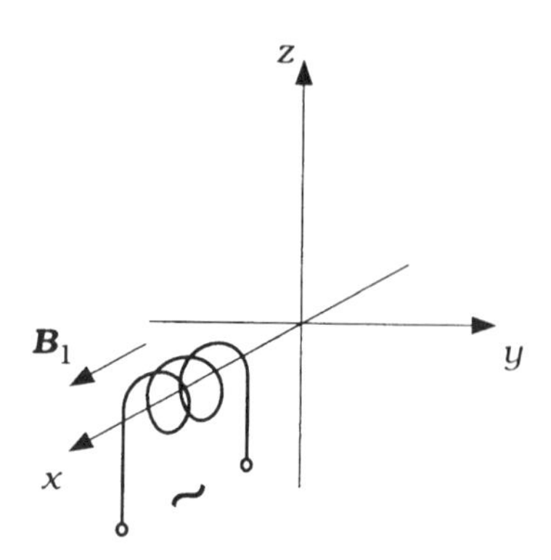

Figure 1.4 The RF coil generates a magnetic field $\boldsymbol{B}_1$ along the x-axis.

The linearly oscillating $\boldsymbol{B}_1$-vector can be rewritten as the sum of two rotating vectors, $\boldsymbol{B}_{1\_cw}$ and $\boldsymbol{B}_{1\_ccw}$, rotating in opposite directions, with the first rotating clockwise with angular frequency $-\Omega$, and the other rotating counter-clockwise with the angular frequency $+\Omega$:

$$\boldsymbol{B}_1 = \boldsymbol{B}_{1\_cw} + \boldsymbol{B}_{1\_ccw} = \begin{bmatrix} B_1 \cos(-\Omega t) \\ B_1 \sin(-\Omega t) \\ 0 \end{bmatrix} + \begin{bmatrix} B_1 \cos(\Omega t) \\ B_1 \sin(\Omega t) \\ 0 \end{bmatrix} = \begin{bmatrix} 2B_1 \cos(\Omega t) \\ 0 \\ 0 \end{bmatrix} \tag{1.17}$$

To describe the motion of the macroscopic magnetization $\boldsymbol{M}$, it is helpful to introduce a new Cartesian coordinate system (x', y', z') rotating around the z-axis of the fixed (x, y, z) system with the angular velocity $-\boldsymbol{\Omega}$, i.e. a coordinate system which follows the rotating $\boldsymbol{B}_{1\_cw}$-vector (see Figure 1.5).

These two coordinate systems are called the 'laboratory frame' (x, y, z) and the 'rotating frame' (x', y', z'), respectively. When viewed from the rotating frame, $\boldsymbol{B}_{1\_cw}$ will appear static, whereas $\boldsymbol{B}_{1\_ccw}$ will rotate with an angular velocity of $2\boldsymbol{\Omega}$.

The dynamics of $\boldsymbol{M}$ in the presence of both the static $\boldsymbol{B}_0$–field and the $\boldsymbol{B}_1$–field is obtained from Equation (1.11) and summation of the individual magnetic moments, as follows:

$$\frac{d\boldsymbol{M}}{dt} = \gamma \boldsymbol{M} \times (\boldsymbol{B}_0 + \boldsymbol{B}_1) \tag{1.18}$$

Let $d\boldsymbol{M}/dt$ denote the time derivative of $\boldsymbol{M}$ when viewed from the laboratory frame, and $\partial\boldsymbol{M}/\partial t$ the time derivative of $\boldsymbol{M}$ when viewed from the rotating frame. The time derivatives are related by the following:

$$\frac{d\boldsymbol{M}}{dt} = \frac{\partial\boldsymbol{M}}{\partial t} + (\boldsymbol{\Omega} \times \boldsymbol{M}) \tag{1.19}$$

The motion of $\boldsymbol{M}$ when viewed from the rotating frame is thus given by:

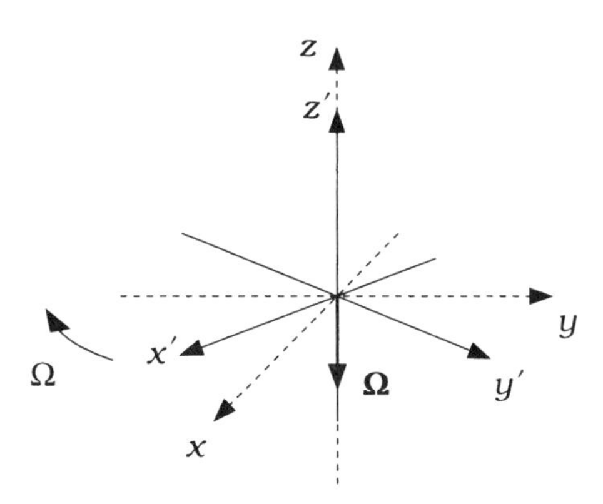

Figure 1.5 The 'rotating frame' (x', y', z'-coordinates).

$$\frac{\partial \boldsymbol{M}}{\partial t} = \frac{d\boldsymbol{M}}{dt} - (\boldsymbol{\Omega} \times \boldsymbol{M})$$
$$= \gamma \boldsymbol{M} \times (\boldsymbol{B}_0 + \boldsymbol{B}_1) + (\boldsymbol{M} \times \boldsymbol{\Omega}) \tag{1.20}$$
$$= \gamma \boldsymbol{M} \times \left(\boldsymbol{B}_0 + \boldsymbol{B}_1 + \frac{\boldsymbol{\Omega}}{\gamma} \right)$$

where:

$$\boldsymbol{\Omega} = \begin{bmatrix} 0 \\ 0 \\ -\Omega \end{bmatrix} \tag{1.21}$$

The end result is thus summarized in the following equation:

$$\frac{\partial \boldsymbol{M}}{\partial t} = \gamma \boldsymbol{M} \times \boldsymbol{B}_{eff} \tag{1.22}$$

where $\boldsymbol{B}_{eff}$ is the 'effective field' given by $\boldsymbol{B}_0 + \boldsymbol{B}_1 + \boldsymbol{\Omega}/\gamma$. Note that the vectors $\boldsymbol{\Omega}$ and $\boldsymbol{B}_0$ have opposite directions. If $\Omega = \gamma B_0$ (i.e. the frequency of the RF field equals the Larmor frequency), the effective field consists only of the $\boldsymbol{B}_1$-field, which in turn is composed of the static $\boldsymbol{B}_{1\_cw}$-field and the rotating $\boldsymbol{B}_{1\_ccw}$-field. Since the $\boldsymbol{B}_{1\_ccw}$-component moves rapidly with the angular frequency of 2Ω relative to $\boldsymbol{M}$, the torque caused by this component will rapidly change direction as well, and thus have no net effect when averaged in time. The result of applying a RF field to the sample is then a precession of $\boldsymbol{M}$ around the $\boldsymbol{B}_{1\_cw}$-component with the angular velocity as follows:

$$\boldsymbol{\omega}_1 = -\gamma \boldsymbol{B}_1 \tag{1.23}$$

If the $\boldsymbol{B}_1$-field is turned on during the time t_p, $\boldsymbol{M}$ will rotate an angle $\beta_1 = \omega_1 t_p$ from its original position down towards the $x'y'$-plane (see Figure 1.6). The angle β_1 is usually called the 'flip angle' of the RF pulse.

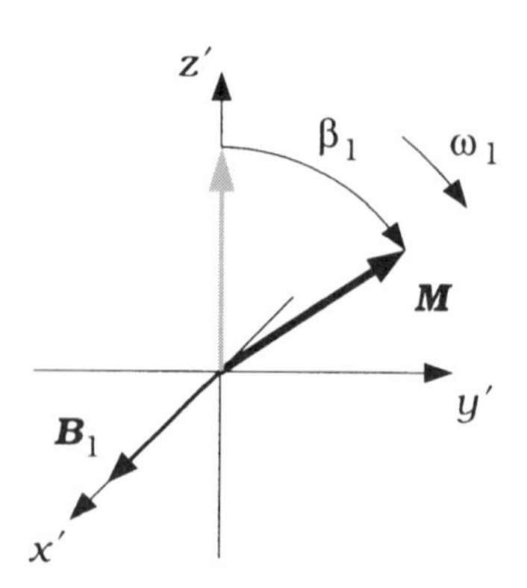

Figure 1.6 *M* is turned by an angle β_1 during the RF-excitation.

After the RF field is turned off, $\boldsymbol{M}$ will thus have a composant in the xy-plane, $\boldsymbol{M}_{xy}$, which is greater than zero. Viewed from the laboratory frame, this component will continue to rotate around $\boldsymbol{B}_0$ according to Equation (1.18). If a sensitive detector coil is placed close to the sample, the rotating magnetization $\boldsymbol{M}_{xy}$ will induce a detectable current in the coil. $\boldsymbol{M}_{xy}$ is the only observable component of $\boldsymbol{M}$ (M_z is not observable because it does not rotate). Usually, the same coil is used both for excitation and detection (but not simultaneously). For further reading about the physical principles of MRI, see references [1–3].

2 RELAXATION PROCESSES

In the previous section we have seen how an RF pulse can turn the macroscopic magnetization away from the z-axis. The magnetization will then, after the $\boldsymbol{B}_1$-field has been turned off, rotate around the $\boldsymbol{B}_0$–field at the Larmor frequency. This far, we have assumed that the magnetization will continue this rotation forever. However, due to relaxation processes, the magnetization will eventually return to its equilibrium position along the z-axis, and the NMR signal will thus vanish.

2.1 SPIN-LATTICE RELAXATION

The mechanism responsible for the returning of the M_z-component to its equilibrium value is known as spin-lattice relaxation or T_1-relaxation, because it is characterized by a time constant T_1. A spin in a high-energy state can make a transition to the low-energy state either via spontaneous emission or stimulated emission. However, the probability for spontaneous emission depends on the third power of the frequency for the energy transition and at radiofrequencies this term is far too small to be significant. Thus, all NMR transitions are stimulated.

In order to undergo a transition, the spin needs stimulation in form of a fluctuating magnetic field at the Larmor frequency. Such a fluctuating magnetic field is formed by the random motions of the molecules in the surrounding medium. As the molecules move around, they carry the nuclear magnetic moments with them and the motion of all magnetic moments generates a magnetic 'noise' at the site of any nucleus. The broad frequency range of this noise will cover the Larmor frequency of the nuclei and stimulate a transition back to the lower energy level. Several processes, which are commonly referred to as T_1–processes, can generate magnetic noise, namely magnetic dipole–dipole interaction, electric quadrupole interaction ($I > 1/2$ nuclei only), chemical shift anisotropy interaction, scalar coupling interaction, and spin-rotation interaction. In general, any mechanism which gives rise to fluctuating magnetic fields is a possible relaxation mechanism. All mechanisms contribute to the observable relaxation time according to the following:

$$\frac{1}{T_1} = \sum_m \frac{1}{T_{1m}} \tag{1.24}$$

where T_{1m} is the time constant of mechanism m. The amount of contribution to the overall relaxation rate from a single mechanism may differ greatly among different chemical groups, and in addition to the chemical environment, also depends on parameters such as temperature and field strength.

2.2 SPIN–SPIN RELAXATION

The magnetization present in the xy-plane after an RF excitation, $\boldsymbol{M}_{xy}$, will decay towards zero due to a process known as spin–spin relaxation or T_2-relaxation, where T_2 is the time constant of this process. Assume an excitation by a 90° RF pulse. Immediately after the pulse, the individual spins rotate coherently around the $\boldsymbol{B}_0$–field at the Larmor frequency, thus creating a macroscopic magnetization in the xy-plane, $\boldsymbol{M}_{xy}$. The fluctuating magnetic field caused by the random motion of the molecules will produce random fluctuations of the angular velocity of the individual spins, thus resulting in a loss of phase coherence. The fluctuating magnetic field will also induce 'flip-flop' transitions, where neighboring spins change energy states. After such a transition, the coherence between the spins is lost. As the total phase dispersion of the spins increases, $\boldsymbol{M}_{xy}$ will decrease to zero. The loss of phase coherence is commonly referred to as 'dephasing' of the spins.

The spins will also dephase if there are bulk inhomogeneities in the $\boldsymbol{B}_0$-field. If the field within one volume element is changed by an amount $\Delta\boldsymbol{B}_0$ due to inhomogeneities, the spins in this volume element will rotate at a different Larmor frequency, as compared to the spins in an adjacent volume element. The dephasing caused by $\Delta\boldsymbol{B}_0$ does not reflect any intrinsic properties of the sample, and these inhomogeneities should therefore be kept as low as possible.

The actual decay rate of the observable NMR signal is characterized by a time constant called T_2^*, rather than T_2. This consists of a term related to the intrinsic T_2-relaxation rate caused by the interactions between spins, and of a term related to the bulk inhomogeneities $\Delta\boldsymbol{B}_0$, as follows:

$$\frac{1}{T_2^*} = \frac{1}{T_2} + \gamma\Delta B_0 \tag{1.25}$$

For imaging experiments, the $\Delta\boldsymbol{B}_0$ term should be interpreted as the inhomogeneity within a single voxel of the image.

2.3 THE BLOCH EQUATION

The relaxation processes acting on the magnetization $\boldsymbol{M}$ can with good accuracy be described by the following three differential equations, as proposed by Felix Bloch in 1946:

$$\frac{dM_x}{dt} = -\frac{M_x}{T_2}, \quad \frac{dM_y}{dt} = -\frac{M_y}{T_2}, \quad \frac{dM_z}{dt} = -\frac{M_z - M_0}{T_1} \tag{1.26}$$

where M_0 is the equilibrium magnetization along the z-axis, given by Equation (1.15). These differential equations are valid both in the laboratory and the rotating frames. The rate of change of $\boldsymbol{M}$ due to relaxation can be superimposed on the motion given by Equation (1.22). Using matrix notation, this leads to the equation:

$$\frac{d\boldsymbol{M}}{dt} = \gamma \boldsymbol{M} \times \boldsymbol{B}_{eff} - \boldsymbol{R}(\boldsymbol{M} - \boldsymbol{M}_0) \tag{1.27}$$

where:

$$\boldsymbol{R} = \begin{bmatrix} \frac{1}{T_2} & 0 & 0 \\ 0 & \frac{1}{T_2} & 0 \\ 0 & 0 & \frac{1}{T_1} \end{bmatrix} \text{ and } \boldsymbol{M_0} = \begin{bmatrix} 0 \\ 0 \\ M_0 \end{bmatrix} \tag{1.28}$$

Equation (1.27) is known as the Bloch equation (in the rotating frame).
Some general comments can be made about the time constants T_1 and T_2:

- T_1 is longer in solids than in liquids, because the greater mobility of the molecules in the latter causes more 'magnetic noise', which is responsible for the T_1-relaxation.
- T_2 is longer in liquids, because the local magnetic field variations tend to average out in mobile media.
- T_1 is always longer than T_2; the differences are very large in solids ($T_1 \sim$ h, $T_2 \sim$ ms), whereas T_1 and T_2 are similar in pure water.
- In biological tissues, T_1 increases with the increasing magnetic field strength B_0, whereas T_2 is relatively independent of B_0.

For further reading on the above, see reference [3].

2.4 RELAXATION AND IMAGE CONTRAST

Following an RF excitation (given that $B_{eff} = 0$ after the RF pulse), the solutions to Equation (1.27) are given by the following.

$$M_z(t) = M_0[1 - \exp(-t/T_1)] + M_z(0)\exp(-t/T_1) \tag{1.29}$$

and:

$$\boldsymbol{M}_{xy}(t) = \boldsymbol{M}_{xy}(0)\exp\left(-t/T_2\right) \tag{1.30}$$

Equations (1.29) and (1.30) form an important basis for all calculations of signal behavior in MRI. From the above, the influence of the time constant T_1 on the NMR signal becomes evident. Assume an NMR experiment where a train of RF pulses are applied to the sample. After the first RF pulse, the full equilibrium magnetization (or a part of it, depending on the flip angle) is turned into the xy-plane. If the time between successive RF pulses is short compared to T_1, the magnetization component along the z-axis does not reach its equilibrium value by the time the next RF pulse is applied. The magnitude of the magnetization turned down in the xy-plane by the next pulse will then be smaller, thus decreasing the amplitude of the signal. A steady-state situation is eventually established where the amount of magnetization brought into the xy-plane by a given RF pulse is constant. The time taken to reach such an equilibrium depends on the flip angle of the RF pulse, but is $\approx 5T_1$ (shorter if the flip angle of the pulses is high).

If we define the parameter TE ('echo time') to be the time interval between the RF pulse and the detection of the NMR signal, and the parameter TR ('repetition time') to be the time between two successive RF pulses, a general expression for the steady-state value of the NMR signal, S, (which is proportional to the transverse magnetization) can be given as follows:

$$S \propto \frac{\sin(\alpha)[1-\exp(-TR/T_1)]}{1-\cos(\alpha)\exp(-TR/T_1)}\exp\left(-TE/T_2'\right) \tag{1.31}$$

where α is the flip angle of the RF pulse; T_2' can be either the intrinsic T_2 of the tissue, or the combination of T_2 and the relaxation due to bulk inhomogeneities (referred to as T_2^*). As will be shown in Chapter 5, the acquisition method can be made sensitive to either T_2^* or just T_2, depending on the way in which the magnetization is manipulated prior to signal detection. Certain assumptions have been made in deriving this expression:

(A) All of the transverse magnetization is lost in the TR interval; this will happen if $T_2 \ll TR$. If this is not the case, the effect of transverse magnetization can actively be neutralized in a process known as 'spoiling'. Equation (1.31) is therefore often referred to as the signal equation for a 'spoiled gradient echo sequence'.

(B) The image must be acquired under steady-state conditions. This means that no significant transient effects occur at the beginning of the RF pulse train. Several 'dummy' RF pulses are often applied prior to the real data acquisition in order to avoid transient signal behavior which tends to create artifacts in the image.

3 IMAGING HARDWARE

Up until this point, a theoretical description of NMR has been given which includes the following:

- the behaviour of atomic nuclei in the $\boldsymbol{B}_0$-field;
- the formation of the macroscopic magnetization $\boldsymbol{M}$;
- the excitation of the nuclear spin system with an RF field $\boldsymbol{B}_1$.

In the MRI scanner, the macroscopic magnetization within an object placed in the static magnetic field is detected and processed to form an image. In order to achieve this, several pieces of hardware are utilized, namely RF coils and RF electronics, magnetic gradients and signal processing devices. The basic concepts for the hardware of a MRI scanner will be outlined in this section.

3.1 THE MAGNET

The most essential part of any NMR system is the magnet generating the B_0-field, at least when considering the cost; the cost of the magnet is probably one half to three quarters of the total cost of the NMR system.

Permanent magnets have no running costs, but are rarely used today in medical scanners. Resistive magnets are found in 'mid-range' systems. These consist of an electrical coil wound around an iron core, with the patient being placed in an air gap of the core. The disadvantages are low field strengths (0.1–0.3 T), heavy weight, high power consumption and poor field stability. The magnet power supply must be controlled in a closed loop, by monitoring the NMR resonance frequency of a fixed sample, in order to achieve sufficient field stability. The major advantage is the lower manufacturing cost compared to superconducting magnets.

Superconducting magnets, where the magnetic field is produced by an electric current flowing in a closed loop through a superconducting coil, are used in 'high-end' systems with field strengths above 0.5 T. The higher field strength gives a superior signal-to-noise ratio when compared to systems with resistive magnets, and the superconducting magnets are also superior with respect to field stability and homogeneity. However, the purchasing and running costs are high, with the latter being due to the need for cooling cryogenic materials (liquid helium and nitrogen). Modern superconducting magnets use a helium refrigerator which eliminates the need for liquid nitrogen.

3.2 RF ELECTRONICS

In Figure 1.7, the RF electronics of the NMR scanner are shown. The frequency synthesizer, which maintains frequency stability by a reference signal

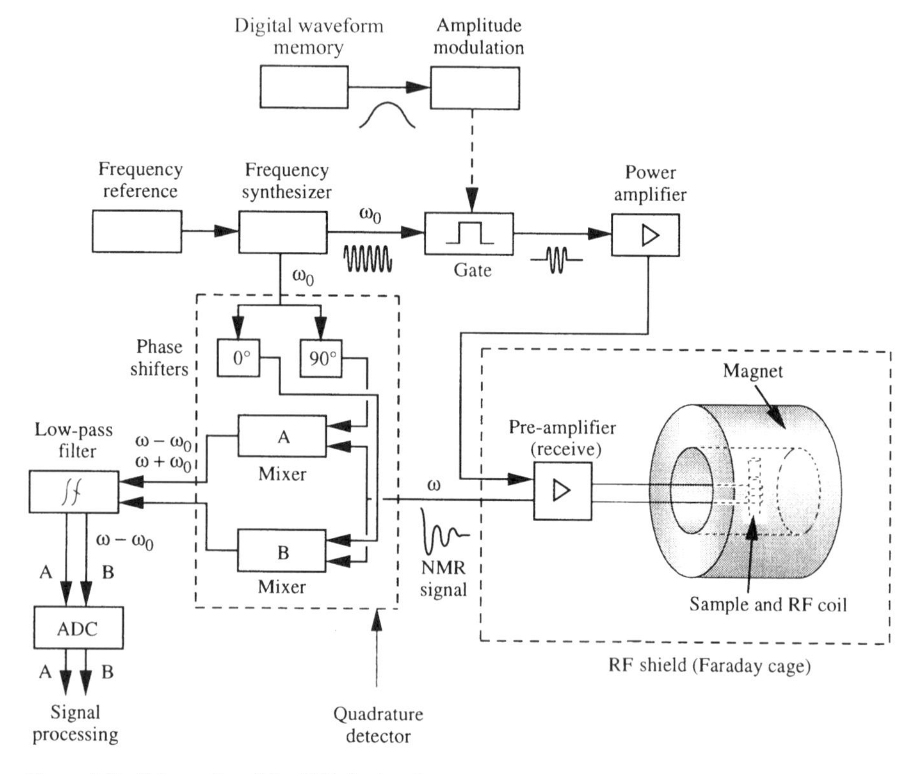

Figure 1.7 Schematic of the RF electronics.

from a quartz crystal, generates a sinusoidal carrier signal with an extremely well defined angular frequency ω_0. The carrier signal is gated to form a short pulse (usually with duration of a few ms or less), which is amplified and fed to the transmitter coil, to produce an excitation of the sample, as explained in the previous sections. A (short) RF pulse without any amplitude modulation is denoted as a 'hard pulse'. Optionally, the RF pulse can be amplitude- and phase-modulated with a shape from a waveform memory. If this is the case, the pulse is denoted as a 'soft pulse' or a 'shaped pulse'. Shaped pulses are essential for the selection of an imaging slice, which will be shown later.

After the excitation, an NMR signal is induced in the receiver coil by the rotating magnetization $\boldsymbol{M}_{xy}$ with an angular frequency ω. The transmitting coil and the receiving coil may physically be the same coil, but can also be two separate coils. The received NMR signal, is in a first step, amplified by a preamplifier. Due to the weakness of the NMR signal, the preamplifiers are mounted as close as possible to the receiver coil, in order to minimize signal losses. Usually, the whole magnet room is RF-shielded (Faraday cage) so as to prevent interference of the NMR signal from radiofrequency signals from the outside world. The amplified NMR signal is then fed into the quadrature detector together with the carrier.

A direct digitizing of the high-frequency NMR signal (60–130 MHz in high-field systems) would be impractical for several reasons. Instead, the signal is converted to a low-frequency signal in an analog mixer, prior to digitization. In the quadrature detector, the NMR signal is split into two parts, or channels. One channel is phase-shifted by 90° before the mixing step. As a result, the signals in the two output channels from the quadrature detector will correspond to cos $[(\omega - \omega_0)t)]$ and sin $[(\omega - \omega_0)t]$, respectively, equivalent to the projection of the magnetization component $\boldsymbol{M}_{xy}$ on to two perpendicular axes in the rotating frame. With this arrangement, it is possible to discriminate both positive and negative frequency differences $(\omega - \omega_0)$.

The two output signals (channels A and B) of the quadrature detector are low-pass filtered, sampled and digitized. The digitized signal is finally stored in a computer for further processing. A pair of samples from the channels A and B can be represented as one complex sample, where channel A corresponds to the real part and channel B corresponds to the imaginary part.

3.3 *GRADIENT AND SHIM COILS*

In order to generate images, it is necessary to encode spatial information into the NMR signal. As will be discussed in the next section, this is accomplished by using time-varying, linear magnetic gradients which changes the magnetic field according to the following relationship:

$$B(x, t) = B_0 + G_x(t)x \tag{1.32}$$

where x is the spatial coordinate along the x-axis, measured from the magnet center, and $G_x(t)$ is the gradient strength. Gradients are generated in the same way also in the y- and z-directions. These gradients are produced by three orthogonal coils placed inside the magnet and surrounding the RF coil and the sample. Each gradient coil is driven by a power amplifier. In clinical systems, the maximum gradient strength is typically 20–25 mT/m, and the rise time from zero to maximum gradient is typically 300–500 μs. In addition to the gradient coils, most systems are equipped with a number of coils (denoted shim coils) generating non-linear gradients, e.g. as given by the following:

$$B(x, y, z) = B_0 + c_s z(x^2 - y^2) \tag{1.33}$$

where c_s is proportional to the current fed through the shim coil. The purpose of the shim coils is to fine-tune the magnetic field for maximal homogeneity across the sample. Since the homogeneity is disturbed by any sample inserted into the magnet, the currents through the shim coils may be adjusted individually for each sample, either manually or automatically.

4 IMAGE GENERATION

4.1 MAGNETIC GRADIENT ACTION

The linear relationship between the magnetic field strength $\boldsymbol{B}$ and the NMR resonance frequency ω, at a position $\boldsymbol{r}$ in space and at time t, is as follows:

$$\boldsymbol{\omega}(\boldsymbol{r}, t) = -\gamma \boldsymbol{B}(\boldsymbol{r}, t), \quad \boldsymbol{r} = \begin{bmatrix} x \\ y \\ z \end{bmatrix} \tag{1.34}$$

This relationship gives the opportunity to map spatial positions to frequencies, given that the magnetic field $\boldsymbol{B}_0$ can be made inhomogeneous 'in a controlled manner'. The gradient $\boldsymbol{G}$ is defined as:

$$\boldsymbol{G}(t) = \frac{\partial B_z(\boldsymbol{r}, t)}{\partial \boldsymbol{r}} = \begin{bmatrix} \frac{\partial B_z(\boldsymbol{r}, t)}{\partial x} \\ \frac{\partial B_z(\boldsymbol{r}, t)}{\partial y} \\ \frac{\partial B_z(\boldsymbol{r}, t)}{\partial z} \end{bmatrix} \tag{1.35}$$

and the magnetic field at position $\boldsymbol{r}$ is thus given by:

$$B_z(\boldsymbol{r}, t) = B_0 + \boldsymbol{G}_r(t) \cdot \boldsymbol{r} \tag{1.36}$$

Note that we assume that the gradient only affects the z-component of the magnetic field. This assumption violates the Maxwell field equations, which tell us that a gradient in the z-component must also create gradients in the x- and y-components. These latter gradients are known as 'concomitant gradients'. However, the influence of these concomitant gradients can be ignored as long as the magnetic field produced by the gradient $\boldsymbol{G}$ is weak when compared to the main field $\boldsymbol{B}_0$. If the gradient coils are perfectly linear, the gradient $\boldsymbol{G}$ is invariant with respect to $\boldsymbol{r}$.

4.2 RF EXCITATION IN THE PRESENCE OF A MAGNETIC GRADIENT

The fact that the magnetic gradient makes the resonance frequency of the spins position-dependent gives the possibility of exciting only parts of the sample. Let the gradient $\boldsymbol{G}$ be given by the following:

$$\boldsymbol{G} = \begin{bmatrix} 0 \\ 0 \\ G_z \end{bmatrix} \tag{1.37}$$

i.e. a time-constant gradient in the z-direction with strength G_z. Assume further that an RF pulse with constant amplitude and frequency Ω generates a $\boldsymbol{B}_1$-field along the x'-axis in the rotating frame. At a position Δz along the z-axis, the effective magnetic field is thus given by:

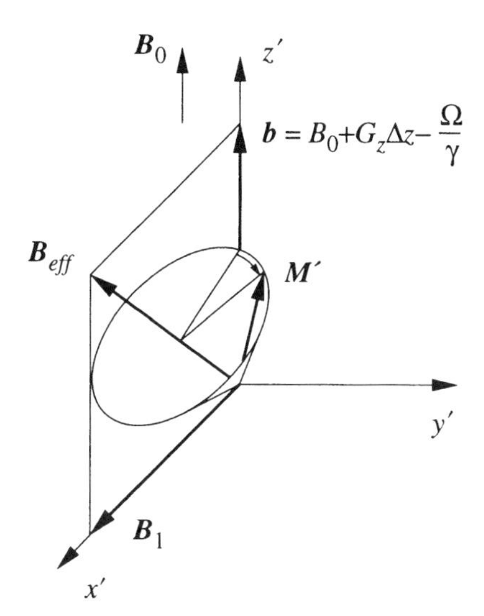

Figure 1.8 RF excitation in the presence of a gradient in the z-direction. Reproduced with permission from [3].

$$\boldsymbol{B}_{eff} = \boldsymbol{B}_0 + \boldsymbol{G}\cdot\boldsymbol{r} + \boldsymbol{B}_1 + \frac{\boldsymbol{\Omega}}{\gamma} = \begin{bmatrix} B_1 \\ 0 \\ B_0 + G_z\Delta z - \dfrac{\Omega}{\gamma} \end{bmatrix} \tag{1.38}$$

According to Equation (1.22), the magnetization $\boldsymbol{M}'$ will rotate around the $\boldsymbol{B}_{eff}$-vector, as shown in Figure 1.8. Here, it is assumed that $\boldsymbol{M}'$ is aligned with the z'-axis at the onset of the RF pulse.

If the vector $\boldsymbol{b}$ in Figure 1.8 is zero, $\boldsymbol{M}'$ will rotate around $\boldsymbol{B}_1$ down towards the $x'y'$-plane. This can be achieved by adjusting Ω to $\gamma(B_0 + G_z\Delta z)$. At the position $z' = \Delta z$, the effective field is thus given by Equation (1.39):

$$\boldsymbol{B}_{eff} =_{|z=\Delta z} \begin{bmatrix} \boldsymbol{B}_1 \\ 0 \\ 0 \end{bmatrix} \tag{1.39}$$

i.e. the $\boldsymbol{B}_{eff}$-field is entirely in the x'-direction, and the magnetization will therefore rotate around the x'-axis towards the $x'y'$-plane.

On the other hand, if $\boldsymbol{b}$ is large compared to $\boldsymbol{B}_1$, $\boldsymbol{B}_{eff}$ will be almost parallel to the z'-axis and $\boldsymbol{M}'$ will remain close to its original position, i.e. the magnetization is not affected by the RF pulse. Thus, it is possible to create transversal magnetization, $\boldsymbol{M}_{xy}$, in an interval (slice) along the z'-axis. The position of the slice can be selected by adjustment of the frequency Ω of the RF pulse. The thickness of the slice can also be selected by adjustment of either the strength of the $\boldsymbol{B}_1$-field or the strength of the gradient G_z. A stronger gradient results in a thinner slice. An RF pulse with longer duration and correspondingly weaker

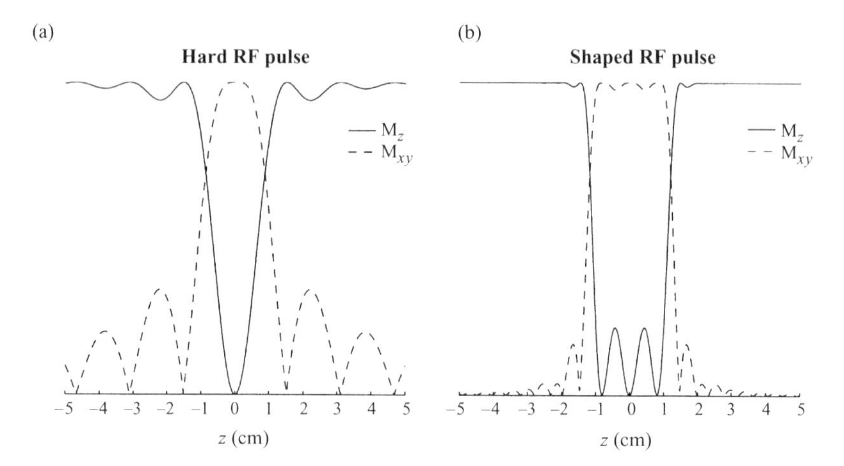

Figure 1.9 Slice profiles for (a) hard and (b) shaped RF pulses.

B_1-field will also result in a thinner slice. A gradient which is active during RF excitation is referred to as a 'slice selection gradient' or just a 'slice gradient'.

Alternatively, the slice thickness can be explained as follows: imagine the B_1-field having a frequency content with bandwidth $\Delta\omega$. The slice thickness is then given by $\Delta\omega/(\gamma G_z)$. Thus, the slice thickness can be made thinner by either increasing G_z or decreasing $\Delta\omega$ (by increasing the RF-pulse duration).

The B_1-field, previously assumed to be constant in time during the RF pulse (hard RF pulse), may also be time-varying (shaped RF pulse). In Figure 1.9, solutions to the differential equation (Equation (1.22)) are plotted versus the z-coordinate, first for a hard RF pulse, and then for a shaped pulse (sinc shape). In both cases, the flip angle of the pulse was 90°.

As shown in this figure, the slice profile after excitation with the shaped RF pulse is much sharper than the profile obtained from the hard pulse.

In the previous example, the excited slice was oriented perpendicular to the z-axis because the slice gradient was directed along the z-axis. However, the orientation of the slice can be chosen arbitrarily by using combinations of the x-, y- and z-gradients for slice selection.

4.3 THE k-SPACE

In the previous section, it was shown that it is possible to excite a plane, or a slice, of the object to be imaged. By applying magnetic gradients perpendicular to the slice gradient, it is possible to affect the phase of the magnetization in the plane, in a way that makes it possible to reconstruct an image from the sampled NMR signal.

Assume again that the z-gradient was used for slice selection during the RF excitation process. Immediately after the RF pulse, we assume that the magnetization in all parts of the slice has been turned down to the xy-plane and is

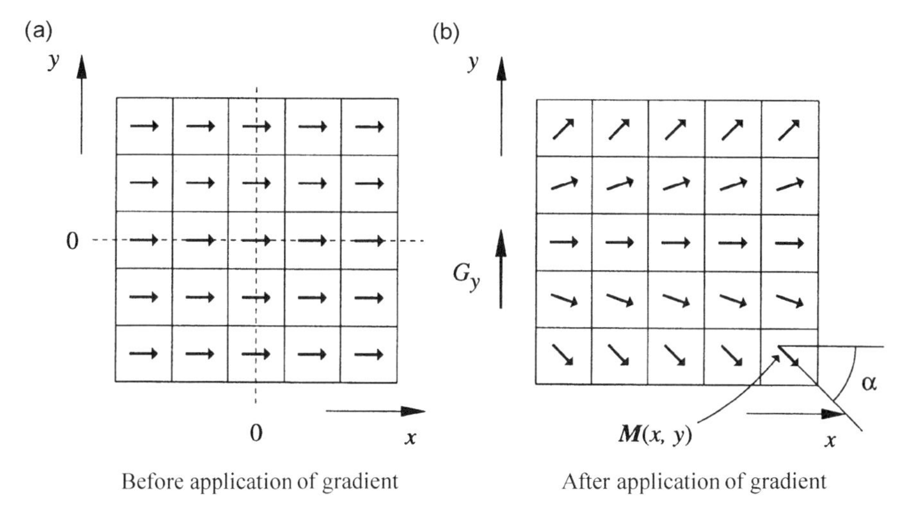

Figure 1.10 The phase angle of the magnetization vectors in the xy-plane, shown before (a) and after (b) the application of a magnetic gradient in the y-direction

aligned with the x-axis. This situation is pictured in Figure 1.10(a), where the slice is divided in 5×5 squares. Each square contains a magnetization vector $\boldsymbol{M}(x, y)$ at the co-ordinates (x, y).

Figure 1.10(b) shows how the magnetization vectors are turned by an angle α (viewed from the rotating frame) after a magnetic gradient, with strength G_y, has been applied in the y-direction. If G_y is constant in time, the phase angle α at the position (x, y) is given by:

$$\alpha(x, y, t) = -\gamma G_y y(t - t_0), t > t_0 \tag{1.40}$$

where $t = 0$ when all of the magnetization vectors have the initial phase $\alpha = 0$ and t_0 is the time when the gradient G_y is turned on. More generally, when the gradient is time-varying in all directions, the phase can be written as follows:

$$\alpha(\boldsymbol{r}, t) = -\gamma \int_0^t \boldsymbol{G}(\tau) \cdot \boldsymbol{r} d\tau \tag{1.41}$$

Let the equilibrium magnetization at position $\boldsymbol{r}$ in the slice be given by a spin density function $\rho(\boldsymbol{r})$. After the slice excitation, the magnetization component in the xy-plane has a length proportional to $\rho(\boldsymbol{r})$ and a phase given by Equation (1.41). If complex notation is introduced, the magnetization $\boldsymbol{M}_{xy}(\boldsymbol{r}, t)$ is given by the following:

$$\boldsymbol{M}_{xy}(\boldsymbol{r}, t) = |\boldsymbol{M}_{xy}(\boldsymbol{r})| \exp[j\alpha(\boldsymbol{r}, t)] \propto \rho(\boldsymbol{r}) \exp\left[-j\gamma \int_0^t G(\tau) \cdot \boldsymbol{r} d\tau\right] \tag{1.42}$$

The magnetization $M_{xy}(\boldsymbol{r}, t)$ gives rise to a signal that is picked up by the receiver coil. Because the receiver coil detects signals from all parts of the sample simultaneously, the signal at time t after the excitation, $S(t)$, is obtained by integrating Equation (1.42) over the whole sample, as follows:

$$S(t) \propto \int\int_{\boldsymbol{r}}\int \rho(\boldsymbol{r})\exp\left[-j\gamma\int_0^t \boldsymbol{G}(\tau)\cdot\boldsymbol{r}d\tau\right]d\boldsymbol{r} \tag{1.43}$$

The 3-dimensional continuous Fourier transform is defined as:

$$F(\boldsymbol{k}) = \int\int_{R^3}\int f(\boldsymbol{r})\exp(-j2\pi\boldsymbol{k}\cdot\boldsymbol{r})d\boldsymbol{r} \tag{1.44}$$

By comparing Equations (1.43) and (1.44), it is seen that the NMR signal at time t is actually given by the Fourier transform of the spin density function $\rho(\boldsymbol{r})$, in the point:

$$\boldsymbol{k} = \frac{\gamma}{2\pi}\int_0^t \boldsymbol{G}(\tau)d\tau = \begin{bmatrix} k_x \\ k_y \\ k_z \end{bmatrix} \tag{1.45}$$

By tradition, the letter k has been used in MRI to represent the coordinate in the Fourier domain. Accordingly, the Fourier domain is denoted as the 'k-space' in MRI. Since the MRI scanner collects successive samples from the k-space during the acquisition of an image, the Fourier domain (or k-space) is also referred to as the 'time domain'. The magnetic resonance image (represented in Equation (1.43) by the function $\rho(\boldsymbol{r})$) is thus reconstructed by using a Fast Fourier Transform (FFT) of the acquired NMR signal.

Data collected near the origin of the k-space represent low spatial frequencies, whereas data collected near the edges of the k-space represent high spatial frequencies. This concept is illustrated in Figure 1.11, which shows various MR images through the human head. In Figure 1.11(a), the full data matrix from the k-space was used for reconstruction of the image. In Figure 1.11(b), only the central parts of the k-space were used and the outer parts (indicated by the gray area) were set to zero prior to the FFT stage. The fine details in the image are lost, although the image contrast is basically preserved. In Figure 1.11(c), the central part of the k-space was set to zero; the resulting image contains information concerning the fine details, but with little contrast.

Note that the pure spin density ρ is of minor interest in biological and medical imaging, because this density is relatively constant and thus gives little information. Preferably, the brightness variation (contrast) of the reconstructed image should reflect the variation of relaxation rates across the sample. It is therefore desirable to sample the k-space with a timing which allows relaxation processes to influence the magnitude of the NMR signal, according to Equation (1.31). Further reading concerning k-space can be found in references [4–5].

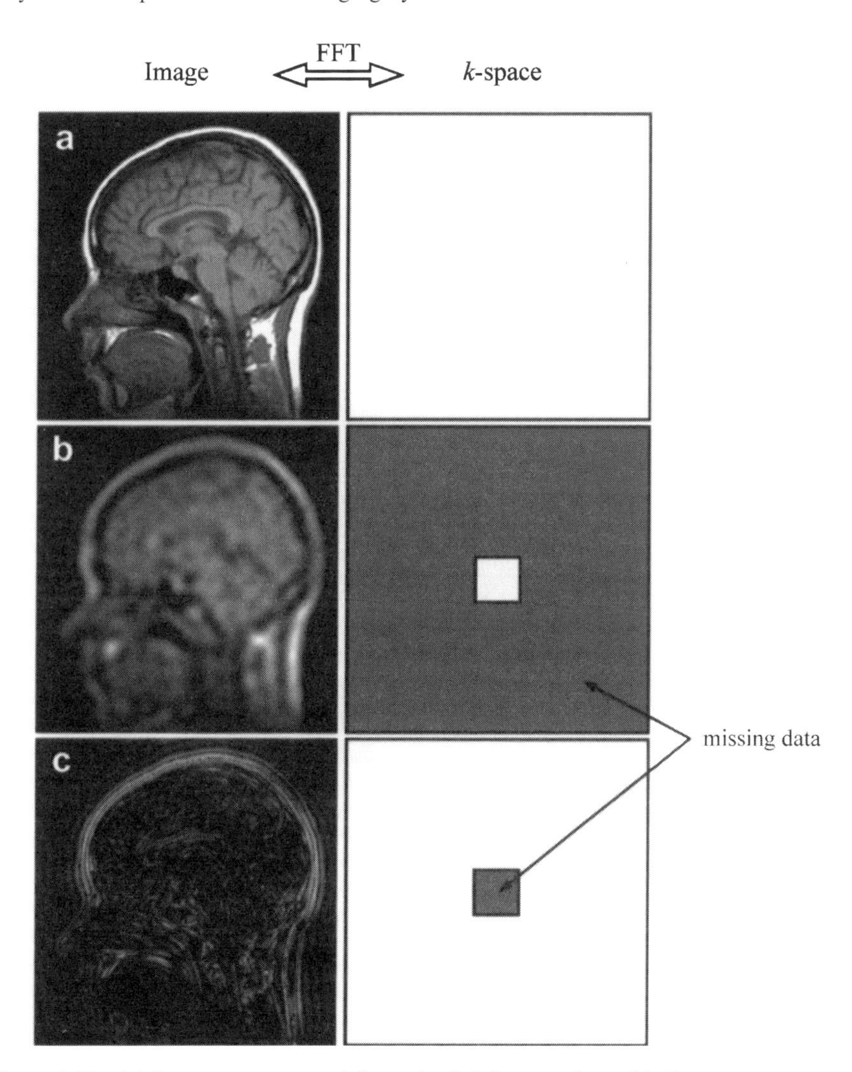

Figure 1.11 (a) Image reconstructed from the full k-space data. (b). Image reconstructed from the central k-space data only. (c) Image reconstructed from the peripheral k-space data only.

5 PULSE SEQUENCES

5.1 k-SPACE TRAJECTORIES

As shown in the previous section, if the NMR signal is sampled after an RF pulse, these samples are equivalent to samples obtained from the Fourier transform of the spin density function $\rho(\boldsymbol{r})$. Thus, it should be possible to arrange the samples in a matrix, which after discrete Fourier transformation (i.e. FFT) would return an image of the spin density. It remains now to explain how this is achieved in practice, and how the gradients should be arranged in order to correctly sample the k-space.

In the following, we assume again that the slice is perpendicular to the z-axis and that the x- and y-gradients are used to define sampling points in the k-space. It is also assumed that the RF pulse has negligible duration and that all spins have zero phase immediately after the RF pulse is applied (time $t = 0$). It then follows from Equation (1.45) that the position in the k-space is at the origin ($\boldsymbol{k} = 0$) at time $t = 0$. Switching of the x- and y-gradients will produce a trajectory in the k-space, which will be denoted by $\boldsymbol{k}(t)$. The arrangement of RF pulses, gradients and sampling intervals are commonly referred to as 'pulse sequences'.

A first attempt to construct a gradient pattern that can be used for imaging is shown in Figure 1.12, together with the corresponding k-space trajectory. After the RF pulse, gradient pulses in the x- and y-directions are switched on for a short time. After the gradients have been switched off, a single datapoint is sampled from the NMR signal. Because the image reconstruction is a digital process, a finite number of complex samples are needed to perform the FFT. In order to obtain an image with, e.g. 256×256 pixels, a total of 64k samples has to be collected. Thus, the sequence illustrated in Figure 1.12 has to be repeated 65 536 times, each time with different values for G_x and G_y. This is indicated in the figure by the dotted lines and arrows for the x- and y-gradients. Note that the k-space trajectory always starts from the origin after each RF excitation.

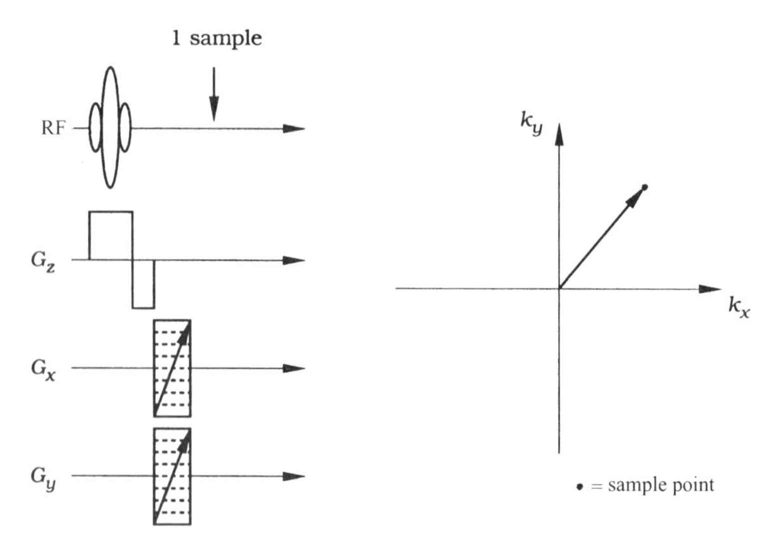

Figure 1.12 A simple approach to generate an MR image.

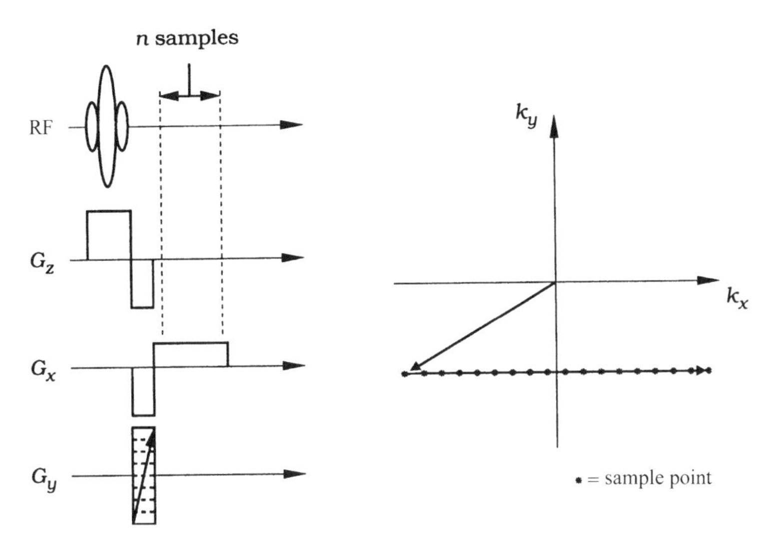

Figure 1.13 The basic gradient-echo sequence.

Although this approach is fully capable of generating an image, it is very time consuming due to the large number of repeated measurements that are required.

The pulse sequence can be made more time efficient by collecting several samples for each RF excitation. Figure 1.13 shows a modified sequence, where a whole k-space line is collected after the RF pulse. The number of points in each line is still determined by the desired matrix size.

After the RF pulse, the negative lobe of the x-gradient moves $\boldsymbol{k}(t)$ to negative k_x values. Simultaneously, $\boldsymbol{k}(t)$ is moved along the k_y-axis by the y-gradient pulse. Thereafter, $\boldsymbol{k}(t)$ again moves towards positive $k_x$values when the x-gradient is switched to a positive value. The imaging sequence described in figure 1.13 is the simplest form of a 'gradient-echo' sequence.

Many other strategies for traversing the k-space have been proposed [5]. For example, in echo-planar imaging (EPI), the k-space is traversed along zig-zag trajectories [3]. The trajectories may also be spiral shaped [6].

5.2 THE GRADIENT-ECHO SEQUENCE

We can use the gradient-echo (GRE) sequence above as a starting point for a generalized discussion concerning image resolution, gradient strengths, sampling rates, etc.

Let $f(\boldsymbol{r})$ denote a discrete two-dimensional function with (N by N) samples, where $\boldsymbol{r}$ is a spatial vector as usual. Let $F(\boldsymbol{k})$ be the (N by N) FFT of $f(\boldsymbol{r})$, where $\boldsymbol{k}$ is the k-space vector. The properties of the fast Fourier transform then governs the following:

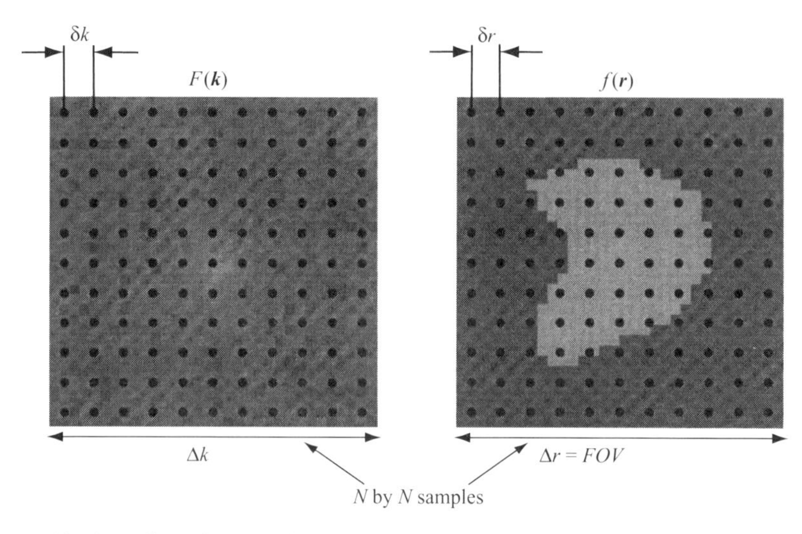

Figure 1.14 Sampling of a two-dimensional function.

$$\delta k = \frac{1}{N\delta r} \tag{1.45}$$

where δk and δr are the sampling distances of $F(\boldsymbol{k})$ and $f(\boldsymbol{r})$, as shown in Figure 1.14. (see reference [7]).

The length Δk covered by the samples in the k-space is $N\delta k$ and the distance Δr covered by the samples in the spatial domain is $N\delta r$. Since $\boldsymbol{r}$ is a spatial vector, the distance Δr is also denoted as the field of view (FOV). We therefore obtain the following:

$$\Delta k = \frac{N}{FOV}$$

$$\delta k = \frac{1}{FOV} \tag{1.47}$$

Equation (1.47) can now be applied to the gradient-echo sequence described in Figure 1.13. In Figure 1.15, a few more variables have been introduced. In this figure, f_s is the sampling frequency of the analog-to-digital converter (ADC), and the sampling time, t_{read}, is therefore N_x/f_s. The time between two samples is $\Delta t = 1/f_s$, G_{x_r} is the gradient strength of the x-gradient during the sampling interval, and G_{x_rew} is the x-gradient strength during the 'rewinding' interval, t_{rew}. In addition, N_x and N_y are the number of pixels in the x- and y-direction, respectively, while FOV_x and FOV_y are respectively the fields of view in the x- and y-directions. For each RF excitation, N_x k-space points are sampled. In order to collect the full number of (N_x by N_y) samples, the sequence must be repeated N_y times. For each turn of the loop, the y-gradient is stepped linearly from $-G_{y\_max}$ to $+G_{y\_max}(N_y - 2)/N_y$.

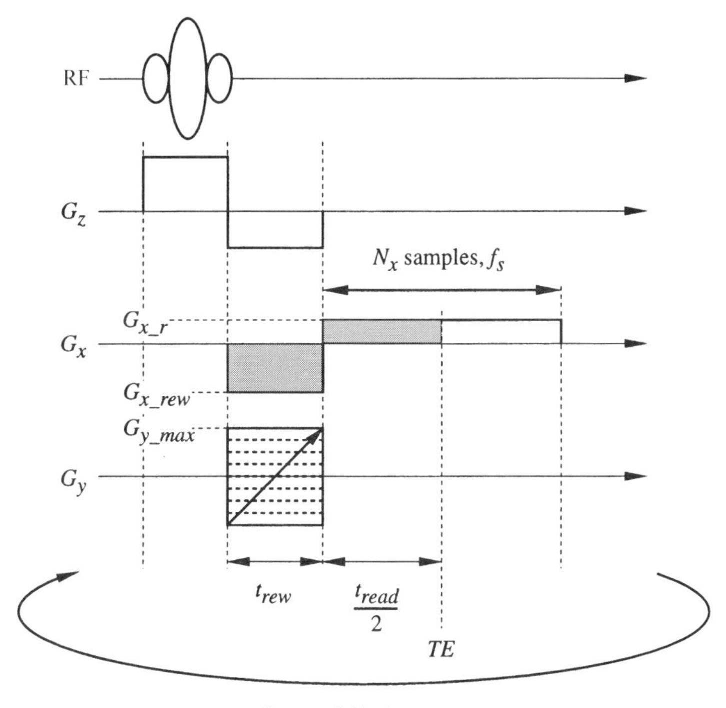

Figure 1.15 The gradient-echo sequence.

The k-space distance between two successive samples is calculated from the following equation:

$$\delta k_x = \frac{\gamma}{2\pi}\int_t^{t+\Delta t} G_x(\tau)d\tau = \frac{\gamma}{2\pi}G_{x\_r}\Delta t = \frac{\gamma}{2\pi}\frac{G_{x\_r}}{f_s} \tag{1.48}$$

Combining Equations (1.47) and (1.48) yields the following:

$$G_{x\_r} = \frac{2\pi}{\gamma}\frac{f_s}{FOV_x} \tag{1.49}$$

In the y-direction, a similar calculation gives:

$$\delta k_y = \frac{\gamma}{2\pi}\frac{2G_{y\_\max}}{N_y}t_{rew} \tag{1.50}$$

With the help of the relationship, $\delta k_y = 1/FOV_y$, we obtain:

$$G_{y\_\max} = \frac{2\pi}{\gamma}\frac{N_y}{2FOV_y t_{rew}} \tag{1.51}$$

Finally, the k-space coordinate k_x should be zero at the center of the sampling interval, thus leading to the condition:

$$\frac{\gamma}{2\pi}\left(G_{x\_rew}t_{rew} + G_{x\_r}\frac{t_{read}}{2}\right) = 0 \tag{1.52}$$

i.e. the integral of the area shaded in gray in Figure 1.15 should equal zero.

In general, a higher image resolution (FOV/N) must be 'paid for' by higher gradient strengths. Faster pulse sequences, where the durations of RF pulses, gradient intervals and sampling intervals have been reduced, also lead to a need for higher gradient strengths.

5.3 THE SPIN-ECHO SEQUENCE

Another example of a pulse sequence is the spin-echo (SE) sequence which utilizes two RF pulses per repeated measurement, the first one with a 90° flip angle and the second with a 180° flip angle. The purpose of this arrangement is to reduce the influence of field inhomogeneities of the static B_0-field. A high sensitivity for field inhomogeneities is desired in some situations when information can be gained from the inhomogeneities. In other situations, the inhomogeneities can cause problems and reduce image quality.

First, we study the influence of the 180° pulse in the absence of gradients. In Figure 1.16, an RF pulse has created transverse magnetization along the x-axis (Figure 1.16(a)). If magnetic inhomogeneities are affecting the magnetization, spins experiencing a higher field will rotate at a higher angular velocity than spins experiencing a lower field. After a time $t = TE/2$, the spins are dephased and spread out in the xy-plane (Figure 1.16(b)). If a 180° pulse is applied along the y-axis in this moment, all magnetization is flipped around the y-axis, but after the 180° pulse will continue to precess with the same angular velocities (Figure 1.16(c)). At the time $t = TE$, all spins will again have the same phase, but now along the negative x-axis (Figure 1.16(d)). A 'spin echo' is thus created at time $t = TE$.

At time $t = TE$ (but *only* at this time), the 180° pulse has 'neutralized' the magnetic inhomogeneities of the B_0-field, i.e. the spins have been 'refocused' to have the same phase in the xy-plane. Thus, the 180° pulse is normally referred to as a 'refocusing pulse'.

The action of the 180° pulse can be super imposed on the action of the magnetic gradients that were used to form the gradient echo. The gradients are still used to create a trajectory through the k-space. Using the k-space formalism, the effect of the 180° pulse can be described as a reflection of the k-space coordinate through the origin:

$$\boldsymbol{k}_{post} = -\boldsymbol{k}_{pre} \tag{1.53}$$

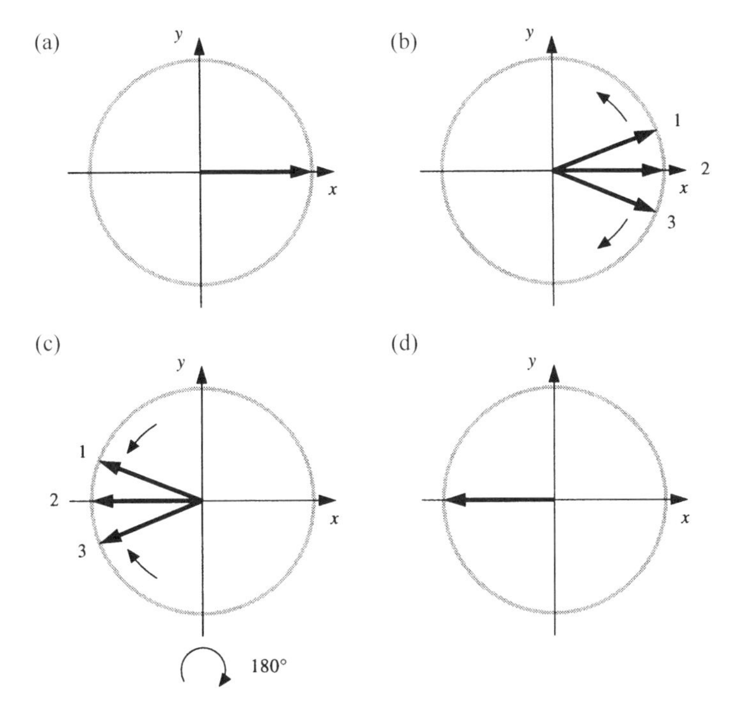

Figure 1.16 (a) Immediately after the excitation pulse, all spins have the same phase. (b) The spins are dephasing due to inhomogeneities. (c) After the 180° pulse, the spins begin to rephase. (d) The spins have the same phase again at time $t = TE$.

where $\boldsymbol{k}_{post}$ and $\boldsymbol{k}_{pre}$ are respectively the k-space coordinates immediately after and before the 180° pulse. In order to optimize the performance of the spin-echo sequence, the spin echo must coincide with the gradient echo, i.e. the echo time TE must be the same for both the spin echo and the gradient echo. This is obtained by placing the refocusing 180° pulse exactly in the middle of the 90° excitation pulse and the midpoint of the echo.

Figure 1.17 shows an illustration of the spin-echo sequence. The same calculations of gradient strengths as were carried out in the previous section for the gradient-echo sequence are also valid for the spin-echo sequence. However, the reflection of the k-space co-ordinate (Equation (1.53) caused by the 180° pulse must be taken into account. As shown in Figure 1.17, this is handled by a reversal of the sign of the rewinding gradient, compared to the gradient-echo sequence in Figure 1.15. This is because the rewinding gradient and the read-out gradient are placed on opposite sides of the 180° pulse in the spin-echo sequence. In addition, the gradient in the y-direction is stepped from positive to negative values, as opposed to the gradient-echo sequence where the steps go from negative to positive values.

The signal equation for the spin-echo sequence is given by the following:

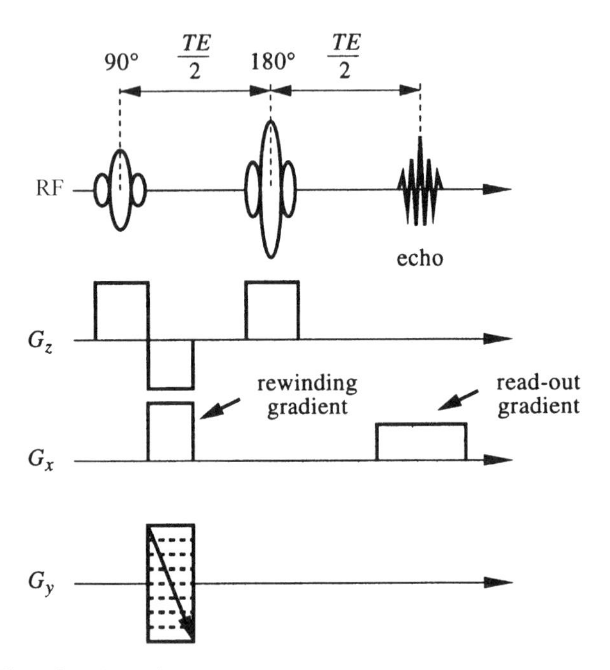

Figure 1.17 The spin-echo pulse sequence.

$$S_{SE} \propto M_0 \left[1 - 2\exp\frac{-(TR - TE/2)}{T_1} + \exp(-TR/T_1) \right] \exp(-TE/T_2) \qquad (1.54)$$

where TR is the interval between two successive 90° RF pulses. If TE is small compared to TR, which is usually the case, Equation (1.54) can be simplified to give:

$$S_{SE} \propto M_0[1 - \exp(-TR/T_1)]\exp(-TE/T_2) \qquad (1.55)$$

By making TR long compared to TE, it is possible to generate several spin echoes in one TR interval. This is successfully utilized in a relatively new derivation of the SE sequence which is called fast spin echo (FSE) or turbo spin echo (TSE) [8]. The FSE sequence can create images with a similar contrast to spin-echo images, but with a much shorter scan time when compared to conventional spin echo. In FSE sequences, a single 90° pulse is followed by several 180° pulses. Each 180° pulse will create an echo, which is individually phase encoded, and several k-space lines (see Figure 1.14) can thus be acquired following a single 90° excitation pulse. Although several spin echoes are generated with a short time interval between them, the so-called *effective TE*, which determines the T_2-weighting in the image, is mainly determined by the time from the 90° excitation pulse and the echo where the phase-encoding gradient is set to zero. By manipulation of the order in which the phase-encoding values are given, the effective *TE* can thus also be manipulated. Both *TR* and the effective *TE* can be made very long in TSE sequences while still maintaining a good signal-to-noise ratio. Since TSE is not a steady-state sequence, it cannot be described mathematically by a single equation like the SE and GRE sequences (for further reading on this subject, see reference [9]).

6 IMAGE CONTRAST IN SPIN-ECHO AND GRADIENT-ECHO IMAGES

We have seen above how the achievable NMR signal intensity is closely related to the scan parameters, *TR*, *TE* and flip angle, as well as the intrinsic tissue parameters, proton density (ρ), T_1 and T_2 (or T_2^*). From Equation (1.31) or in the case of spin-echo sequences, Equation (1.54), it is thus possible to predict the resulting image contrast between tissues with known relaxation properties for a given pulse sequence. The image contrast in MRI is typically classified according to its sensitivity to the three tissue parameters, ρ, T_1 and T_2. A sequence which is mainly sensitive to proton density is called a proton-density-weighted image. Similarly, image sequences which are mainly sensitive to T_1 or T_2 relaxation times are called T_1- or T_2-weighted images, respectively.

6.1 CONTRAST- AND SIGNAL-TO-NOISE RATIO

The final MR image is made up of a matrix of pixels with individual intensities on a gray-level scale. The numerical value of the signal intensity in a given pixel is somewhat arbitrary and has no real physical meaning, unlike in computenized tomography (CT) where the pixel intensity can be directly quantified in Hounsfield units. The ability to detect disease or to discern different anatomical structures relies on a difference in pixel signal intensity in the structures of interest. Furthermore, this difference in signal intensity needs to be large compared to the image noise, which will give rise to a random signal intensity component in all of the pixels. The noise present in the MR image is a complicated function of many parameters. To some extent, such noise is generated in the receiver coil and the electronic circuits of the MR scanner, although the predominant source of noise is the patient being imaged.

The signal-to-noise ratio (SNR) of an MR image is defined as the signal intensity of a given region of interest divided by the standard deviation of the noise in the image, i.e. $\mathrm{SNR} = \mathrm{SI}_{\mathrm{tissue}}/\mathrm{SD}_{\mathrm{noise}}$. The standard deviation of the noise is usually measured in an area of the image where the signal ideally should be zero (i.e. in the air surrounding the patient). Contrast-noise ratio (CNR) can thus be defined as the difference in signal intensity between two areas of interest divided by the standard deviation of the noise, i.e. $\mathrm{CNR} = (\mathrm{SI}_{\mathrm{tissue1}} - \mathrm{SI}_{tissue2})/\mathrm{SD}_{\mathrm{noise}}$. From this, it follows that an optimal

SNR for a given tissue does not always give the most diagnostically useful image. In most clinical situations, the sequence needs to be optimized to give the best possible CNR between the tissues of interest.

6.2 SPIN-ECHO (SE) CONTRAST BEHAVIOR

It is evident from Equation (1.55) that the T_1-weighting in the image can be minimized by making $TR >> T_1$ for all tissues of interest, i.e. TR should be four to five times longer than the longest T_1 value. Since the T_1 of tissues ranges from <300 to >2000 ms, TR must be of the order of several seconds to completely eliminate T_1-weighting in the image. Such a long TR is not always practical since the scan time increases with increasing TR. In practice, therefore, there is always a certain degree of T_1-weighting in the image. However, the main rule is that long TR values (>2 s) will minimize the influence of T_1 relaxation times on image contrast. By minimizing the T_1-dependence, T_2-weighting can be introduced by making the echo time, (TE), comparable to the T_2-times of the target tissues. This means that TE should be in the range 70–300 ms in most situations. The longer the TE, then the more the image contrast is weighted towards tissues with long T_2 relaxation times. However, the signal-to-noise ratio will go down with increasing TE since the NMR signal intensity falls off exponentially with TE. This limits the length to which TE can be made in practice. Proton-density weighting can be achieved by combining a long TR ($TR >> T_1$) and a short $TE(TE << T_2)$. In such a sequence, both T_1- and T_2-weighting is minimized, with the main contrast-determining tissue property being proton density.

In Figures 1.18, the relative signal intensity is plotted versus (a) TR and (b) TE for three different combinations of T_1, T_2 and ρ. Notice that in Figure 1.18(a) the magnetization of the tissue type with the longest T_1 (e.g.

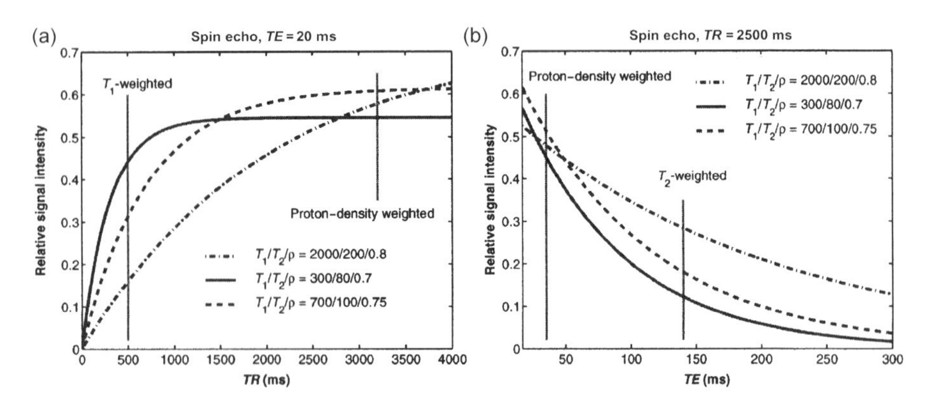

Figure 1.18 Relative signal intensity versus (a) TR and (b) TE in spin-echo sequences.

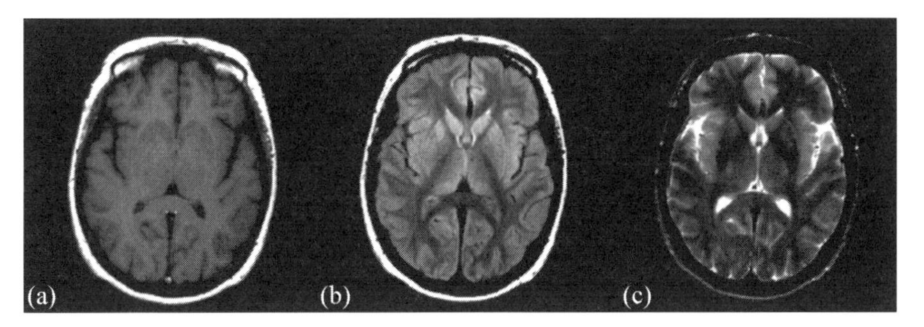

Figure 1.19 Spin-echo images of the brain: (a) T_1-weighted image ($TR/TE = 505/20$ ms); (b) proton-density weighted image ($TR/TE = 2200/20$ ms); (c) T_2-weighted image ($TR/TE = 2200/90$ ms).

cerebrospinal fluid (CSF)) has not fully recovered during the TR interval, and the image is thus somewhat T_1-weighted for these long T_1 times, even at a long TR time of 3000 ms.

The contrast in the SE sequences can be summarized as follows:

- short TR and short TE = T_1-weighted image;
- long TR and long TE = T_2-weighted image;
- long TR and short TE = proton-density weighted image.

The combination of short TR and long TE is not generally used because it gives a poor signal-to-noise ratio and a mixed T_1/T_2 contrast which is not very useful clinically.

Figure 1.19 shows three spin-echo images of the brain with different contrast weightings. Figure 1.19(a) is a T_1-weighted image ($TR/TE = 505/20$ ms). Notice that the tissue with the shortest T_1, i.e. fat, has the brightest signal. Furthermore, white matter has a higher signal intensity than gray matter due to its shorter T_1-value. Conversely, the tissue with the longest T_1, i.e. the CSF, has the lowest signal intensity. Figure 1.19(b) is the proton-density weighted image ($TR/TE = 2200/20$ ms). The contrast in this image is weighted towards proton-density differences. Notice, however, that the fat is still very bright, suggesting that the TR is not long enough to completely eliminate the T_1-weighting in the image. Figure 1.19(c) shows the T_2-weighted image ($TR/TE = 2200/90$ ms), in which the CSF has the highest signal intensity due to its very long T_2 relaxation time.

6.3 GRADIENT-ECHO (GRE) CONTRAST BEHAVIOR

There are two main differences in the contrast behavior of a GRE sequence when compared to an SE sequence. First, the TR values are generally much shorter in GRE sequences, and secondly 'true' T_2-weighting is not possible since the magnetic field inhomogeneities are not refocused. Consequently, the image contrast is modulated by T_2^* rather than T_2. Furthermore, a third contrast-determining parameter, the flip angle, is introduced (the flip angle is almost always 90° in spin-echo sequences). The fact that *TR* is generally shorter in GRE than in SE sequences is not an absolute requirement, but rather one of the main advantages of GRE sequences over SE sequences. Since no RF-refocusing (180°) pulse is applied, the *TR* can be made much shorter, which will speed up the acquisition time considerably. On modern scanners, *TR* can be made as short as 2–5 ms, which allows for very rapid image acquisitions. In addition, the echo time can be made much shorter in a GRE sequence compared to a SE sequence, because no additional time is needed for a refocusing pulse. It is therefore possible to reduce, or even eliminate, the influence of T_2^* or signal intensity, in spite of T_2^* being generally much shorter than T_2.

GRE sequences are typically used to generate highly T_1-weighted images, by using both a short *TR* and a short *TE*. A reduction of the flip angle can be used to introduce some degree of T_2-weighting, or more correctly, to increase the sensitivity for longer T_1 relaxation times.

It can easily be shown from Equation (1.31) that, for a given TR/T_1 ratio, the signal intensity reaches a maximum at a given flip angle α. This flip angle is called the Ernst angle (α_{ernst}), and is given by the following:

$$\cos(\alpha_{ernst}) = \exp(-TR/T_1) \tag{1.56}$$

From Equation (1.56), it is seen that, for a given *TR*, the Ernst angle is larger for tissues with a short T_1, compared to tissues with a longer T_1. A large flip angle increases the T_1-weighting in GRE images, and conversely a small flip angle reduces the T_1-weighting.

Exactly what is meant by a 'small' and a 'large' flip angle depends on the *TR* used and on the T_1 of the target tissues. Figure 1.20 shows a plot of relative signal intensity versus flip angle for three different T_1 values. It is assumed that $TE << T_2^*$. Notice how the Ernst angle is shifted to the left as T_1 increases.

6.4 OTHER PARAMETERS INFLUENCING IMAGE CONTRAST

One of the real strengths of MRI is that the image contrast can be made sensitive to many physiologically important parameters. The tissue parameters discussed this far, i.e. T_1, T_2 and proton density, can all be significantly altered in pathological tissue relative to normal tissue. Furthermore, the image contrast can also be made sensitive to other physiologically important parameters such as flow. Both microscopic (diffusion) and macroscopic flow can readily be observed in MRI. This is a result of the fact that an MR image is generated by applying field gradients to the imaged object. One can intuitively understand that, since a magnetic gradient will alter the Larmor frequency as a function of

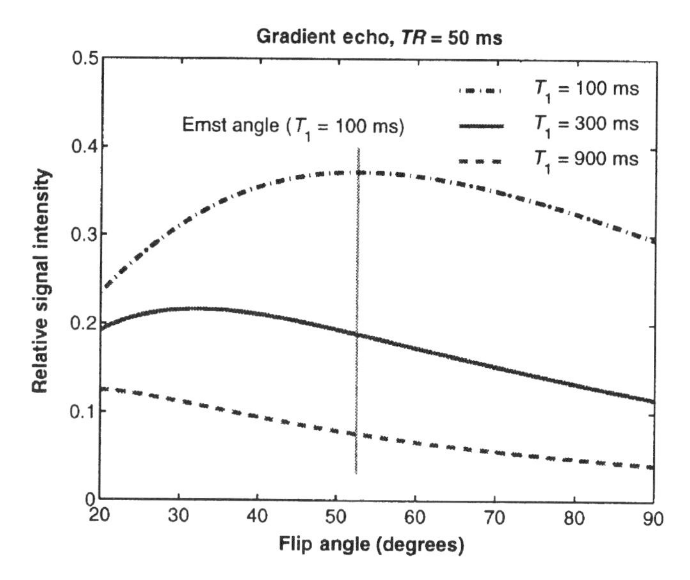

Figure 1.20 Relative signal intensity versus flip angle in a 'spoiled' gradient-echo sequence.

position, flowing spins will behave differently to stationary spins under the influence of a magnetic field gradient. A thorough discussion of flow-related effects in MRI is beyond the scope of this present book, but the reader should be aware of the fact that MRI is a very powerful technique for visualizing and quantifying flow phenomena *in vivo*.

Another way to improve the contrast in MRI is to introduce contrast agents which alter the T_1- and/or T_2-relaxation times of the target tissue. A general overview of the MR contrast agents in use today will be given in the next section, while detailed discussions of contrast-agent mechanisms and contrast-agent design can be found in future chapters of this book.

7 APPLICATION OF CONTRAST AGENTS IN MRI

7.1 CLASSIFICATION OF MR CONTRAST AGENTS

The use of contrast-enhancing agents has become an integral part of MR imaging for many applications. Currently available MR contrast agents can be categorized according to their magnetic behavior, biodistribution and effect on the image, as shown in Figure 1.21.

All MR contrast agents work by reducing the T_2 and/or T_1 relaxation times of the target tissue, and are thus commonly described as either 'T_1-agents' or 'T_2-agents' depending on whether the relative reduction in relaxation times

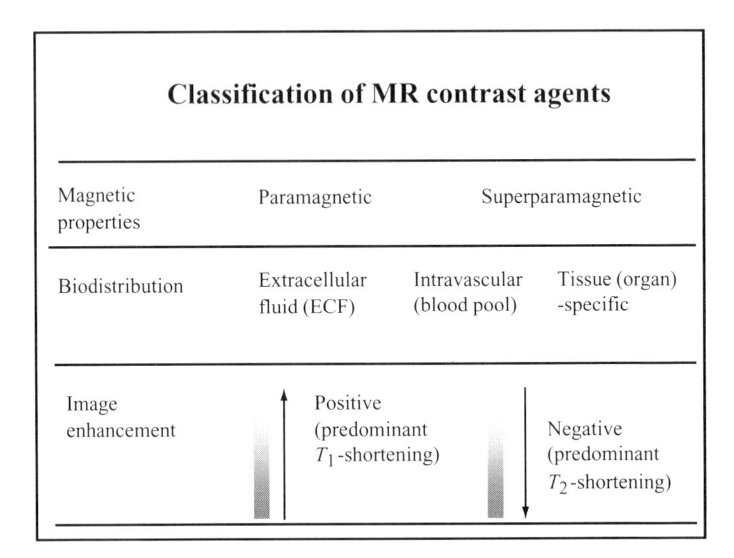

Figure 1.21 Classification of MR contrast agents according to magnetic behavior and biodistribution.

caused by the contrast agent is greater for the longitudinal (T_1) or the transversal (T_2 or T_2^*) relaxation times. The ability of the agent to reduce the T_1 and T_2 relaxation times are respectively described by the r_1 and r_2 *relaxivity* values of the agent. These parameters will be discussed in more detail in Section 7.2.

7.1.1 Paramagnetic Agents

Metal ions with one or more unpaired electrons are paramagnetic, and therefore have a permanent magnetic moment. Organic free radicals are also paramagnetic because of the odd number of valence electrons in these compounds. In aqueous solution, there is a dipolar magnetic interaction between the electronic magnetic moment of the paramagnetic atom and the much smaller magnetic moments of the protons of the nearby water molecules. Random fluctuations in this dipolar magnetic interaction, mainly a result of molecular motions, reduces both the longitudinal (T_1) and transverse (T_2) relaxation times of the water protons. Gadolinium (Gd^{3+}) and manganese (Mn^{2+}) are examples of paramagnetic ions which are used in MR contrast agents. Paramagnetic metals can not be used as contrast agents in their ionic form due to an undesirable biodistribution and the relatively high toxicities of the metal ions. By binding the metal to a ligand, a more desirable biodistribution and safety profile can be achieved. The resulting metal-ion complex is called a chelate. The relaxation theory of paramagnetic substances is well understood and has been extensively described in the literature [10–12].

7.1.2 Superparamagnetic Agents

Iron oxide particles (typically 5–200 nm in diameter) are made up of several thousand magnetic ions and are said to have superparamagnetic properties if the magnetic ions are mutually aligned. Iron oxides are divided into two classes according to the overall size of the crystal: if the diameter (d) is > 50 nm, they are called super paramagnetic iron oxide (SPIO) particles, while if $d < 50$ nm, they are known as ultra-small SPIO (USPIO) particles.

The magnetic moments of the individual iron ions that make up the SPIO and USPIO particles do not cancel out; thus, the particles have a permanent magnetic moment which is very large when exposed to a magnetic field, i.e. much larger than that of a single molecule of a Gd chelate. As a consequence of their large size and magnetic moment, SPIOs were initially developed as T_2-agents [13]. However, a new generation of USPIOs with crystal sizes less than 10 nm have also been reported to have excellent T_1-enhancing properties [14–19].

7.1.3 Extracellular Fluid (ECF) Agents

Contrast agents which are distributed to the extracellular fluid (ECF) are often referred to as ECF agents. Such agents leak rapidly from the blood into the interstitium with a distribution half-life of about 5 min and are cleared by the kidneys with an elimination half-life of about 80 min [20]. The presence of the blood brain barrier (BBB) prevents extravascular leakage of ECF agents into the brain in areas with an intact BBB. Since many neuro-pathological conditions are associated with an altered capillary permeability, such conditions can often be better visualized after contrast administration due to the selective accumulation of contrast agent in these regions. ECF agents have also been extensively used in extra-cranial applications. Notably, the use of gadolinium-based ECF agents has shown great utility in MR angiography (MRA). By selectively reducing the T_1 relaxation time of blood, high-quality MR angiograms can be obtained after administration of Gd-ECF agents. Since these agents clear rapidly from the blood, the images are typically acquired in the early phase following a bolus injection, when used in conjunction with MRA.

7.1.4 Blood-Pool Agents

Blood-pool agents (BPAs, or intravascular agents) belong to a new class of MR contrast agents. These compounds are significantly larger in 'size' than ECF agents, thus preventing leakage into the interstitium. Two main approaches are currently pursued with MR blood-pool agents. One approach is to use ultra-small iron oxide particles with a low r_2/r_1 ratio [14, 15]. The second approach is to use gadolinium-based water-soluble macromolecules with a vascular residence time and relaxivity which is higher than those of ECF agents [21]. The macromolecule can either be a biological entity (e.g. albumin) [22] or a chemical entity (e.g. a polymer) [23]. These agents have been primarily developed for use in MR angiography, where the aim is to have an agent with a relatively long vascular half-life and the highest possible r_1-relaxivity. At the same time, the r_2-relaxivity must be low enough to avoid excessive signal loss due to T_2 or T_2^* relaxation. By selectively reducing the T_1 relaxation time of blood, high-quality MR angiograms can be generated. The potential advantage of using a blood-pool agent in MRA is the prolonged imaging window. Since a longer image acquisition time translates into higher image resolution and/or signal-to-noise ratio, higher-quality angiograms can potentially be attained compared to when gadolinium-based ECF agents are used. The disadvantage of this 'steady-state' approach is that both arteries and veins are equally enhanced in the image, thus making their differentiation more challenging.

7.1.5 Organ-specific agents

An organ-specific agent can be defined as a contrast agent which is selectively taken up by a particular type of cell (e.g. Kupffer cells or hepatocytes) and thereby only enhances organs where these cells are present (e.g. the liver, the spleen and the lymph nodes). All contrast agents are 'organ-specific' to some extent in that they are excreted either by the liver or by the kidneys. Iron-oxide-particle-based agents (SPIOs and USPIOs) accumulate in the reticuloendothelial system in the liver. They can therefore be described as liver specific agents even if they have a long vascular half-life prior to liver accumulation.

Other contrast agents are specifically designed to accumulate in the liver, e.g. Mn-DPDP [24] or Gd-EOB-DTPA [25]. Another type of organ-specific agents are the oral contrast agents which are distributed in the gastrointestinal tract to improve delineation of the bowel from surrounding tissues [26].

7.2 BASIC PRINCIPLES OF MR CONTRAST AGENTS

7.2.1 Inner-and Outer-Sphere Relaxation

As mentioned previously, the relaxation of water protons is enhanced by a contrast agent due to a time-dependent magnetic dipolar interaction between the magnetic moments of the paramagnetic ions and water protons. An MR contrast agent can thus be thought of as a catalyst that increases the R_1 $(= 1/T_1)$ and the $R_2 (= 1/T_2)$ relaxation rates of tissue protons. The efficiency of a given contrast agent to shorten the relaxation times of protons is defined in terms of the *relaxivity* of the agent, r_1 and r_2, as follows:

$$\begin{aligned} &\text{For } r_1,\ R_1 = r_1[CA] + R_{1\_tissue} \\ &\text{For } r_2,\ R_2 = r_2[CA] + R_{2\_tissue} \end{aligned} \tag{1.57}$$

where $[CA]$ is the concentration of the contrast agent in mmol/liter (mM). Consequently, r_1 and r_2 have the dimensions of $mM^{-1}s^{-1}$. The quantities r_1, r_2, R_{1_tissue}, and R_{2_tissue}, depend on both temperature and magnetic field strength. For gadolinium-based ECF agents at 0.5 T and 37 °C, $r_1 \cong 4\ mM^{-1}s^{-1}$ and $r_2 \cong 4.7\ mM^{-1}s^{-1}$ when measured in water. Proton relaxation due to r_1 and r_2 requires the protons to come into close contact with the contrast agent. For small-molecular-weight paramagnetic agents, half of the r_1 and r_2 relaxation is due to an inner-sphere effect [10, 12], where a water molecule binds in the inner co-ordination sphere of the paramagnetic ion and exchanges rapidly with the bulk water protons. The other half is due to water molecules diffusing in the outer-sphere environment of the paramagnetic ion (outer-sphere relaxation). The outer-sphere relaxation mechanism is entirely responsible for determining the relaxivity of superparamagnetic particles. An important result from the outer-sphere theory is that the ratio of r_2/r_1 increases with increasing particle size, and smaller particles are much better T_1-shortening agents than larger [11, 14–16].

7.2.2 Susceptibility-induced relaxation

On a more 'macroscopic' scale, long-range interactions can be a very dominant source of T_2 or T_2^* relaxation for all types of contrast agents. This type of mechanism is related to the magnetization of a contrast agent, and is usually referred to as 'susceptibility-induced relaxation' [27]. In a magnetic field, the magnetic moments of the paramagnetic metal become partially aligned along the direction of the magnetic field. When a contrast agent becomes compartmentalized, such as when superparamagnetic particles are taken up by Kupffer cells, or within a vessel, the compartment containing the contrast agent functions as a secondary contrast agent. This secondary agent is essentially a magnetic bulk material from the point of view of water protons located outside of the compartment. Water protons on the outside of the compartment are therefore relaxed by an outer-sphere mechanism. The magnitude of this long-range T_2/T_2^* relaxation effect depends on a number of factors, with the most important ones being as follows:

(a) the magnetic moment and the local concentration of the contrast agent;
(b) the dimensions and local geometry of the contrast-agent-containing compartment;
(c) the diffusion constant of water;
(d) the type of imaging sequence employed.

At MRI fields, superparamagnetic agents are magnetized much more than paramagnetic agents due to their larger magnetic moment (by about a factor of 1000). Consequently, superparamagnetic agents have a more potent long-range T_2/T_2^* effect than paramagnetic agents.

From the above, it follows that the classification of MR contrast agents as either a 'T_1-agent' or a 'T_2-agent' is not always accurate, since any contrast agent which reduces T_1 must also reduce the T_2. However, any agent that reduces T_2 does not have to reduce T_1, at least at MRI field strengths. Whether the contrast agent functions as a 'T_1-agent' or a 'T_2-agent' for MRI is a function of the imaging sequence used, the field strength, the correlation time (essentially the size of the agent) and how the contrast agent is compartmentalized in the tissue.

7.3 IMAGING CONSIDERATIONS

In order to achieve significant T_1-contrast, the relative effect of the contrast agent on the T_1-relaxation time of the target tissue must be larger than its relative effect on T_2/T_2^*-relaxation. This is easily achieved with paramagnetic agents at typical clinical dosages. Given that $T_1 \gg T_2$ for most tissues, the *relative* effect of the contrast agent on T_1 is therefore much larger than the relative effect on T_2.

Figure 1.22 shows a clinical example of the use of a gadolinium-based ECF agent (Gd-DTPA-BMA, Omniscan®) [28] in a patient with a brain lesion. The T_1-shortening effect is clearly visible on the right-hand image (b) which was acquired after administration of a standard dose of 0.1 mmol/kg of this contrast agent. As discussed in Section 7.1.3, selective T_1-shortening is often obtained in brain lesions due to the presence of the blood brain barrier on normal brain

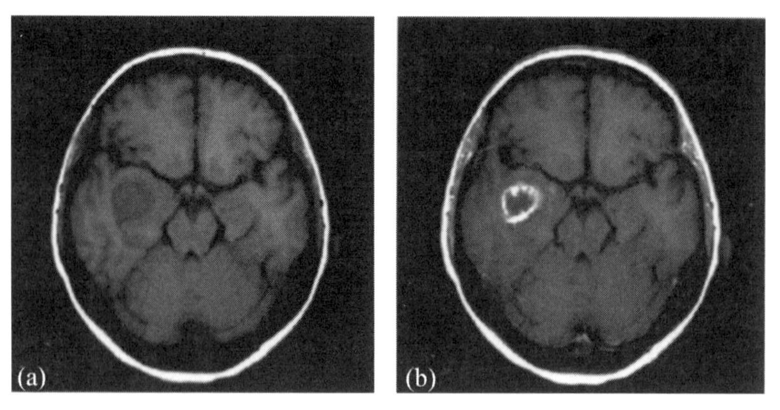

Figure 1.22 Positive-contrast enhancement in a brain lesion obtained by using a spin-echo T_1-weighted sequence (TR/TE = 64/15 ms): (a) pre-contrast; (b) post-contrast. (Images provided by the courtesy of G. Myhr, MR-Centre, Trondheim, Norway.)

tissues. The achievable T_1-enhancement can be estimated by making some assumptions about the contrast-agent concentration in the enhancing tissue.

The T_1 of brain white matter is approximately 800 ms at 1.5 T. It is evident from Figure 1.22(a) that the brain lesion has a slightly longer T_1 than the white matter in the non-enhanced image (lower signal intensity), say 1000 ms. The standard dose of gadolinium-ECF agents is 0.1 mmol Gd/kg body weight. Given that the ECF agents are distributed in the extracellular volume (plasma and interstitium), their distribution volume is about 200 ml/kg. The initial equilibrium concentration in tissue is therefore 0.1 mmol/kg per 200 ml/kg = 0.5 mM. Assuming that the contrast-agent concentration in the enhancing part of the lesion equals the tissue equilibrium concentration, the T_1 of the lesion after contrast administration can be estimated by using Equation (1.57) as follows:

$$R_{1\_post} = 1\,\mathrm{s}^{-1} + 0.5\,mM \times 4\,mM^{-1}\,s^{-1} = 3\,s^{-1} \tag{1.58}$$

and $T_{1\_\mathrm{post}} = 1/R_{1\_post} = 330$ ms. The T_1 relaxation time of the tumour is thus reduced from about 1000 to about 330 ms, i.e. by more than 60 %. The relative reduction in the T_2 relaxation time is much smaller, since T_2 is significantly shorter than the T_1 before contrast administration. Assuming the T_2 of the lesion to be 120 ms pre-contrast, the T_2 of the lesion after contrast administration can similarly be found to be about 90 ms (using Equation (1.57), with $r_2 =$ 5 $\mathrm{mM}^{-1}\mathrm{s}^{-1}$). Thus, the relative reduction in T_2 is only about 25 %. Given that all T_1-weighted sequences use a short TE (typically 15–20 ms for spin-echo and 3–10 ms for gradient-echo sequences), the effect of the contrast agent on T_2 is negligible in T_1-weighted images. The achievable contrast enhancement is thus determined mainly by the repetition time (TR) which is used. Figure 1.23 shows

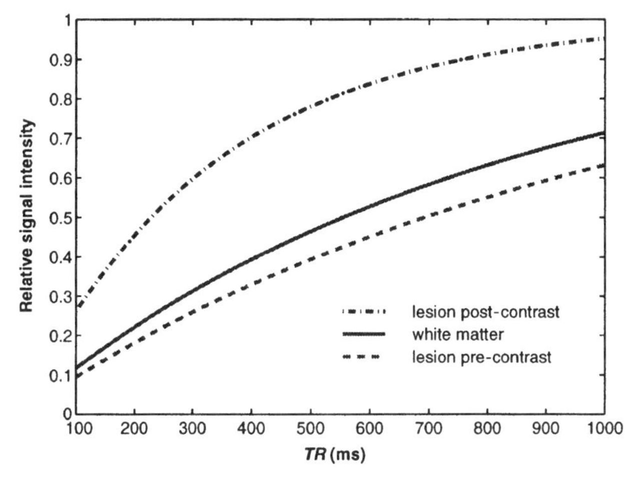

Figure 1.23 Effect of a varying TR on the relative signal difference between white matter and lesion by using a spin-echo sequence (see text for details).

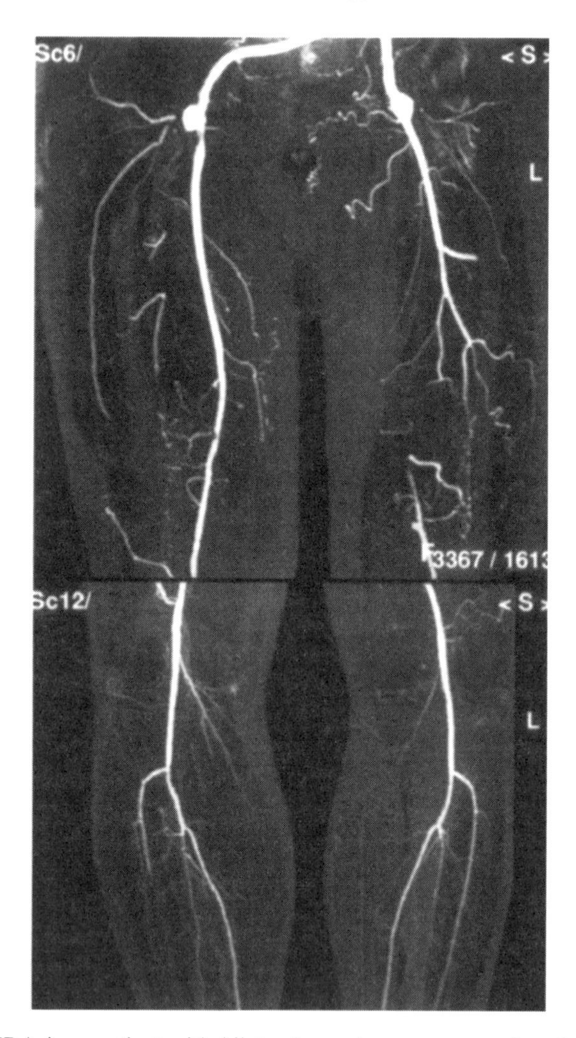

Figure 1.24 MRA in a patient with bilateral pseudo-aneurysm and occlusion of the left femoral artery following bolus administration of GdDTPA-BMA (Omniscan®). (Images provided by the courtesy of H. Ahlström and L. Johansson, Uppsala University Hospital, Sweden.)

the effect of a varying *TR* on the relative brain-lesion contrast by using the above numerical example. Notice that an optimal *TR* exists which gives the largest brain-lesion contrast for a given T_1-difference (see Section 6.1).

Contrary to tumor imaging, contrast-enhanced MR angiography usually requires a rapid bolus injection of a paramagnetic ECF agent, since these agents rapidly redistribute from the vascular space into the interstitium. Following such a bolus injection, the initial gadolinium concentration in blood can easily be ten times higher than the steady-state concentration. The benefit of such a

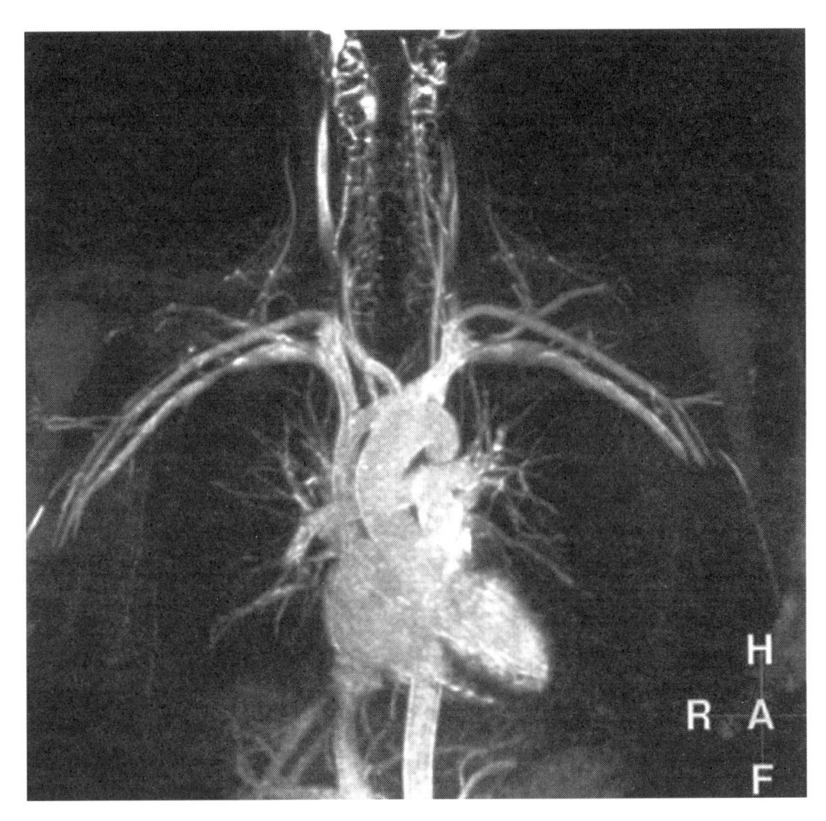

Figure 1.25 MRA of the thoracic region following administration of the UPSIO blood-pool agent, NC100150 Injection (Clariscan™). Image courtesy of H. Ahlström and L. Johansson, Uppsala University Hospital, Sweden.)

high gadolinium concentration is that the resulting T_1 of blood is extremely short, typically less than 100 ms, thus giving rise to a very large blood-background contrast in heavily T_1-weighted images. This effect is successfully utilized in contrast-enhanced MR angiography [29, 30]. However, T_2/T_2^* will also be very short in the bolus phase, and it therefore becomes a sequence question as to whether an additional decrease in T_1 will give a corresponding increase in signal intensity.

Figure 1.24 shows an MR angiogram obtained following a rapid bolus administration of GdDTPA-BMA. High-quality MR angiograms have also been obtained by using UPSIO-type blood-pool agents.

Figure 1.25 shows an image obtained in a healthy volunteer following the administration of an NC100150 Injection (Clariscan^{TM}, Nycomed Imaging, Oslo, Norway). This image was acquired in the 'steady-state', which is evident from the fact that arteries and veins are equally enhanced. The advantage of this approach over Gd-ECF-enhanced MRA is, however, that a much larger imaging window is available, and that images of a higher signal to-noise ratio and/or resolution can be obtained. The relative utility of the Gd-ECF approach versus the blood-pool-agent approach in MR angiography has not yet been established.

REFERENCES

1. Abragam A. *Principles of Nuclear Magnetism*, Oxford University Press, Oxford, 1961, pp. 19–57.
2. Canet D. *Nuclear Magnetic Resonance: Concepts and Methods*, John Wiley & Sons, Chichester, 1996, pp. 1–102.
3. Mansfield P. and Morris P. G. *NMR Imaging in Biomedicine*, Academic Press, Orlando, FL, 1982, pp. 10–65.
4. Wehrli F. W. *Fast-Scan Magnetic Resonance: Principles and Applications*, Raven Press Ltd, New York, 1991, pp. 3–9.
5. Ljunggren S. *J. Magn. Reson.* 1983, **54**, 338.
6. Meyer C. H., Hu B. S., Nishimura, D. G. and Macovski, A. *Magn. Reson. Med.* 1992, **28**, 202.
7. Gonzales R. C. and Wintz P. *Digital Image Processing*, Addisson-Wesley Publishing Company, Reading, MA, 1987, pp. 61–100.
8. Listerud J., Einstein S., Outwater E. and Kressel H. Y. *Magn. Reson. Q.* 1992, **8**, 199.
9. Vlaardingerbroek. M. T. and den Boer J. A. *Magnetic Resonance Imaging: Theory and Practice*, Springer-Verlag, Berlin, 1996.
10. Solomon I. *Phys. Rev. 1955*, **99**, 559.
11. Freed J. H. *J. Chem. Phys.* 1978, **68**, 4034.
12. Koenig S. H. and Brown R. D. III *Prog. Nucl. Magn. Reson. Spectr. OSC.*, 1991, **22**, 487.
13. Vogl T. J., Hammerstingl R., Schwarz W., Mack M. G., Muller P. K., Pegios W., Keck H., Eibl-Eibesfeldt A., Hoelzl J., Woessmer B., Bergman C. and Felix R. *Radiology*, 1996, **198**, 881.
14. Koenig S. H. and Kellar K. E. *Magn. Reson. Med.* 1995, **34**, 227.
15. Kellar K. E., Fujii D. K., Gunther W. H. H., Briely-sæbø K., Bjørnesud A., Spiller M. and Koenig S. H. J. *Magn. Reson. Imaging* 2000, **11**, 488.
16. Roch A., Muller R. N and Gillis P. J. *Chem. Phys.* 1999, **110**, 5403.
17. Ahlström K. H., Johansson L. O., Rodenburg J. B., Ragnarsson A. S., Åkeson P. and Børseth A. *Radiology* 1999, **211**, 865.
18. Wikstrom L. J., Johansson L. O., Ericsson B. A., Børseth A., Åkeson P. A and Ahlström K. H. *Acad. Radiol.* 1999, **6**, 292.
19. Wagenseil J. E., Johansson L. O. and Lorenz C. H. *J. Magn. Reson. Imaging* 1999, **10**, 784.
20. Van Wagoner M. and Worah D. *Invest. Radiol.* 1993, **28**, S44.
21. Koenig S. H. and Kellar K. E. *Acad. Radiol.* 1998, **5**, S200.
22. Lauffer R. B., Parmelee D. J., Dunham S. U., Ouellet H. S., Dolan R. P., Witte S., McMurry T. J. and Walovitch R. C. *Radiology* 1998, 207, 529.
23. Dong Q., Hurst D. R., Weinmann H. J., Chenevert T. L., Londy F. J. and Prince M. R. *Invest. Radiol.* 1998, **33**, 699.
24. Hemmingsson A. (Ed.) *Acta Radiol.* 1997, **38**.
25. Reimer P., Rummeny E. J., Daldrup H. E., Hesse T., Balzer T., Tombach B. and Peters P. E. *Eur. Radiol.* 1997, **7**, 275.

26. Jacobsen T. F., Laniado M., Van Beers B. E., Dupas B., Boudghene F. P., Rummeny E., Falke T. H., Rinck P. A., MacVicar D. and Lundby B. *Acad. Radiol.* 1996, **3**, 571.
27. Yablonskiy D. A. and Haacke E. M. *Magn. Reson. Med.* 1994, **32**, 794.
28. Aslanian V., Lemaignen H., Bunouf P., Svaland M. G, Børseth A. and Lundby B. *Neuroradiology* 1996, **38**, 537.
29. Meaney J. F., Prince M. R., Nostrant T. T. and Stanley J. C. *J. Magn. Reson. Imaging.* 1997, **7**, 171.
30. Neimatallah M. A., Dong Q., Schoenberg S. O., Cho K. J. and Prince M. R. *J. Magn. Reson. Imaging* 1999, **10**, 357.

2 Relaxivity of Gadolinium(III) Complexes: Theory and Mechanism

ÉVA TÓTH, LOTHAR HELM and ANDRÉ E. MERBACH
University of Lausanne, Switzerland

1 INTRODUCTION

The aim of using a contrast agent in Magnetic Resonance Imaging (MRI) is to accelerate the relaxation of water protons in the surrounding tissue. This objective can be achieved by paramagnetic substances. In 1948, Bloch *et al.* reported the use of the paramagnetic ferric nitrate salt to enhance the relaxation rates of water protons [1]. Some 30 years later, Lauterbur *et al.* applied a Mn(II) salt to distinguish between different tissues based on the differential relaxation times and thus produced the first MR image [2]. Nowadays, Gd(III) complexes are by far the most widely used contrast agents in clinical practice. The choice of Gd(III) is explained by its seven unpaired electrons which makes it the most paramagnetic stable metal ion. Besides this, Gd(III) has another significant feature: owing to the symmetric S-state, its electronic relaxation is relatively slow which is relevant to its efficiency as an MRI contrast agent.

As most of the current contrast agent applications in MRI concern Gd(III) complexes, the discussion of the relaxation theory and of the experimental results in this present chapter will be focused on Gd(III)-based agents. Several reviews, one of which is rather recent, have already been published on this topic [3–5]. In addition to Gd(III) compounds, a Mn(II) chelate, i.e. Mn(II)DPDP, is the only metal complex which has been approved as an MRI contrast agent (DPDP $= N, N'$-dipyridoxylethylenediamine-N, N'-diacetate 5, 5'-bis(phosphate)) [6]. Mn(II)DPDP is a weak chelate that dissociates *in vivo* to give free manganese which is taken up by hepatocytes [7]. The presence of the ligand is necessary because it facilitates a slower release of manganese than would have been the case had manganese been administered in the form of a simple salt, such as manganese chloride.

The relaxation mechanism of paramagnetic particles will be treated separately in Chapter 10. A discussion of free radicals [8, 9], or of the new and rapidly emerging field of hyperpolarized noble gases [10] as MRI contrast agents is beyond the scope of this present survey.

The Chemistry of Contrast Agents in Medical Magnetic Resonance Imaging
Edited by A. E. Merbach and É. Tóth.

The general theory of solvent nuclear relaxation in the presence of paramagnetic substances was developed by the groups of Bloembergen, Solomon, and others [11–16]. A Gd(III) complex induces an increase of both the longitudinal and transverse relaxation rates, $1/T_1$ and $1/T_2$, respectively, of the solvent nuclei. The observed solvent relaxation rate, $1/T_{i,obs}$, is the sum of the diamagnetic ($1/T_{i,d}$) and paramagnetic relaxation rates ($1/T_{i,p}$):

$$\frac{1}{T_{i,obs}} = \frac{1}{T_{i,d}} + \frac{1}{T_{i,p}}, \text{where} \quad i = 1, 2 \tag{2.1}$$

Although Equation (2.1) is widely used as a general description, strictly speaking it is only valid for dilute paramagnetic solutions, which condition is generally fulfilled. The diamagnetic term, $1/T_{i,d}$, corresponds to the relaxation rate of the solvent (water) nuclei in the absence of a paramagnetic solute. The paramagnetic term, $1/T_{i,p}$, gives the relaxation rate enhancement caused by the paramagnetic substance, which is linearly proportional to the concentration of the paramagnetic species, [Gd]:

$$\frac{1}{T_{i,obs}} = \frac{1}{T_{i,d}} + r_i[\text{Gd}], \text{where} \quad i = 1, 2 \tag{2.2}$$

In Equation (2.2), the concentration is usually given in mmol/L; for non-dilute systems, the linear relationship is valid only if the concentration is expressed in mmol/kg of solvent (millimolality). According to Equation (2.2), a plot of the observed relaxation rates versus the concentration of the paramagnetic species gives a straight line and its slope defines the relaxivity, r_i (in units of $\text{mM}^{-1}\ \text{s}^{-1}$). If we consider the relaxation of water protons, which is the basis of imaging by magnetic resonance, and is consequently significant from the practical point of view, we can introduce the corresponding term *proton relaxivity*. Proton relaxivity directly refers to the efficiency of a paramagnetic substance to enhance the relaxation rate of water protons, and thus to its efficiency to act as a contrast agent. It has to be noted that the simple term 'relaxivity' is very often used in the context of MRI contrast agents and always refers to 'longitudinal proton relaxivity', even if the adjectives are omitted. In the following, we primarily deal with the theory of proton relaxation in the presence of Gd(III)-containing paramagnetic species.

The paramagnetic relaxation of the water protons originates from the dipole–dipole interactions between the nuclear spins and the fluctuating local magnetic field caused by the unpaired electron spins. This magnetic field around the paramagnetic center vanishes rapidly with distance. Therefore, specific chemical interactions that bring the water protons into the immediate proximity of the metal ion play an important role in transmitting the paramagnetic effect towards the bulk solvent. For Gd(III) complexes, this specific interaction corresponds to the binding of the water molecule(s) in the first coordination sphere of the metal ion. These inner-sphere water protons then

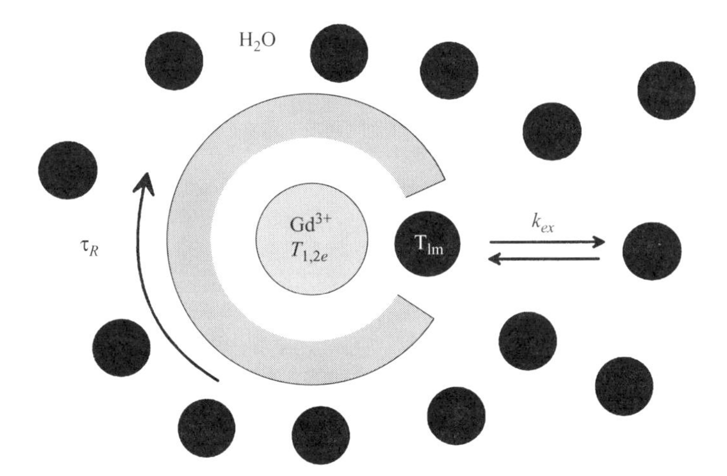

Figure 2.1 Schematic representation of a Gd(III) chelate with one inner-sphere water molecule, surrounded by bulk water; τ_R refers to the rotational correlation time of the molecule, k_{ex} to the water/proton exchange rate and $1/T_{1,2e}$ to the relaxation rate of the Gd(III) electron spin.

exchange with bulk solvent protons and in this way the paramagnetic influence is propagated to the bulk. This mechanism is depicted as the inner-sphere contribution to the overall proton relaxivity (Figure 2.1). Solvent molecules of the bulk also experience the paramagnetic effect when they diffuse in the surroundings of the paramagnetic center. The effect of the random translational diffusion is defined as outer-sphere relaxation. Thus, we separate the inner-and outer-sphere contributions based on the intra-and intermolecular nature of the interaction, respectively. This separation is also useful in explaining the observed proton relaxivities in terms of existing theories. The total paramagnetic relaxation rate enhancement due to the paramagnetic agent is therefore given as in Equation (2.3), or expressed in relaxivities in Equation (2.4):

$$\left(\frac{1}{T_{i,p}}\right) = \left(\frac{1}{T_{i,p}}\right)^{IS} + \left(\frac{1}{T_{i,p}}\right)^{OS} \tag{2.3}$$

$$r_i = r_i^{IS} + r_i^{OS} \tag{2.4}$$

where the superscripts 'IS' and 'OS' refer to the inner and outer sphere, respectively.

For certain agents, solvent molecules that are not directly bound in the first coordination sphere may also remain in the proximity of the paramagnetic metal for a relatively long time, e.g. due to hydrogen bridges to the ligand (e.g. to its carboxylate or phosphonate groups) or to the solvent molecule(s) in the first coordination sphere. The relaxivity contribution originating from these

Figure 2.2 Three different types of water molecules around a Gd(III) complex as obtained by a molecular dynamics simulation of $[Gd(DOTA)(H_2O)]^-$ in aqueous solution. The *inner-sphere water* is directly coordinated to the metal (its oxygen is black). *Second-sphere water* molecules are on the hydrophilic side of the complex, close to the carboxylate groups. These are oriented with their hydrogens towards the carboxylate oxygens to form hydrogen bonds (their oxygens are represented as gray balls). *Outer-sphere or bulk water* molecules have no preferential orientation (shown in white). The bulk water molecules in front and behind the complex have been removed for clarity.

interactions is called second-sphere relaxivity (even if the symmetry is not spherical), and can be described by the same theory as the inner-sphere term. However, very often this contribution is neglected or its effect is taken into account in the outer-sphere term. The three different types of water (inner, second and outer sphere) are represented in Figure 2.2.

For the currently used, monomeric Gd(III)-based contrast agents, the outer- and inner-sphere relaxation mechanisms contribute approximately to the same extent to the observed proton relaxivity at the imaging fields. It is the inner-sphere term that can be considerably augmented, whereas the outer-sphere contribution can hardly be modified. Consequently, for the new-generation agents of higher efficiency (higher relaxivity), the inner-sphere term becomes relatively much more important and represents the major contribution to the overall proton relaxation rate.

In this chapter, first we will discuss the inner-sphere relaxivity by analyzing each factor that determines this relaxation mechanism. This will be followed by a survey of the second-and outer-sphere contributions to proton relaxivity. We will also discuss in detail proton Nuclear Magnetic Resonance Dispersion (NMRD), a technique that is widely used and is rather specific for the characterization of MRI contrast agents, and about which is little known outside of this field. In the last section of this chapter, we will draw conclusions from the existing results and give indications on how to optimize the parameters that govern proton relaxivity, with the aim of designing more efficient MRI contrast agents.

2 INNER SPHERE PROTON RELAXIVITY

The inner-sphere contribution to the overall proton relaxation rate enhancement results from the chemical exchange of the coordinated water protons with the bulk, and thus represents a two-site exchange problem. One site (the bulk) is much more populated than the other site (coordinated water), and the observed signal corresponds to that of the free water. The longitudinal and transverse inner-sphere relaxation rates are given by Equations (2.5) [17] and (2.6) [18], respectively:

$$\left(\frac{1}{T_1}\right) = \frac{cq}{55.5}\left(\frac{1}{T_{1m} + \tau_m}\right) = P_m \frac{1}{T_{1m} + \tau_m} \tag{2.5}$$

$$\left(\frac{1}{T_2}\right)^{IS} = \frac{P_m}{\tau_m}\left[\frac{T_{2m}^{-2} + \tau_m^{-1} T_{2m}^{-1} + \Delta\omega_m^2}{\left(\tau_m^{-1} + T_{2m}^{-1}\right)^2 + \Delta\omega_m^2}\right] \tag{2.6}$$

where c is the molal concentration, q is the number of bound water nuclei per Gd (hydration number), P_m is the mole fraction of the bound water nuclei, τ_m is the lifetime of a water molecule in the inner sphere of the complex (equal to the reciprocal water exchange rate, $1/k_{ex}$), $1/T_{1m}$ and $1/T_{2m}$ are, respectively, the longitudinal and transverse proton relaxation rates in the bound water, and $\Delta\omega_m$ is the chemical shift difference between the bound water and the bulk water.

The relaxation of the bound water protons is governed by the dipole–dipole (DD) and scalar (SC) (or contact) mechanisms, where both are dependent on the magnetic field. The dipole–dipole part of the interaction is modulated by the reorientation of the nuclear spin – electron spin vector, by changes in the orientation of the electron spin (electron spin relaxation) and by water (proton) exchange. The scalar interaction remains unaffected by reorientation of the molecule and is only modulated by electron spin relaxation and water exchange. Although later on in this chapter we will limit the detailed discussion to longitudinal relaxation, which today largely controls NMR image enhancement, here we give the expressions for transverse relaxation as well. The relaxation rates are generally expressed by the modified Solomon–Bloembergen equations:

$$\frac{1}{T_{im}} = \frac{1}{T_i^{DD}} + \frac{1}{T_i^{SC}}, \text{ where } \quad i = 1,2 \tag{2.7}$$

$$\frac{1}{T_1^{DD}} = \frac{2}{15}\left(\frac{\gamma_I^2 g^2 \mu_B^2}{r_{GdH}^6}\right) S(S+1)\left(\frac{\mu_0}{4\pi}\right)^2 \left(7\frac{\tau_{c2}}{1+\omega_s^2\tau_{c2}^2} + 3\frac{\tau_{c1}}{1+\omega_I^2\tau_{c1}^2}\right) \tag{2.8}$$

$$\frac{1}{T_1^{SC}} = \frac{2S(S+1)}{3}\left(\frac{A}{\hbar}\right)^2 \left(\frac{\tau_{e2}}{1+\omega_s^2\tau_{e2}^2}\right) \tag{2.9}$$

$$\frac{1}{T_2^{DD}} = \frac{1}{15}\left(\frac{\gamma_I^2 g^2 \mu_B^2}{r_{GdH}^6}\right) S(S+1)\left(\frac{\mu_0}{4\pi}\right)^2 \left(13\frac{\tau_{c2}}{1+\omega_s^2\tau_{c2}^2} + 3\frac{\tau_{c1}}{1+\omega_I^2\tau_{c1}^2} + 4\tau_{c1}\right) \tag{2.10}$$

$$\frac{1}{T_2^{SC}} = \frac{S(S+1)}{3}\left(\frac{A}{\hbar}\right)^2 \left(\frac{\tau_{e2}}{1+\omega_s^2\tau_{e2}^2} + \tau_{e1}\right) \tag{2.11}$$

In Equations (2.7)–(2.11), γ_I is the nuclear gyromagnetic ratio, g is the electron g-factor, μ_B is the Bohr magneton, r_{GdH} is the electron spin – proton distance, ω_I and ω_s are the nuclear and electron Larmor frequencies, respectively ($\omega = \gamma B$, where B is the magnetic field), and $A/\hbar$ is the hyperfine or scalar coupling constant between the electron of the paramagnetic center and the proton of the coordinated water. The correlation times which are characteristic of the relaxation processes are depicted as:

$$1/\tau_{ci} = 1/\tau_R + 1/T_{ie} + 1/\tau_m, \text{ where } \quad i = 1,\ 2 \tag{2.12}$$

$$1/\tau_{ei} = 1/T_{ie} + 1/\tau_m, \text{ where } \quad i = 1,,\ 2 \tag{2.13}$$

where τ_R is the rotational correlation time or, more precisely, the reorientational correlation time of the metal – proton vector, and T_{1e} and T_{2e} are, respectively the longitudinal and transverse electron spin relaxation times of the metal ion.

Since Gd(III) forms highly ionic bonds and the proton is relatively far from the paramagnetic center, the scalar coupling must be very weak, although the exact value of the coupling constant, $A/\hbar$, is not known. Moreover, the $\tau_{e2}/(1 + \omega_s^2\tau_{e2}^2)$ term completely vanishes at frequencies above 10 MHz. Consequently, at least for $1/T_{1m}$, the scalar mechanism represents only a small contribution to the overall proton relaxation (Equation (2.7)).

The modified Solomon–Bloembergen equations are based on several assumptions. First of all, they are valid within the Redfield limit, i.e. the motions in the lattice must occur on a much faster time scale than the motions in the spin system. In addition to the Redfield relaxation theory, two main assumptions are involved, namely that the electron spin relaxation is uncorrelated with molecular reorientation, and that the electron spin system is dominated by the electron Zeeman interaction and other interactions result only in electron spin relaxation. Further assumptions are that (i) the electron spin can be considered as a point dipole centered at the metal ion, (ii) the electron g-tensor is isotropic, (iii) the reorientation of the nuclear–electron spin is isotropic and can be characterized with a single correlation time, τ_R, and (iv) the chemical exchange is not correlated with the motions of the lattice.

The electronic relaxation rates, as described by Bloembergen and Morgan [15], and McLachlan [19], also depend on the magnetic field. For Gd(III) complexes they are usually interpreted in terms of a zero-field-splitting (ZFS) interaction. The electronic relaxation rates can be described by Equations (2.14)–(2.16), which are often referred to as the Bloembergen–Morgan theory of paramagnetic electron spin relaxation:

$$\left(\frac{1}{T_{1e}}\right)^{\mathrm{ZFS}} = 2C\left(\frac{1}{1+\omega_S^2\tau_v^2} + \frac{4}{1+4\omega_s^2\tau_v^2}\right) \tag{2.14}$$

$$\left(\frac{1}{T_{2e}}\right)^{\mathrm{ZFS}} = C\left(\frac{5}{1+\omega_S^2\tau_v^2} + \frac{2}{1+4\omega_s^2\tau_v^2} + 3\right) \tag{2.15}$$

$$C = \frac{1}{50}\Delta^2\tau_v[4S(S+1)-3] \tag{2.16}$$

Here Δ^2 is the mean-square zero-field-splitting energy and τ_v is the correlation time for the modulation of the zero-field-splitting interaction. This modulation results from the transient distortions of the complex. Some authors express C with the low-field limiting value of the electronic relaxation rate, τ_{s0}:

$$C = \frac{1}{10\tau_{s0}}$$

where:

$$\frac{1}{\tau_{s0}} = \frac{1}{5}\Delta^2\tau_v[4S(S+1)-3] = 12\Delta^2\tau_v \tag{2.17}$$

The validity of Equations (2.14)–(2.16) is restricted to conditions where $\omega_0\tau_\nu \ll 1$ is fulfilled.

The combination of the modified Solomon–Bloembergen equations (Equations 2.7–2.11) with the equations for electron spin relaxation (Equations 2.14–2.16) constitutes a complete theory for relating the observed paramagnetic relaxation rate enhancement to the microscopic properties, and is generally referred as to the Solomon–Bloembergen–Morgan (SBM) theory. Detailed discussions of this relaxation theory can be found in the literature [20, 21].

It is clear from the above equations that proton relaxivity is influenced by numerous parameters. In one limiting case, when the proton exchange is very slow ($T_{1m} \ll \tau_m$, the exchange rate will be the only determining factor for relaxivity (Equation (2.5)). If the proton exchange rate is fast enough such that $\tau_m \ll T_{1m}$ holds, the observed proton relaxivity will be exclusively determined by the relaxation rate of the coordinated protons, T_{1m}, which itself depends on the rate of proton exchange, rotation and electronic relaxation. In all cases, further variables are the Gd–proton distance (r_{GdH}) and the hydration number (q). These latter parameters influence proton relaxivity in an obvious manner. It is easy to recognize that doubling q (two inner-sphere water molecules instead of one) will double the inner-sphere relaxivity (Equation (2.5)), or that a slightly longer Gd–H distance results in a significant decrease in r_1 due to the sixth-power dependence. However, it is less evident to predict the influence of other parameters, such as the rotational correlation time, the proton exchange rate or the electronic relaxation parameters. Any prediction is further complicated by the fact that the influence of the above parameters strongly depends on the magnetic field. In order to visualize these effects, Figure 2.3 shows simulated inner-sphere proton relaxivities as a function of the proton exchange rate and of the rotation at different fields and at different values of the electronic relaxation rate, as determined by fixed τ_v–Δ^2 pairs.

The relaxivity maximum is attained when the correlation time, τ_{c1}, equals the inverse proton Larmor frequency ($1/\tau_{c1} = 1/\tau_R + 1/\tau_m + 1/T_{1e} = \omega_I$). However, τ_m enters also into Equation (2.5), which predicts a higher relaxivity for a shorter τ_m value. Consequently, τ_m should be optimally decreased but not as much so that it starts to limit T_{1m}. The optimal relationship is given as follows:

$$\frac{1}{T_{1m}} < \frac{1}{\tau_m} < \frac{1}{\tau_R}; \frac{1}{T_{1e}} \tag{2.18}$$

Figure 2.3 clearly shows that all three main parameters (rotational correlation time, proton exchange rate and electronic relaxation rate) have to be optimized at the same time in order to attain maximum relaxivities. Another important point is that the theoretical relaxivity maximum decreases with increasing magnetic field. In practice, however, this negative effect of the higher magnetic fields can be compensated for by the greater spatial resolution and sensitivity.

Contrast agents that are currently used in clinical practice are situated at the bottom of the theoretical surfaces (see Figure 2.3). The reason is that neither their rotational correlation time ($\sim 10^{-10}$s), nor their proton exchange rate

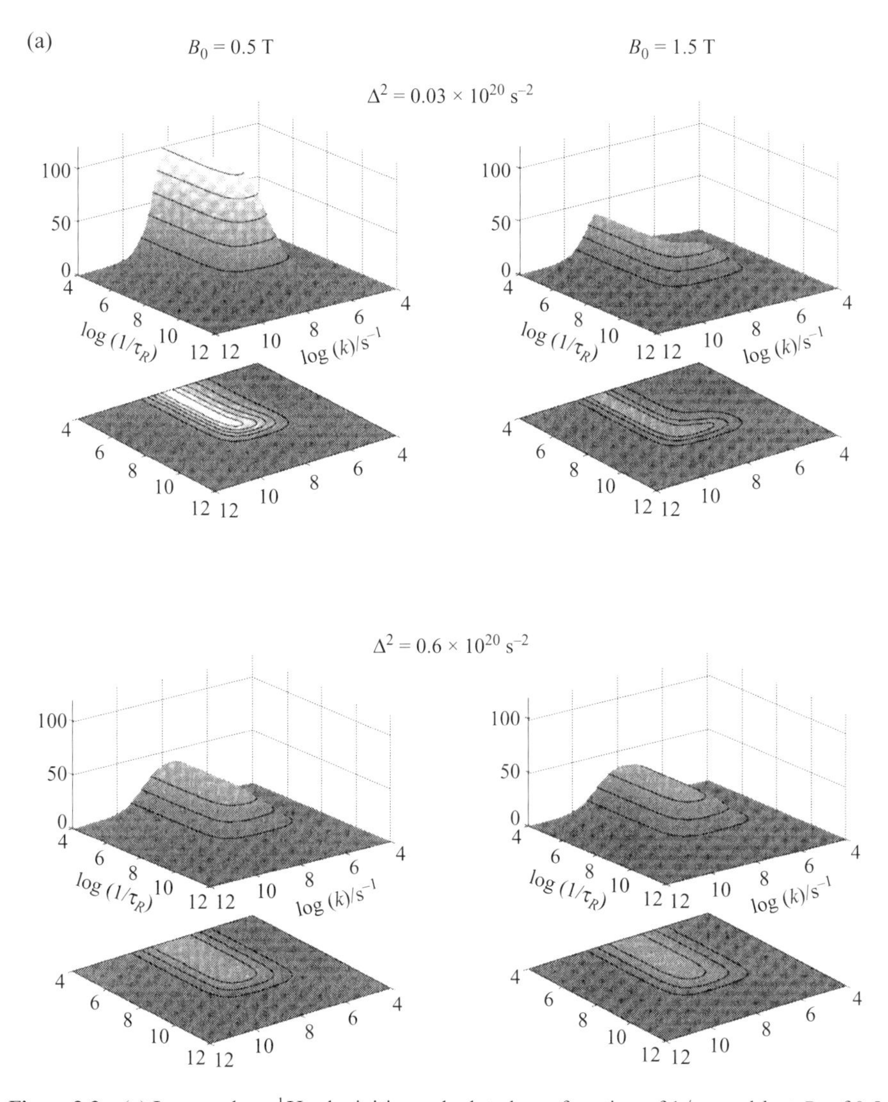

Figure 2.3 (a) Inner-sphere ^{1}H relaxivities, calculated as a function of $1/\tau_R$ and k at B_0 of 0.5 T and 1.5 T ($\tau_v = 1 \times 10^{-12}$ s^{-1}), and plotted as three-dimensional and contour plots. The electron spin relaxation times, T_{1e} are 2.9×10^{-8} s (top) and 2.9×10^{-9} s (bottom).

($\sim 10^6$s^{-1}) is optimal. These simulated surfaces give a clear answer on how to improve the efficiency of the approved agents: on the one hand, their rotational motion has to be slowed down, while on the other hand their proton exchange has to be accelerated. At the end of this chapter, a section will be devoted to the optimization of these parameters and to the design of high-relaxivity MRI contrast agents.

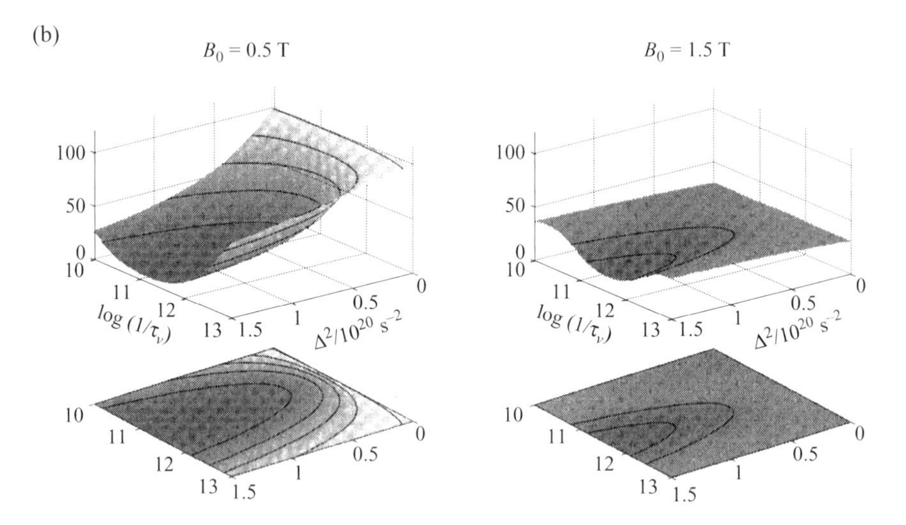

Figure 2.3 (b) Inner-sphere ^{1}H relaxivities, calculated as a function of $1/\tau_v$ and Δ^2 at B_0 of 0.5 T and 1.5 T ($\tau_R = 1 \times 10^{-8}\ s^{-1}$ and $k = 1 \times 10^8\ s^{-1}$), and plotted as three-dimensional and contour plots.

2.1 HYDRATION NUMBER AND HYDRATION EQUILIBRIA

Inner-sphere proton relaxivity is linearly proportional to the number of water molecules directly coordinated to the Gd(III) ion, as shown by Equation 2.5. In the Gd^{3+} aqua ion there are eight inner-sphere water molecules which would result in a proportionally high relaxation enhancement. However, the aqua ion itself is toxic, since, under physiological conditions, it has a strong tendency to precipitate in the form of gadolinium-hydroxide or to form stable complexes with endogenously available anions. Therefore, as will be discussed in detail in Chapter 6, the metal ion has to be incorporated into a multidentate ligand which will protect it. All approved Gd(III)-based contrast agents are poly (amino carboxylate) complexes where the chelating ligand occupies eight coordination sites and leaves enough room for the coordination of one water molecule in the inner sphere. Attempts have been made to increase this low hydration number; however, this apparently obvious way of increasing the proton relaxivity is not easily applicable to the Gd(III)-based contrast agents due to toxicity concerns. The thermodynamic and kinetic stabilities of the complex, which are guarantees of non-toxicity, can considerably decrease when more than one coordination site is occupied by water.

Equation (2.5) shows clearly that proton relaxivities cannot be analyzed without knowing the hydration number. Although solid-state structures provide q, there are examples where the hydration state changes from the solid to the solution state [22]. Moreover, metal complexes in general may show coordination and/or hydration equilibria in aqueous solution. Therefore, there is an inevitable need to determine q in aqueous solution. Several experimental techniques are available to obtain the hydration number of lanthanide complexes, although none of these can be directly applied to Gd(III) chelates.

One method uses lanthanide-induced shifts (LISs) [23]. For paramagnetic lanthanide(III) ions, the contact contribution to the Ln(III)-induced shift of a Ln(III)-bound ^{17}O nucleus is independent of the nature of the ligand and of the other ligands coordinated to the metal. For Dy(III), the induced ^{17}O shift is dominated by the contact contribution (> 85 %), and thus the separation of contact and pseudo-contact terms can be avoided. In a general case, the plot of the observed ^{17}O shifts in a Dy(III) chelate solution versus the complex concentration gives a straight line, with the slope being proportional to the hydration number of the Dy(III) complex. Hence, the ratio of the slopes obtained for different Dy(III) complexes gives the ratios of their hydration numbers. If one has a reference complex of a known q (generally, $[Dy(H_2O)_8]^{3+}$ is used), the shift ratio directly provides the water coordination number for the complex in question. The applicability of ^{17}O NMR spectroscopy for studying the hydration state of Dy(III) complexes has been demonstrated by using several examples and will be discussed in detail in Chapter 8.

Hydration numbers can also be determined by laser-induced luminescence. In this technique, the difference in luminescence lifetimes measured in H_2O and D_2O solutions is related to the hydration number, q, of the complex via the following simple equation [24, 25]:

$$q = A_{Ln}\left(\tau_{H_2O}^{-1} - \tau_{D_2O}^{-1}\right) \tag{2.19}$$

where τ is the luminescence lifetime and A_{Ln} is a proportionality constant specific to a given lanthanide ion (for further details, see Chapter 11). This method is widely used for Eu(III) and Tb(III) complexes. Since these two lanthanides neighbor gadolinium in the periodic table, and consequently have very similar ionic radii, the results of this technique can give a good estimate for the hydration number of the corresponding Gd(III) chelate. Some hydration numbers obtained for selected Ln(III) complexes are given in Table 2.1, with the structures of the corresponding ligands being shown in Charts 2.1 and 2.2.

A common drawback to both of the above methods, although often neglected, is that they provide an average hydration number with an error of ±10 to ± 20 %, and thus we can not tell with any certainty whether different hydration states, e.g. $[LnL(H_2O)n]$ and $[LnL(H_2O)_{n+1}]$, coexist in solution (hydration equilibrium), or if the complex has a single hydration state. An ultimate answer to this question is provided by UV–Vis spectrophotometric measurements on the Eu(III) chelates. The Eu(III) has an absorption band in the visible spectrum (578–582 nm), at a wavelength which is very sensitive to small changes in the coordination environment. Although the intensity of this $^7F_0 \rightarrow ^5D_0$ transition is low, the bands are relatively narrow, which allows us to

Table 2.1 Average hydration numbers (per Ln(III) ion), q, for poly(amino carboxylate) complexes derived from ^{17}O NMR or Eu(III) and Tb(III) luminescence measurements. For comparison, the corresponding numbers of bound water molecules found by using solid-state X-ray crystallography are also shown.

Ligand[a]	^{17}O NMR	Luminescence	X-ray	Reference
DTPA	1.3	1.1	1	23, 66, 151, 152
DTPA-BPA	1.0	1.0	—	57, 153
15-DTPA-en	1.2	2.3	—	154, 155
30-DTPA-en-DTPA-en	0.8	1.2	—	154, 155
TTHA	0.2	0.2	0	23, 156, 157
NOTA	2.5	3.3	—	23, 57
DOTA	1.0	1.0, 1.1	1	151, 158, 159
TACI	2.3	—	2	49

[a] Structures are shown in Charts 2.1 and 2.2.

distinguish different coordination states of the metal. This transition can be used to determine the number of species present in solution [26], and, in particular, to characterize hydration equilibria for Eu(III) complexes [27–30].

A Eu(III) complex which has two different hydrated forms in aqueous solution presents two absorption bands belonging to the two species and separated by about 0.5 nm (Figure 2.4). In all cases, the band at the lower wavelength decreases, whereas the other band increases with temperature, thus indicating that there is a dynamic equilibrium between the two species of different coordination number:

$$\left[\mathrm{EuL(H_2O)}_{q+1}\right] \rightleftharpoons \left[\mathrm{Eu(L)(H_2O)}_q\right] + \mathrm{H_2O} \tag{2.20}$$

Based on the temperature behavior, the band at the higher wavelength can be attributed to the chelate with the higher hydration number, and the band at the lower wavelength to the chelate with the lower hydration number. This assignment is in accordance with the findings of Frey and Horrocks [31]. The latter studied the luminescence for a great number of Eu(III) complexes, and correlated the frequency of the $^7F_0 \rightarrow {}^5D_0$ transition with the nephelauxetic effects of the ligating atoms. Their empirical rule also predicts a higher wavelength for the complex of higher hydration (and higher overall coordination) number than for the one of lower hydration number. Hence, although the UV-Vis spectrum does not directly provide the hydration number for the Eu(III) complex, the attribution of the transition bands to total coordination numbers is quite straightforward. In general, the total coordination number observed for Eu(III) complexes can be either eight or nine, consequently the hydration equilibrium, if it exists, is always between eight-and nine-coordinated complexes. If one knows how many donor atoms of the ligand(s) participates in the chelation, it is easy to calculate

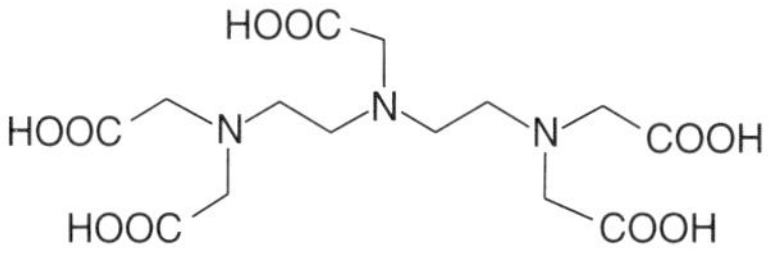

DTPA

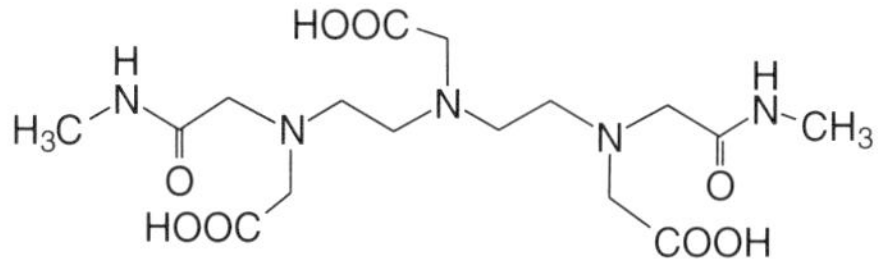

DTPA-BMA

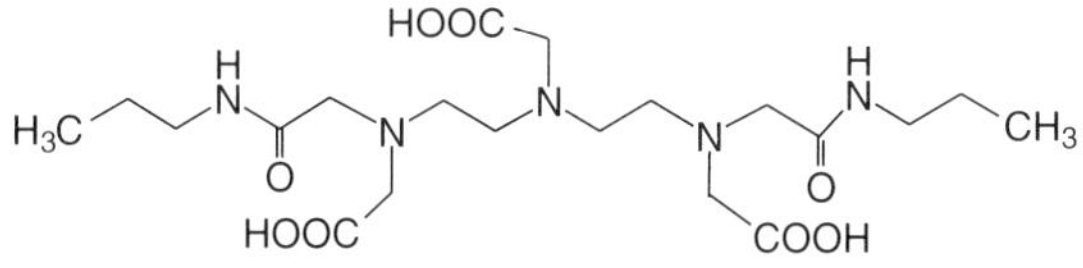

DTPA-BPA

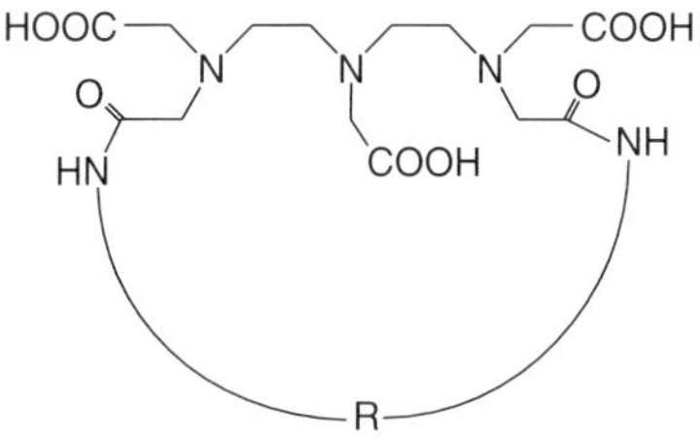

15-DTPA-en $R = CH_2CH_2$

30-DTPA-en-DTPA-en ($n = 2$)

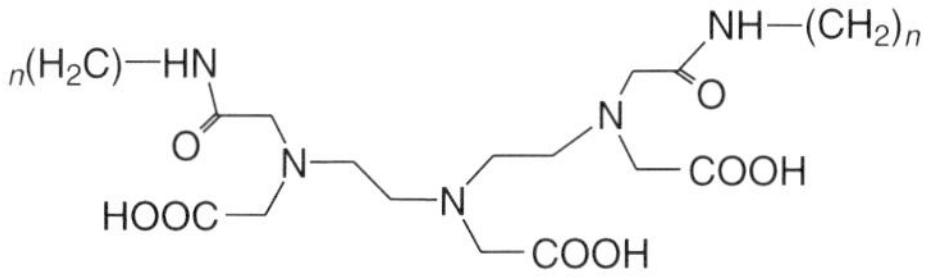

[**Chart 2.1**]

the possible hydration number for the two species in equilibrium. If the number of the coordinating atoms from the ligand is unknown, it is necessary to perform one of the techniques discussed above that will provide the absolute number for q (however, they do not necessarily indicate if there is a hydration equilibrium or not).

A temperature dependence of the UV-Vis spectra allows the determination of the thermodynamic parameters for the hydration equilibrium, whereas the pressure dependence yields the reaction volume. In the analysis, the ratio of the peak integrals (*Int*), obtained from a least-squares fit of the experimental spectrum, is related to the equilibrium constant (K_{eq}) [28]:

TTHA

DOTA

NOTA

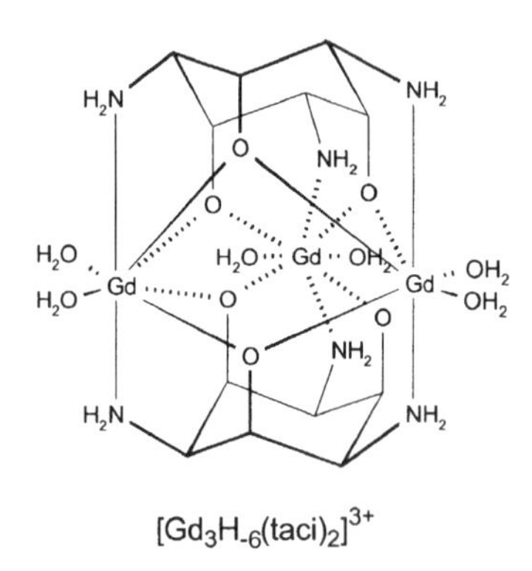

$[Gd_3H_{-6}(taci)_2]^{3+}$

[Chart 2.2]

$$\frac{Int_q}{Int_{q+1}} = K_{eq}\frac{I_q}{I_{q+1}} \tag{2.21}$$

where I is the integrated intensity of the band. Equations (2.22) and (2.23) then give the temperature and pressure dependence of this ratio:

$$\ln\left(\frac{Int_q}{Int_{q+1}}\right) = -\frac{\Delta H^0}{RT} + \frac{\Delta S^0}{R} + \ln\left(\frac{I_q}{I_{q+1}}\right) \tag{2.22}$$

$$\ln\left(\frac{Int_q}{Int_{q+1}}\right) = \ln K_{eq,0} - \frac{\Delta V^0 P}{RT} + \ln\left(\frac{I_q}{I_{q+1}}\right) \tag{2.23}$$

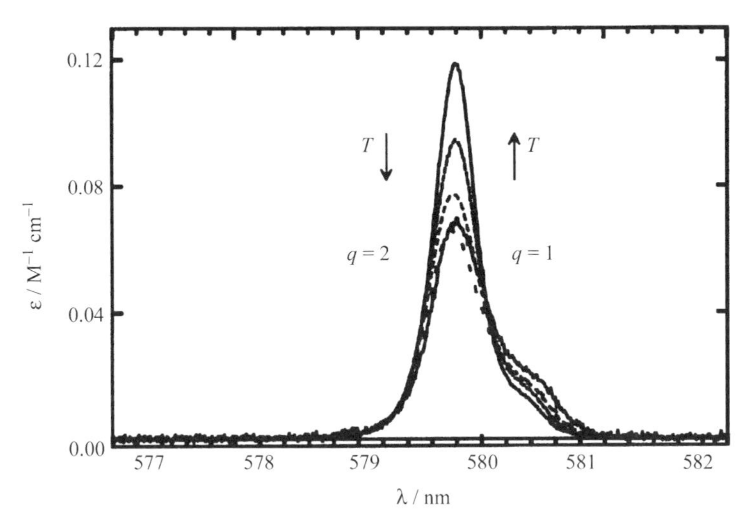

Figure 2.4 Typical example of the $^7F_0 \rightarrow {}^5D_0$ transition in UV-Vis spectra of a Eu(III) chelate with a hydration equilibrium. The spectra were recorded in a $[Eu(DO3A)(H_2O)_q]$ solution as a function of the temperature (T = 275.2, 292.7, 311.6, 333.7 and 348.8 K) [29] (1999). © John Wiley & Sons, Ltd. Reproduced with permission.

Provided that the ratio of the integrated intensities of the peaks, I_q/I_{q+1}, does not change significantly with temperature and pressure (i.e. the molar absorbances of the two complexes are the same), the slope of the dependence of $\ln(Int_q/Int_{q+1})$ with respect to inverse temperature yields the reaction enthalpy, ΔH^0, and that with respect to pressure gives the reaction volume, ΔV^0. With the additional hypothesis that the integrated intensities of the two peaks are identical ($\ln(I_q/I_{q+1}) = 0$ in Equations (2.22) and (2.23), the temperature dependence also yields the reaction entropy, ΔS^0, and thus the equilibrium constant at any temperature. This hypothesis is not necessarily valid, especially if the symmetry is not the same for the two differently hydrated species; however, the ΔH^0 and ΔV^0 values will be in any case correct. The thermodynamic parameters characterizing the hydration equilibrium, as well as the average hydration numbers are summarized for different Eu(III) complexes in Table 2.2, with the structures of the corresponding ligands being shown in Chart 2.3.

In conclusion, as there is no direct method to measure the hydration number on a Gd(III) complex, one has to extrapolate from the q values directly determined for the Dy, Eu or Tb analogues. Neither the dysprosium-induced shift, nor the luminescence method, can tell if there is an equilibrium between differently hydrated species, and therefore UV-Vis spectrophotometric measurements performed on the corresponding Eu(III) chelate are necessary to unambiguously characterize the hydration state.

It has to be noted that the value of the ^{17}O scalar coupling constant, $A/\hbar$, obtained for Gd(III) complexes also gives a good hint if the hydration number

Table 2.2 Thermodynamic parameters and average hydration numbers (q_{av}) determined by UV-Vis spectrophotometry for the equilibrium, $[Eu(L)(H_2O)_q] \leftrightarrow [Eu(L)(H_2O)_{q-1}] + H_2O$, for various selected ligands, L.

L	$EDTA^{4-a}$	$CDTA^{4-b}$	$DO3A^{3-c}$	$DO2A^{2-d}$
K_{eq}^{298}	0.59	0.11	0.13	0.25
ΔH^0 (kJ mol^{-1})	+17.7	+8.3	+12.6	+12.1
ΔS^0 (J mol^{-1} K^{-1})	+55	+19.7	+25.2	+28.9
ΔV^0 (cm mol^{-1})		+3.0	+7.5	—
q_{av}	2.6	2.9	1.9[e]	2.8

[a] Reference [28].
[b] CDTA = cyclohexane-1, 2-diamine-*N*, *N*, *N'*, *N'*-tetraacetate [28].
[c] Reference [29].
[d] Reference [30].
[e] $q = 1.8 \pm 0.2$ was found by luminescence for $[Tb(DO3A)(H_2O)_q]$ [160].

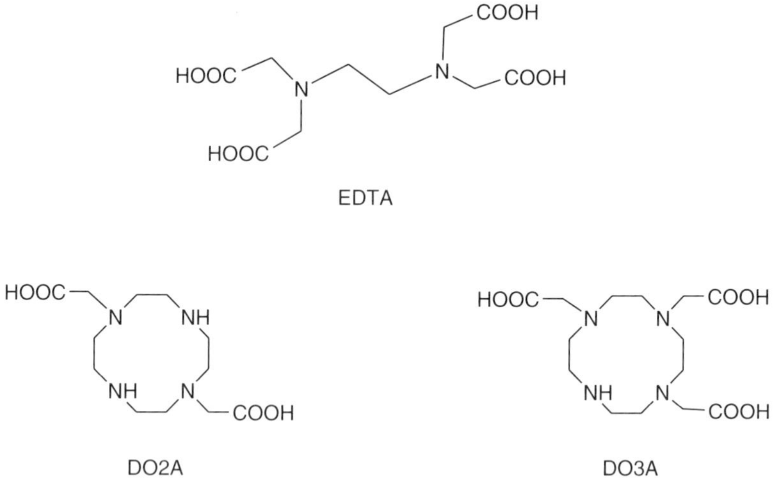

EDTA

DO2A

DO3A

[Chart 2.3]

used to calculate the coupling constant from the experimental chemical shifts is correct or not. For similar Gd(III) chelates, the values of the scalar coupling constant are within a certain range [32]; when the coupling constant is outside this usual range, the hydration number can be suspected to be wrong.

2.2 GD–H DISTANCE

The dipole–dipole relaxation term (Equation (2.8)) which determines inner-sphere proton relaxivity has a sixth-order dependence of the distance between the coordinated water proton and the Gd electron spin (r_{GdH}). A rapid calculation shows that a decrease of 0.1 Å in the Gd–H distance corresponds to a 20 %

increase in inner-sphere proton relaxivity, while a decrease of 0.2 Å results in as much as a 50 % increase. In principle, two possibilities could be imagined to shorten the metal–H distance with the aim of increasing proton relaxivity. First, higher tilt angles between the plane of the bound water and the metal–O bond could be induced by hydrogen bonding of the coordinated water to an appropriate side group of the chelate, which could result in a significant decrease of the metal–proton distance. However, it is not likely that such subtle structural features can be controlled. The second possibility for increasing relaxivity through changes in the metal–H distance could be the electron delocalization towards the ligand. Indeed, an anomalously high proton relaxivity was found for the Mn(III)TPPS4 complex and this was attributed to the anisotropy of the ground-state wavefunction of Mn^{3+} in the phorphyrin complex, thus effectively bringing the spin density of the metal ion closer to the coordinated water protons than would a spherically symmetric S-state ion [33]. However, for Gd(III) with its redox-stable and highly symmetric S-state, this route is not realizable.

Despite their great importance, the r_{GdH} values used in relaxivity analyses of Gd(III) complexes are in most cases only estimations. The reason is that the metal–coordinated water hydrogen distance is a difficult parameter to obtain experimentally. Generally, it is deduced from the Gd–coordinated water oxygen distance, which is easier to obtain. However, the calculation of the Gd–H distance from the Gd–O distance is complicated by the fact that the tilt angle of the plane of the bound water molecule with respect to the Gd–O bond is not well defined in solution.

To a first approximation, the Gd–O distances obtained from solid-state X-ray structure determination give a good estimation for the solution case as well. So far, there has been only one recent study that reports the Gd–O distances for Gd(III) poly(amino carboxylate) chelates measured directly in solution [34]. The significance of this X-ray absorption fine structure (XAFS) study performed on $[Gd(DOTA)(H_2O)]^-$ and $[Gd(DTPA)(H_2O)]^{2-}$ is that it proved experimentally that the local environment around the gadolinium ions, including the Gd–O distance, is indeed conserved in solution when compared to the solid-state X-ray structure. The structure of the complexes was found to be very rigid and the distances were practically independent of temperature (25–90 °C) and pH (1.5–7.0). The Gd–water oxygen distance was similar for the two contrast agents, $r_{GdO} = 2.46$ and 2.47 Å for $[Gd(DOTA)(H_2O)]^-$ and $[Gd(DTPA)(H_2O)]^{2-}$, respectively.

There are considerably more solution data available on lanthanide(III) aqua ions, with some of them on Gd^{3+}_{aq} itself. An X-ray diffraction (XRD) study in highly concentrated $GdCl_3$ aqueous solution resulted in 2.37 Å for the Gd–O distance [35]. Extended X-ray fine structure (EXAFS) measurements on perchlorate solutions of a series of lanthanide(III) materials revealed that the metal–oxygen distance changes along the series from 2.51 Å (Nd) to 2.31 Å (Lu), which is accompanied by a change in the hydration number (9.5 for Nd and 7.7 for Lu) [36]. This study reports $r_{GdO} = 2.41$ Å and $q = 7.6$ for Gd_{aq}^{3+}. A neutron scattering first-order difference study with isotopic substitution in D_2O confirmed the change of coordination number from nine for the light to eight for the heavy lanthanides, with a hydration equilibrium at Sm_{aq}^{3+} ($q = 8.5$) [37, 38]. Both lanthanide–oxygen and lanthanide–deuterium distances have been obtained, namely $r_{LnO} = 2.46$ and 2.50 Å and $r_{LnD} = 3.11$ and 3.03 Å for Sm_{aq}^{3+} and Dy_{aq}^{3+}, respectively, which are the closest metals to Gd_{aq}^{3+} in the periodic table that were investigated. Much shorter metal–deuterium distances (around 2.7 Å) have been calculated for frozen solutions of $[Gd(EDTA)(H_2O)_x]^-$ and $[Gd(DTPA)(H_2O)]^{2-}$ from Electron Spin Echo Envelope Modulation (ESEEM) data [39]. This method provides an effective value for r_{GdD}, which represents the magnitude of the electron nuclear interaction as if it were entirely due to a point-dipole interaction. However, the approximations made in the analysis, such as the use of a point-dipole model, may introduce large errors which can lead to an underestimation of the Gd–D distance as compared to the values determined by neutron diffraction. The r_{LnD} values obtained from the neutron diffraction and from the ESEEM study represent the only experimentally determined numbers for lanthanide–coordinated water hydrogen distances in solution.

The r_{GdH} values generally used in the analysis of relaxivity data for Gd(III) poly(amino carboxylate) complexes vary between 3.0 and 3.2 Å, according to various authors. Some workers have attempted to obtain the Gd–H distance from a fit of the proton relaxivity data. As generally there are many unknown parameters that are simultaneously fitted, this is a rather uncertain way to obtain r_{GdH} values. However, if all other influencing parameters, namely the rates of proton exchange, rotation and electronic relaxation, and the outer-sphere effect, are determined from independent measurements, in principle it is possible to obtain the effective Gd–H distance from the measured proton relaxivities (for further details of the analysis of proton relaxivities see Section 2.4 on 'Relaxivity and NMRD profiles').

2.3 PROTON/WATER EXCHANGE

The residence lifetime of protons, τ_M, plays a dual role in determining proton relaxivity. It modulates the efficiency of chemical exchange from the inner sphere of the metal to the bulk (Equation (2.5)), and it also contributes to the overall correlation time, τ_c, that governs the dipole–dipole interaction between the electron and nuclear spin (Equations (2.8) and (2.12)).

The exchange of coordinated water protons can occur in two ways, i.e. independently of the exchange of the entire water molecule on which it resides, or via the exchange of the water molecule itself. At around neutral pH, which is of interest for practical applications, the overall proton exchange rate is generally equal to the exchange rate of the entire water molecules, i.e. each proton exchanges with the bulk in the form of intact H_2O molecules. On increasing the acidity or basicity of the solution, the proton exchange may become considerably faster than the water exchange due to acid-or base-catalyzed pathways [40, 41] (see below). In any case, the water exchange rate represents a lower limit for the proton exchange rate.

When Gd(III) complexes started to be applied as MRI contrast agents, their water exchange rate was believed to be faster than that of the Gd aqua ion itself ($k_{ex} \approx 10^9$ s^{-1}), which would be too fast to influence the proton relaxivity. This assumption was based on the analogy to d transition metal complexes, where multidentate ligands are known to labilize the remaining water molecules [42]. In 1993, Merbach and co-workers measured directly the water exchange rates by ^{17}O NMR for $[Gd(DTPA)(H_2O)]^{2-}$ and $[Gd(DOTA)(H_2O)]^{-}$ and found that they were almost three orders of magnitude lower than that of $[Gd(H_2O)_8]^{3+}$ [43]. This result made it evident that for lanthanide poly(amino carboxylate) complexes, the water exchange can be considerably slowed down when compared with the aqua ion, and consequently can affect proton relaxivity, even for small, fast rotating agents.

2.3.1 Determination of the Water Exchange Rate: ^{17}O NMR

The water exchange rate can be directly obtained from variable temperature transverse ^{17}O relaxation rates measured on the Gd(III) complex solution. The reduced transverse ^{17}O relaxation rate, $1/T_{2r}$, can be calculated from the measured ^{17}O NMR relaxation rate of the paramagnetic solution, $1/T_2$, and of the reference, $1/T_{2A}$, and can be expressed as shown in Equation (2.24) [17], where $1/T_{2m}$ is the relaxation rate of the bound water ^{17}O and $\Delta\omega_m$ is the chemical shift difference between bound and bulk water:

$$\frac{1}{T_{2r}} = \frac{1}{P_m}\left(\frac{1}{T_2} - \frac{1}{T_{2A}}\right) = \frac{1}{\tau_m}\frac{T_{2m}^{-2} + \tau_m^{-1}T_{2m}^{-1} + \Delta\omega_m^2}{\left(\tau_m^{-1} + T_{2m}^{-1}\right)^2 + \Delta\omega_m^2} + \frac{1}{T_{2OS}} \tag{2.24}$$

In the ^{17}O NMR measurements an external reference is used, which is a solution of an analogous diamagnetic complex of the same concentration and pH as the Gd(III) sample. For this purpose, Y(III) is the best choice as its size is very similar to that of Gd(III). For low-molecular-weight complexes, if the pH of the Gd(III) sample is 4.0–6.5, acidified water can also be used as reference.

Outer-sphere contributions to both transverse and longitudinal ^{17}O relaxation rates are negligible [43]. Due to the small chemical shifts observed for Gd(III) complexes, $\Delta\omega_m$ is also negligible in Equation (2.24) ($\Delta\omega_\mathrm{m} \ll 1/\mathrm{T}_{2\mathrm{m}};1/\tau_\mathrm{m}$), and thus the latter simplifies to give Equation (2.25):

$$\frac{1}{T_{2r}} = \frac{1}{T_{2m} + \tau_m} \tag{2.25}$$

Since the oxygen is directly coordinated to Gd(III), in the transverse relaxation the scalar contribution, $1/T_{2SC}$, is the most important one (Equation 2.26) [43]. In this equation, $1/\tau_{e1}$ is the sum of the water exchange rate constant and the longitudinal electron spin relaxation rate:

$$\frac{1}{T_{2m}} \cong \frac{1}{T_{2SC}} = \frac{S(S+1)}{3}\left(\frac{A}{\hbar}\right)^2 \tau_{el}; \frac{1}{\tau_{el}} = \frac{1}{\tau_m} + \frac{1}{T_{1e}} \tag{2.26}$$

The chemical shift of the coordinated water oxygen, $\Delta\omega_m$, is determined by the hyperfine interaction between the Gd^{3+} electron spin and the ^{17}O nucleus [44]. Therefore, the scalar or hyperfine coupling constant, $A/\hbar$, can be directly obtained from the chemical shifts measured for the paramagnetic sample, ω, referred to the chemical shift of the reference, ω_A, through Equations (2.27) and (2.28):

$$\Delta\omega_r = \frac{1}{P_m}(\omega - \omega_A) = \frac{\Delta\omega_m}{\left(1+\tau_m T_{2m}^{-1}\right)^2 + \tau_m^2 \Delta\omega_m^2} + \Delta\omega_{OS} \tag{2.27}$$

$$\Delta\omega_m = \frac{g_L \mu_B S(S+1) B}{3 k_B T} \frac{A}{\hbar} \tag{2.28}$$

In Equations (2.27) and (2.28), $\Delta\omega_r$ is the reduced ^{17}O chemical shift, B is the magnetic field, S is the electron spin and g_L is the isotropic Landé g-factor. The outer-sphere contribution to the ^{17}O chemical shift, $\Delta\omega_{os}$, is assumed to be proportional to $\Delta\omega_m$, where C_{os} is an empirical constant [43, 45]:

$$\Delta\omega_{OS} = C_{OS}\Delta\omega_m \tag{2.29}$$

The binding time ($\tau_m = 1/k_{ex}$) of water molecules in the inner sphere is assumed to obey the Eyring equation (Equation (2.30), where $\Delta S^{\ddagger}$ and $\Delta H^{\ddagger}$ are the entropy and enthalpy of activation for the exchange process, and k_{ex}^{298} is the exchange rate at 298.15 K:

$$\frac{1}{\tau_m} = k_{ex} = \frac{k_B T}{h}\exp\left(\frac{\Delta S^{\ddagger}}{R} - \frac{\Delta H^{\ddagger}}{RT}\right) = \frac{k_{ex}^{298} T}{298.15}\exp\left[\frac{\Delta H^{\ddagger}}{R}\left(\frac{1}{298.15} - \frac{1}{T}\right)\right] \tag{2.30}$$

In the correlation time that governs the bound water transverse ^{17}O relaxation (τ_{e1} in Equation 2.26), the electron spin relaxation term, $1/T_{1e}$, is field-dependent, whereas the exchange contribution, $1/\tau_m$, is not. Consequently, variable-field ^{17}O NMR measurements can be useful in separating the two terms, i.e. the exchange and the electronic relaxation effect, which contribute to τ_{e1}. A typical plot of the paramagnetic relaxation rate enhancement versus the inverse temperature is represented by the [Gd(DTPA-BMA)(H_2O)] complex, as shown in Figure 2.5(a).

At low temperatures, the reduced transverse relaxation rates increase with temperature. In this *slow kinetic region* $1/T_{2r}$ is directly determined by the exchange rate, k_{ex}. At high temperatures, in the *fast exchange region*, the

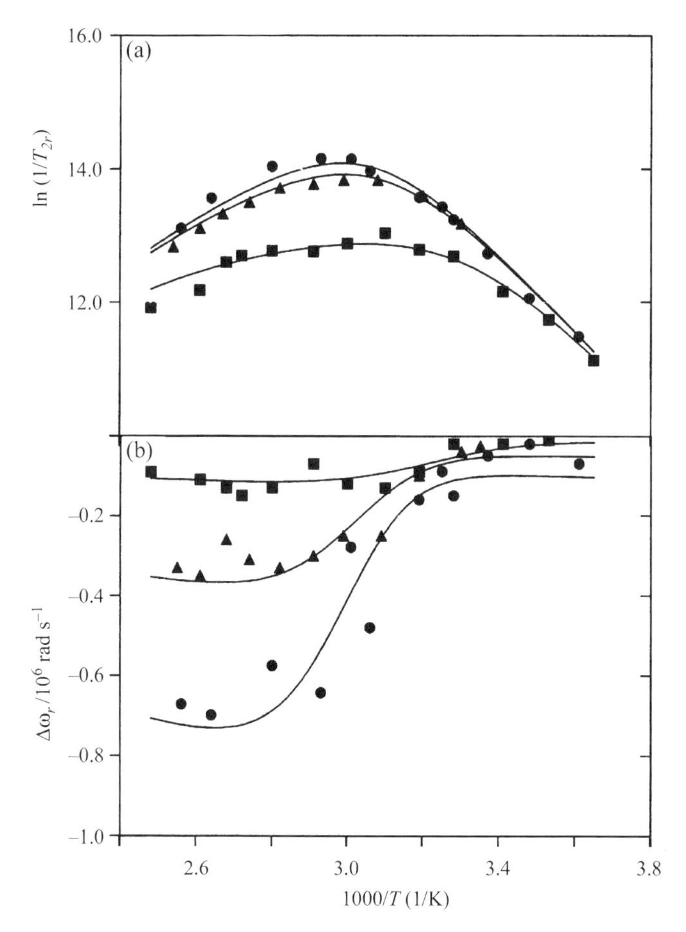

Figure 2.5 Temperature dependence of (a) the reduced transverse ^{17}O relaxation rates, and (b) the reduced chemical shifts for $[Gd(DTPA\text{-}BMA)(H_2O)]$ at three different magnetic fields; $B = 9.4$ T (circles), 4.7 T (triangles) and 1.41 T (squares).

reduced transverse relaxation rates decrease with temperature and are determined by the transverse relaxation rate of the coordinated water oxygen, $1/T_{2m}$. The change-over from the slow to the fast exchange region corresponds to the maxima on the $1/T_{2r}$ curves and to the inflection point on the plot of the reduced chemical shifts, $\Delta\omega_r$ (Figure 2.5(b)). If the exchange rate is very fast, i.e. $k_{ex} >> 1/T_{2m}$, no slow exchange region is observed. Under such conditions, it is absolutely necessary to have additional information on the electron spin relaxation (from electron paramagnetic resonance (EPR) measurements) and to measure ^{17}O relaxation rates at variable fields in order to be able to calculate an exact water exchange rate.

2.3.2 Variable Pressure ^{17}O NMR: Assessment of the Water Exchange Mechanism

The mechanism of the water exchange process can be determined from variable-pressure ^{17}O transverse relaxation measurements which give access to the volume of activation, $\Delta V^{\ddagger}$, which is a potent tool in assigning the mechanism [46]. In general terms, $\Delta V^{\ddagger}$ is defined as the difference between the partial molar volume of the transition state and the reactants and is related to the pressure dependence of the exchange rate constant through Equation (2.31):

$$\frac{1}{\tau_m} = k_{ex} = (k_{ex})_0^T \exp\left(-\frac{\Delta V^{\ddagger}}{RT} P\right) \tag{2.31}$$

where $(k_{ex})_0^T$ is the water exchange rate at zero pressure and temperature T. Thus, the exchange reaction is either slowed down or accelerated by increasing pressure when $\Delta V^{\ddagger}$ is positive or negative, respectively. The term $\Delta V^{\ddagger}$ is assumed to be a direct measure of the degree of bond making and bond breaking occurring in the transition state and the concurrent lengthening and shortening of the non-exchanging ligands to metal center distances. A continuous variation of transition state configurations may be envisaged, as shown in Figure 2.6.

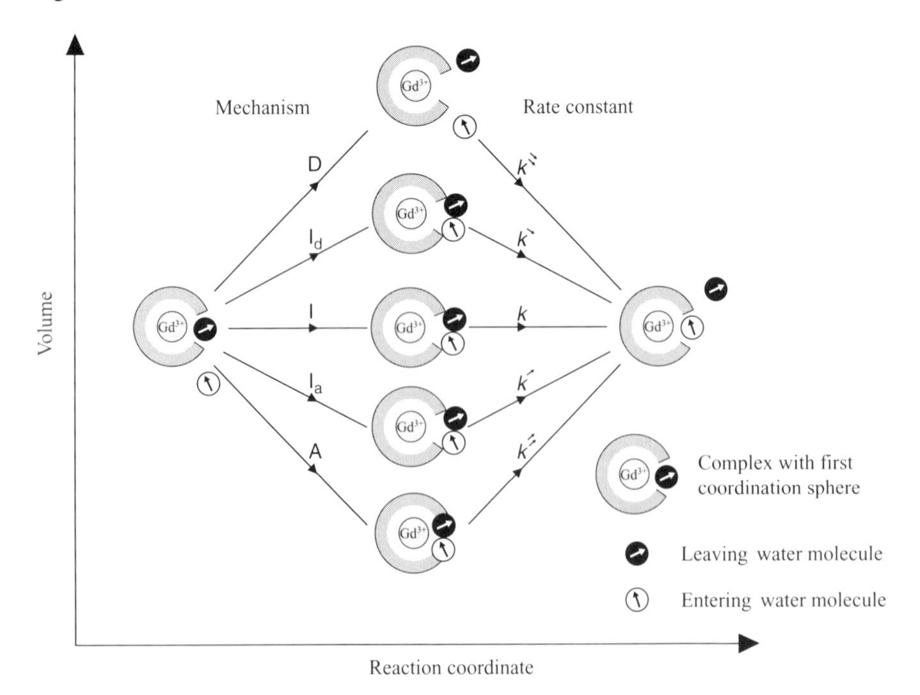

Figure 2.6 Volume profiles for the spectrum of water exchange processes on Gd(III) chelates.

At one extreme the dissociative (D) mechanism is characterized by a greatly expanded transition state, and thus $\Delta V^{\ddagger}$ has a large positive value. At the other extreme, the associative (A) mechanism is characterized by a greatly contracted transition state, and $\Delta V^{\ddagger}$ is large and negative. In between these two mechanisms is the interchange (I) mechanism, in which the bond breaking and bond making compensate each other in contributing to $\Delta V^{\ddagger}$. On either side of I are the I_d and I_a mechanisms, which are characterized by positive and negative $\Delta V^{\ddagger}$ values corresponding to a greater or smaller bond breaking contributions, respectively.

In the most simple case (and if possible), the variable-pressure measurements are performed in the slow-exchange region. Here the reduced relaxation rate directly corresponds to the water exchange rate, and thus the experimentally observed pressure dependence is that of the water exchange rate itself. Things become somewhat more complicated if there is no slow-exchange region. Under these conditions, one has to make certain assumptions, such as the pressure independence of the scalar coupling constant, $A/\hbar$, and of the electron spin relaxation. The $A/\hbar$ term has indeed been found to be independent of pressure for different lanthanide(III) aqua ions [47]. On the other hand, nothing is known about the pressure dependence of the electronic relaxation. Therefore, in the analysis of the variable-pressure ^{17}O NMR data, a check is usually made on what effect a reasonable pressure variation of $\tau_v(|\Delta V_v^{\ddagger}| \leq 4\ \mathrm{cm^3\ mol^{-1}})$ could have on the water exchange rate [48, 49]. Certainly, if it is not possible to perform the variable-pressure study in the slow-exchange region, because it cannot be observed, one has to find the optimum conditions (by possibly varying the temperature and magnetic field) where $1/T_{1e}$ contributes the least to τ_{e1}.

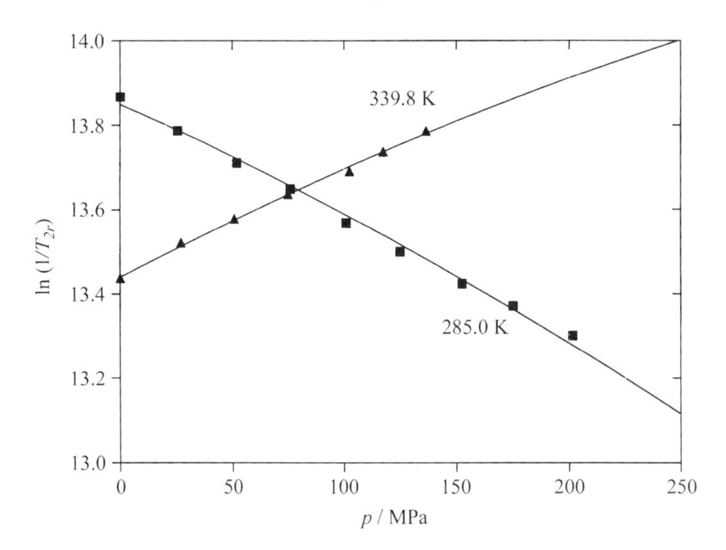

Figure 2.7 Pressure dependence of the reduced transverse ^{17}O relaxation rates of an aqueous solution of $[Gd(DOTA)(H_2O)]^-$ at two different temperatures, i.e. $T = 339.8$ K (fast-exchange region) and $T = 285.0$ K (slow-kinetic region).

Table 2.3 Kinetic parameters for water exchange on selected small-molecular-weight Gd(III) chelates, as measured by ^{17}O NMR

Ligand[a]	Coordinating unit	q	k_{ex}^{298} (10^6 s^{-1})	$\Delta H^{\neq}$ (kJ mol^{-1})	$\Delta S^{\neq}$ (J mol^{-1} K^{-1})	$\Delta V^{\neq}$ (cm^3 mol^{-1})	Mechanism	Reference
aqua	—	1	804	15.3	−23.1	−3.3	A	51
DTPA	DTPA	1	3.30	51.6	53	12.5	D	32
BOPTA	DTPA	1	3.45	—	—	—	—	74
EOB-DTPA	DTPA	1	3.60	49.1	45	12.3	D	60
MP-2269	DTPA	1	4.20	51.6	56	—	D	72
COPTA	DTPA	1	3.40	—	—	—	—	71
DTPA-N-MA	DTPA-monoamide	1	1.30	48.6	36	12.7	D	60
DTPA-N'-MA	DTPA-monoamide	1	1.90	50.6	40	10.6	D	60
DTPA-BMA	DTPA-bisamide	1	0.45	47.6	23	7.3	D	32
DTPA-BMEA	DTPA-bisamide	1	0.39	49.8	27	7.4	D	59
DTPA-BGLUCA	DTPA-bisamide	1	0.38	47.6	22	6.8	D	58
DTPA-BENGALAA	DTPA-bisamide	1	0.22	42.5	0	5.6	D	58
DOTA	DOTA	1	4.1	49.8	49	10.5	D	32
DO3A	DO3A	1.9	11	33.6	2	—	—	29
DO2A	DO2A	2.8	10	21	−39	—	—	30
DOTASA	DOTA+COO$^-$	1	6.3	45.4	38	—	I_d–D	69
DO3A-bz-NO_2	DOTA-monoamide	1	1.6	40.9	11	7.7	D	62
pip$(DO3A)_2$	DOTA-monoamide	1	1.5	34.2	−12	—	I_d	32
bisoxa$(DO3A)_2$	DOTA-monoamide	1	1.4	38.5	2	2.3	I_d	32
DOTA-C_{12}	DOTA-monoamide +COO$^-$	1	4.8	42.7	27	—	I_d–D	81
DOTAM	DOTA-tetraamide	1	0.053	—	—	—	—	50
DOTTA	DOTA-tetraamide	1	0.128	—	—	—	—	50
DTMA	DOTA-tetraamide	1	0.059	—	—	—	—	50
BO$(DO3A)_2$	DO3A+OH	1	1.0	30.0	−29	0.5	I_d	61

Table 2.3 *Cont.*

Ligand[a]	Coordinating unit	q	k_{ex}^{298} (10^6 s^{-1})	$\Delta H^{\neq}$ (kJ mol^{-1})	$\Delta S^{\neq}$ (J mol^{-1} K^{-1})	$\Delta V^{\neq}$ (cm^3 mol^{-1})	Mechanism	Reference
PCTP-[13]	—	1	125	58	105	—	—	76
PCTP-[12]	—	1	170	14	−40	—	—	77
PCTA-[12]	—	2	14	45	43	—	—	77
EGTA	—	1	31	42.7	42	10.5	D	73
PDTA	—	2	102	11	−55	−1.5	I_a	51
TTAHA	—	2	8.6	40.4	23	2.9	I_d	75
taci	—	2	11.0	59.8	−89	−12.7	A	49

[a] Structures are shown in Charts 2.1–2.8.

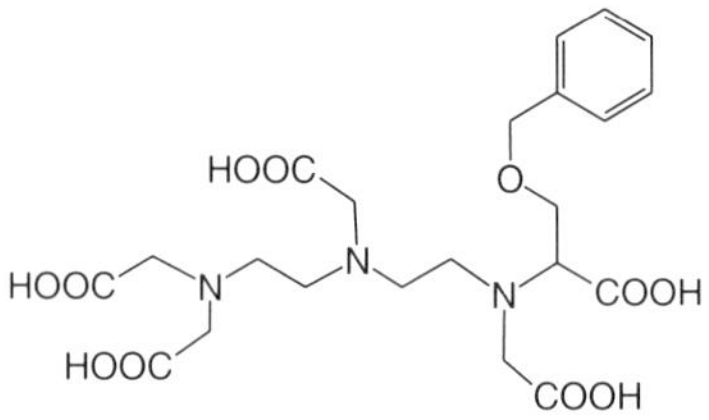

BOPTA

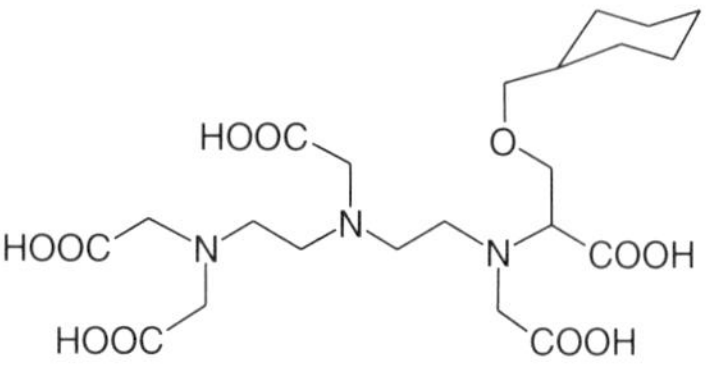

COPTA

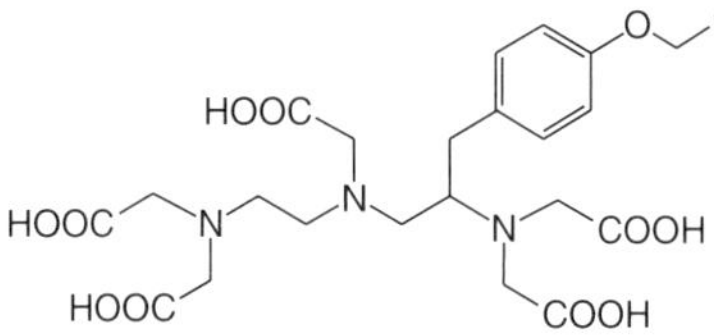

EOB-DTPA

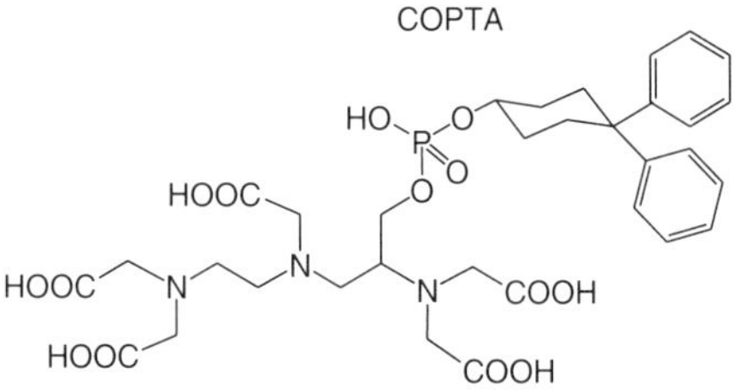

MS-325-L

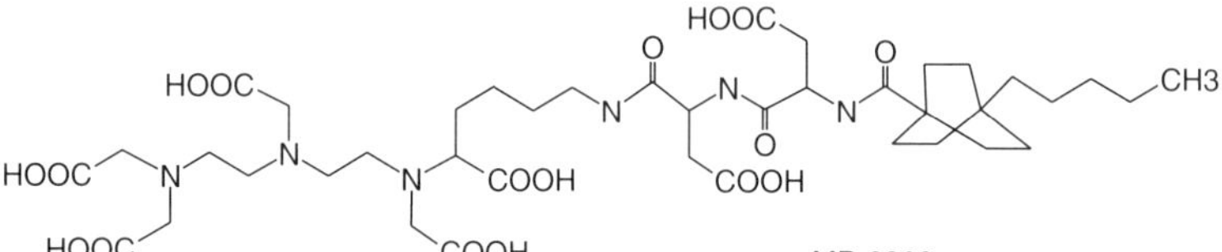

MP-2269

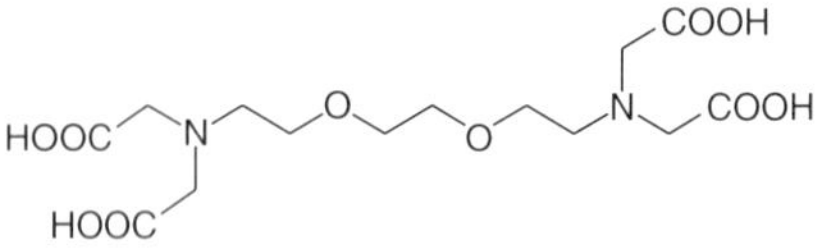

EGTA

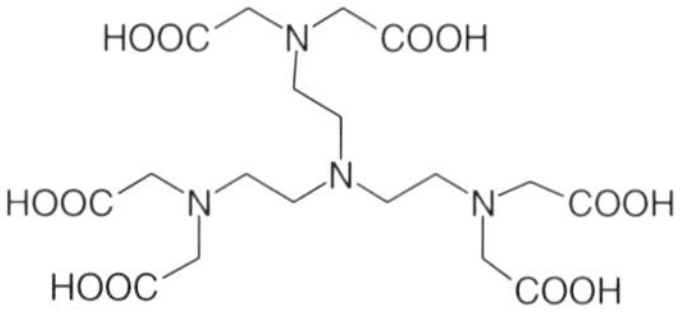

TTAHA

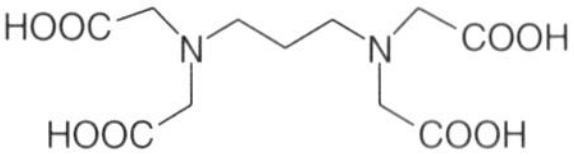

PDTA

[**Chart 2.4**]

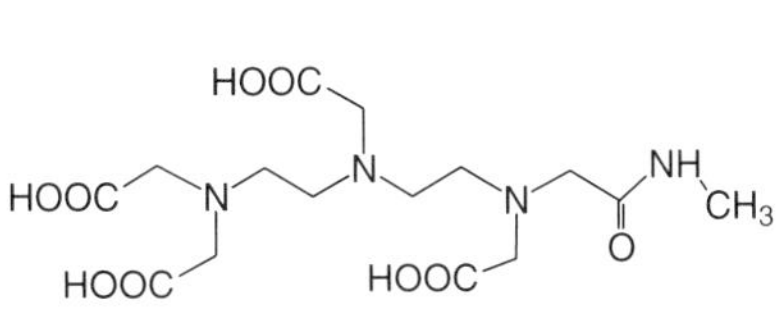

DTPA-N-MA

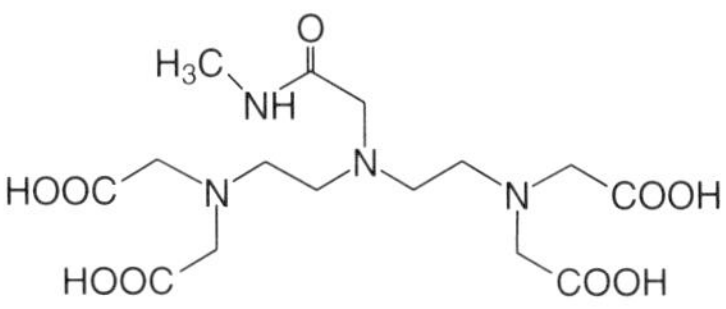

DTPA-N'-MA

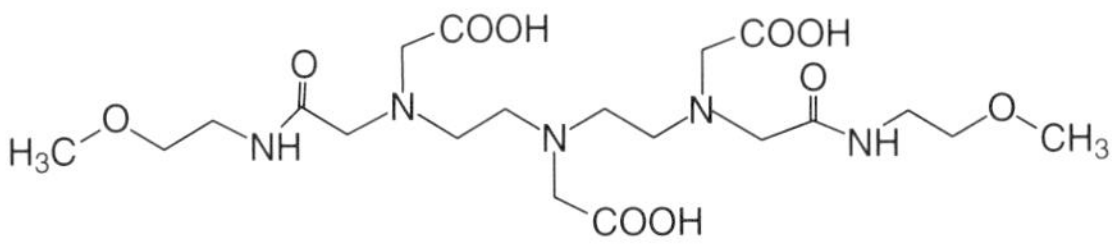

DTPA-BMEA

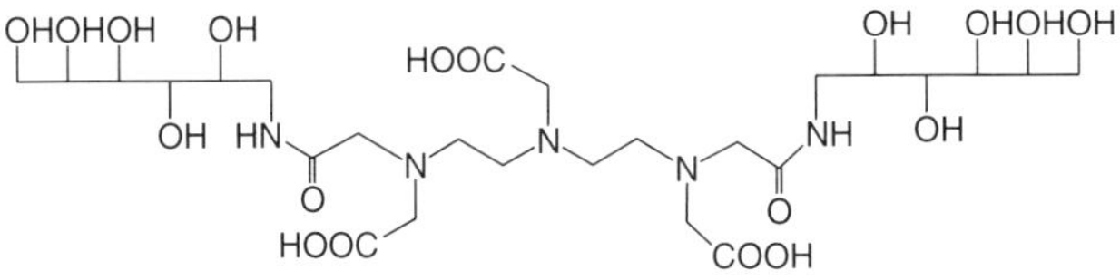

DTPA-BGLUCA

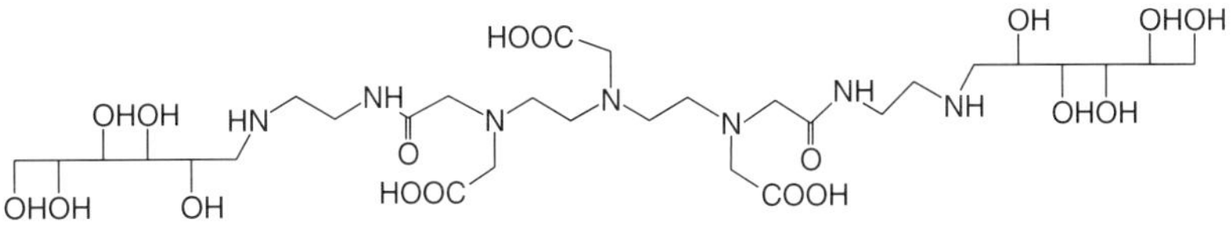

DTPA-BENGALAA

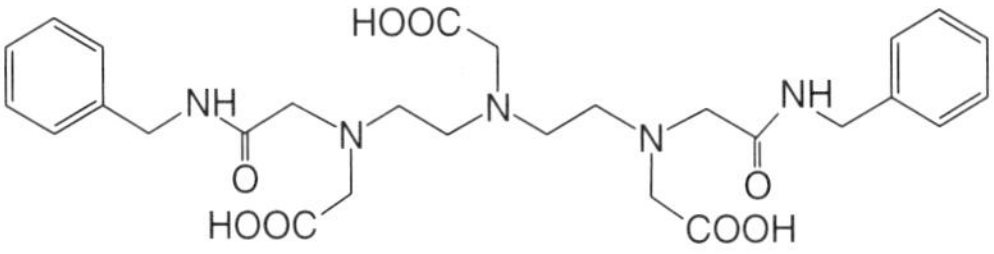

DTPA-BBA

[Chart 2.5]

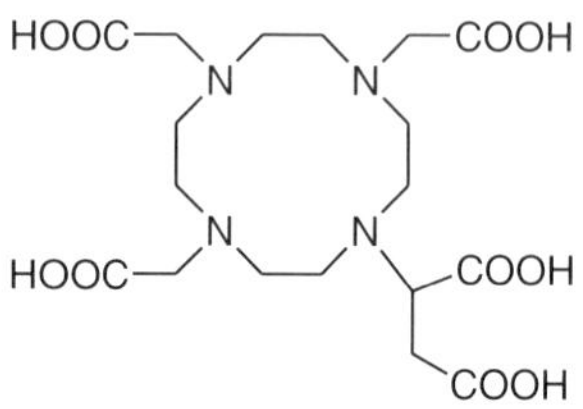

DOTASA

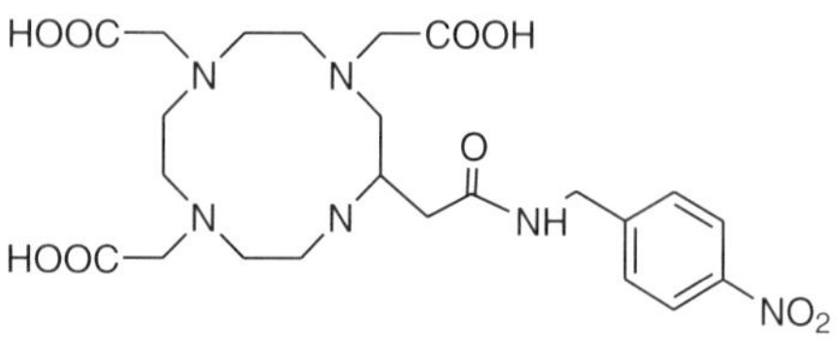

DO3A-bz-NO_2

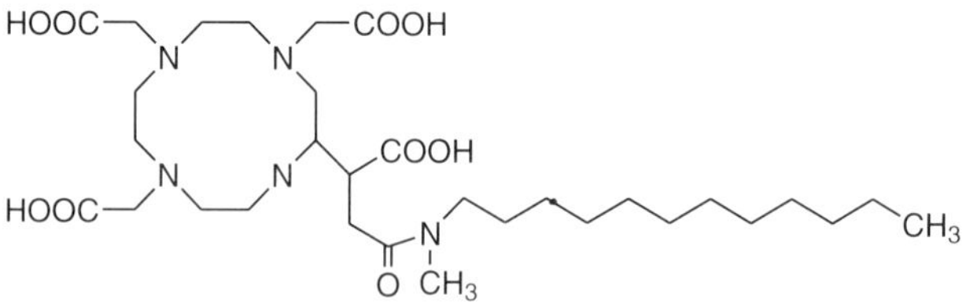

DOTA-C_{12}

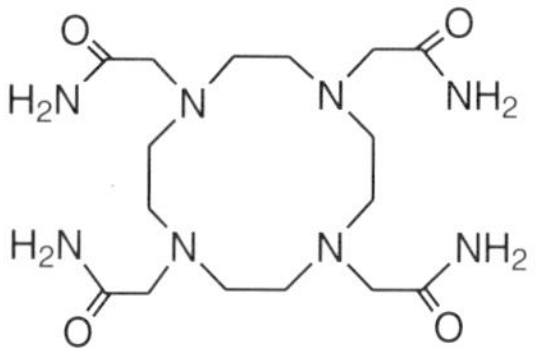

DOTAM

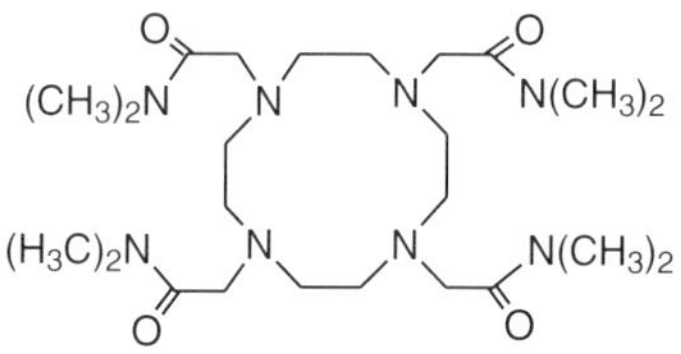

DOTTA

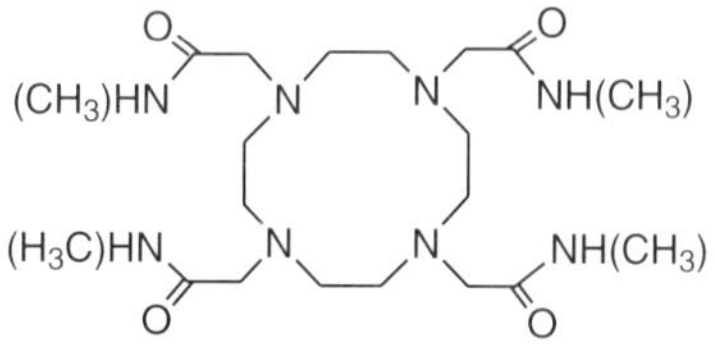

DTMA

[**Chart 2.6**]

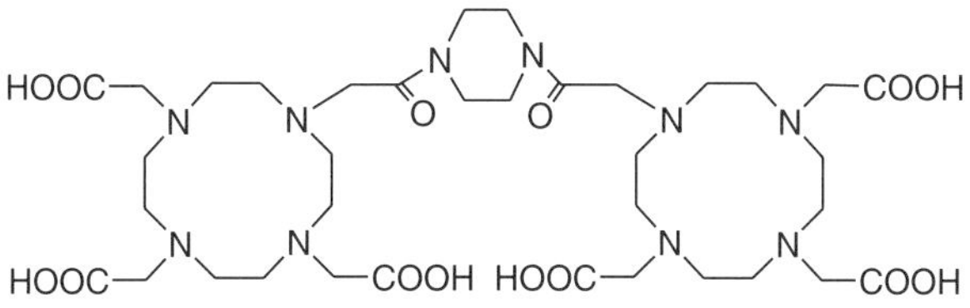

pip(DO3A)$_2$

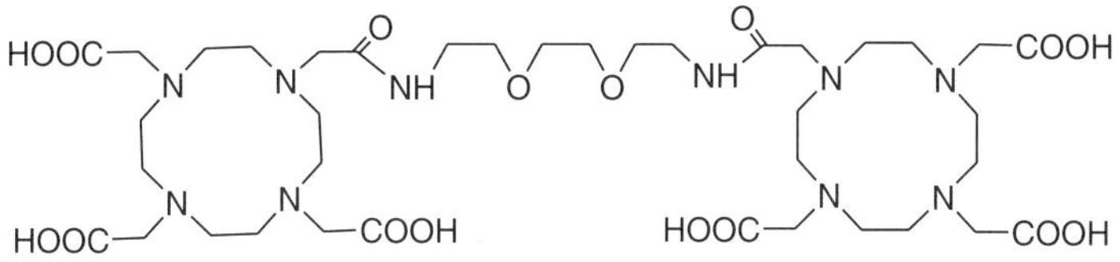

bisoxa(DO3A)$_2$

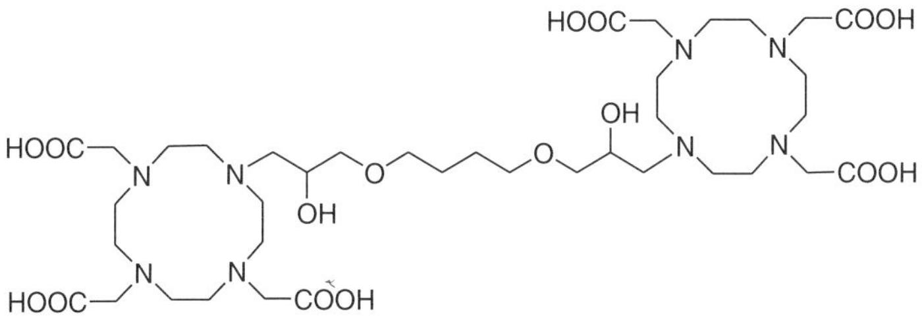

BO-(DO3A)$_2$

[**Chart 2.7**]

Figure 2.7 shows two representative cases of variable-pressure measurements: both of these were performed on $[Gd(DOTA)(H_2O)]^-$ but at different temperatures, i.e. one in the slow and the other in the fast-exchange regime [43]. The opposite slopes of the two curves is due to the opposite k_{ex} dependences of the reduced transverse ^{17}O relaxation rates, T_{2r}, in the slow and fast kinetic region (Equations (2.25) and (2.26)). The activation volumes determined for different Gd(III) complexes are presented in Table 2.3, with the structures of the corresponding ligands being shown in Charts 2.1–2.8.

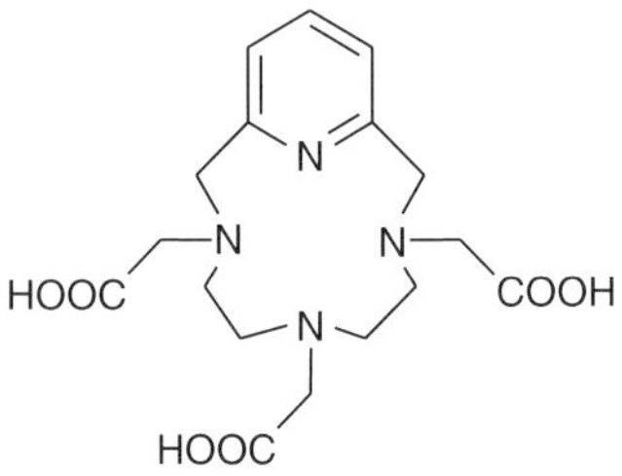

PCTA-[12]

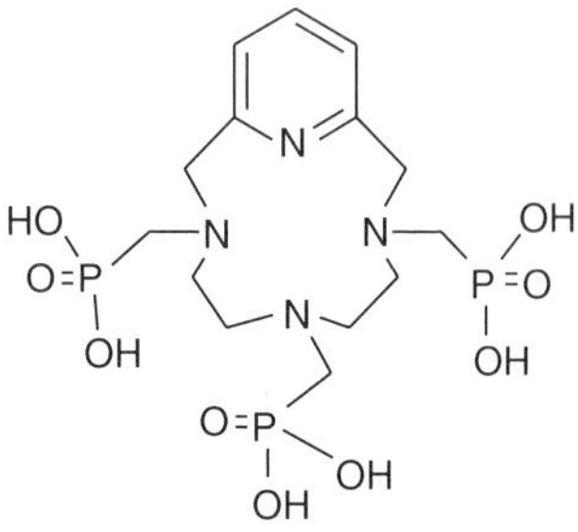

PCTP-[12]

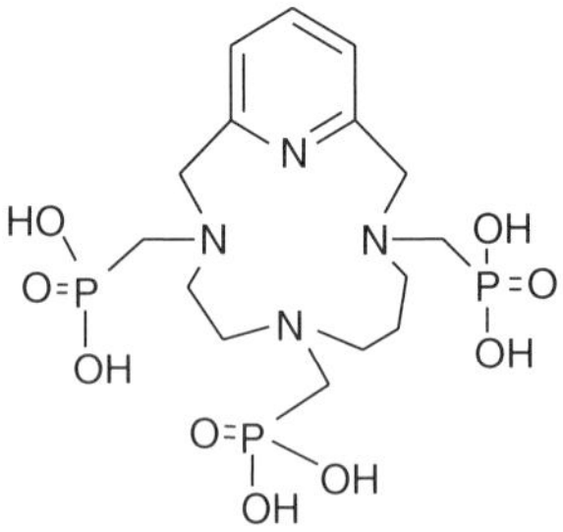

PCTP-[13]

[Chart 2.8]

2.3.3 Water Exchange on Monomer Gd(III) Complexes

The water exchange rates reported for different Gd(III) complexes cover a range of more than four orders of magnitude, from the lowest for DOTA-tetraamide complexes [40, 50] ($k_{ex}^{298} = 4.5 \times 10^4 s^{-1}$ for $[Gd(DOTMA)(H_2O)]^{3+}$) to the highest, $k_{ex}^{298} = 8 \times 10^8 s^{-1}$ for the aqua ion itself [51] (see Table 2.3). (Recently, an even lower value, i.e. $k_{ex}^{298} = 8.3 \times 10^3 s^{-1}$, has been found for the individual exchange rate of one isomer of the $[Eu(DOTAM)(H_2O)]^{3+}$ complex [50, 52, 53], which is likely to be similar for the corresponding Gd(III) isomer as well.)

Both the rate and the mechanism of water exchange are intimately related to the inner-sphere solution structure of the Gd(III) complexes. For lanthanide(III) aqua ions, the water exchange rates decrease by more than one order of magnitude between $[Gd(H_2O)_8]^{3+}$ and $[Yb(H_2O)_8]^{3+}$ [47, 54]. From neutron diffraction measurements, it is known that as the ionic radius decreases, the coordination number of the lanthanide aqua ions changes from nine at the beginning of the series to eight at the end, with the Sm^{3+} having an average coordination number of 8.5 [38]. The activation volumes indicate associatively activated water exchange processes for all the octaaqua ions from $[Gd(H_2O)_8]^{3+}$ to $[Yb(H_2O)_8]^{3+}$. The fast water exchange on $[Gd(H_2O)_8]^{3+}$ can therefore be interpreted in terms of activation energy: being relatively close to an equilibrium state between eight-and nine-coordinated species, for the $[Gd(H_2O)_8]^{3+}$ ion little energy is required to reach the transition state (coordination number of nine) in an associatively activated process.

The nine-coordinate Gd(III) poly(amino carboxylate)s all have positive activation volumes which are indicative of dissociatively activated water exchange (see Table 2.3). This is expected, considering that in a nine-coordinate lanthanide complex there is no longer space for a second water molecule to enter before the subsequent departure of the bound water molecule. On the other hand, for several of these complexes the eight-coordinate transition state is energetically unstable, since for them only the coordination number of nine is observed all through the lanthanide series [55–57]. The instability of the transition state, and thus the high activation energy needed, results in a decreased rate constant. Another important factor is the rigidity of the inner coordination sphere. Whereas in the aqua ion the rearrangement of the flexible coordination sphere occurs easily, the poly(amino carboxylate) complexes have a much more rigid inner-sphere structure whose rearrangement requires higher energy. In conclusion, the difference in the inner-sphere structure, and hence the difference in the mechanism, is the reason why water exchange on nine-coordinate lanthanide(III) poly(amino carboxylate) complexes is generally much slower when compared to the eight-coordinate $[Gd(H_2O)_8]^{3+}$ or to the nine-coordinate early lanthanide aqua ions.

Let us now consider the differences between different nine-coordinate linear or macrocyclic poly(amino carboxylate)s of Gd(III) with one inner-sphere water molecule. Although the mechanism is always dissociative, there is a tenfold decrease in k_{ex} going e.g. from the pentacarboxylate $[Gd(DTPA)(H_2O)]^{2-}$ to the bisamide derivatives [45, 58, 59] ($[Gd(DTPA\text{-}BMA)(H_2O)]$ or $[Gd(DTPA\text{-}BMEA)(H_2O)]$), with the k_{ex} values for monoamide complexes being within this range (see Table 2.3) [60]. The same diminution of the exchange rate is found for macrocyclic complexes when one carboxylate of DOTA is substituted by an amide or OH coordinating group [32, 61, 62]. Numerous amide derivatives in the DOTA or DTPA family have been synthesized in the last few years, with the aim of diminishing the total complex charge, or, at the same time, forming dimeric complexes or linking the chelate to various macromolecules. On the basis of all of the k_{ex} values now available for amide or OH derivatives of either DTPA or DOTA, we can generally state that the replacement of one carboxylate group by an amide or OH moiety decreases the water exchange rate of the Gd(III) complex by a factor of between three and four.

An amide or alcoholic OH group is coordinated less strongly towards the lanthanide ion than a carboxylate, which is exhibited by smaller stability constants for the amide complexes when compared to carboxylates in solution

[63–65] and by longer Gd–amide oxygen distances in the solid state, when compared to Gd–carboxylate oxygen distances (e.g. the average Gd–carboxylate oxygen distance in $Na_2[Gd(DTPA)(H_2O)]$ is 0.240 nm, [66], and the Gd–amide oxygen distance, e.g. in the Gd(III) bis(benzylamide)-DTPA complex, is 0.244 nm [67]. As a consequence, the inner sphere is less crowded in amide than in carboxylate complexes. In dissociatively activated water exchange processes, the steric crowding is of primary importance, i.e. a tightly coordinating ligand forces the water molecule to leave, and thus favours the dissociative activation step. The significance of crowding at the water binding site was also demonstrated by an ^{17}O NMR study on the whole lanthanide series of DTPA-BMA complexes [68]. On progressing from the middle to the end of the series, the eight-coordinate transition state becomes more and more accessible since the radius of the lanthanide ion decreases, and the result is a large increase in the water exchange rate from $[Eu(DTPA\text{-}BMA)(H_2O)]$ to $[Ho(DTPA\text{-}BMA)(H_2O)]$.

Besides steric crowding, charge effects can also explain these differences in the water exchange rate. A higher negative overall charge favors the leaving of the water molecule in a dissociative process, and thus accelerates the exchange. Indeed, about a 50 % higher water exchange rate was found for the pentacarboxylate DOTA derivative, $[Gd(DOTASA)(H_2O)]^{2-}$ as compared to $[Gd(DOTA)(H_2O)]^-$ [69]. In an analogous way, the observed water exchange rate on $[Gd(DOTA)(H_2O)]^-$ decreases significantly with increasing extent of protonation, and thus with decreasing negative charge, and at 1.0 M H^+ concentration it is about ten times lower than in neutral media [70].

While the water exchange rate is strongly affected when coordinating groups or the overall charges are changed, it remains relatively constant on introducing different substituents which do not directly interfere in the inner coordination sphere. Very similar exchange rates have been reported for all of the different bisamide DTPA-derivatives DTPA-BMA, DTPA-BMEA, DTPA-BENGALAA, etc [45, 58, 59]. Likewise, even bulky substituents on the carbon backbone of the DTPA have almost no influence on the water exchange kinetics: the rates are all similar for the Gd(III) complexes of DTPA, EOB-DTPA, COPTA, and MP-2269 [43, 60, 70–71]. The two dimeric complexes, $[pip\{Gd(DO3A)(H_2O)\}_2]$ and $[bisoxa\{Gd(DO3A)(H_2O)\}_2]$, which differ in the bridge between the two DO3A-amide units, also have identical water exchange rates [32].

Despite its lower negative charge, the nine-coordinate $[Gd(EGTA)(H_2O)]^-$ species exchanges ten times faster than $[Gd(DTPA)(H_2O)]^{2-}$ [73]. This can be explained again in terms of steric contraints: in $[Gd(EGTA)(H_2O)]^-$, the ethyl group bridging the two coordinating oxygens causes a steric compression of these atoms around the site occupied by the water molecule. This destabilizes the bound water molecule and accelerates the water exchange. The increasing steric crowding also explains the increasing water exchange rates for Gd(DTPA)-derivatives with one, two and three benzyloxymethylenic substituents on the carboxymethyl group, i.e. $[Gd(BOPTA)(H_2O)]^{2-}$, $[Gd(DTPA\text{-}(BOM)_2)(H_2O)]^{2-}$ and $[Gd(DTPA\text{-}(BOM)_3)(H_2O)]^{2-}$ [74]. The nine-coordinate tripod $[Gd(TTAHA)(H_2O)_2]^{2-}$ complex also exchanges about twice as fast as $[Gd(DTPA)(H_2O)]^{2-}$, via a dissociatively activated interchange (I_d) mechanism [75]. In this case, the presence of two inner-sphere water molecules decreases the stereorigidity of the system. Moreover, the partial participation of the entering water molecule is also favored, as shown by the low value of the activation volume ($\Delta V^{\ddagger} = 2.9$ cm^3 mol^{-1}). These two factors explain the relatively fast water exchange.

Gd(III) complexes of some pyridine-based macrocycles have been reported to have, in general, remarkably higher water exchange rates than that of the parent $[Gd(DOTA)(H_2O)]^-$ species [76, 77]. The exchange is particularly fast for the eight-coordinate phosphonate derivative $[Gd(PCTP\text{-}[12])(H_2O)]$, and was accounted for by a probable associatively activated mechanism. A comparison of the water exchange rates on the nine-coordinate acetate derivative $[Gd(PCTA\text{-}[12])(H_2O)_2]$ and the eight-coordinate phosphonate derivative $[Gd(PCTP\text{-}[12])(H_2O)]$ is an illustrative example of the effect of the ground-state coordination number: the change in the water exchange mechanism from a dissociative ($[Gd(PCTA\text{-}[12])(H_2O)_2]$) to an associative ($[Gd(PCTP\text{-}[12])(H_2O)]$) activation mode is accompanied by a one order of magnitude increase in the exchange rate.

A few other eight-coordinate complexes have also been investigated. The water exchange on $[Gd(PDTA)(H_2O)_2]^-$ is extremely fast and only somewhat slower than that on the aqua ion, and proceeds via an associative interchange (I_a) mechanism [51]. The trimer, $[Gd_3(H_{-3}\text{ taci})_2(H_2O)_6]^{3+}$, undergoes a slightly slower water exchange than $[Gd(PDTA)(H_2O)_2]^-$, although the mechanism is much more associatively activated (limiting A mechanism), as shown by the large negative activation volume [49]. The lower exchange rate was attributed to the particularly rigid structure of this trimer complex, which slows down the transition from the eight-coordinate reactant to the nine-coordinate transition state.

The effect of the varying lanthanide ion size on the water exchange has been reported for two poly(amino carboxylate) ligands. For the eight-coordinate $[Ln(PDTA)(H_2O)_2]^-$ complexes (Ln = Gd, Tb, Dy and Er), the mechanism remains associatively activated; however, the rate decreases by three orders of magnitude with decreasing metal ion size [28]. An equilibrium between nine-and eight-coordinate species occurs for complexes with the early lanthanides; the nearer the particular ion is to the position of this equilibrium, then the easier it is to reach the nine-coordinate transition state, and hence the faster is the associatively activated exchange reaction. As a further result of the decreasing size, the exchange becomes dissociatively activated for $[Yb(PDTA)(H_2O)_2]^-$.

For the series of $[Ln(DTPA\text{-}BMA)(H_2O)]$ complexes (Ln = Nd, Eu, Gd, Tb, Dy and Ho), the hydration number does not vary. While the water exchange rate is nearly constant for the complexes of Nd, Eu and Gd, it takes a steep rise for Tb, Dy and Ho [68]. This trend is rationalized in terms of a change in the mechanism. The volumes of activation depict a change-over from an interchange mode for the Nd(III) complex to a limiting dissociative one for the heavier lanthanides.

These results on the water exchange rates and mechanisms of different Gd(III) complexes have important implications for the design of MRI contrast agents. First, they show that the low water exchange rates observed for commercial MRI contrast agents, which could limit relaxivity improvements in the new generation of high-molecular-weight contrast agents, are not an intrinsic property of nine-coordinate Ln(III) complexes. Secondly, they provide some indications for ligand design on how to tune the water exchange rates to attain maximum efficiency. For the eight-coordinate complexes with associatively activated water exchange, changes of the ligand that decrease steric crowding will decrease the free energy necessary to reach the nine-coordinate transition state, and consequently they will increase the exchange rate. For dissociatively activated exchange on nine-coordinate MRI contrast agents, structural changes of the ligand that induce greater steric crowding will decrease the activation energy needed to reach the eight-coordinate transition state, and thus increase the exchange rate. Charge effects should also be considered: higher negative overall charge will always favor the leaving of the water molecule in a dissociatively activated water exchange process, and thus accelerate the exchange.

2.3.4 Water Exchange on Macromolecular Gd(III) Complexes

While rotation is obviously slowed down to a smaller or higher extent when a monomer Gd(III) chelate is attached to a macromolecule, it is less straightforward to predict how the water exchange is affected. In the last few years, several macromolecular systems have been studied by variable-temperature and-pressure ^{17}O NMR in order to determine the rate and mechanism of water exchange. The kinetic parameters characterizing the water exchange on macromolecular Gd(III) complexes are presented in Table 2.4, with the structures of the various complexes and ligands being shown in Charts 2.9 and 2.10.

Three different generations (gen. 3, 4 and 5) of PAMAM dendrimers functionalized with the same DO3A-monoamide Gd(III) chelate have been compared to the corresponding monomer chelate [62]. The water exchange rates, k_{ex}^{298}, on the generation-5 [G5(N{CS}N-bz-Gd{DO3A}{H_2O})$_{52}$], generation-4 [G4(N{CS}N-bz-Gd{DO3A}{H_2O})$_{30}$], generation-3 [G3(N{CS}N-bz-Gd{DO3A}{H_2O}]$_{23}$], and the monomer [Gd(DO3A-bz-NO_2)(H_2O)] complexes are very similar. It can be noted that the k_{ex} value for the generation-3 derivative, which has 96 % of its terminal amine groups functionalized, is slightly smaller than that of the monomer complex or of the higher-generation dendrimer complexes, where the substitution is only 64 % (gen. 4) and 53 % (gen. 5). Although it cannot be completely excluded that the difference in the

Table 2.4 Kinetic parameters for water exchange on selected macromolecular Gd(III) chelates, as measured by ^{17}O NMR. Values obtained for the corresponding monomer units (in italics) are also shown for comparison.

Complex[a]	k_{ex}^{298} (10^6 s^{-1})	$\Delta H^{\neq}$ (kJ mol^{-1})	$\Delta S^{\ddagger}$(J mol^{-1} K^{-1})	$\Delta V^{\ddagger}$ (cm^3 mol^{-1})	Reference
[Gd(DO3A – bz – NO2)(H2O)]	*1.6*	*40.9*	*11*	*7.7*	*62*
[G3(N{CS}N-bz-Gd{DO3A}{H2O})23]	1.0	28.8	−30	3.1	62
[G4(N{CS}N-bz-Gd{DO3A}{H2O})30]	1.3	27.7	−31	—	62
[G5(N{CS}N-bz-Gd{DO3A}{H2O})52]	1.5	24.0	−43	—	62
[Gd(DTPA-BMA)(H2O)]	*0.45*	*47.6*	*23*	*7.3*	*32*
$[Gd(DTPA\text{-}BA)(H_2O)(CH_2)_n]_x$ $n = 6$	0.43	50.2	32	9.6	79
$[Gd(DTPA\text{-}BA)(H_2O)(CH_2)_n]_x$ $n = 10$	0.66	40.0	2	—	79
$[Gd(DTPA\text{-}BA)(H_2O)(CH_2)_n]_x$ $n = 12$	0.50	49.8	31	—	79
$[Gd(DTPA\text{-}BA)(H_2O)\text{-}PEG]_x$	0.48	47.0	22	9.2	78

[a] Structures of the complexes and ligands are shown in Charts 2.9 and 2.10.

relative number of free amines affect the water exchange rate, it does not seem to be significant in the present case. The positive activation volumes indicate a dissociatively activated mechanism, even if it is not as accentuated for the dendritic complex [G3(N{CS}N-bz-Gd{DO3A}{H_2O})$_{23}$] ($\Delta V^{\ddagger} = +3.1$ cm^3 mol^{-1}) as it is for the monomer ($\Delta V^{\ddagger} = +7.7$ cm^3 mol^{-1}). All four complexes exhibit rate constants which are smaller by a factor of three when compared to that of $[Gd(DOTA)(H_2O)]^-$, which fits well to the empirical rule about the diminution of the exchange rate when amide donors are substituted for carboxylates, as discussed above. This study proved experimentally for the first time that the attachment of a macrocyclic unit to a large dendrimer molecule did not significantly influence the kinetics of water exchange on the Gd(III) chelate.

Water exchange parameters have been assessed on linear copolymers containing DTPA-bisamide chelators with poly(ethylene glycol) (PEG) [78] or polyalkyl ($-(CH_2)_n-$; $n = 6$, 10 and 12) spacers [79]. For both $[Gd(DTPA\text{-}BA)(H_2O)\text{-}PEG]_x$ and $[Gd(DTPA\text{-}BA)(H_2O)\text{-}(CH_2)_n]_x$, the rates and the mechanisms of the exchanges were identical to those of [Gd(DTPA-BMA)(H_2O)], where the latter can be considered as the monomer unit of

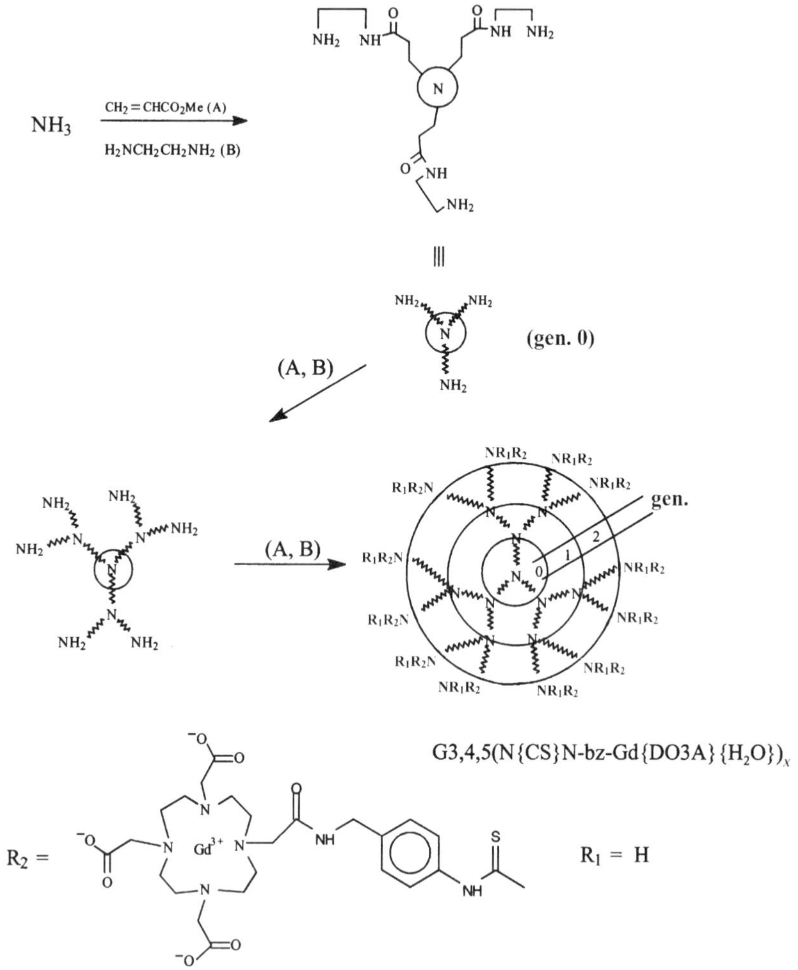

NH_3 $\xrightarrow[H_2NCH_2CH_2NH_2\ (B)]{CH_2=CHCO_2Me\ (A)}$

(gen. 0)

(A, B)

(A, B)

gen.

G3,4,5(N{CS}N-bz-Gd{DO3A}{H_2O})$_x$

R_2 =

R_1 = H

[Chart 2.9]

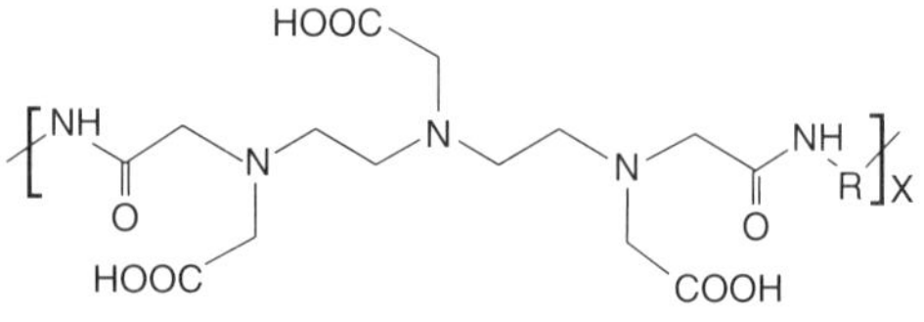

R = $(CH_2)_n$ n = 6,10,12   DTPA-BA-$(CH_2)_n$

R = CH_2-CH_2-$(OCH_2)_{32}$ —— DTPA-BA-PEG

[Chart 2.10]

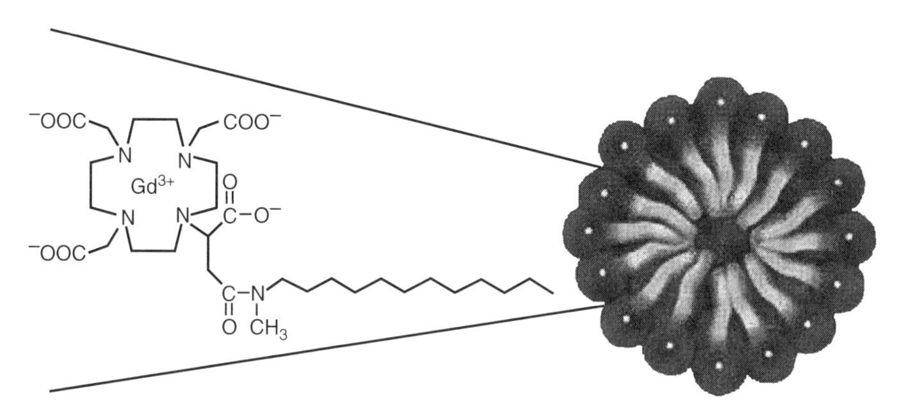

Figure 2.8 Schematic representation of the micellar structure formed in aqueous solution of the $[Gd(DOTA\text{-}C_{12})(H_2O)]^-$ complex. Reproduced by permission of Wiley–VCH [81].

these polymeric Gd(III) complexes. The Gd(DTPA-bisamide)–alkyl copolymers form rigid intramolecular micelle-like structures in aqueous solutions [80]. The polymeric chains consist of a non-ionic 'hydrophilic' polar part (the Gd chelate) and a 'hydrophobic' part (the methylene groups) and in this respect they behave as typical non-ionic polymeric surfactants which are known to form intramolecular micelle-like domains. The ^{17}O NMR measurements on these polymers clearly showed that their micellar structure does not affect the water exchange. The same conclusion was drawn from a ^{17}O NMR study performed on the amphiphilic DOTA derivative, $[Gd(DOTA\text{-}C_{12})(H_2O)]$, which is capable of self-organization by forming micelles in aqueous solution [81]. In the micelles (intra- or intermolecular for the linear polymers and the aggregated monomers, respectively) the Gd(III) chelates point towards the hydrophilic exterior, and thus there is easy access from the bulk water to the paramagnetic center (Figure 2.8). Although the degree to which water is present in the deep interior of the micelles has been the subject of some controversy, it is generally accepted that water molecules penetrate one or two CH_2 groups toward the center and the head-group is always fully hydrated [82]. Consequently, the parameters describing water exchange can not be much influenced by the micellar structure.

Another type of macromolecular agent is represented by Gd(III) complexes that non-covalently bind to biological molecules, particularly to proteins (see Chapter 5) [72, 76, 83–86]. Water exchange rates have been estimated or directly obtained for some protein-bound chelates. Aime and co-workers reported a significant decrease in the water exchange rate on two Gd(III) complexes when they bind to albumin via electrostatic or hydrophobic interactions. In these cases, the exchange rate was deduced from the analysis of NMRD profiles. The slow water exchange was explained in terms of a reduced accessibility of the water coordination site induced by the binding. This behavior occurs if the interaction between the Gd(III) chelate and the protein involves the bound

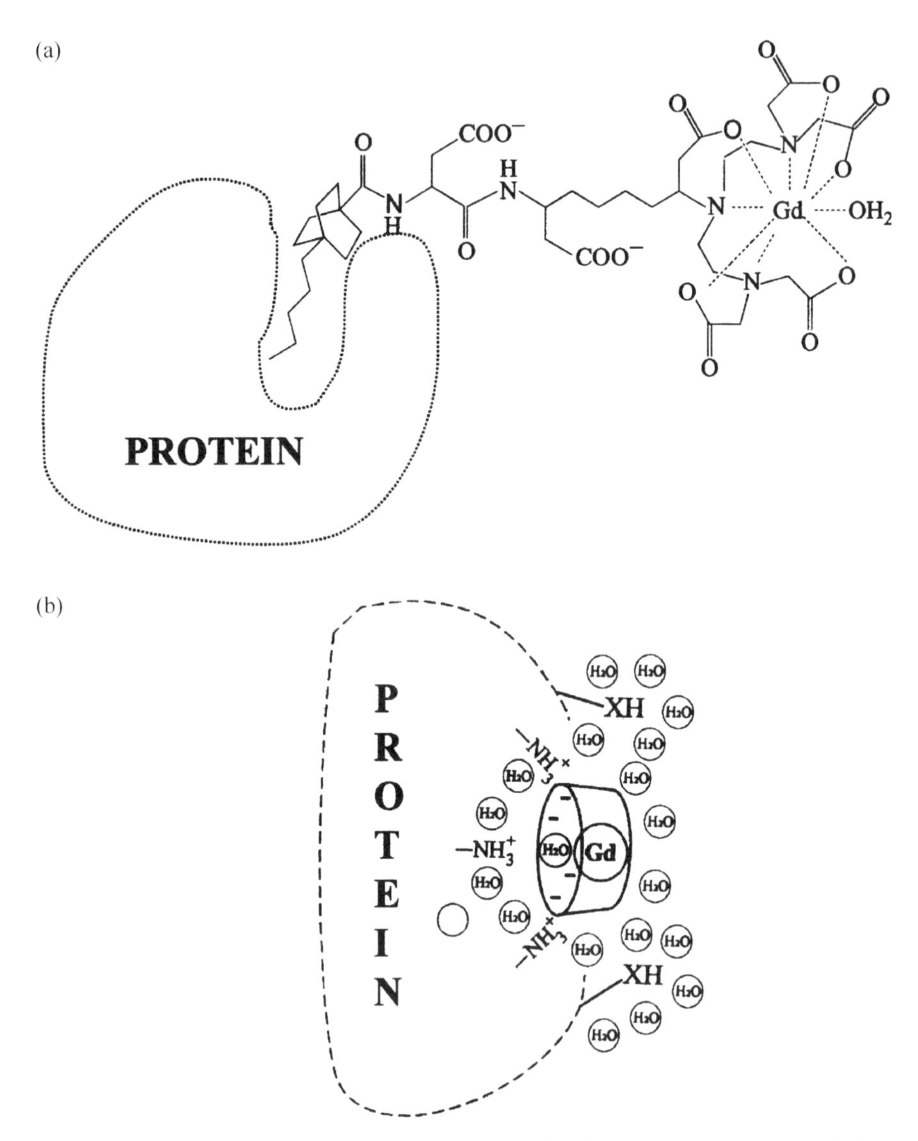

Figure 2.9 Schematic representations of the non-covalent interactions between (a) a hydrophobic side-chain of the ligand and a protein, and (b) a negatively charged paramagnetic chelate and a protein. In the case of (a), the protein binding site is too far from the Gd(III) chelate to influence the water exchange, whereas in (b) the exchange of the inner-sphere water is slowed down.

water side of the complex (Figure 2.9(b)). Similarly, the fitting of NMRD profiles of the protein-bound Gd(III) chelate suggested a ca. 50 % decrease in the water exchange rate on MS-325 when bound to human serum albumin [87, 88].

The effect of albumin-binding on water exchange has been directly evaluated by ^{17}O NMR on MP-2269, a DTPA-derivative Gd(III) complex which binds serum albumin through hydrophobic interactions [72]. Transverse ^{17}O relaxation rates of MP-2269 were measured in the absence and in the presence of bovine serum albumin (BSA). Optimum concentrations of the Gd(III) chelate and of the protein have been chosen so that the maximum effect in measured relaxation rates can be observed. The reduced ^{17}O relaxation rates measured in the absence and in the presence of BSA were found to be very similar. Since at low temperatures (below 300 K), the system is in the slow-exchange regime, the measured relaxation rates are determined directly by the water exchange rate, and the relaxation rate of the coordinated water oxygen ($1/T_{2m}$) has no significant contribution. The observed ^{17}O relaxation rates are not influenced either by the residence time of the whole GdL complex on the protein. The similarity of the transverse relaxation rates, measured both without and with protein, clearly indicated that the water exchange rate on the Gd(III) complex did not change considerably when it was bound to the protein: as a lowest limit, $k_{ex,GdL-BSA} \geq k_{ex,GdL}/2$ was determined. The finding that the water exchange on MP-2269 is not affected by the protein can be accounted for by the relatively large distance between the metal- and protein-binding sites which facilitates the access from the bulk to the coordinated water (Figure 2.9(a)).

2.3.5 Effect of Hydration Equilibrium on the Water Exchange

We have discussed above how the rate and the mechanism of the water exchange is closely related to the inner-sphere structure of the Gd(III) chelate. Since it is usually admitted that only the coordination numbers of eight or nine are available for Gd(III) in aqueous solution, nine-coordinated complexes will exchange in dissociatively activated processes, i.e. via an eight-coordinate transition state, whereas eight-coordinated complexes exchange in associatively activated processes, i.e. via a nine-coordinate transition state. The rate of the water exchange depends on the accessibility of the transition state. If eight- and nine-coordinated species, ($[GdL(H_2O)_n]$ and $[GdL(H_2O)_{n+1}]$), are both present in the solution in a dynamic equilibrium, one can expect that the energy gap between them, i.e. between the nine-coordinate ground state and the eight-coordinate transition states (or vice versa), is relatively small, and thus the exchange will be fast. As there is no direct means to check the presence of a hydration equilibrium for a Gd(III) complex, one has to use the $^7F_0 \rightarrow {}^5D_0$ absorption band of Eu(III) in the visible spectrum which is sensitive to changes in the coordination environment. This transition can be used to characterize hydration equilibria for Eu(III) complexes, as discussed above in Section 2.1., and the results can then be extrapolated to the corresponding Gd(III) chelate. Since the size difference between Gd^{3+} and Eu^{3+} is very small, the assumption of identical equilibrium parameters for the complexes of the two adjacent lanthanide ions is quite reasonable.

When two differently hydrated Gd(III) complexes are present in a dynamic equilibrium, the rigorous analysis of the observed ^{17}O transverse relaxation rates becomes rather complicated since there are two sites ([GdL$(H_2O)_n$] and [GdL$(H_2O)_{n+1}$]) that can both exchange with the bulk and between themselves, and thus the situation has to be described as a three-site exchange problem. In the absence of spin–spin coupling, the Kubo–Sack formalism allows for a correct calculation of the NMR lineshape with chemical exchange between different sites [89]. The ^{17}O NMR data for [Gd(DO3A)$(H_2O)_{1,2}$] have been evaluated by using this rigorous method. The parameters calculated for the water exchange have been compared to those obtained by simplifying the system to two parallel two-site exchange problems which can be analyzed by the usual Swift–Connick equations [29]. In this latter case, an average hydration number is used which depends on the temperature as calculated from a UV-Vis spedroscopic study on the Eu(III) analogue. For the [Gd(DO3A)$(H_2O)_{1,2}$] system, the equilibrium is strongly shifted towards the bishydrated species, with the average coordination number being 1.88 at 298 K.

Although the Kubo–Sack formalism allows for a theoretically correct analysis of the ^{17}O NMR data, several approximations have to be made, namely that identical electronic relaxation parameters and scalar coupling constants are assumed for the two differently hydrated species. As far as electron spin relaxation is concerned, very little, in general is known about the relationship between complex structure and relaxation rate. For the scalar coupling constants, the real values are probably somewhat different for the two complexes. However, the concentration of [Gd(DO3A)(H_2O)] is small compared to [Gd(DO3A)$(H_2O)_2$], which would make the separation of the contributions of the two species difficult. Consequently, introducing independent parameters for the contributions of the mono-and bisaqua complexes to the observed relaxation rate and chemical shift would certainly ameliorate the quality of the fit; however, the resulting values could be less reliable. Therefore, the hypothesis of identical τ_v, Δ^2, E_v and $A/\hbar$ values for the two complexes is acceptable. It implies that the [Gd(DO3A)$(H_2O)_2$] and [Gd(DO3A)(H_2O)] sites are differentiated only by the exchange rate of their coordinated water molecule(s) which, through the scalar relaxation mechanism, also affects the $1/T_{2m}$ relaxation term.

As shown in Table 2.5, the parameters obtained by the Kubo–Sack treatment are very similar to those calculated with the more simple Swift–Connick approach. This means that, despite the crude approximation of simplifying the system to two-site exchange problems, the Swift–Connick analysis which uses a temperature-dependent average hydration number, leads to a sufficiently precise value for the water exchange rate, and thus the use of the much more complicated Kubo–Sack formalism is not necessary in the case of Gd(III) complexes with a hydration equilibrium shifted towards one species.

Table 2.5 Water exchange parameters obtained from ^{17}O transverse relaxation rates and chemical-shift data for $[Gd(DO3A)(H_2O)_n]$ ($n = 1, 2$), by using the simple Swift–Connick analysis (two two-site exchange problems), or the theoretically correct Kubo–Sack formalism (three-site exchange).

$[Gd(DO3A)(H_2O)_n]$	Two-site exchange Swift–Connick approach $n = 1.7$–1.9	Three-site exchange Kubo–Sack approach
k_{ex}^{298} ($10^6 s^{-1}$)	11 ± 1	10 ± 2
$\Delta H^{\ddagger}$ (kJ mol^{-1})	33.6 ± 1.2	34.1 ± 3.0
$\Delta S^{\ddagger}$ (J mol^{-1} K^{-1})	$+ 2.4 \pm 0.4$	$+9.4 \pm 8.0$

Water exchange parameters have also been obtained on the $[Gd(DO2A)(H_2O)_{2,3}]^+$ complex which represents another example of hydration equilibrium [30]. Contrary to the expectations, for both $[Gd(DO2A)(H_2O)_{2,3}]^+$ ($k_{ex}^{298} = 10 \times 10^6 s^{-1}$) and $[Gd(DO3A)(H_2O)_{1,2}]$ ($k_{ex}^{298} = 11 \times 10^6 s^{-1}$), only about a twofold gain is observed in the water exchange rate in comparison to $[Gd(DOTA)(H_2O)]^-$ ($k_{ex}^{298} = 4.8 \times 10^6 s^{-1}$). The reason is that besides the favorable influence of the equilibrium state on increasing the water exchange rate, one has to consider another, i.e. negative, factor, which is the increased overall positive charge of the complex. The interplay of the opposite effects of the hydration equilibrium and of the increasing positive charge results in the limited water exchange gain.

2.3.6 Water Exchange on Different Isomers of Ln(III) Tetraazamacrocyclic Complexes

It is now well known that Ln(III) DOTA-type complexes exist in two diastereoisomeric forms (**m** and **M**) that differ by the layout of their acetate arms: the **M** isomer has an antiprismatic geometry, whereas the **m** isomer has a twisted antiprismatic geometry (Figure 2.10) [90].

Basically, the two structures display a different orientation of the two square planes formed by the four cyclen nitrogens and the four binding oxygens, making an angle of ca. 40° in **M**-type structures, whereas this situation is reversed and reduced to ca. 20° in the **m**-type derivatives. In solution, the two isomers, **m** and **M**, may exist in equilibrium. In the solid state for $[Ln(DOTA)(H_2O)]^-$ complexes, **m**-type structures have been observed for La and Ce, and **M**-type structures for Eu, Gd, Dy, Ho, Lu and Y [91–94]. For DOTA-tetraamide Ln(III) complexes, an **m**-type structure has been found in the case of $[Eu(DOTAM)(H_2O)]^{3+}$ [95], whereas the Dy complexes with ligands where one or both amide protons have been substituted by methyl groups, gave **M**-type X-ray structures [50].

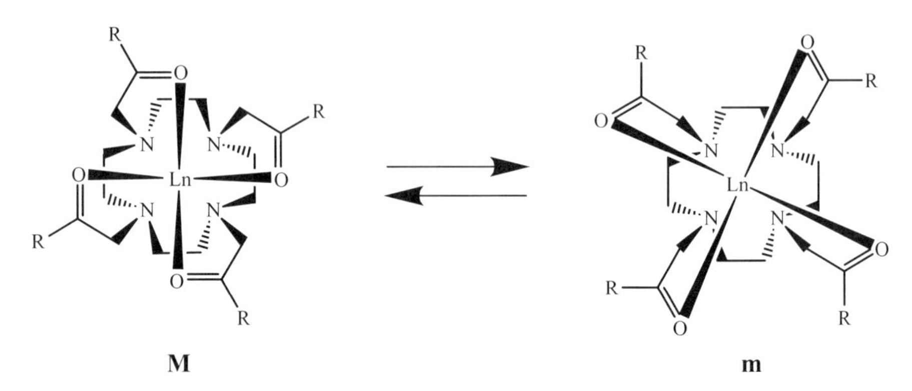

Figure 2.10 Two diastereoisomeric forms, **m** and **M**, of Ln(III) complexes with DOTA-type ligands.

The bound-water signal of $[Ln(DOTA)(H_2O)]^-$ complexes has never been observed due to the fast exchange, and thus it is impossible to determine the contribution of each isomer to the overall exchange with the bulk water. Moreover, for any Gd(III) complex the slow electronic relaxation prevents the observation of the NMR signal of the bound-water molecule. Therefore, in order to see the coordinated water signal one has to choose a different lanthanide ion to gadolinium, and then find a complex with a sufficiently slow exchange. Coordinated-water ^1H and ^{17}O NMR signals have been detected for both isomers of $[Eu(DOTAM)(H_2O)]^{3+}$ in an acetonitrile–water solvent system [50, 52, 53]. (It has to be noted that acetonitrile has no influence on the water exchange, i.e. the low exchange rate is not due to the presence of this solvent.) The water and proton exchange rates, determined from the ^1H and ^{17}O NMR linewidths, respectively, have been found to be identical. The water exchange on **m** is about 50 times faster than on **M**, and even though the equilibrium constant $K = [M]/[m]$ equals 4.5, the contribution of **m** to the overall exchange rate still represents 90 %. The interconversion between the **M** and **m** isomers, which happens mainly through a rotation of the amide arms in an interchange mechanism, has been found to be related to the water exchange process. A non-hydrated complex is proposed as a common intermediate for both the water exchange and the isomer interconversion processes, but only one **M** $\rightarrow$ **m** interconversion occurs while two to three water exchanges take place. This study has important implications to contrast agent design within the family of DOTA-type ligands, since, for the first time, it shows the necessity to synthesize complexes which mainly exist in the fast-exchanging **m**-form in solution.

Similar results have been obtained on the Gd(III) complex of another DOTA derivative, where two carboxylic moieties are transformed into *N*-bis-methyl-carboxamide functionalities [74]. The transverse ^{17}O relaxation rates had to be analyzed in terms of the sum of the contributions arising from two isomeric species which are characterized by a large difference in their water exchange rates. The isomeric ratio obtained in this way is temperature-dependent and corresponds well to that determined by ^{1}H NMR for the analogous Eu(III) complex.

2.3.7 Proton Exchange versus Water Exchange

Water exchange rates give only a lower limit for the proton exchange rate. The distinction between proton and water exchange can be made by means of ^{1}H relaxometric measurements, once the water exchange has been independently determined from ^{17}O transverse relaxation rates. Reliable kinetic parameters can only be obtained for the proton exchange if the exchange of the entire water molecules is considerably slower than that of the protons. Proton exchange becomes accelerated in acidic or basic media due to H^{+-} or OH^{-}-catalyzed processes. The general form of the proton exchange rate is expressed as shown in Equation 2.32, where $k_{ex}^{H_2O}$ is the water exchange rate as obtained from ^{17}O NMR measurements, with k^{H} and K^{OH} being the rate constants for the acid- and base-catalyzed prototropic exchange processes, respectively [74]:

$$k = k_{ex}^{H_2O} + k^{H}[H^{+}] + k^{OH}[OH^{-}] \tag{2.32}$$

For the Gd(III) chelates used as MRI contrast agents, at physiological pH values the proton exchange rate equals the water exchange rate (the second and third terms are negligible in Equation (2.32)). The acceleration of proton exchange on Gd(III) complexes due to H^{+} or OH^{-} catalysis is an interesting phenomenon from the practical point of view, since it represents a possibility to avoid the situation where slow water/proton exchange limits the proton relaxivity. Consequently, in current contrast agent research there is a big effort to displace the domain of H^{+} or OH^{-} catalysis and to take advantage of the accelerated proton exchange around the physiological pH.

The H^{+}-and OH^{-}-catalyzed proton exchange was evidenced for a tetra-amide-DOTA-derivative Gd(III) complex, $[Gd(DTMA)(H_2O)]^{3+}$, from relaxivity measurements as a function of pH (Figure 2.11) [40]. The inner-sphere water of this complex is so inert ($k_{ex}^{298} = 4.5 \times 10^{4} s^{-1}$) that the relaxivities measured around a neutral pH value correspond only to the outer-sphere contribution. Below a pH of 2 and above a pH of 8, the relaxivities start to increase significantly. This phenomenon can be accounted for in terms of proton-and base-catalyzed proton exchange which results in an inner-sphere contribution to the overall relaxivity. It has to be noted that the $[Gd(DTMA)(H_2O)]^{3+}$ complex shows a remarkable kinetic stability toward acid-catalyzed dissociation which makes it possible to investigate the H^{+} catalyzed proton exchange even at very acidic pH values. Further examples for

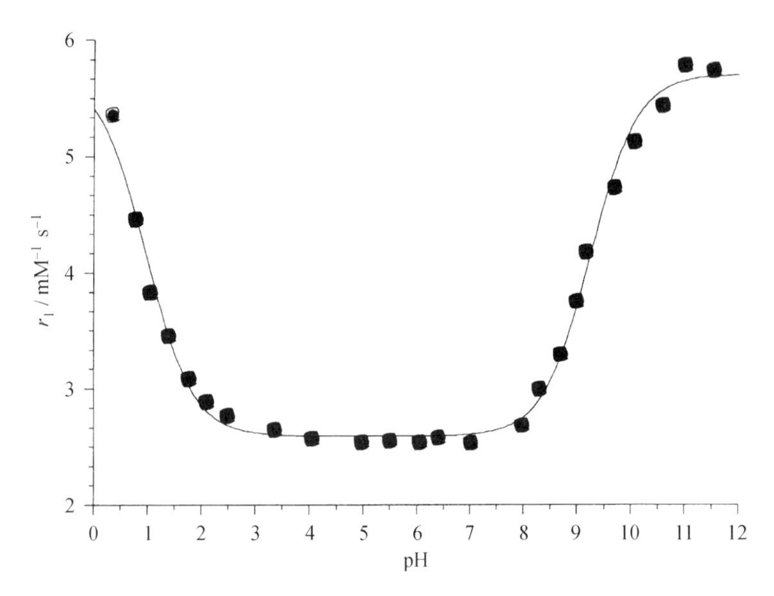

Figure 2.11 Plot of the pH dependence of the solvent longitudinal relaxivity for an aqueous solution of $[Gd(DTMA)](CF_3SO_3)_3$ at 298 K and 20 MHz. Reprinted with permission from [40]. Copyright (1997) *American Chemical Society*.

H^+-or OH^--catalyzed prototropic exchange have been given by the DTPA-bisamide-derivative complex, $[Gd(DTPA\text{-}BBA)(H_2O)]$ [41], and by DO3A-triamide Gd(III) complexes [96].

Certainly, the effect of prototropic exchange can only be observed on relaxivity if at neutral pH the relaxivity is limited by the water exchange lifetime, i.e. the inner-sphere proton relaxation rate, $1/T_{1m}$, is comparable to the water exchange rate, k_{ex} (see Equation (2.5)). This condition can be fulfilled if the chelate is attached to a slowly tumbling macromolecule. This was exploited by studying the prototropic exchange for the amphiphilic $[Gd(COPTA)(H_2O)]^{2-}$ species which is capable of forming an adduct with cyclodextrin [71]. While the exchange has no influence on the relaxivity of the monomer itself, it affects the relaxivity of the slowly rotating cyclodextrin conjugate, and thus the acid- or base-catalyzed prototropic exchange processes can be evaluated from the pH dependence of the proton relaxivities.

2.4 ROTATION

For small-molecular-weight Gd(III) chelates, it is the rotational correlation time, τ_R, that mainly determines the effective correlation time of proton relaxation, τ_c, in Equation (2.12), i.e. the fast rotation is the limiting factor for proton relaxivity at magnetic fields relevant to MRI (see Figure 2.3). This is well

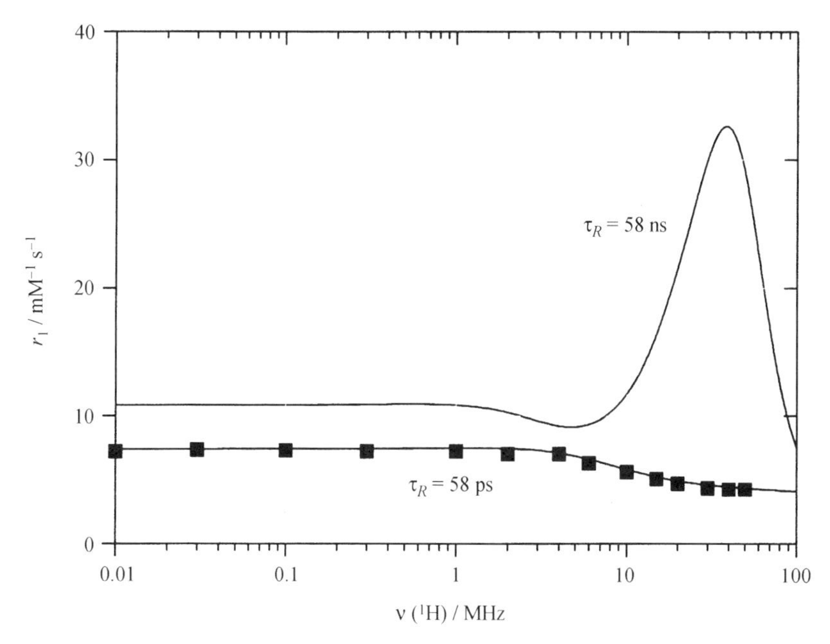

Figure 2.12 An experimental NMRD profile of $[Gd(DTPA)(H_2O)]^{2-}$ (25 °C; bottom curve) and a calculated NMRD profile (upper curve) obtained by using the same parameters as obtained for this species, except for the rotational correlation time, τ_R, which was taken at 1000 times of its value.

demonstrated by Figure 2.12 which shows the nuclear magnetic resonance dispersion (NMRD) profile of the low-molecular-weight, and thus fast-rotating, $[Gd(DTPA)(H_2O)]^{2-}$ species (τ_{R1} = 58 ps) and a curve calculated with the same parameters except for the rotational correlation time ($\tau_{R2} = 1000 \times \tau_{R1}$ = 58 ns, which is of the order of rotational correlation times measured, e.g. for serum albumins. The fact that the relaxivity of the small Gd(III) chelates is limited by rapid tumbling was recognized at an early stage of contrast agent development and has led to a wide variety of approaches to slow down rotation by increasing the molecular weight of the complex.

2.4.1 Techniques for Determining the Rotational Correlation Time

Although a knowledge of the rotational correlation time is crucial for the analysis of NMRD curves, it is not an easy parameter to determine in an independent way. In fact, it is the NMRD technique which is most widely used to determine rotational correlation times, with the majority of the τ_R values reported in the literature being obtained from proton relaxation measurements.

The Debye–Stokes equation (Equation (2.33)) provides an estimation of τ_R for spherical molecules:

$$\tau_R = \frac{4\pi\eta r_{eff}^3}{3k_B T} \tag{2.33}$$

The drawback of this method is that the value of the radius, r_{eff}, is never known exactly, and η, which stands for the microviscosity, can be considerably different from the measurable macroviscosity of the solution. However, rotational correlation times for similar systems with different molecular weights can be suitably compared on the basis of Equation (2.33), by supposing that the microviscosity, as well as the density, of the molecules are the same, and thus the ratio of the r_{eff}^3 values can be expressed by the ratio of the molecular weights.

Nuclear relaxation of different nuclei other than the proton can also be used to determine the rotational correlation time. The longitudinal relaxation of the ^{17}O, which is present in the water molecule coordinated to the Gd(III), is governed by quadrupolar [97] and dipolar [98] mechanisms, $1/T_{1q}$ and $1/T_{1d}$, respectively, with the quadrupolar term being the major contributor:

$$\frac{1}{T_{1q}} = \frac{3\pi^2}{10}\left(\frac{2I+3}{I^2(2I-1)}\right)\chi^2(1+\eta^2/3)\left[0.2\left(\frac{\tau_R}{1+\omega_I^2\tau_R^2}\right)+0.8\left(\frac{\tau_R}{1+4\omega_I^2\tau_R^2}\right)\right] \tag{2.34}$$

$$\frac{1}{T_{1d}} = \frac{2}{15}\left(\frac{\gamma_I^2 g^2 \mu_B^2}{r_{\mathrm{GdO}}^6}\right)S(S+1)\left(\frac{\mu_0}{4\pi}\right)^2\left[7\left(\frac{\tau_c}{1+\omega_s^2\tau_c^2}\right)+3\left(\frac{\tau_c}{1+\omega_I^2\tau_c^2}\right)\right] \tag{2.35}$$

In Equations (2.34) and (2.35), I is the nuclear spin, χ^2 is the quadrupolar coupling constant, η is an asymmetry parameter, and r_{GdO} is the Gd–O distance. (The transverse ^{17}O relaxation rates in Gd^{3+} solutions are determined by a scalar relaxation mechanism and contain no information on the rotational motion of the system.) The difficulty in the evaluation of the longitudinal ^{17}O relaxation rates is that the $\chi^2(1+\eta^2/3)$ term is unknown for Gd(III) chelates, while in addition, the Gd–O distance in solution can only be estimated. Despite these difficulties, rotational correlation times obtained from longitudinal ^{17}O relaxation rates can provide a good comparison for similar Gd(III) complexes. It is usually supposed that neither the Gd–O distance, nor the quadrupolar coupling constant, will change significantly from one complex to the other. A clear advantage of this method is that the rotational correlation time is measured directly on the Gd(III) complex and no metal substitution is needed (see below). Furthermore, the τ_R determined in this way corresponds to the rotation of the Gd(III)–coordinated water oxygen vector which is probably analogous to the rotation of the Gd(III)–coordinated water proton, which itself determines the parameter of practical importance, namely the proton relaxivity.

It has to be noted that rotational correlation times obtained from ^{1}H NMRD measurements are generally up to 50 % lower than those determined from ^{17}O longitudinal relaxation rates. This discrepancy can be partly explained by the incompatible Gd–O and Gd–H distances used in the two cases. Another source of the difference is that the quadrupolar coupling constant is unknown for Gd(III) complexes, and generally the χ-value determined for water is used. Furthermore, this discrepancy may also arise in part from the contribution of second-sphere water to the ^{17}O longitudinal relaxation rates, although usually this is considered to be negligible. In conclusion, we can say that even if the absolute values of τ_R measured by ^{17}O NMR are not correct, their comparison for similar Gd(III) complexes is always reliable.

Recently, there have been certain concerns that the coordinated water molecule may rotate around the Gd–O axis independently of the rotation of the complex, which would result in a lower rotational correlation time for the Gd–H than that measured for the Gd–O vector. This phenomenon could be a serious limiting factor for the attainment of high proton relaxivities for immobilized systems. However, so far there is no experimental evidence for this hypothesis.

Deuterium or ^{13}C relaxation measurements on the ligand can also be useful for determining τ_R [58, 99–101]. The big advantage is that they are very direct methods in the sense that the measured relaxation rates are directly proportional to the rotational correlation time, and no separation of different contributions is needed. The disadvantage of both deuterium or ^{13}C relaxation measurements is that they require the use of a diamagnetic analogue instead of the Gd(III) complex itself (Y(III), La(III) or Lu(III)). Furthermore, it is not the rotation of the metal–coordinated water vector, which is important from the practical point of view, that is monitored (this can be especially problematic in the case of large molecules). Another drawback is the low sensitivity of ^{2}D or ^{13}C at natural abundance levels. This problem can be overcome by enrichment of the ligands, which is, however, not always easily realizable.

EPR spectra are often extremely sensitive to rotational motion, which makes it possible to determine τ_R from simulation of the spectral lineshape. Unfortunately, Gd(III) does not exhibit rotationally modulated EPR spectra at conventional magnetic fields. This problem can be circumvented by substituting Gd(III), which presents motion-insensitive EPR spectra, by a similar-sized and similar-shaped cation which has a motion-sensitive EPR signal. Vanadyl (VO^{2+}) was proposed due to its anisotropic g-factor [102–105]. An additional advantage of this technique is that vanadyl EPR line shapes allow for the distinction of isotropic and anisotropic motions. However, substitution of the triply charged gadolinium ion with the doublly charged vanadyl species may mainly modify the number of water molecules which are hydrogen-bonded to the ligand, and hence the overall size of the tumbling entity, and this consequently also results in modified rotational dynamics.

Rotational diffusion of fluorophores is a dominant cause of fluorescence depolarization [106]. Based on this phenomenon, fluorescence polarization spectroscopy can also be used to determine τ_R, provided that the chelate contains a fluorophore, which, moreover, must have a fluorescence lifetime comparable to the rotational correlation time. This technique, mainly due to the synthetic challenge of introducing a fluorescent group into the ligand, has not found wide application. Another problem may be that labeling with suitable fluorophores can modify the size and shape of the complex, and hence alter the rotational dynamics.

2.4.2 Rotation of Monomers, Dimers and Macromolecules–Internal Flexibility

A selection of rotational correlation times, τ_R, determined by different techniques for a series of monomer, dimer and macromolecular Gd(III) chelates are presented in Table 2.6, with the structures of the various ligands being shown in Charts 2.1, 2.2, 2.4, and 2.7–2.10. For comparative purposes, this table also contains the molecular weight and proton relaxivity values of the complexes. The examples shown have been chosen in order to represent general trends and not to cover all data available in the literature. A more complete database has been published recently [5].

As expected, the τ_R values generally increase with increasing molecular weight; however, this relationship is far from the linearity predicted by the Debye formula (Equation (2.33)). The reason for this is the internal flexibility of the molecules. The rotational correlation time as obtained from proton or ^{17}O longitudinal relaxation rates corresponds to the rotational motion of the Gd–coordinated water H, or Gd–coordinated water O, vectors, and thus represents an effective correlation time and not one for the whole molecule. Since the molecules are not completely rigid, this motion, which then determines proton relaxivity, can be considerably faster than the motion of the whole molecule, which is itself related to the molecular weight. This phenomenon is particularly important for several types of macromolecular agents, either due to the general flexibility of the whole molecule or to the flexibility of the linking group that is used to attach the Gd(III) chelate to the macromolecule. The first example is well illustrated by certain linear polymers whose proton relaxivities can be completely independent of the molecular weight (usually above $\approx$ 10 kDa) for a family of analogous polymers [107, 108]. Since their relaxivity values are limited by rotation, the invariance of the relaxivity reflects the invariance of the rotational correlation time. Similarly, although not in the context of MRI contrast agents, molecular-weight-independent rotational correlation times were obtained from EPR measurements on nitroxide-labeled linear polymers [109]. The reason is that the rotational correlation times of linear polymers are dominated by segmental motions which, for large polymers, are independent of the molecular weight. Consequently, linear polymers in general are known to

Table 2.6 Rotational correlation times, obtained by different methods, and proton relaxivities for selected Gd(III) complexes.

Type	Ligand[a]	τ_R^{298} (10^{-12}s)	Method	M_w	r_1 (mM^{-1} s^{-1}) 20 MHz; 37A°C	Reference
Monomers	aqua	41	NMRD/$^{17}O^b$	301	—	32
	DTPA	58	NMRD/$^{17}O^b$	563	4.02	32
		103	$^{17}O^c$	—	—	32
	EOB-DTPA	84	NMRD	696	5.3	99, 161
		93	^{2}H NMR	—	—	99
		178	$^{17}O^c$	—	—	60
	BOPTA	88	NMRD	683	4.39	162
	DTPA-BMA	66	NMRD/$^{17}O^b$	587	3.96	32
		167	$^{17}O^c$	—	—	32
	MP-2269	139	NMRD/$^{17}O^b$	1069	5.64	72
	DOTA	77	NMRD/$^{17}O^b$	575	3.83	32
		90	$^{17}O^c$	—	—	32
	PCTA-[12]	70	NMRD	552	6.9 (25 °C)	77
	PCTP-[12]	106	NMRD	657	7.5 (25 °C)	77
Dimers	pip$(DO3A)_2$	171	NMRD/$^{17}O^b$	1202	5.79	32
	bisoxa$(DO3A)_2$	106	NMRD/$^{17}O^b$	1264	4.94	32
Linear polymers	$[DTPA\text{-}BA\text{-}PEG]_x$	232	NMRD/$^{17}O^b$	20.2 kDa	6.31	78
	$[DTPA\text{-}BA(CH_2)_n]_x$	801	$^{17}O^c$	19.4 kDa	9.8	79, 80
	$n = 6$					
	$n = 10$	$\tau_g = 2900^d$	$^{17}O^b$	10.3 kDa	15.4	79, 80
	$n = 10$	$\tau_1 = 460^e$	Lipari–Szabo	—	—	79, 80
	$n = 12$	$\tau_g = 4400^d$	$^{17}O^b$	15.7 kDa	19.6	79, 80
	$n = 12$	$\tau_1 = 480^e$	Lipari–Szabo	—	—	79, 80

Table 2.6 (*Continued*)

Type	Ligand[a]	τ_R^{298} (10^{-12}s)	Method	M_w	r_1 (mM^{-1} s^{-1}) 20 MHz; 37A°C	Reference
Dendrimers	[G3(N{CS}N-bz-{DO3A})$_{23}$]	580	$^{17}O^c$	22.1 kDa	14.6	62
	[G4(N{CS}N-bz-{DO3A})$_{30}$]	700	$^{17}O^c$	37.4 kDa	15.9	62
	[G5(N{CS}N-bz-{DO3A})$_{52}$]	870	$^{17}O^c$	61.8 kDa	18.7	62
Protein-bound	PCTP-[13]-HSA	30 000	NMRD	69 kDa	45.0	76
	MS-325-HSA	3000–4000	NMRD	69 kDa	48.9	87
		6000–7000	^2H NMR	—	—	87
	MP-2269–BSA	1000	NMRD	66 kDa	24.5	72

[a] Structures are shown in Charts 2.1, 2.2, 2.4, and 2.7–2.10.
[b] In the fitting procedure, the Gd–H distance, r_{GdH}, was fixed at 0.31 nm, and the Gd–O distance, r_{GdO}, was left variable (0.21–0.24 nm).
[c] The τ_R values were calculated from ^{17}O NMR data alone, with the Gd–O distance, r_{GdO}, fixed at 0.25 nm.
[d] Global rotational correlation time.
[e] Local rotational correlation time.

have a considerable internal flexibility, and thus relatively low proton relaxivities.

The flexibility of the linking group was found to be responsible for the much lower effective rotational correlation times, and consequently the lower relaxivities of many different types of macromolecular agents. Dendrimers are inherently rigid macromolecular systems. The proton relaxivities attained by Gd(III) chelates attached to dendrimers increased with molecular weight, but, however, were lower than expected solely on the basis of molecular weight. A part of the reason is that the linking group between the Gd(III) chelate and the rigid dendrimer molecule has some flexibility and consequently the Gd(III) chelate experiences a more rapid motion when compared to the rotation of the dendrimer itself. (In addition, slow water exchange also limits proton relaxivity.) Other examples for internal flexibility can be found in the field of non-covalently bound Gd(III) chelate–protein adducts. For MP-2269 bound to bovine serum albumin, a τ_R of around 1.0 ns has been determined from both ^{17}O and ^{1}H longitudinal relaxation rates [72]. This value is more than one order of magnitude lower than the rotational correlation time of the entire protein molecule. This remarkable internal flexibility has been attributed to the relatively long distance between the Gd–binding and the protein-binding sites of the molecule. Contrary to this, a much longer rotational correlation time (of the order of τ_R for the protein) has been found for other protein-bound Gd(III) chelates, where the binding to the protein is ensured by phosphonate groups which are also participating in the coordination of Gd(III), thus precluding all flexibility between the metal- and protein-binding sites [76] (see Figure 2.9).

2.4.3 Separation of Fast Local and Slow Global Motions: the Lipari–Szabo Approach

It is becoming standard in NMR relaxation studies of macromolecular solutions to treat the dynamics in terms of a spectral density function involving an orientational order parameter and two correlation times describing overall and internal motions. Most of these studies are based on ^{13}C or ^{15}N nuclear spin relaxation measurements. Although the number of macromolecular systems studied in the context of contrast agent research is increasing, the problem of rotational dynamics has been rarely addressed in detail. Clarkson and co-workers investigated the molecular dynamics of vanadyl-EDTA and DTPA complexes in sucrose solution or attached to Polyamidoamine (PAMAM) dendrimers by using EPR [103, 104]. The motion-sensitive EPR data of the dendrimeric system have been fitted to an anisotropic model which is described by an overall spherical rotation, combined with a rotation around the axis of the arm branching out of the central core. The motions around the axis of the branch connecting the chelate to the central core were found to be very rapid, whereas the overall tumbling motions were slow.

Information on motional dynamics of macromolecular Gd(III) chelates can also be obtained from the longitudinal ^{17}O relaxation rates by using the Lipari–Szabo approach [79, 110]. The 'model-free' Lipari–Szabo approach is a widely used method for the evaluation of relaxation data for polysaccharides [111, 112], proteins [113], micellized surfactants [114] or calixarenes [115]. In this approach, two kinds of motion are assumed to modulate the interaction causing the relaxation, namely a rapid, local motion which lies in the extreme narrowing limit and a slower, global motion. If the two motions are statistically independent and if the global molecular reorientation is isotropic (it can be described by a single rotational correlation time), the reduced spectral density function can be written as follows:

$$J(\omega) = \left(\frac{S^2 \tau_g}{1 + \omega^2 \tau_g^2} + \frac{(1 - S^2)\tau}{1 + \omega^2 \tau^2} \right), \text{where} \quad \tau^{-1} = \tau_g^{-1} + \tau_l^{-1} \tag{2.36}$$

where τ_g is the correlation time for the global motion (common to the whole molecule), and τ_l is the correlation time for the fast local motion, which is specific for the individual relaxation axis, and thus related to the motion of the Gd(III) chelate unit. The generalized order parameter, S, is a model-independent measure of the degree of spatial restriction of the local motion, with $S = 0$ if the internal motion is isotropic, $S = 1$ if the motion is completely restricted. (In the original paper [110] and in the literature in general, τ_M is used for the overall, and τ_e for the local motion. However, in the context of contrast agent research τ_M is reserved for the binding time of the coordinated water molecule; hence, here we prefer a different notation.)

The dipolar term for the longitudinal ^{17}O relaxation rates in non-extreme narrowing conditions is given by the following equation [97]:

$$\frac{1}{T_{1dd}} = \frac{2}{15} \frac{\gamma_I^2 g^2 \mu_B^2}{r_{GdO}^O} S(S+1) \left(\frac{\mu_0}{4\pi} \right)^2 [3J(\omega_I; \tau_{d1}) + 7J(\omega_s; \tau_{d2})] \tag{2.37}$$

The Lipari–Szabo spectral density function is expressed by the following:

$$J(\omega_I; \tau_{d1}) = \left[\frac{S^2 \tau_{d1g}}{1 + \omega_I^2 \tau_{d1g}^2} + \frac{(1 - S^2)\tau_{d1}}{1 + \omega_I^2 \tau_{d1}^2} \right] \tag{2.38}$$

$$\frac{1}{\tau_{d1g}} = \frac{1}{\tau_m} + \frac{1}{\tau_g} + \frac{1}{T_{1e}}; \quad \frac{1}{\tau_{d1}} = \frac{1}{\tau_m} + \frac{1}{\tau} + \frac{1}{T_{1e}} \tag{2.39}$$

$$\tau^{-1} = \tau_g^{-1} + \tau_l^{-1} \tag{2.40}$$

In the corresponding expression for $J(\omega_s; \tau_{d2})$, we have:

$$\frac{1}{\tau_{d2g}} = \frac{1}{\tau_m} + \frac{1}{\tau_g} + \frac{1}{T_{2e}}; \quad \frac{1}{\tau_{d2}} = \frac{1}{\tau_m} + \frac{1}{\tau} + \frac{1}{T_{2e}} \tag{2.41}$$

In addition, the quadrupolar relaxation term is given by the following [98]:

$$\frac{1}{T_{1q}} = \frac{3\pi^2}{10}\frac{2I+3}{I^2(2I-1)}\chi^2(1+\eta^2/3)\times[0.2J_1(\omega_I)+0.8J_2(\omega_I)] \tag{2.42}$$

Contrary to the dipolar part, for the quadrupolar relaxation the spectral density function contains, directly, the rotational correlation times, and is given by the following:

$$J_n(\omega_I) = \left[\frac{S^2\tau_g}{1+n\omega_I^2\tau_g^2} + \frac{(-S^2)\tau}{1+n\omega_I^2\tau^2}\right], \text{where} \quad n = 1, 2 \tag{2.43}$$

This analysis has been applied for linear [Gd(III)DTPA-bisamide]–alkyl copolymers with $(CH_2)_n$ spacers of different length between the Gd(III) chelates, $[Gd(DTPA\text{-}BA)(H_2O) - (CH_2)_n]_x$ ($n = 10$ and 12) [79]. Contrary to the usual case of linear polymers, these copolymers have relatively high proton relaxivities as a result of slow rotation. The latter can be attributed to their special structure: they behave as non-ionic surfactants and form rigid intramolecular micelle-like aggregates due to hydrophobic interactions between the alkyl domains. The analysis of their rotational dynamics has led to significant results concerning the factors that determine relaxivity. The local rotational correlation times and the extent to which they contributed to relaxivity were found to be similar for the two polymers (the same as for much less rigid systems, e.g. PEG-based copolymers; Table 2.6). The global rotational correlation times, however, were different for the polymers with different $-CH_2$-groups and reflected their molecular weights. The difference in proton relaxivity could be interpreted with the different global rotational correlation times which clearly shows that, for these polymers, the overall motion also contributes to the relaxivity.

2.5 ELECTRONIC RELAXATION

Since a complete chapter in this volume will discuss the electron spin relaxation of Gd^{3+} (Chapter 9), here we will deal only very briefly with this problem. Electronic relaxation is a crucial and rather difficult issue in the analysis of proton relaxivity data. The difficulty resides, on the one hand, in the lack of a theory which is valid under all real conditions, and, on the other hand, by the technical problems of independent and direct determination of electronic relaxation parameters. Equations (2.14)–(2.15), which describe a mono-exponential electronic relaxation, are valid only in the limit of extreme narrowing ($\omega_s^2\tau_v^2 \ll 1$), while outside of this limit the relaxation becomes multi-exponential.

Electronic relaxation can be indirectly assessed by NMRD studies. However, due to the problems associated with the fit of NMRD profiles (see Section 4.), the parameters obtained in this way are not always reliable. Transverse relaxation rates can be directly determined from EPR measurements. Here the general problem is the limited validity of the Bloembergen–Morgan equations (Equations (2.14)–(2.16) which are used to describe the field dependence of the electronic relaxation rates and to obtain the experimentally non-accessible longitudinal electronic relaxation rates from the transverse relaxation rates. Variable-field EPR measurements performed in a relatively large field range are extremely useful and often indispensable; however, high-field EPR spectrometers are rarely accessible. Furthermore, improvements in the theory currently available for Gd(III) electron spin relaxation could also be necessary.

3 SECOND- AND OUTER-SPHERE RELAXATION

After describing the relaxivity enhancement due to the inner-sphere water molecule(s) by using well-established theory, we will now describe the theoretical model for the relaxation of water molecules outside of the first coordination shell of the gadolinium complex (see Figure 2.2). Very often the contribution of these to the overall relaxation enhancement is summarized in the term r_i^{OS} (Equation (2.4)), without distinguishing between different types of the so-called outer-sphere water molecules. The interaction between the water proton nuclear spin, I, and the gadolinium electron spin, S, is supposed to be a *dipolar intermolecular interaction* whose fluctuations are governed by random translational motion of the molecules [116]. For unlike spins, the relaxation rate, $1/T_1$ is given by the following [117–119]:

$$\frac{1}{T_1} = \frac{32\pi}{405}\left(\frac{\mu_0}{4\pi}\right)^2 \frac{N_A[\mathrm{M}]}{dD}\gamma_I^2\gamma_S^2\hbar^2 S(S+1)[j_2(\omega_I - \omega_s) + 3j_1(\omega_I) + 6j_2(\omega_I + \omega_s)] \tag{2.44}$$

where N_A is the Avogadro number, d is the closest distance of approach of spins I and S, D is the diffusion coefficient for relative diffusion, ($D = D_I + D_S$), and [M] is the molar concentration of the metal bearing the spin S. The spectral densities, $j(\omega)$, are the Fourier transforms of the time correlation functions, $g(t)$, which depend on the conditional probabilities for the relative diffusion of spins I and S, i.e. $P(r_0|r, t)$. This latter term can be approximated as being the solution of the Smoluchowski equation [119, 120]:

$$\frac{\delta P(\mathbf{r}_0 \mid \mathbf{r}, t)}{\delta t} = D\nabla\left[\nabla P(\mathbf{r}_0 \mid \mathbf{r}, t) + \frac{1}{k_B T} P(\mathbf{r}_0 \mid \mathbf{r}, t)\nabla U(r)\right] \tag{2.45}$$

Here, $\mathbf{r}_0$ and $\mathbf{r}$ are the separations of spins I and S at times of 0 and t, respectively, while $U(r)$ is the potential of the averaged forces between the spin-bearing molecules, which may be obtained from the radial distribution function, $g(r)$, by using $\ln[g(r)] \equiv -U(r)/k_B T$. Freed and other workers [120–

122] showed that an analytical expression for $1/T_1$ exists for the simplest case of a force-free model. The spectral densities are given in this case by the following:

$$j_k(\omega) = Re\left(\frac{1 + z/4}{1 + z + 4z^2/9 + z^3/9}\right)$$

with:

$$z = \sqrt{i\omega\tau + \tau/T_{ke}\tau} = d^2/D;\ k = 1, 2 \tag{2.46}$$

where T_{ke} is the longitudinal or transverse electron spin relaxation time. If the S spin is not located in the center of the molecule, various correction terms have to be added [119, 123].

The force-free model is certainly only a rough approximation for the interaction of outer-sphere water molecules with poly(amino carboxylate) or phosphonate Gd(III) complexes. It was found, for example, that phosphonate moieties are especially capable of tightly binding water molecules. According to the strength of this binding of the water molecules to functional groups of the chelating ligand, we can apply two limiting models.

A weak binding of water molecules to the outer part of a Gd(III) complex can be described by using Freed's theory for translational motion, together with a potential of averaged forces for an attraction of H_2O molecules by, for example, COO^- groups of the ligand. Such a potential could be obtained from integral equations of the statistical mechanics of liquids [124] or from classical molecular dynamics simulations.

A strong binding of water molecules to functional groups of the chelating ligand does not allow a treatment using translational diffusion. Similarly to the treatment of the inner-sphere term, the relaxation of the second-sphere water molecules has then to be calculated with equations derived from rotational diffusion. Aime *et al.* [76] and Botta [125] have proposed to replace Equation (2.4) by the following:

$$r_i = r_i^{IS} + r_i^{2nd} + r_i^{OS} \tag{2.47}$$

where r_1^{2nd} is the contribution from water molecules in the second coordination sphere and r_1^{OS} deals with the contribution from H_2O molecules which diffuse in the proximity of the paramagnetic complex. The term 'second coordination sphere' can be misleading in the sense that r_1^{2nd} concerns only a few water molecules close to the functional groups of the ligand which do not form a complete spherical shell around the complex, in contrast to what is observed, for example, around $[Cr(H_2O)6]^{3+}$ [126]. Moreover, the orientation of these 'second-sphere' water molecules is different from that of the first-coordination-sphere water molecules: in contrast to the first sphere, in the 'second sphere' the protons of the water molecules are closer to the Gd(III) than the oxygens (see Figure 2.2).

The contribution r_1^{2nd} can be calculated from expressions similar to Equation (2.5) applied for the first sphere relaxation. A more realistic model should consider that several distinct binding sites, characterized with a different type of binding mode, can be present on the chelating ligand. Consequently, we have to sum up the interactions due to each of these particular binding sites, and then write the following for the contribution to longitudinal relaxation:

$$r_1^{2nd} = P_m \sum_{j=1}^{M} \frac{1}{T_{1j} + \tau_{mj}} \tag{2.47}$$

where τ_{mj} is the lifetime of a specific water molecule j in the second shell, and the summation is up to the number of specially bound water molecules in the second sphere, M. The term $1/T_{1j}$ can be estimated from equations similar to those used for inner-sphere water molecules (Equations (2.7)–(2.13)). We have furthermore to keep in mind that the water–^{1}H–gadolinium distance, introduced as $(r_{GdH}{}^{2nd})^{-6}$, is different for the two water protons of hydrogen-bonded H_2O in the second shell. From these considerations, it is understandable why the second-sphere contribution is mostly included in the outer-sphere term, thus leading to d and D values (and sometimes also inner-sphere water r_{GdH}) which are only fitting parameters without any significant physical meaning.

More detailed information can be obtained from gadolinium complexes without any inner-sphere water. Clarkson and co-workers [105] showed that for TTHA the second-sphere contribution can be substantial, i.e. more than 30 % (Figure 2.13). Taking into account only one type of second-shell binding site,

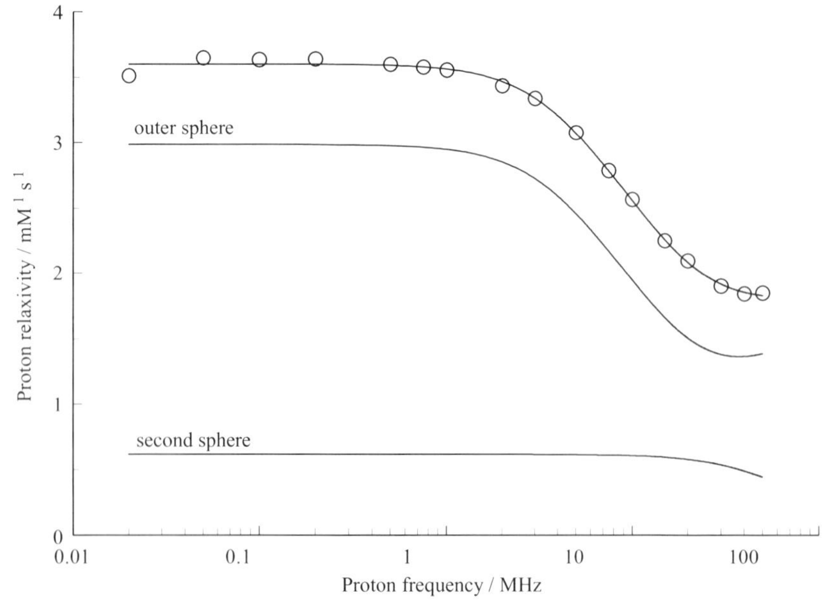

Figure 2.13 Gd(TTHA) NMRD profile at 293 K and its best-fits under the combined model utilizing both the translational outer-sphere model and the second-sphere model. Reprinted with permission from [105]. Copyright (1998) *American Chemical Society*.

they obtained values of $q_{2nd} = 11.7$, $r_{GdH}^{2nd} = 4.40$ Å and $\tau_m^{2nd} = 4.48$ ps. This lifetime obtained for the second-sphere water seems to be too short for a hydrogen-bonded water molecule; however, the value is strongly correlated with the improbably high number of second-sphere water molecules. Furthermore, one has to consider that there is an array of hydrogen-bonded solvent molecules, which are characterized by different distances and residence lifetimes [125].

4 RELAXIVITY AND NMRD PROFILES

The relaxation of a nuclear magnetic spin in general is a function of experimental parameters such as temperature, pressure, sample composition and magnetic field, B_0. Changing thermodynamic parameters, like the temperature, influences the physical or chemical state of the sample under investigation. Variation of the magnetic field, however, has normally no influence on the chemistry of the sample. This is therefore a valuable tool for separation of different interaction mechanisms and dynamic processes influencing the relaxation behavior. Measuring the relaxation rates of an abundant nuclear species as a function of the magnetic field over a wide range is called relaxometry. A relaxometry profile is a plot of the nuclear magnetic relaxation rate, usually $1/T_1$, as a function of the Larmor frequency or the magnetic field on a logarithmic scale (see Figure 2.12). This profile is also called a Nuclear Magnetic Relaxation Dispersion (NMRD) profile.

The measurement of relaxation rates is a routine task in the range of 0.47 T (20 MHz ^{1}H) to 18.8 T (800 MHz ^{1}H) where commercial NMR spectrometers are available. Laboratory-built spectrometers can extend the range of available fields down to about 0.023 T (1 MHz). In principle, there is no theoretical limitation on further decreasing the field, but the dramatic decrease in sensitivity sets a practical limit. However, NMRD profiles of paramagnetic solutions very often show interesting features at frequencies below 1 MHz. This led to the development of a special experimental technique which uses fast cycling of the magnetic field. Although the general principle of the technique was developed in the early 1960s, the first laboratory-built fast-field-cycling (FFC) relaxometers appeared only about 10 years later [127–129]. A commercially available FFC relaxometer has existed since the mid 1990s (STELAR s.n.c., Mede, Italy), mainly due to the increasing research demand in the field of MRI contrast agents.

Two working schemes of FFC relaxometers are shown in Figure 2.14. At a very low relaxation field, B_r, (typically, $B_r < B_d$; B_d = detection field) the nuclear spins are polarized at a polarisation field, B_p during a time which is long compared to the relaxation time T_1 at that field (Figure 2.14(a)). Then B_0 is switched rapidly to the relaxation field B_r. The switching time has to be shorter than T_1, otherwise most of the polarization created is lost and the

(a)

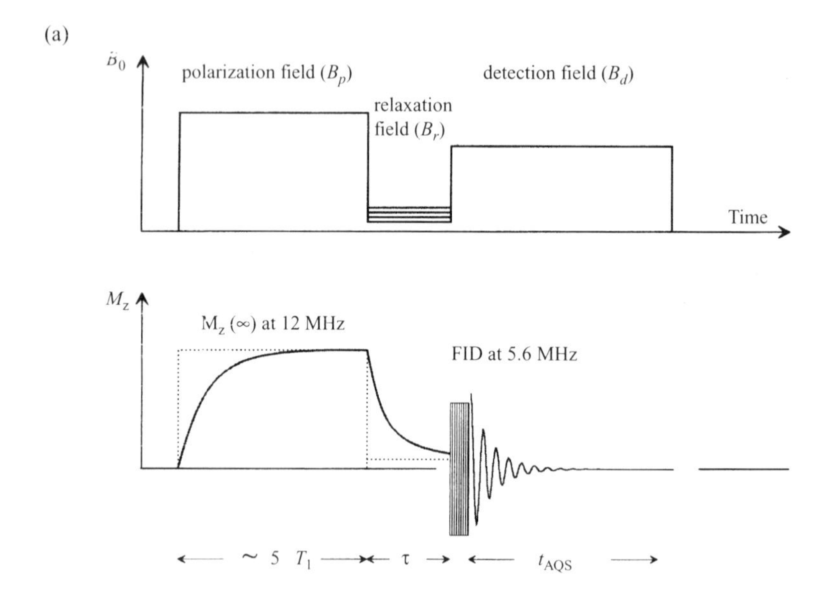

(b)

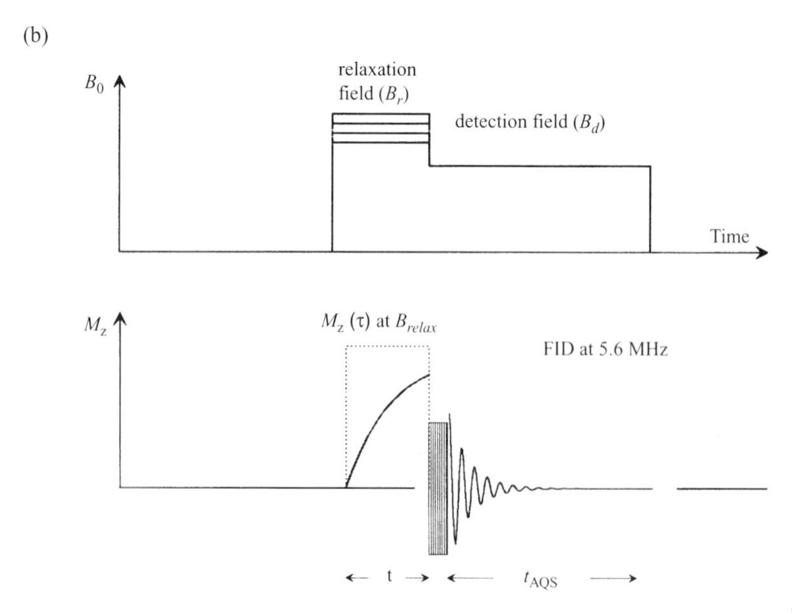

Figure 2.14 Working schemes of FFC relaxometers, showing measurements (a) with and (b) without prepolarization. In both cases, the upper schemes show the variation of the magnetic field, whereas the bottom schemes represent the evaluation of the magnetization during the measurement.

sensitivity becomes poor. The spin system then relaxes towards its new equilibrium value. After a time τ, the magnetization is monitored by an RF pulse. In order to simplify the procedure, the NMR signal is always recorded at the same frequency (corresponding to B_d), and no tuning of the probe is necessary for different relaxation fields. At higher magnetic fields, B_r, ($B_d < B_r < B_{max}$) the increase of the magnetization after switching the magnetic field to the B_r-value is recorded directly (Figure 2.14(b) as a function of τ. Since the field induction decay (FID) is also recorded at the same frequency, a whole NMRD profile can be measured automatically without tuning the probe.

4.1 FITTING OF NMRD PROFILES

The field dependence of proton relaxivities is determined by the numerous physicochemical parameters appearing in Equations (2.5)–(2.16) which are discussed above. Consequently, the NMRD method can be of great help in determining these parameters and has therefore played a central role in the development of our understanding of proton relaxivity. The underlying complexity, on the other hand, represents an important drawback: there are too many influencing parameters, and hence they are often ill-defined by the–very often featureless–NMRD curves alone. In particular, the relative contributions from outer-, inner-, and eventually, second-sphere relaxivity to the NMRD curve can not be easily separated for a given complex. Even in a relatively simple situation, where second-sphere relaxation can be neglected and the outer-sphere part can be estimated by Equation (2.44) using reasonable parameters, and with furthermore the hydration number, q, being known, the Solomon–Bloembergen–Morgan theory for the inner-sphere contribution contains several parameters to be fit, i.e. τ_m, τ_R, τ_v, Δ^2 and r_{GdH}. In certain cases it is possible to separate these parameters to some extent, since they affect the profile at different magnetic fields. Thus, electronic relaxation usually dominates the dipole-dipole correlation time for inner-sphere relaxivity at low field, and produces the first dispersion at around a 3–4 MHz proton Larmor frequency, whereas rotation dominates at higher fields and determines the dispersion at around 30 MHz. However, as stressed by Koenig a long time ago [130], an accurate interpretation of NMRD profiles can only be made by reference to independent information from other techniques, as otherwise the parameters obtained by fitting only the profiles can be completely meaningless. Two additional techniques, namely EPR and ^{17}O NMR, have proved especially useful as probes for a number of the parameters of importance to proton relaxivity. EPR linewidths give direct access to transverse electronic relaxation rates, whereas ^{17}O NMR relaxation rates and chemical shifts, over a range of magnetic fields and as a function of temperature and pressure, permit estimates of the number of inner-sphere water molecules, the rotational correlation time, and the longitudinal electronic relaxation rate of the Gd(III) complexes. Most importantly, the ^{17}O NMR technique allows accurate determination of the water exchange rate. Although EPR and mainly ^{17}O NMR are often used to characterize Gd(III)-based MRI contrast agents, the data are usually analyzed separately for each method and the parameters obtained in this way are then implemented to the NMRD fit. Since the results of the three techniques, EPR, ^{17}O NMR and NMRD, are influenced by a number of common parameters, it seems more reasonable–where possible–to subject them to a simultaneous least-squares fitting procedure. This will allow a more reliable determination of the set of parameters governing proton relaxivity, provide a more stringent test of the relaxation theories applied to the three techniques, and permit a validation of current models for the dynamics in paramagnetic solutions.

Variable-temperature data, including NMRD profiles, as well can also be very useful, and assuming physically reasonable exponential or Eyring behavior for the different correlation times, rather then fitting independent values at each temperature, has also proved helpful in the analysis. The temperature dependence of the NMRD profiles already gives some indication as to which parameter limits the relaxivity, especially at high fields. Rotation and water exchange, the two main parameters that can limit proton relaxivity in a general case, lead to an opposite temperature behavior of the relaxivity: when rotation is the principal factor governing relaxivity, the r_1 values decrease with increasing temperature, whereas an exchange limitation results in an opposite temperature effect.

A simultaneous analysis of EPR, ^{17}O NMR and NMRD data has been carried out for several, mainly low-molecular-weight, Gd(III) complexes [32, 49, 59, 78]. Certainly, this integrated approach can not be applied when the different experimental conditions (concentration) used, e.g. in the ^{17}O NMR and NMRD measurements, affect one of the parameters fitted (usually the rotational correlation time), due to concentration-dependent associations, such as micelle formation or protein binding.

4.2 RELAXIVITY OF LOW-MOLECULAR-WEIGHT GD(III) COMPLEXES

NMRD profiles of low-molecular-weight Gd(III) complexes with one inner-sphere water can be typified by that of the commercialized agents, $[Gd(DTPA)(H_2O)]^{2-}$, $[Gd(DTPA\text{-}BMA)(H_2O)]$ or $[Gd(DOTA)(H_2O)]^-$ and have the general forms shown in Figure 2.15.

The main feature of these profiles is that the relaxivity is limited by fast rotation, especially at high frequencies (> 10 MHz). As a consequence, the high-field relaxivities of the three agents are practically the same, since their sizes, and thus their rotational correlation times, are also very similar. The one-order-of-magnitude lower water exchange rate determined for $[Gd(DTPA\text{-}BMA)(H_2O)]$ has no influence on the high-field relaxivities, as the latter are exclusively limited by rotation. The different low-field relaxivities reflect the

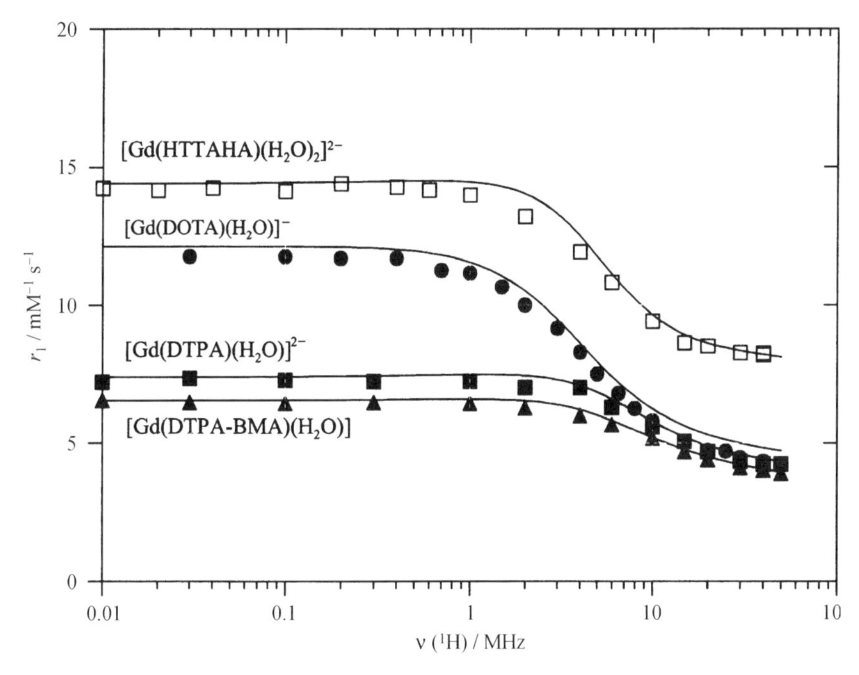

Figure 2.15 Typical NMRD profiles of monomer Gd(III) chelates, showing three examples of commercially available contrast agents, i.e. [Gd(DOTA)(H_2O)]$^-$, [Gd(DTPA)(H_2O)]$^{2-}$ and [Gd(DTPA-BMA)(H_2O)], plus, for comparison, a chelate with two inner-sphere water molecules, [Gd(HTTAHA)(H_2O)$_2$]$^{2-}$.

remarkably slower electronic relaxation of the symmetric [Gd(DOTA)(H_2O)]$^-$ species, as compared to the linear chelates, which has no more influence at higher fields due to the dispersion of the electronic relaxation term. For these low-molecular-weight agents, the contributions from inner- and outer-sphere relaxation contribute more or less to the same extent to the overall effect. According to this, a considerable relaxivity increase can be attained with a bishydrated Gd(III) chelate. Indeed, the relaxivities over the whole frequency range are about 50 % higher for the [Gd(TTAHA)(H_2O)$_2$] complex. Other Gd(III) complexes with two inner-sphere water molecules also have relaxivities of the same magnitude. For the [Gd(TREN-Me-3, 2-HOPO)(H_2O)$_2$] complex, an even higher relaxivity was measured (r_1 = 10.5 mM^{-1} s^{-1}; 37 °C, 20 MHz). This can be only partly explained by the two inner-sphere water molecules, and one has to assume a relatively fast water exchange rate and slower rotation as well, which result in a higher inner-sphere contribution per coordinated water molecule than that found for DTPA- or DOTA-type complexes (TREN-Me-3, 2-HOPO = tris[(3-hydroxy-1-methyl-2-oxo-1, 2-didehydropyridine-4-carboxamido)ethyl]amine) [131].

Table 2.7 Proton relaxivities of some low-molecular-weight Gd(III) chelates.

Ligand[a]	r_1(mM^{-1}s^{-1}) 20 MHz; 25 °C	Reference
DTPA	4.69	32
BOPTA	5.2	163
MS-325	6.6 (37 °C)	164
DTPA-BMA	4.39	32
DTPA-BPA	4.86	67
DTPA-BBA	4.80	41
DOTA	4.74	32
DOTASA	5.93	69
HP-DO3A	3.7 (40 °C)	165
DO3A	4.8 (40 °C)	165
TTAHA	8.50	75
EGTA	4.73	73

[a] Structures are shown in Charts 2.1–2.6.

A series of relatively small-molecular-weight multimeric Gd(III) complexes have been synthesized and these were found to have relaxivities which were linearly proportional to the molecular weight [132]. For those chelates of molecular weight less then 5 kDa, the effective rotational correlation time that determines proton relaxivity increases with molecular weight as the internal flexibility, which becomes important for macromolecular agents, is still negligible. Since the chelates studied contain the same complexing units, their water exchange rate and electronic relaxation is the same. Consequently, the linear dependence of the relaxivities on molecular weight reflects their rotational limitation.

A non-negligible contribution to the overall relaxivity can arise from the so-called second-sphere relaxation. This effect was mainly observed for ligands which have highly charged, oxygen-containing groups, such as phosphonates [76].

Table 2.7 shows proton relaxivity values obtained for a series of different low-molecular-weight Gd(III) chelates, with the structures of the various ligands being shown in Charts 2.1–2.6.

4.3 RELAXIVITY OF MACROMOLECULAR MRI CONTRAST AGENTS

Macromolecular conjugates of Gd(III) chelates are widely investigated as MRI contrast agents. In addition to the potential increase in relaxivity due to their slower rotation, they have another advantage over the Gd(III) chelates, namely an extended lifetime in the blood pool which is necessary for magnetic resonance angiography applications. In the last few years several approaches have been attempted to increase the molecular weight. Generally, two main groups of macromolecular agents are distinguished, based on the covalent or non-covalent nature of the binding between the monomeric agent and the macromolecule. Covalent binding may involve conjugation of functionalized Gd(III) chelates to polymers, dendrimers, or biological molecules, or the synthesis of poly(amino carboxylate)-containing copolymers. Non-covalent binding is largely represented by protein-bound chelates.

Another hope associated with polymeric agents is receptor targeting. These molecules can contain a high number of spatially accumulated Gd(III) ions, and thus they could be expected to have sufficiently high relaxivities for targeting of low-concentration receptors, provided a suitable targeting moiety is also present in the molecule. However, so far these hopes have not become a reality.

In the following, we will survey the most typical classes of high-molecular-weight agents from the point of view of their proton relaxivities.

4.3.1 Linear Polymers

Gd(III) chelates have been attached to linear polymer chains as side chains or incorporated into the linear polymer chain itself in the form of copolymers. The first group is typically represented by polylysine derivatives, where the Gd(III) poly(amino carboxylate) moiety can be easily conjugated to the ϵ-amino group of the lysine backbone [133–137] (Chart 2.11). For this type of linear polymeric agent the relaxivities have been found to be much lower than expected solely on

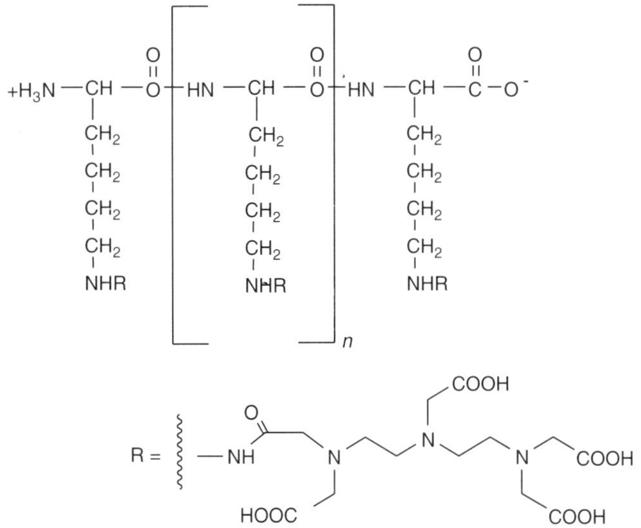

[Chart 2.11]

the basis of the increased molecular weight, and more particularly, they do not change with the molecular weight of the polymer. As an example, Gd(DTPA)-loaded polylysine chains of considerably different molecular weights (in the range of 36–480 kDa) all had relaxivities between 10.4–11.9 mM^{-1} s^{-1} [136]. Similarly, the relaxivities of PEG-modified polylysine chains with Gd(DTPA-monoamide)$^-$ chelates were constant (6.0 mM^{-1} s^{-1} at 20 MHz and 37 °C) for molecular weights of 10.8–83.4 kDa [135]. This was explained by the highly flexible nature of the polylysine backbone and of the side-chain containing the Gd(III) unit.

Gd(DTPA-bisamide) complexes copolymerized with PEG moieties are representative of the second group of linear polymers (see Chart 2.10) [78]. This group have also shown the same flexibility, which explains their very low relaxivity. A combined ^{17}O NMR, EPR and NMRD study performed on $[Gd(DTPA\text{-}BA)\text{-}PEG]_x$ concluded that the effective rotational correlation time in the long polymer chain is not higher than that of the same Gd(III) monomer unit, [Gd(DTPA-BMA)], which is restricted to rotate around a single axis (τ_R for the polymer equals three times the value of τ_R for the monomer; Table 2.6) [78]. An interesting feature of this compound is that its relaxivity is practically independent of temperature, which makes it a good candidate for use as a standard to calibrate MRI signal intensities and T_1 values. Provided that the T_1 of the standard is not so long that the temperature-dependence of the solvent makes a significant contribution, this parameter will not change with positioning from the patient, where large ranges in temperature (body temperature to room temperature) are possible. Detailed analysis of the parameters determining relaxivity has proved that this temperature independence is the consequence of the interplay between the different temperature behaviors of the fast rotation and relatively slow water exchange.

Considerably higher proton relaxivities have been found for another class of DTPA-bisamide copolymers, $[Gd(DTPA\text{-}BA)\text{-}(CH_2)_n]_x$ ($n = 6$, 10 and 12), which contain hydrophobic sections in the polymer chain (Chart 2.10; Table 2.6) [79]. The slowly tumbling intramolecular aggregates that form in solution are the origin of the high relaxivities (see Section 2.4.3). The NMRD curves of the polymers where $n =$ 10 and 12 (Figure 2.16) show the typical high-field peaks which are characteristic of slowly rotating Gd(III) complexes.

4.3.2 Dendrimer-Based MRI Contrast Agents

Dendrimers represent a unique class of synthetic polymers in the sense that, in contrast to linear polymers, highly rigid and almost monodisperse systems can be produced [138]. Their synthesis consists of repetitive reaction steps (leading to the so-called 'generations') starting from a small core molecule and resulting in a three-dimensional structure, very often with a quasi-spherical or spherical shape for higher generations. The surface groups of the molecule, whose

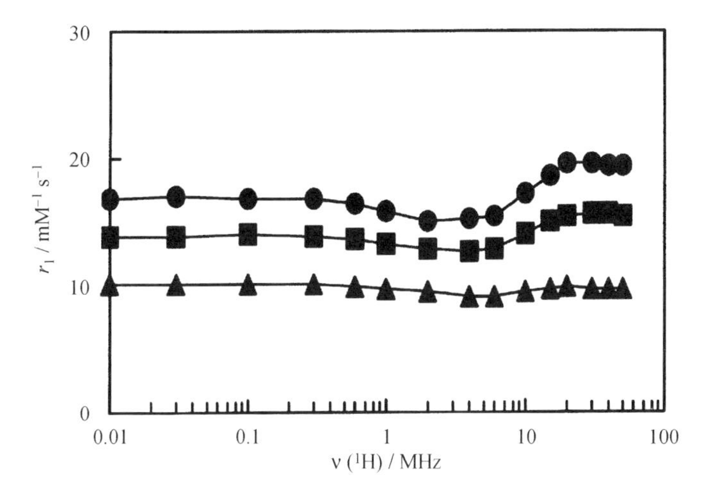

Figure 2.16 NMRD profiles of the linear polymer Gd(III) complexes, [Gd(DTPA-BA)-$(CH_2)_n]_x$ ($n = 6$ (triangles), 10 (squares) and 12 (circles)) at $T = 308$ K.

number is largely increasing with increasing generation, can be used for the conjugation of Gd(III) chelates (see Chart 2.9). This method has made possible the accumulation of as many as 1860 gadolinium chelates within one molecule, as has been reported for a generation-10 PAMAM (polyamidoamine) dendrimer [139]. As far as the proton relaxivity of the dendrimeric compounds is concerned, a key issue–besides the sufficiently rapid water exchange–is the right choice of the linking group between the macromolecule and the Gd(III) complex. This has to be rigid enough so that the slow rotation of the rigid dendrimer molecule is transmitted to the surface chelate itself. Relaxivities up to 36 mM^{-1} s^{-1} (20 MHz; 23 °C) have been reported for dendrimeric Gd(III) complexes.

PAMAM dendrimers of different generations have been loaded with DOTA- [139, 140] or DTPA-type [141–143] chelates, most often by using the *p*-NCS-benzyl functional group as the linking group. For these types of dendrimer-based Gd(III) complexes, the relaxivity increases with increasing generation before reaching a plateau for the high-generation compounds (above generation 7, e.g. for [Gd(*p*-NCS-bz-DOTA)]$^-$-loaded dendrimers) (Table 2.8, and Charts 2.9 and 2.12). The relaxivity profiles show the typical high-field peak at around 20 MHz which is characteristic of slow rotation. The relaxivities for the high-generation dendrimers ($G = 5$–10) decrease as the temperature decreases, thus indicating that slow water exchange of the bound water molecules limits the relaxivity. In such circumstances, further increases in the rotational correlation time of the macromolecules associated with higher generation of dendrimers do not result in significant improvement in proton relaxivity. However, even if the relaxivity for each Gd ion does not considerably increase with generation, the total molecular relaxivities increase from 2880 to 66960

Table 2.8 Relaxivities of some dendrimeric contrast agents.

Dendrimer[a]	M_W (kDa)	r_1[b] (mM^{-1} s^{-1})	r_1[c] (mM^{-1} s^{-1})	$\nu_X t$	Reference
G5(N{CS}N-bz-Gd{DOTA})$_{96}$	118	30	2880	20 MHz 23 °C	139
G7(N{CS}N-bz-Gd{DOTA})$_{380}$	375	35	13 300	"	"
G9(N{CS}N-bz-Gd{DOTA})$_{1320}$	1600	36	47 520	"	"
G10(N{CS}N-bz-Gd{DOTA})$_{1860}$	3000	36	66 960	"	"
G3(N{CS}N-bz-Gd{DO3A})$_{23}$	22.1	14.6	336	20 MHz 37 °C	62
G4(N{CS}N-bz-Gd{DO3A})$_{30}$	37.4	15.9	477	"	"
G5(N{CS}N-bz-Gd{DO3A})$_{52}$	61.8	18.7	972	"	"
G2(N{CS}N-bz-Gd{DTPA})$_{11}$	8.5	21.3	234	25 MHz 20 °C	141
G6(N{CS}N-bz-Gd{DTPA})$_{170}$	139	34	5800	"	"
Gadomer 17 (24 Gd)	17.5	17.3	415	20 MHz 39 °C	144

[a] Structures are shown in Charts 2.9 and 2.12.
[b] per Gd ion.
[c] Per molecule.

mM^{-1} s^{-1} for the $G = 5$ to 10 $[Gd(p\text{–NCS–bz–DOTA})]^-$-loaded PAMAM dendrimers.

The same conclusion, i.e. a water-exchange limitation of the proton relaxivities, has been drawn from a combined ^{17}O NMR and NMRD study on Gd(DO3A-monoamide)-functionalized dendrimers (generation-5 [G5(N{CS}N-bz-Gd{DO3A}{H_2O})$_{52}$], generation-4 [G4(N{CS}N-bz-Gd{DO3A}{H_2O})$_{30}$] and generation 3 [G3(N{CS}N-bz-Gd{DO3A}{H_2O})$_{23}$]) [62]. These results have important implications for the design of optimized macromolecular contrast agents. They indicate that high-molecular-weight complexes such as these dendrimers have rotational correlation times that are long enough for the water exchange to influence the overall relaxivity. Therefore, besides slowing down rotation, a further improvement in the efficiency of the contrast agents can only be achieved with higher water exchange rates.

A different type of dendrimer has been synthesized by workers at Schering AG (Berlin, Germany) [144, 145]. Here, the core is a 1, 3, 5 benzene-tricarboxylic acid which is coupled to various lysines, and the generation-4 dendrimer, called Gadomer-17, contains 24 Gd(III)-DOTA-monoamide chelates on the surface (see Chart 2.12).

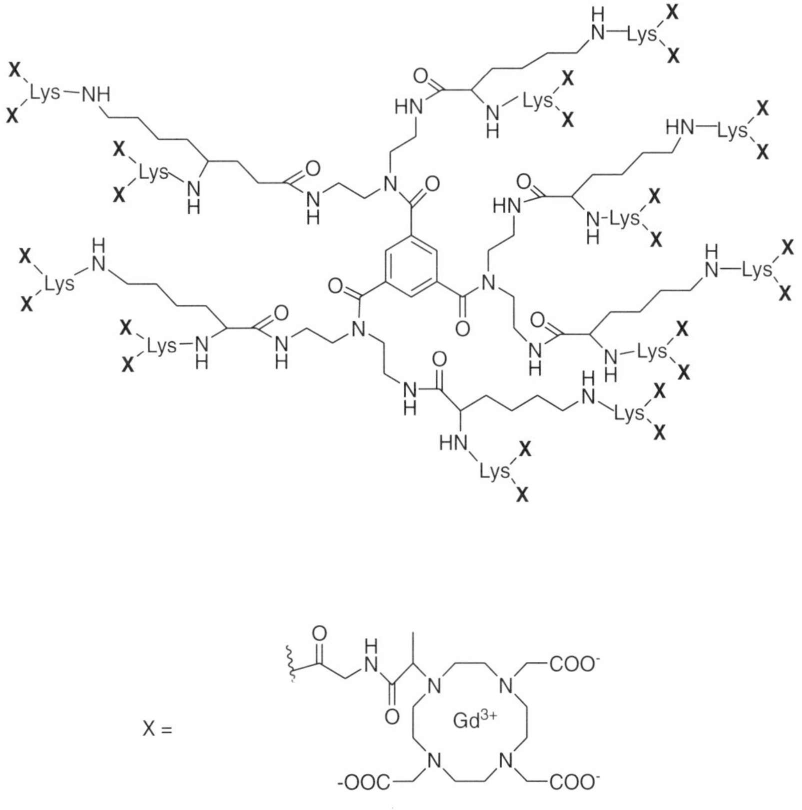

[Chart 2.12]

4.3.3 Dextran-Based Agents

Dextrans as macromolecular carriers of paramagnetic chelates can be very useful, due to their hydrophilicity, clinical use as plasma expanders, and their different available molecular weights plus narrow polydispersity, which can be used to control the biodistribution and the versatility of the activation methods applicable to them. Several DTPA- or DOTA-loaded carboxymethyl dextran (CMD) derivatives have been prepared and tested in blood pool MR imaging [146–149]. The highest relaxivities reported for the magnetic fields of practical importance do not exceed 16 mM^{-1} s^{-1} [148].

4.3.4 Micellar Structures

An interesting way of increasing the rotational correlation time, and thus the proton relaxivity, is to use amphiphilic monomer Gd(III) chelates which are capable of self-organization by forming micellar aggregates in aqueous solution. The Gd(III) complex of the DOTA-derivative ligand containing a dodecyl side-chain represents a good example (see Figure 2.8). Its 20 MHz relaxivity in saline solution ($r_1 = 18.0$ mM^{-1} s^{-1}; 25 °C) is comparable to those attained so far only with macromolecular Gd(III) chelates [81].

4.3.5 Protein-Bound Chelates

An exhaustive discussion of protein-bound Gd(III) chelates is presented in Chapter 5.

4.3.6 Metallofullerenes

Fullerenes that contain a paramagnetic lanthanide ion such as Gd^{3+} could have a number of advantages as MRI contrast agents. The Gd^{3+} ion provides the unpaired electron density, while the fullerene cage protects the lanthanide from chemical attack and sequesters the toxicity of a naked Gd^{3+} ion. Once trapped inside the fullerene cage, the metal ion is protected from the external environment, and the resulting metallofullerene is stable with respect to dissociation in even the most extreme chemical environments. In addition, similarly to 'empty' fullerenes, metallofullerenes can be derivatized via chemical reactions. For example, a water-soluble polyhydroxylated Gd^{3+} metallofullerene, [Gd@$C_{82}(OH)_x$], has been prepared and its proton relaxivity was $r_1 = 20$ mM^{-1} s^{-1} (at 20 MHz; 40 °C) [150]. One hypothesis to explain this surprisingly high value for a caged Gd^{3+} ion is that the relaxivity is due to the electronic structure of the paramagnetic ($S = 1/2$) metallofullerene cage and not directly to the f-orbitals of the Gd^{3+} ion. In such a relaxation process, the [Gd@$C_{82}(OH)_x$] species could simultaneously relax the protons of many hydrogen-bonded water molecules on the 200 Å^2 paramagnetic metallofullerene surface (Figure 2.17). This relaxation mechanism is unique when compared to conventional Gd(III) chelates, suggesting that metallofullerenes could form the basis of an entirely new class of contrast agents.

5 DESIGN OF HIGH RELAXIVITY AGENTS: SUMMARY

As discussed in the previous sections, the most obvious approach to increase proton relaxivity, as compared to the commercialized MRI contrast agents, is

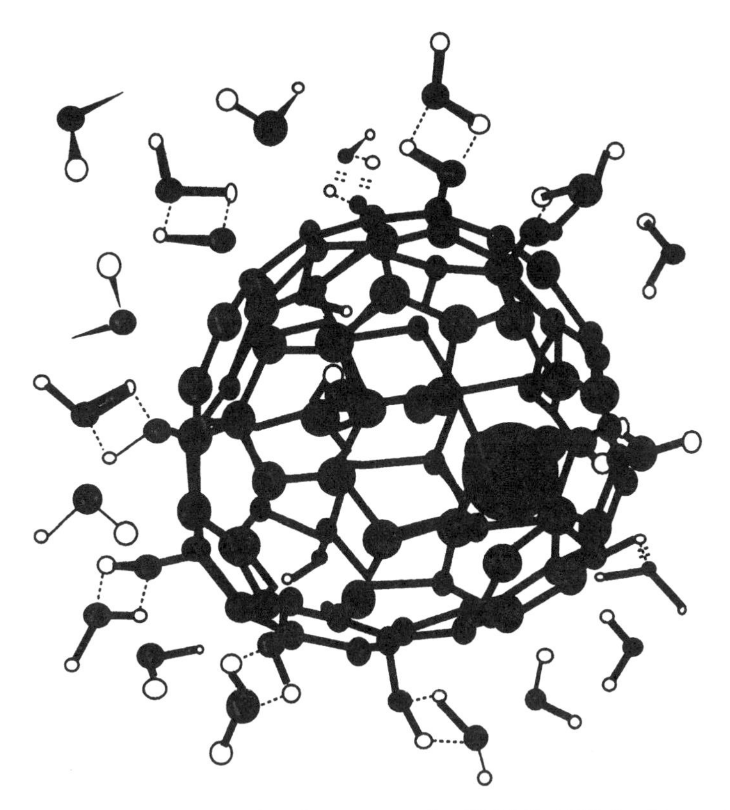

Figure 2.17 A water proton relaxation process for [Gd@$C_{82}(OH)_x$] with an $S = 1/2$ fullerene cage. The affected water molecules are shown as being hydrogen bonded to the cage-hydroxyl groups [150]. Reproduced by permission of The Electrochemical Society, Inc.

to slow down the rotation. This approach has brought a three- to four fold relaxivity gain for the new-generation agents which are currently under clinical trial. This improvement is much less than that expected on the basis of the molecular-weight increase. One reason is the lack of rigidity for the macromolecular agents which makes the effective rotational correlation time of the Gd–water H vector, which is important from the practical point of view, much shorter than that of the whole macromolecule. Consequently, one way of development is the optimization of the rotational correlation time by synthesizing more rigid molecules.

It has been proved for several types of macromolecular agents that even if the rotational correlation time is not yet of optimal length, slow water exchange starts to limit relaxivity. Therefore, in an effort which is parallel to the increase of the rotational correlation time, one has to increase the water exchange rate as well. The introduction of further types of ligands other than poly(amino carboxylate)s can lead to significant improvement in this field. Another possibility for avoiding water exchange limitations can be the acceleration of the proton exchange via H^+- or OH^--catalyzed processes.

Doubling the inner-sphere contribution to the proton relaxivity could be easily achieved with bishydrated Gd(III) chelates. However, the high thermodynamic stabilities and kinetic inertness of these complexes has to be maintained in order not to increase the toxicity of the contrast agent.

With appropriate ligand design, the relaxivity contribution form the so-called second sphere can also be important. As the same theory applies for the inner-and for the second-sphere relaxation, slow rotation of the molecule is beneficial for the second-sphere contribution as well.

REFERENCES

1. Bloch, F., Hansen, W. W. and Packard, M *Phys. Rev.* 1948, **70**, 474.
2. Lauterbur, P. C., Mendoca-Dias, M. H., and Rudin, A. M. in *Frontiers of Biological Energetics*, Dutton, P. L., Leigh, L. S., and Scarpa, A. (Eds), Academic Press, New York, 1978, p. 752.
3. Lauffer, R. B. *Chem. Rev.* 1987, **87**, 90.
4. Peters, J. A., Huskens, J. and Raber, D. J. *Prog. Nucl. Magn. Reson. Spectrosc.* 1996, **28**, 283.
5. Caravan, P., Ellison, J. J., McMurry, T. J. and Lauffer, R. B. *Chem. Rev.* 1999, **99**, 2293.
6. Rocklage, S. M., Cacheris, W. P., Quay, S. C., Hahn, F. E. and Raymond, K. N. *Inorg. Chem.* 1989, **28**, 477.
7. Gallez, B., Bacic, G. and Swartz, H. M. *Magn. Reson. Med.* 1996, **35**, 14.
8. Bennett, H. F., Brown, R. D. III, Koenig, S. H. and Swartz, H. M. *Magn. Res. Med.* 1987, **4**, 93.
9. Vallet, P., Van Haverbeke, Y., Bonnet, P. A., Subra, G., Chapat, J.-P. and Muller, R. N. *Magn. Res. Med.* 1994, **32**, 11.
10. Kauczor, H., Surkau, R. and Roberts, T. *Eur. Radiol.* 1998, **8**, 820.
11. Bloembergen, N., Purcell, E. M. and Pound, R. V. *Phys. Rev.* 1948, **73**, 678.
12. Solomon, I. *Phys. Rev.* 1955, **99**, 559.
13. Solomon, I. and Bloembergen, N. *J. Chem. Phys.* 1956, **25**, 261.
14. Bloembergen, N. *J. Chem. Phys.* 1957, **27**, 572.
15. Bloembergen, N. and Morgan, L. O. *J. Chem. Phys.* 1961, **34**, 842.
16. Connick, R. E. and Fiat, D. *J. Chem. Phys.* 1966, **44**, 4103.
17. Swift T. J. and Connick, R. E. *J. Chem. Phys.* 1962, **37**, 307.
18. Luz, Z. and Meiboom, S. *J. Chem. Phys.* 1964, **40**, 2686.
19. McLachlan, A. D. *Proc. R. Soc. London* 1964, **A280**, 271.
20. Kowalewski, J., Nordenskiöld, L., Benetis, N. and Westlund, P.-O. *Prog. Nacl. Magn. Reson. Spectrosc.* 1985, **17**, 141.
21. Kowalewski, J. in *Encyclopedia of Nuclear Magnetic Resonance*, Grant, D. M. and Harris, R. K. (Eds), Wiley, Chichester, 1996, pp. 3456–3462.
22. Cossy, C. and Merbach, A. E. *Pure Appl. Chem.* 1988, **60**, 1785.
23. Alpoim, M. C., Urbano, A. M., Geraldes, C. F. G. C. and Peters, J. A. *J. Chem. Soc. Dalton Trans.* 1992, 463.
24. Horrocks, W. D. Jr and Sudnick, D. R. *J. Am. Chem. Soc.* 1979, **101**, 334.
25. Parker, D. and Williams, J. A. G. *J. Chem. Soc. Dalton Trans.* 1996, 3613.
26. Albin, M., Farber, G. K. and Horrocks, W. D. Jr *Inorg. Chem.* 1984, **23**, 1648.

27. Geier, G. and Jorgensen, C. K. *Chem. Phys. Lett.*, 1971, **9**, 263.
28. Graeppi, N., Powell, D. H., Laurenczy, G., Zékány, L. and Merbach, A. E. *Inorg. Chim. Acta* 1994, **235**, 311.
29. Tóth, É., Ni Dhubhghaill, O. M., Besson, G., Helm, L. and Merbach, A. E. *Magn. Res. Chem.* 1999, **37**, 701.
30. Yerly, F., Dunand, F. A., Tóth, É., Figueirinha, A., Kovács, Z., Sherry, A. D., Geraldes C. F. G. C. and Merbach, A. E. *Eur. J. Inorg. Chem.*, 2000, 1001.
31. Frey, S. T. and Horrocks, W. D. Jr *Inorg. Chim. Acta* 1995, **229**, 383.
32. Powell, H. D., Ni Dhubhghaill, O, M., Pubanz, D., Helm, L., Lebedev, Y., Schlaepfer, W., and Merbach, A. E. *J. Am. Chem. Soc.*, 1996, **118**, 9333.
33. Koenig, S. H. and Brown, R. D. III. and Spiller, M. *Magn. Res. Med.* 1987, **4**, 252.
34. Bénazeth, S., Purans, J., Chalbot, M.-C., Mguyen-van-Duong, M. K., Nicolas, L., Keller, F. and Gaudemer, A. *Inorg. Chem.* 1998, **37**, 3667.
35. Steele, M. L. and Wertz, D. L. *J. Am. Chem. Soc.* 1976, **98**, 4424.
36. Yamaguchi, T., Nomura, M., Wakita, H. and Ohtaki, H. *J. Chem. Phys.* 1988, **89**, 5153.
37. Cossy, C., Barnes, A., Enderby, J. E. and Merbach, A. E. *J. Chem. Phys.* 1989, **90**, 3254.
38. Cossy, C., Helm, L., Powell, D. H. and Merbach, A. E. *New J. Chem.* 1995, **19**, 27.
39. Clarkson, R. B., Hwang, J.-H. and Belford, R. L. *Magn. Res. Med.* 1993, **29**, 521.
40. Aime, S., Barge, A., Botta, M., Parker, D. and De Sousa, A. S. *J. Am. Chem. Soc.* 1997, **119**, 4767.
41. Aime, S., Botta, M., Fasano, M., Paoletti, S. and Terreno, E. *Chem. Eur. J.* 1997, **3**, 1499.
42. Margerum, D. W., Cayley, G. R., Weatherburn, D. C. and Pagenkopf, G. K. in *Coordination Chemistry*, Martell, A. E. (Ed.), Vol. 2, American Chemical Society, Washington, DC, 1978, pp. 1–220.
43. Micskei, K., Helm, L., Brücher, E. and Merbach, A. E. *Inorg. Chem.* 1993, **32**, 3844.
44. Brittain, H. G. and Desreux, J. F. *Inorg. Chem.* 1984, **23**, 4459.
45. Gonzalez, G., Powell, D. H., Tissières, V. and Merbach, A. E. *J. Phys. Chem.* 1994, **48**, 53.
46. Lincoln, S. F. and Merbach, A. E. *Adv. Inorg. Chem.* 1995, **42**, 1.
47. Cossy, C., Helm, L. and Merbach, A. E. *Inorg. Chem.*, 1989, **28**, 2699.
48. Caravan, P., Tóth, É., Rockenbauer, A. and Merbach, A. E. *J. Am. Chem. Soc.* 1999, **121**, 10403.
49. Tóth, É., Helm, L., Merbach, A. E., Hedinger, R., Hegetschweiler, K. and Jánossy, A. *Inorg. Chem.* 1998, **37**, 4104.
50. Aime, S., Barge, A., Bruce, J. I., Botta, M., Howard, J. A. K., Moloney, J. M., Parker, D., De Sousa, A. S. and Woods, M. *J. Am. Chem. Soc.* 1999, **121**, 5762.
51. Micskei, K., Powell, D. H., Helm, L., Brücher, E. and Merbach, A. E. *Magn. Res. Chem.* 1993, **31**, 1011.
52. Aime, S., Barge, A., Botta, M., De Sousa, A. S. and Parker, D. *Angew. Chem. Int. Ed. Engl.* 1998, **37**, 2673.
53. Dunand, A. F., Aime, S. and Merbach, A. E. *J. Am. Chem. Soc.* 2000, **122**, 1506.
54. Cossy, C., Helm, L. and Merbach, A. E. *Inorg. Chem., 1988,* **27***, 1973.*
55. Peters, J. A. *Inorg. Chem.*, 1988, **27**, 4686.
56. Rizkalla, E. M., Choppin, G. R. and Cacheris, W. *Inorg. Chem.*, 1993, **32**, 582.
57. Geraldes, C. F. G. C., Sherry, A. D., Cacheris, W. P., Kuan, K.-T., Brown, R. D. III., Koenig, S. H. and Spiller, M. *Magn. Res. Med.*, 1988, **8**, 191.
58. Lammers, H., Maton, F., Pubanz, D., Van Laren, M. W., Van Bekkum, H., Merbach, A. E., Muller, R. N. and Peters, J. A. *Inorg. Chem.* 1997, **36**, 2527.

59. Tóth, É., Connac, F., Helm, L., Adzamli, K and Merbach, A. E. *Eur. J. Inorg. Chem.* 1998, 2017.
60. Tóth, É., Burai, L., Brücher, E. and Merbach, A. E. *J. Chem. Soc. Dalton Trans.* 1997, 1587.
61. Tóth, É., Vauthey, S., Pubanz, D. and Merbach, A. E. *Inorg. Chem.* 1996, **35**, 3375.
62. Tóth, É., Pubanz, D., Vauthey, S., Helm, L. and Merbach, A. E. *Chem. Eur. J.*, 1996, **2**, 1607.
63. Cacheris, W. P., Quay, S. C. and Rocklage, S. *Magn. Res. Imag.* 1990, **8**, 467.
64. White, D. H., deLearie, L. A., Moore, D. A., Wallace, R. A., Dunn, T. J., Cacheris, W. P., Imura, H. and Choppin, G. R. *Invest. Radiol.*, 1991, **26** S226.
65. Paul-Roth. C. and Raymond, K. N. *Inorg. Chem.* 1995, **34**, 1408.
66. Gries, H. and Miklautz, H. *Physiol. Chem. Phys. Med. NMR* 1984, **16**, 105.
67. Bligh, S. W. A., Chowdhury, A. H. M. S., McPartlin, M., Scowen, I. J. and Bulman, R. A. *Polyhedron* 1995, **14**, 567.
68. Pubanz, D., Gonzalez, G., Powell, D. H. and Merbach, A. E. *Inorg. Chem.* 1995, **34**, 4447.
69. André, J. P., Maecke, H. R., Tóth, É and Merbach, A. E. *J. Biol. Inorg. Chem.* 1999, **4**, 341.
70. Szilágyi, E., Tóth, É., Brücher, E. and Merbach, A. E. *J. Chem. Soc. Dalton Trans.* 1999, 2481.
71. Aime, S., Crich, S. G., Gianolio, E., Terreno, E., Beltrami, A. and Uggeri, F. *Eur. J. Inorg. Chem.* 1998, 1283.
72. Tóth, É., Connac, F., Helm, L., Adzamli, K. and Merbach, A. E. *J. Biol. Inorg. Chem.* 1998, **3**, 606.
73. Aime, S., Barge, A., Borel, A., Botta, M., Chemerisov, S., Merbach, A. E., Muller, R. N. and Pubanz, D. *Inorg. Chem.* 1997, **36**, 5104.
74. Aime, S., Botta, M., Fasano, M. and Terreno, E. *Acc. Chem. Res.* 1999, **32**, 941.
75. Ruloff, R., Muller, R. N., Pubanz, D. and Merbach, A. E. *Inorg. Chim. Acta*, , 1998, **275–276**, 15.
76. Aime, S., Botta, M., Crich, S. G., Giovenzana, G., Pagliarin, R., Piccinini, M., Sisti, M. and Terreno, E. *J. Biol. Inorg. Chem.* 1997, **2**, 470.
77. Aime, S., Botta, M., Crich, S. G., Giovenzana, G., Pagliarin, R., Sisti, M. and Terreno, E. *Magn. Res. Chem.* 1998, **36**, S200.
78. Tóth, É., van Uffelen, I., Helm, L., Merbach, A. E., Ladd, D., Briley-Saebo, K. and Kellar, K. E. *Magn. Res. Chem.* 1998, **36**, S125.
79. Tóth, É., Helm, L., Kellar, K. E. and Merbach, A. E. *Chem. Eur. J.* 1999, **5**, 1202.
80. Kellar, K. E., Henrichs, P. M., Hollister, R., Koenig, S. H., Eck, J. and Wei, D. *Magn. Res. Med.* 1997, **38**, 712.
81. André, J. P., Tóth, É., Maecke, H. R. and Merbach, A. E. *Chem. Eur. J.* 1999, **5**, 2977.
82. Hall, D. G. and Tiddy, G. J. in *Anionic Surfactants*, (Surfactant Science Series) Lucassen-Reynders, E. H., (Ed.), Marcel Dekker, New York, 1981, pp. 55–108.
83. Lauffer, R. B. *Magn. Res. Med.* 1991, **22**, 339.
84. Aime, S., Botta, M., Fasano, M., Crich, S. G. and Terreno, E. *J. Biol. Inorg. Chem.* 1996, **1**, 312.
85. Lauffer, R. B., Parmelee, D. J., Dunham, S. U., Ouellet, H. S., Dolan, R. P., Witte, S., McMurry, T. J. and Walowitch, R. C. *Radiology* 1998, **207**, 529.
86. Bertini, I., Luchinat, C., Parigi, G., Quacquarini, G., Marzola, P. and Cavagna, F. M. *Magn. Res. Med.* 1998, **39**, 124.
87. Muller, R. N., Radüchel, B., Laurent, S., Platzek, J., Pierart, C., Mareski, P. and Vander Elst, L. *Eur. J. Inorg. Chem.* 1999, 1949.

88. Aime, S. , Chiaussa, M., Digilio, G., Gianolio, E. and Terreno, E. *J. Biol. Inorg. Chem.* 1999, **4**, 766.
89. (a) Kubo, R. *J. Phys. Soc. Jpn* 1954, **9**, 888; (b) Sack, R. A. *Mol. Phys.* 1958, **1**, 163; (c) Reeves L. W. and Shaw, K. N. *Can. J. Chem.* 1970, **48**, 3641.
90. Aime, S., Botta, M. and Ermondi, G. *Inorg. Chem.* 1992, **31**, 4291.
91. Aime, S., Barge, A., Botta, M., Fasano, M., Ayala, J. D. and Bombieri, G. *Inorg. Chim. Acta* 1996, **246**, 423.
92. Parker, D., Pulukkody, K., Smith, F. C., Batsanov, A. and Howard, J. A. K. *J. Chem. Soc. Dalton Trans.* 1994, 689.
93. Benetollo, F., Bombieri, G., Aime, S. and Botta, M. *Acta Crystallogr. Sect. C-Cryst. Struct. Commun.* 1999, **55**, 353.
94. Dubost, J. P., Leger, J. M., Langlois, M. H., Meyer, D. and Schaefer, M. *C. R. Acad. Sci. Paris* 1991, **312**, 349.
95. Amin, S., Morrow, J. R., Lake, C. H. and Churchill, M. R. *Ang. Chem. Int. Ed. Engl.* 1994, **33**, 773.
96. Aime, S., Barge, A., Botta, M., Howard, J. A. K., Kataky, R., Lowe, M. P., Moloney, J. M., Parker, D. and de Sousa, A. S. *J. Chem. Soc. Chem. Commun.* 1999, 1047.
97. Halle B. and Wennerstrom, H. *J. Magn. Res.* 1981, **44**, 89.
98. Banci, L., Bertini, I. and Luchinat, C. in *Nuclear and Electron Relaxation*, VCH, Wenheim, 1991, p. 95.
99. Vander Elst, L., Maton, F., Laurent, S., Seghi, F., Chapelle, F. and Muller, R. N. *Magn. Reson. Med.* 1997, **38**, 604.
100. Aime, S. and Nano, R. *Invest. Radiol.* 1988, **23**, S264.
101. Vander Elst, L., Laurent, S. and Muller, R. N. *Invest. Radiol.* 1998, **33**, 828.
102. Chen, J. W., Auteri, F. P., Budil, D. E., Belford, R. L. and Clarkson, R. B. *J. Phys. Chem.* 1994, **98**, 13452.
103. Chen, J. W., Clarkson, R. B. and Belford, R. L. *J. Phys. Chem.* 1996, **100**, 8093.
104. Wiener, E. C., Auteri, F. P., Chen, J. W., Brechbiel, M. W., Gansow, O. A., Schneider, D. S., Clarkson, R. B. and Lauterbur, P. C. *J. Am. Chem. Soc.* 1996, **118**, 7774.
105. Chen, J. W., Belford, R. L. and Clarkson, R. B. *J. Phys. Chem. A* 1998, **102**, 2117.
106. Lakowicz, J. R. *Principles of Fluorescence Spectroscopy*, Plenum, New York, 1983, p. 187.
107. Vexler, V. S., Clement, O., Schmitt-Willich, H. and Brasch, R. C. *J. Magn. Res. Imaging* 1994, **4**, 381.
108. Desser, T. S., Rubin, D. L., Muller, H. H., Qing, F., Khodor, S., Zanazzi, G., Young, S. U., Ladd, D. L., Wellons, J. A., Kellar, K. E., Toner, J. L. and Snow, R. A. *J. Magn. Res. Imaging* 1994, **4**, 467.
109. Bullock, A. T., Cameron, G. G. and Smith, P. M. *J. Phys. Chem.* 1973, **77**, 1635.
110. (a) Lipari, G. and Szabo, A. *J. Am. Chem. Soc.* 1982, **104**, 4546; (b) Lipari, G. and Szabo, A. *J. Am. Chem. Soc.* 1982, **104**, 4559.
111. Maler, L., Widmalm, G. and Kowalewski, J. *J. Biomol. NMR* 1996, **7**, 1.
112. Maler, L., Widmalm, G. and Kowalewski, J. *J. Phys. Chem.* 1996, **100**, 17103.
113. Kemple, M. D., Buckley, P., Yuan, P. and Prendergast, F. *Biochemistry*, 1997, **36**, 1678.
114. Elbayed, K., Canet, D. and Brondeau, J. *Mol. Phys.* 1989, **68**, 295.
115. Antony, J. H., Dölle, A., Fliege, T. and Geiger, A. *J. Phys. Chem. A* 1997, **101**, 4517.
116. McConnell, J. *The Theory of Nuclear Magnetic Relaxation in Liquids*, Cambridge University Press, Cambridge, 1987, pp. 67–77.

117. Abragam, A. *The Principles of Nuclear Magnetism*, Oxford University Press, Oxford, 1961, pp. 294–297.
118. Fries, P. H. and Belorizky, E. *J. Phys. (Paris)* 1978, **39**, 1263.
119. Albrand, J. P., Taieb, M. C., Fries, P. H. and Belorizky, E. *J. Chem. Phys*. 1983, **78**, 5809.
120. Hwang, L. -P. and Freed, J. H. *J. Chem. Phys*. 1975, **63**, 1975.
121. Freed, J. H. *J. Chem. Phys* 1978, **68**, 4034.
122. Polnaszek, C. F. and Bryant, R. G. *J. Chem. Phys*. 1984, **81**, 4038.
123. Ayant, Y., Belorizky, E., Fries, P. H. and Rosset, J. *J. Phys. (Paris)* 1977, **38**, 325.
124. Vigouroux, C., Bardet, M., Belorizky, E., Fries, H. P. and Guillermo, A. *Chem. Phys. Lett*. 1998, **186**, 93.
125. Botta, M. *Eur. J. Inorg. Chem*. 2000, 399.
126. Bleuzen, A., Foglia, F., Furet, E., Helm, L., Merbach, A. E. and Weber, J. *J. Am. Chem. Soc*. 1996, **118**, 12777.
127. Noack, F. *Prog. Nucl. Magn. Reson. Spectrosc*. 1986, **18**, 171.
128. Koenig, S. H. and Brown, R. D. *Prog. Nucl. Magn. Reson. Spectrosc*. 1990, **22**, 487.
129. Kimmich, R. *NMR Tomography, Diffusometry, Relaxometry*, Spinger, Berlin, 1997, pp. 138–158.
130. Koenig, S. H. *J. Magn. Reson*. 1978, **31**, 1.
131. Xu, J., Franklin, S. J., Whisenhunt, D. W. Jr and Raymond, K. N. *J. Am. Chem. Soc*. 1995, **117**, 7245.
132. Ranganathan, R. S., Fernandez, M. E., Kang, S. I., Nunn, A. D., Ratsep, P. C., Pillai, R., Zhang, X. and Tweedle, M. F. *Invest. Radiol*. 1998, **33**, 779.
133. Sieving, P. F., Watson, A. D. and Rocklage, S. M. *Bioconjugate Chem*. 1990, **1**, 65.
134. Schuhmann-Giampieri, G., Schmitt-Willich, H., Frenzel, T., Press, W. R. and Weinmann, H. J. *Invest. Radiol*. 1991, **26**, 969.
135. Desser, T., Rubin, D., Muller, H., Qing, F., Khodor, S., Zanazzi, Y., Ladd, D., Wellons, J., Kellar, K., Toner, J. and Snow, R. *J. Magn. Reson. Imaging* 1994, **4**, 467.
136. Vexler, V. S., Clement, O., Schmitt-Willich, H. and Brasch, R. C. *J. Magn. Reson. Imaging* 1994, **4**, 381.
137. Aime, S., Botta, M., Crich, S. G., Giovenzana, G., Palmisano, G. and Sisti, M. *Bioconjugate Chem*. 1999, **10**, 192.
138. Zeng, F. and Zimmerman, S. C. *Chem. Rev*. 1997, **97**, 1681.
139. Bryant, L. H. Jr, Brechbiel, M. W. Wu, C., Bulte, J. W. M., Herynek, V. and Frank, J. A. *J. Magn. Reson. Imaging* 1999, **9**, 348.
140. Margerum, L. D., Campion, B. K., Koo, M., Shargill, N., Lai, J. -J., Marumoto, A. and Sontum, P. C. *J. Alloys Compd*. 1997, **249**, 185.
141. Wiener, E. C., Brechbiel, M. W., Brothers, H., Magin, R. L., Gansow, O. A., Tomalia, D. A. and Lauterbur, P. C. *Magn. Reson. Med*. 1994, **31**, 1.
142. Tacke, J., Adam, G., Claben, H., Muhler, A., Prescher, A. and Gunther, R. W. *Magn. Reson. Imaging* 1997, **7**, 678.
143. Roberts, H. C., Saeed, M., Roberts, T. P. L., Muhler, A., Shames, D. M., Mann, J. S., Stiskal, M., Demsar, F. and Brasch, R. C. *Magn. Reson. Imaging* 1997, **7**, 331.
144. Dong, Q., Hurst, D. R., Weinmann, H. J., Chenevert, T. L., Londy, F. J. and Prince, M. R. *Invest. Radiol*., 1998, **33**, 699.
145. Schmitt-Willich, H., Platzek, J., Radüchel, B., Ebert, W., Franzel, T., Misselwitz, B. and Weinmann, H. -J. *Abstracts of the 1st Dendrimer Symposium*, 3–5 October 1999, Frankfurt, p. 25.
146. Corot, C., Schaefer, M., Beaute, S., Bourrinet, P., Zehaf, S., Benize, V., Sabatou, M. and Meyer, D. *Acta Radiol*. (Suppl.), 1997, **412** (Suppl.), 91.

147. Casali, C., Janier, M., Canet, E., Obadia, J. F., Benderbous, S., Corot, C. and Revel, D. *Acad. Radiol.* 1998, **5** (Suppl. 1), S214.
148. Meyer, D., Schaefer, M., Bouillot, A., Beaute, S. and Chambon, C. *Invest. Radiol.*, 1991, **26**, S50.
149. Rebizak, R., Schaefer, M. and Dellacherie, E. *Eur. J. Pharm. Sci.* 1999, **7**, 243.
150. Wilson, L. J. *The Electrochemical Society Interface*, The Electrochemical Society, Pennington, NJ, 1999, (Winter), 24.
151. Beeby, A., Clarkson, I. M., Dickins, R. S., Faulkner, S., Parker, D., Royle, L., de Sousa, A. S., Williams, J. A. G. and Woods, M. *J. Chem. Soc. Perkin Trans. 2* 1999, 493.
152. Stekowski, J. J. and Hoard, J. L. *Isr. J. Chem.* 1984, **24**, 323.
153. Geraldes, C. F. G. C., Urbano, A. M., Hoefnagel, M. A. and Peters, J. A. *Inorg. Chem.*, 1993, **32**, 2426.
154. Bovens, E., Hoefnagel, M. A., Boers, E., Lammers, H., van Bekkum, H. and Peters, J. A. *Inorg. Chem.* 1996, **35**, 7679.
155. Frey, S. T., Chang, C. A., Carvalho, J. F., Varadarajan, A., Schultze, L. M., Pounds, K. L. Horrocks, W. DeW. Jr *Inorg. Chem.* 1994, **33**, 2882.
156. Chang, C. A., Brittain, H. G., Telser, J. and Tweedle, M. F. *Inorg. Chem.* 1990, **29**, 4468.
157. Ruloff, R., Prokop, P., Sieler, J., Hoyer, E. and Beyer, L. *Z. Naturforsch. B* 1996, **51**, 963.
158. Bryden, C. C., Reilley, C. C. and Desreux, J. F. *Anal. Chem.* 1981, **53**, 1418.
159. Dubost, J. P., Leger, J. M., Langlois, M. H., Meyer, D. and Schaefer, M. *C. R. Acad. Sci. Ser. II Paris* 1991, **312**, 329.
160. Zhang, X., Chang, C. A., Brittain, H. G., Garrison, J. M., Telser, J. and Tweedle, M. F. *Inorg. Chem.* 1992, **31**, 5597.
161. Weinmann, H. J., Schuhmann-Giampieri, G., Schmitt-Willich, H., Vogler, H., Frenzel, T. and Gries, H. *Magn. Res. Med.* 1991, **22**, 233.
162. Uggeri, F., Aime, S., Anelli, P. L., Botta, M., Brochetta, M., de Haen, C., Ermondi, G., Grandi, M. and Paoli, P. *Inorg. Chem.* 1995, **34**, 633.
163. Aime, S., Botta, M., Panero, M., Grandi, M. and Uggeri, F. *Magn. Res. Chem.* 1991, **29**, 923.
164. Siauve, N., Clement, O., Cuenod, C. -A., Benderbous, S. and Frija, G. *Magn. Reson. Imaging*, 1996, **14**, 381.
165. Kang, S. I., Ranganathan, R. S., Emswiler, J. E., Kumar, K., Gougutas, J. Z., Malley, M. F. and Tweedle, M. F. *Inorg. Chem.* 1993, **32**, 2912.

3 Synthesis of MRI Contrast Agents I. Acyclic Ligands

PIER LUCIO ANELLI and LUCIANO LATTUADA
Milano Research Center, Milano, Italy

1 INTRODUCTION

The whole story of complexes of paramagnetic metal ions as MRI contrast agents started with the publication in the patent [1] and scientific [2] literature of the first reports dealing with the use of the Gd(III) complex of diethylenetriaminepentaacetic acid (pentetic acid, DTPA) (Figure 3.1) for such a purpose. Since then, many research groups belonging to academic and industrial establishments have been devoting their efforts to the search for different, sometimes improved, chelating agents for the complexation of paramagnetic metal ions. To date, Gd(III) has been the preferred 'paramagnetic probe' for the preparation of MRI contrast agents. Indeed, among the complexes which have gone through the long process of being approved for clinical use, all but one are Gd(III) complexes. Nonetheless, other cations, e.g. Mn(II), Fe(III) and Dy(III), have been more or less successfully investigated for the preparation of MRI contrast agents.

In this chapter, we have reviewed the synthesis of acyclic chelating agents which, after complexation with paramagnetic metal ions, lead to compounds of interest as MRI contrast agents. Since such a claim is mentioned in a very large number of papers and patents, we have tried to be selective taking into account only those compounds for which certain studies have somehow proved their

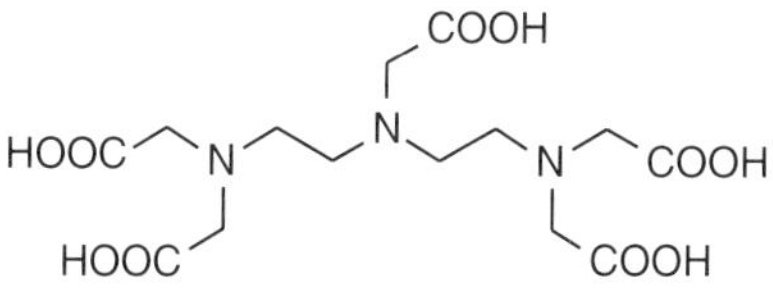

DTPA

Figure 3.1 Chemical structure of diethylenetriaminepentaacetic acid (DTPA).

The Chemistry of Contrast Agents in Medical Magnetic Resonance Imaging
Edited by A. E. Merbach and É. Tóth.

potential application in the MRI context. We apologize to those authors whose contributions have been inadvertently overlooked.

To date, by and large, polyaminopolycarboxylic ligands have had a predominant role over other classes of chelating agents. Some of these ligands (e.g. DTPA derivatives) are also excellent chelating agents for cations of radionuclides, e.g. ^{90}Y(III), ^{212}Bi(III) and ^{111}In(III) and the resulting complexes have been and still are under scrutiny for the preparation of radiopharmaceuticals. In this respect, advances in the chemistry of polyaminopolycarboxylic ligands have been achieved in both of the areas of contrast agents for MRI and of radiopharmaceuticals for nuclear medicine applications. Therefore, overlaps between the chelating agents to be used in either context are quite common in the literature. In our selection, we have tried to stick as much as possible to ligands used in MRI applications. Nonetheless, we believe that some breakthroughs achieved while looking for new radiopharmaceuticals have had, or still could have, very positive feedback for the synthesis of future MRI contrast agents. Therefore, some of the results that, in our opinion, display such a potential are also discussed in this review.

2 POLYAMINOPOLYCARBOXYLIC LIGANDS

2.1 OCTADENTATE LIGANDS (DTPA AND DTPA-MODIFIED LIGANDS)

The first synthesis of DTPA, which is now a commercially available product, appeared in the scientific literature in 1956 and was performed by reacting diethylenetriamine with formaldehyde and sodium cyanide under alkaline conditions [3]. Cyanomethylation of polyamines and subsequent hydrolysis of the nitriles is still the principal manufacturing process for making chelating agents such as DTPA [4], whereas carboxymethylation of diethylenetriamine with chloro- or bromoacetic acid under alkaline conditions [5], which was formerly used for industrial production, is no longer employed. However, carboxymethylation with α-haloacids is still a classical step in the preparation of a variety of polyaminopolycarboxylic ligands on a laboratory scale (see below).

DTPA forms a very stable complex with Gd(III) (log K_{ML} 22.46) [6]. The solid-state structure of Gd-DTPA [7], obtained by X-ray crystallography, shows that the gadolinium ion is nine-coordinate, with the DTPA ligand involved in the complexation of the ion through the three nitrogen atoms and the five oxygen atoms of the carboxylate residues. The ninth position is occupied by a molecule of water and is crucial in determining the effectiveness of the complex as an MRI contrast agent.

In 1988, Gd-DTPA bis(*N*-methylglucammonium) salt (Gd-DTPA dimeglumine, Magnevist®) became the first approved complex for clinical applications. Gd-DTPA dimeglumine is commercially available as a quite concentrated (i.e.

0.5 M) aqueous solution which is administered to the patient by intravenous injection just before the MRI procedure, usually at a dose level of 0.1 mmol/kg of body weight. Soon after administration, the complex starts diffusing from the blood into the extracellular space. Therefore, it crosses blood vessel walls but it does not enter cells. Gd-DTPA dimeglumine is quickly excreted from the human body through the renal route [8].

The success of this compound has led to extensive studies which are aimed at the modification of DTPA by the introduction of suitable residues into either the acetic side-arms or the diethylenetriamine backbone. The purpose of these modifications was to obtain, after complexation with Gd(III), complexes featuring improved properties in terms of selectivity and/or tolerability.

The introduction of a benzyloxymethyl residue into one of the two iminodiacetic moieties of DTPA led to **1** (BOPTA, see Scheme 3.1) [9]. The synthesis of such a modified DTPA skeleton does not require a protection/deprotection strategy but takes advantage of the regioselective alkylation of diethylenetriamine with secondary α-haloacids. Indeed, reaction of 2-chloro-3-(phenylmethoxy)propanoic acid **2** (R = Ph) with an excess of diethylenetriamine in water exclusively leads to the monoalkylated isomer **3** (R = Ph). Subsequent carboxymethylation with bromoacetic acid at pH 10 yields ligand **1**. The general applicability of this procedure is also proved by the preparation of the cyclohexylmethyl analogue **4** [10].

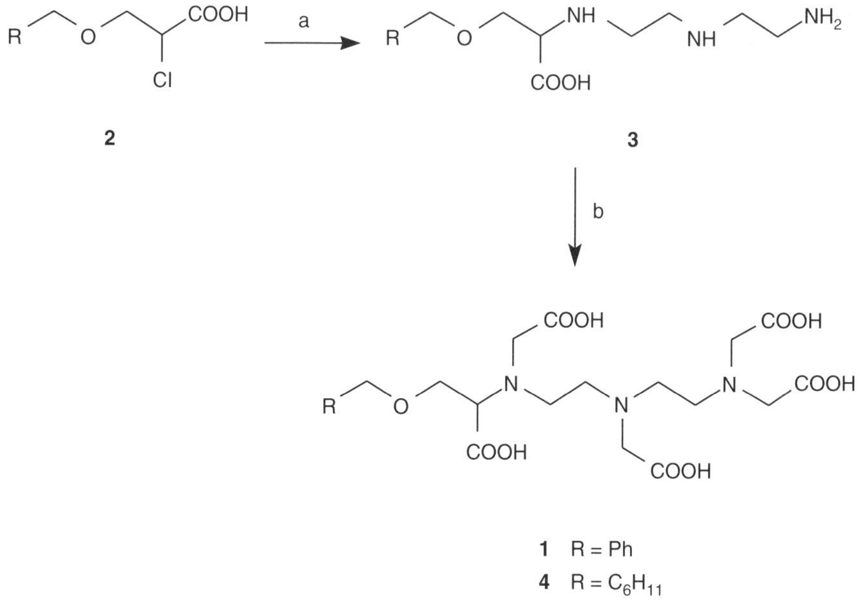

1 R = Ph
4 R = C_6H_{11}

(a) diethylenetriamine; (b) $BrCH_2COOH$, NaOH

Scheme 3.1

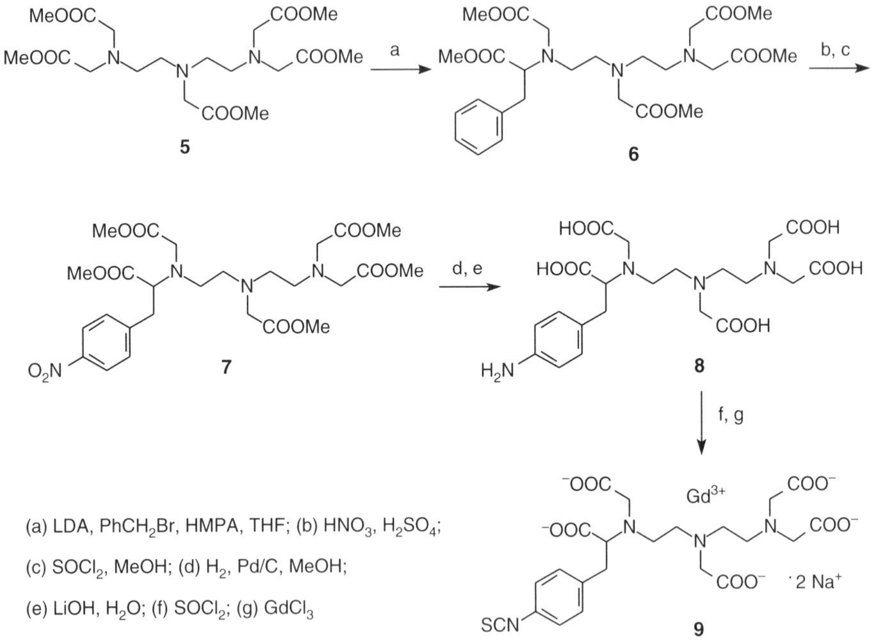

(a) LDA, $PhCH_2Br$, HMPA, THF; (b) HNO_3, H_2SO_4;
(c) $SOCl_2$, MeOH; (d) H_2, Pd/C, MeOH;
(e) LiOH, H_2O; (f) $SOCl_2$; (g) $GdCl_3$

Scheme 3.2

The bis(*N*-methylglucammonium) salt of the gadolinium complex of **1** (i.e. MultiHance®, US Adopted Name: gadobenate dimeglumine) has been developed as a hepatospecific contrast agent because, unlike Gd-DTPA dimeglumine, it is in part eliminated from the body through the hepatic route.

Keana and Mann developed an original methodology to introduce a residue into one of the iminodiacetic moieties of DTPA (Scheme 3.2) [11]. The pentamethyl ester **5**, which is easily obtained by esterification of DTPA with $SOCl_2$/MeOH, is converted into the corresponding enolate with lithium diisopropylamide in THF/HMPA at $-78\,^{\circ}C$ and alkylated with benzyl bromide to give **6** in fair yield. Nitration of **6** with HNO_3/H_2SO_4 gives predominantly the *p*-nitro isomer and, after re-esterification with $SOCl_2$/MeOH, the pentaester **7** is isolated. Subsequently, **7** is reduced and deprotected with LiOH to give the lithium salt of ligand **8**. Keana and Mann continued the synthesis with the conversion of the amino group of **8** into an isothiocyanate by the action of thiophosgene [11]. Complexation with $GdCl_3$ yields **9**, which is a functionalized complex suitable for the conjugation to a variety of carriers.

The synthesis of **8** was also approached by Johnson and co-workers while investigating conjugates of ^{111}In complexes to antibodies [12]. The crucial step in their synthetic path (Scheme 3.3) is the low-yielding reductive alkylation of triamine **10**, which is prepared in two steps from diethylenetriamine, with (4-nitrophenyl)pyruvic acid. The monoacid **11** is carboxymethylated with

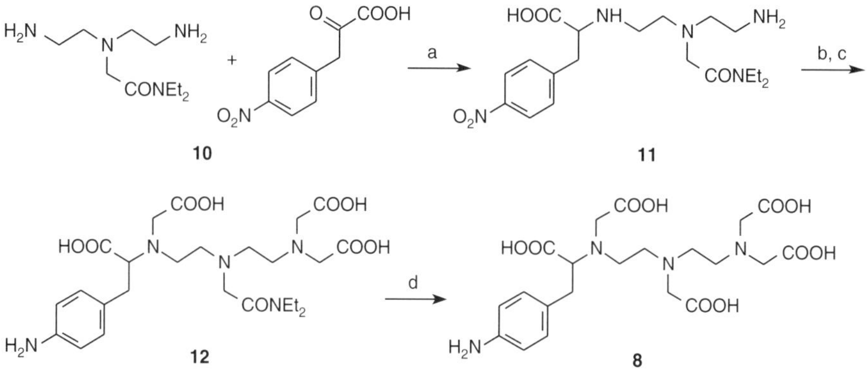

(a) $NaBH_3CN$, MeOH; (b) $BrCH_2COOH$, pH 12; (c) H_2, Pd/C, H_2O, HCOOH; (d) 7 M NaOH, reflux

Scheme 3.3

bromoacetic acid and aqueous NaOH and then reduced by hydrogenolysis to **12**. The amide group is finally removed under strongly alkaline conditions to afford **8**.

An innovative synthesis of a DTPA derivative bearing a *p*-nitrobenzyl group on the central acetic moiety, **13**, was reported in 1993 by Williams and Rapoport (Scheme 3.4) [13]. The key step is the dialkylation of the 4-nitro-L-phenylalanine benzyl ester **14** with di(*t*-butyl) *N*-(bromoethyl)iminodiacetate **15**, which is performed in a two-phase mixture of MeCN and pH 8 phosphate

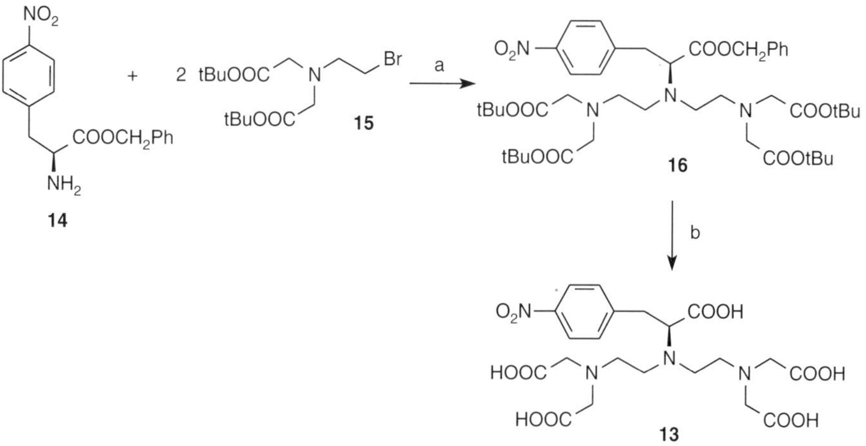

(a) MeCN, 2M phosphate buffer, pH 8; (b) HCl, reflux

Scheme 3.4

buffer. The bromoderivative **15** is easily prepared by dialkylation of ethanolamine with *t*-butyl bromoacetate in DMF in the presence of $KHCO_3$, followed by bromination of the hydroxy group with *N*-bromosuccinimide and triphenylphosphine. The pentaester **16**, under acidic conditions, affords ligand **13**.

The above synthetic pathway is of general applicability provided that a suitable α-amino ester is available. Accordingly, starting from L-glutamic acid and L-lysine and using an appropriate protection/deprotection strategy, the pentaesters **17** and **18**, respectively, were prepared (Chart 3.1) [14]. Both of these are bifunctional synthons which can be used in the preparation of conjugates of Gd complexes. Use of a lysine-derived DTPA, obtained by the Rapoport procedure, recently allowed the synthesis of ligand **19** (Chart 3.1) [15]. The corresponding gadolinium complex (i.e. MP-2269) shows a very strong binding to plasma proteins, which makes it a promising candidate as a contrast agent for the blood pool. Indeed, strong protein binding should prevent, or at least reduce, the extravasation of the contrast agent from plasma into the extracellular fluid after administration. The arrival of the so-called 'blood pool' contrast agents is expected to have a noticeable impact on the future of MRI angiographic procedures, especially coronarography.

Taking advantage of the Rapoport procedure a great number of mono- [16] (e.g. **20**) as well as di- (e.g. **21**) and tri-substituted [17] (e.g. **22**) DTPA ligands have been prepared (Chart 3.2). In order to apply the procedure to the synthesis of ligands such as **21** and **22**, a couple of simple methodologies for the preparation of bromo derivatives such as **23** from simple α-amino acids have been developed [17]. Gadolinium complexes of ligands such as **20–22**, due to their strong binding to plasma proteins, have been evaluated as potential candidates for MRI angiography.

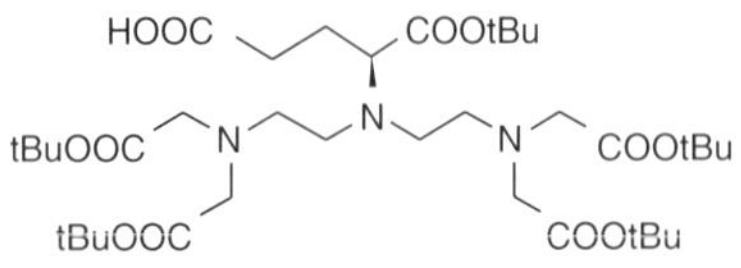

17

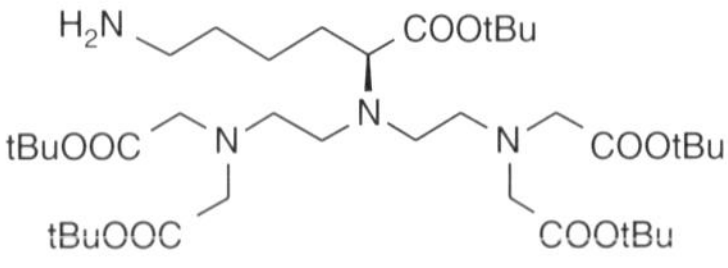

18

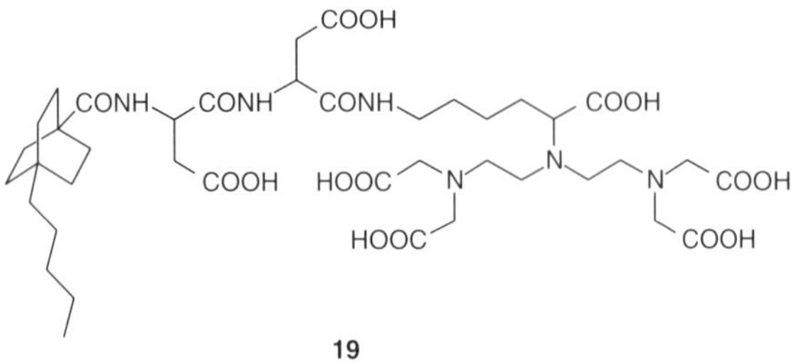

19

[Chart 3.1]

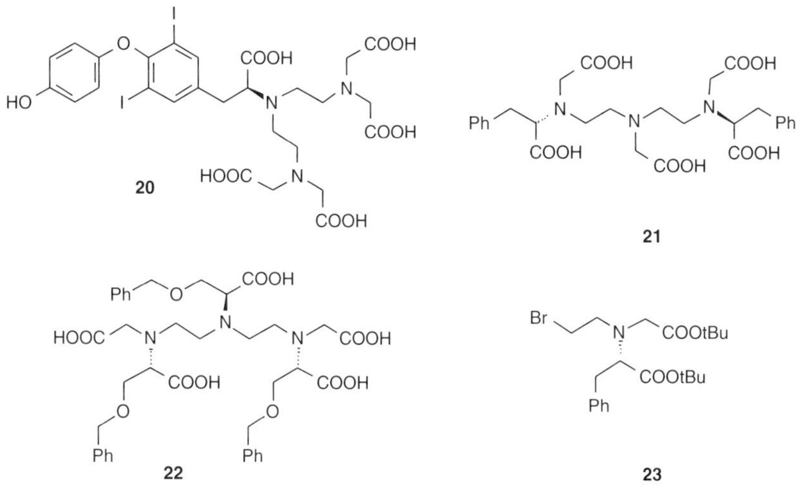

[Chart 3.2]

Modification of the DTPA skeleton with the insertion of a group linked to one of the carbon atoms of the diethylenetriamine moiety has attracted the attention of several research groups. Once again, the results of certain studies aimed at the preparation of bifunctional chelating agents for radiolabeling of monoclonal antibodies with ^{111}In(III) proved of capital importance in the field of MRI contrast agents. Indeed, in 1986 Gansow and co-workers reported the synthesis of the *p*-aminobenzyl-substituted DTPA **24** according to the path described in Scheme 3.5 [18]. The methyl ester of *p*-nitrophenylalanine **25** is

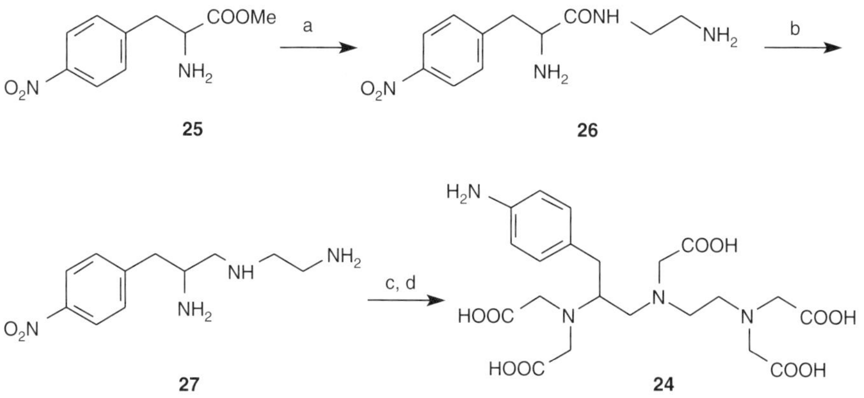

(a) $H_2NCH_2CH_2NH_2$; (b) BH_3, THF; (c) $BrCH_2COOH$, KOH; (d) H_2, Pd/C

Scheme 3.5

reacted with an excess of ethylenediamine to give amide **26**, which is then reduced with diborane in THF to give the substituted diethylenetriamine **27**. Carboxymethylation under classical conditions and reduction of the nitro group by hydrogenation in the presence of Pd/C afforded ligand **24**.

The Gansow strategy has been applied to the synthesis of several MRI contrast agents. One of these, namely the Gd(III) complex of **28** (EOB-DTPA) sodium salt [19] (international non-proprietary name: gadoxetic acid, disodium salt), due to the presence of the lipophilic ethoxybenzyl residue, proved very promising as a hepatospecific MRI contrast agent. Its clinical development has reached the stage of phase-III studies. Two synthetic routes to **28** are depicted in Scheme 3.6. According to path A, the phenol group of the (*Z*)-protected L-tyrosine methyl ester **29** is alkylated with ethyl iodide in DMF in the presence of K_2CO_3 to give **30**, which is then reacted with an excess of

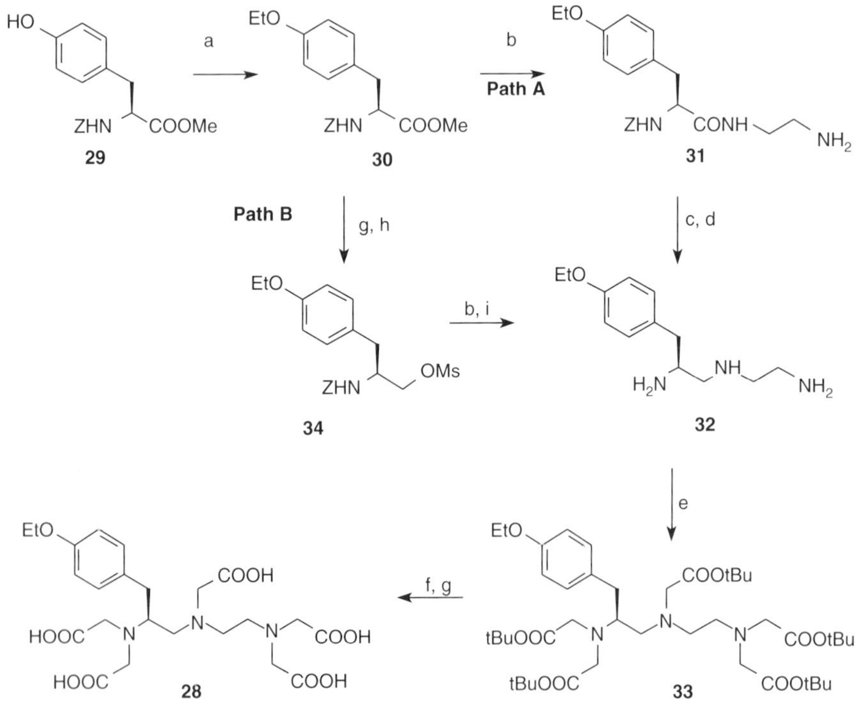

(a) EtI, K_2CO_3, DMF; (b) $H_2NCH_2CH_2NH_2$; (c) H_2, Pd/C, toluene/aq. KOH; (d) BH_3, THF;

(e) $BrCH_2COOtBu$, THF/H_2O, K_2CO_3; (f) NaOH, H_2O/MeOH, reflux then Amberlite IR 120 (H^+ form);

(g) $NaBH_4$ THF; (h) MsCl, Et_3N, THF; (i) H_2, Pd/C, MeOH

Scheme 3.6

ethylenediamine to yield **31**. Removal of the (*Z*)-protection by hydrogenolysis and reduction of the amide moiety with diborane leads to triamine **32**. Differently from the Gansow procedure, the triamine is alkylated with *t*-butyl bromoacetate in THF/H_2O in the presence of K_2CO_3 to afford the pentaester **33**[1]. Deprotection of **33** is not carried out under acidic conditions, as it would be normally expected for *t*-butyl esters, but with sodium hydroxide in water/methanol. By using this methodology, ligand **28** is obtained from **33** in almost quantitative yield. Alkaline deprotection of poly(*t*-butyl esters) such as **33** is becoming quite popular [21] and, when compared to structurally different *t*-butyl esters, is likely to be facilitated by the presence of a nitrogen atom in a position α to the ester function. Path B in Scheme 3.6 has been developed to circumvent the use of diborane in large-scale preparations of **28**. Reduction with sodium borohydride of ester **30** leads to an intermediate alcohol which is mesylated with methanesulfonyl chloride and triethylamine in THF to give **34**. Reaction of **34** with an excess of ethylenediamine and subsequent hydrogenolysis of the (*Z*)-protection affords the triamine **32**. The synthetic path A to ligand **28** has been used by the same research group for the preparation of a variety of similarly substituted DTPA ligands [19a]. A structurally similar ligand in which the 4-ethoxybenzyl group is replaced by a 4-butylbenzyl group has been similarly prepared [22] and its Gd(III) complex (i.e. MS264) investigated to some extent as a hepatospecific contrast agent.

The Gansow procedure has been also used in the first steps of the synthesis of ligand **35**. Starting from the L-serine methyl ester and with a reaction sequence (Scheme 3.7) similar to those described above, the pentaester **36** [23] can be easily obtained. Coupling of the pentaester **36** to 4,4-diphenylcyclohexanol [24] through a phosphate bridge is achieved by using 2-cyanoethyl *N,N*-diisopropylchlorophosphoramidite, a reagent which is commonly utilized in the synthesis of oligonucleotides. Subsequent oxidation of the phosphorus atom with *t*-butyl hydroperoxide affords **37**. Cleavage of the cyanoethyl phosphoric ester with ammonia in MeOH and of the *t*-butyl carboxylic esters under acidic conditions leads to ligand **35**.

From the pentaester **36**, the same authors prepared a large series of ligands which were structurally similar to **35** and differing only in the lipophilic residue attached to the phosphate moiety [25]. Among the Gd(III) complexes of such ligands, many of which presumably feature strong binding to human serum albumin (i.e. the main component of the plasma protein pool), the Gd(III) complex of **35** (i.e. MS325, AngioMark®) was selected as a candidate contrast

[1] It must be stressed that alkylation with *t*-butyl bromoacetate at this stage of the synthesis of the ligands of interest has become very popular as an alternative to carboxymethylation with bromoacetic acid. Despite the introduction of an additional step (i.e. the deprotection), the pentaester usually offers the advantage of an easy purification by silica gel chromatography, as opposed to the ligand which normally has to be purified by reverse-phase chromatography. It is evident that on the laboratory scale the former methodology is by far the preferred approach. As pointed out by several authors, the use of *t*-butyl bromoacetate instead of other (e.g. methyl or benzyl) bromoacetates allows the avoidance of competitive lactamization reactions during the peralkylation of the triamine intermediates [13, 20].

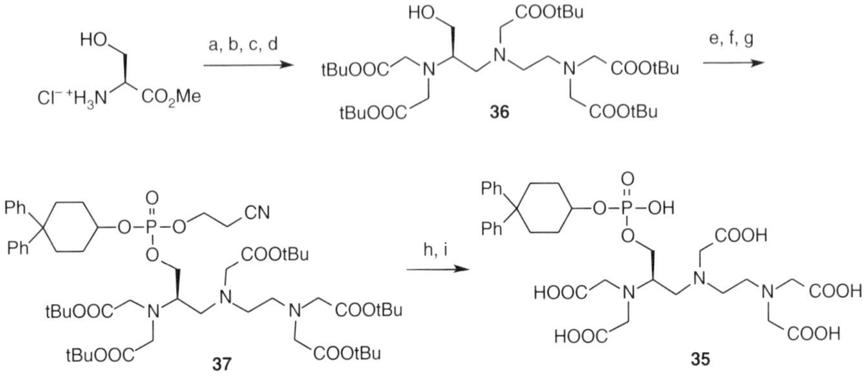

(a) BOC_2O, Et_3N, toluene; (b) $H_2NCH_2CH_2NH_2$; (c) BH_3, THF, then aq. HCl;

(d) $BrCH_2COOtBu$, DIPEA, DMF; (e) 2-cyanoethyl diisopropylchlorophosphoramidite, DIEA, CH_2Cl_2;

(f) 4,4-diphenylcyclohexanol, tetrazole, MeCN; (g) *t*-BuOOH; (h) NH_3, MeOH; (i) conc. HCl, Et_2O

Scheme 3.7

agent for MR angiography/coronarography and is presently under clinical development.

The combination of the Gansow procedure with an original protection/deprotection strategy (i.e. using Cu(II) as the 'protecting group') allowed the preparation of a DTPA ligand functionalized with an ω-aminoalkyl residue (Scheme 3.8) [26]. Amidation of L-lysine methyl ester with ethylenediamine and subsequent reduction leads to tetramine **38**. The diethylenetriamine moiety of **38** is temporarily protected as the Cu(II)-complex by reaction with copper

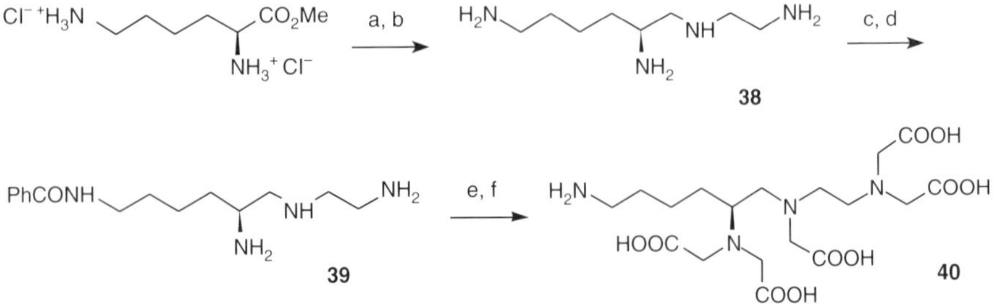

(a) $H_2NCH_2CH_2NH_2$; (b) BH_3, THF; (c) $CuCO_3$, PhCOCl, KOH;

(d) H_2S; (e) $BrCH_2$ COOH, KOH; (f) 6 M HCl, 110 °C

Scheme 3.8

carbonate, thus allowing the selective benzoylation of the amino group which is not involved in chelation of copper. After removal of the Cu(II) ions by treatment with H_2S, **39** is alkylated with bromoacetic acid and, eventually, the benzamide linkage is hydrolyzed under strongly acidic conditions to yield **40**.

It has been previously shown that the introduction of a lipophilic group into the structure of DTPA can be used to tune the specificity of the corresponding Gd(III) complex. To further exploit this issue, Sajiki and Ong have developed a synthetic route which allows the insertion of two residues into the diethylenetriamine backbone in a stereospecific way [27]. The phenylalaninol **41** (Scheme 3.9) is protected with Boc_2O and then brominated with triphenylphosphine and *N*-bromosuccinimide to afford **42**. Azidation with sodium azide in dimethylformamide and reduction with hydrogen on Lindlar catalyst gives **43**. Coupling between **43** and Boc-L-phenylalanine *p*-nitrophenyl ester is the key step in the assembly of the diethylenetriamine moiety. Ligand **45** is obtained from intermediate **44** through a classical sequence of (i) deprotection of the Boc groups, (ii) amide reduction, (iii) alkylation with *t*-butyl bromoacetate, (iv) hydrolysis of the *t*-butyl ester groups. Both the (*S,S*)- and the *meso*-diastereomeric forms of ligand **45** have been prepared from (*S*)- and (*R*)-phenylalaninol, respectively. The strategy shown in Scheme 3.9 appears of general application for the preparation of both symmetrically and asymmetrically disubstituted DTPA ligands. This can be simply achieved by replacing phenylalaninol (**41**) by any amino-acid-derived aminoalcohol and using the activated ester of the amino acid in the coupling step.

In the early 1990s, a few research groups have investigated the insertion of suitable residues in the diethylene moiety of DTPA ligands to achieve metal

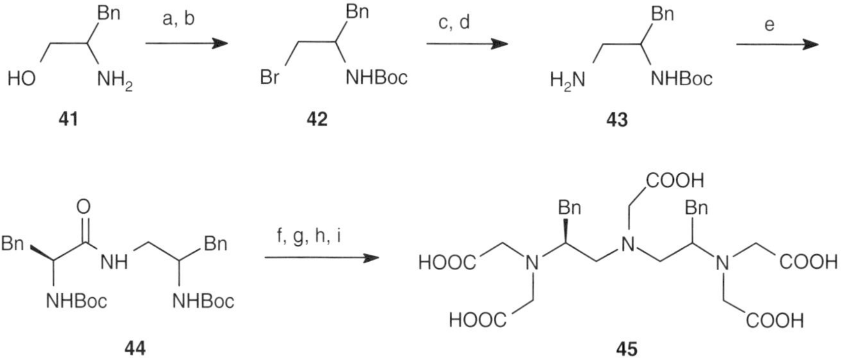

(a) Boc_2O; (b) Ph_3P, NBS; (c) NaN_3; (d) H_2, Lindlar catalyst; (e) Boc-L-Phe 4-$NO_2C_6H_4$ ester; (f) CF_3COOH; (g) BH_3, THF then HCl; (h) $BrCH_2COOtBu$; (i) HCl

Scheme 3.9

complexes whose *in vivo* dissociation is depressed [20, 28]. This issue was put forward as an important requirement in the preparation of conjugates of radionuclides to monoclonal antibodies. Conformationally constrained ligands were conceived with a twofold aim, i.e. (i) a certain degree of preorganization should favor the complexation process, and (ii) decomplexation should be disfavored due to the presence of steric barriers. Insertion in the diethylenetriamine backbone of DTPA of simple moieties such as methyl groups [20a, 28a, 28b] or replacement of one of the ethylene moieties with a cyclohexylene [20b, 28c] have been successfully studied in this context.

More recently, hindered DTPA ligands have been proposed for the preparation of Gd(III) complexes characterized by increased stability (Chart 3.3) [29]. New synthetic routes have been developed to achieve monomethyl (**46a**) and dimethyl (**46b**) derivatives [29b], as well as ligands incorporating one (**47**) and two units (**48**) of pyrrolidine [29a]. Increased rigidity of the DTPA skeleton has also been obtained by incorporating the central nitrogen atom into a five- or six-membered heterocyclic ring as in **49** [30]. The ligand **50**, which contains two cyclohexylene moieties, has been also prepared [31]. The corresponding Gd(III) complex (i.e. WIN 70197), due to its noticeable lipophilicity and its favorable tolerability, has been considered as a candidate for development as a hepatospecific contrast agent. In the synthesis of **50** (Scheme 3.10), special attention must be payed to the stereochemical issues since the ligand contains four stereogenic carbon atoms. *Trans*-2-aminocyclohexanol (**51**) is bis-tosylated with *p*-toluenesulfonyl chloride in pyridine and the ditosyl derivative is converted into *N*-tosylaziridine (**52**) by reaction with sodium hydride in THF. Aziridine ring opening with *trans*-1,2-cyclohexanediamine leads to a 4:1 diastereomeric mixture in favor of the desired diastereomer. After detosylation and column-chromatography separation, triamine **53** is isolated in 33 % yield from

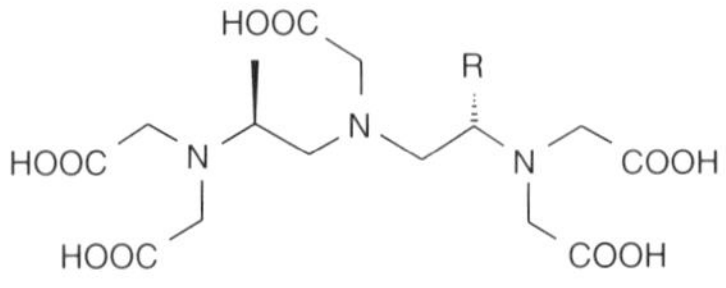

46a R = H
46b R = Me

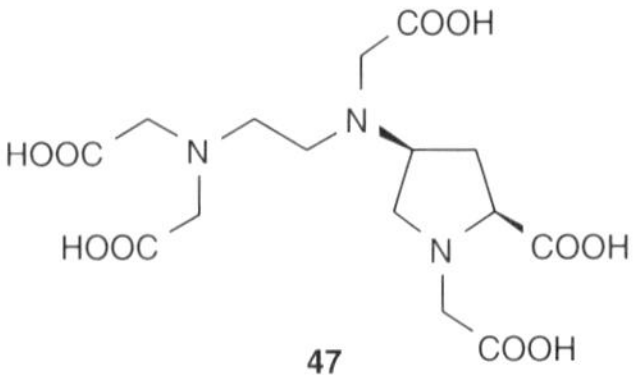

47

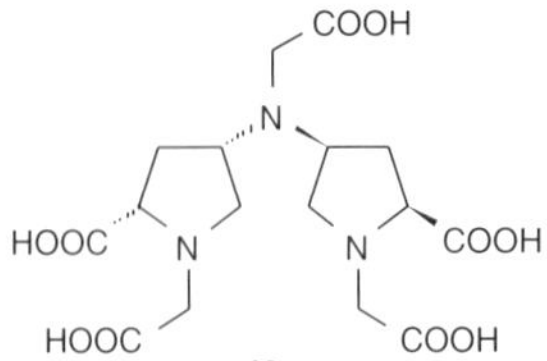

48

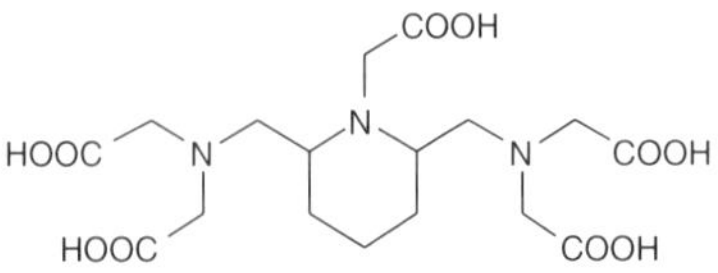

49

[Chart 3.3]

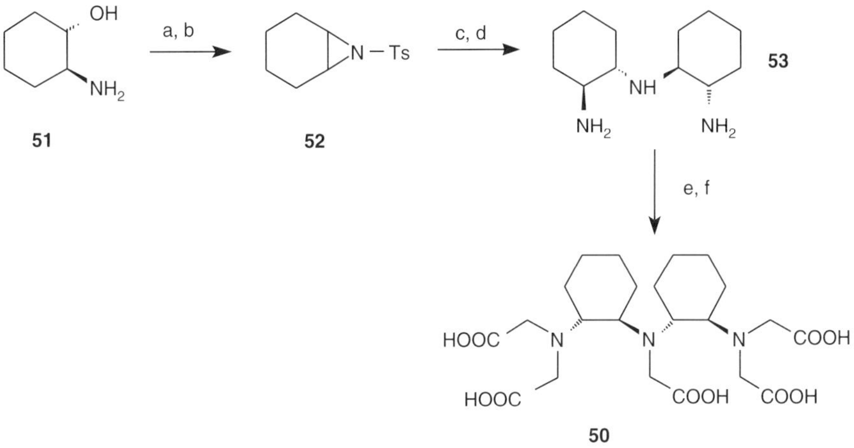

(a) TsCl, pyridine; (b) NaH, THF; (c) *trans*-1,2-cyclohexanediamine, MeCN, reflux;

(d) H_2SO_4, 110 °C; (e) $BrCH_2COOtBu$, K_2CO_3, MeCN, 50 °C, (f) CF_3COOH

Scheme 3.10

52. Classical alkylation of **53** with *t*-butyl bromoacetate, followed by deprotection with trifluoroacetic acid, gives ligand **50**. The stereochemical configuration of **50** was confirmed by the solid-state structure obtained by X-ray diffraction analysis [31].

The Gd(III) complex of a modified DTPA, in which one of the two ethylene bridges is replaced by a 1,3-propylene moiety, has been very recently studied. Ligand **54** (Chart 3.4) is easily prepared from commercially available *N*-(2-aminoethyl)-1,3-propanediamine by alkylation with *t*-butyl bromoacetate followed by acidic deprotection [32]. The Gd(III) complex of **54** does not show any significant advantage over the corresponding complex of DTPA in terms of stability constant and relaxivity.

2.2 HEXADENTATE LIGANDS (EDTA AND EDTA-MODIFIED LIGANDS)

In the early 1980s, the Gd(III) complex of EDTA was studied as a potential candidate for MRI [2]. In principle, such a complex shows several appealing features, i.e. (i) the relaxivity is quite favourable [2] due to the presence of three water molecules in the first coordination sphere of the metal ion, as evidenced by the X-ray crystal structure [33], (ii) the thermodynamic stability of the complex is quite high (log K_{ML} 17.35) [34] and comparable to that of complexes which are routinely used in clinical practice (see below), and (iii) EDTA is the

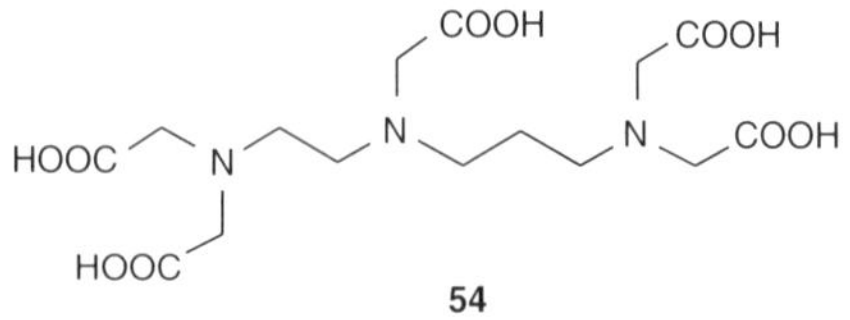

54

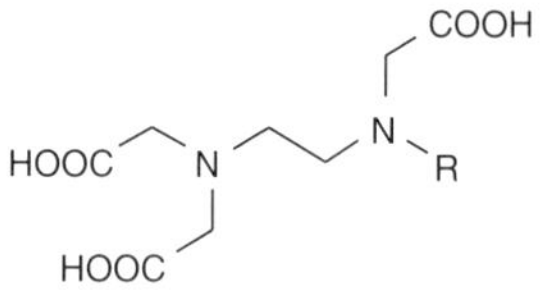

	R
EDTA	CH_2COOH
55a	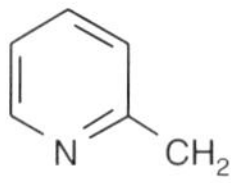 (2-pyridyl)CH_2
55b	$CH_2C_6H_5$
55c	CH_2CH_2OH

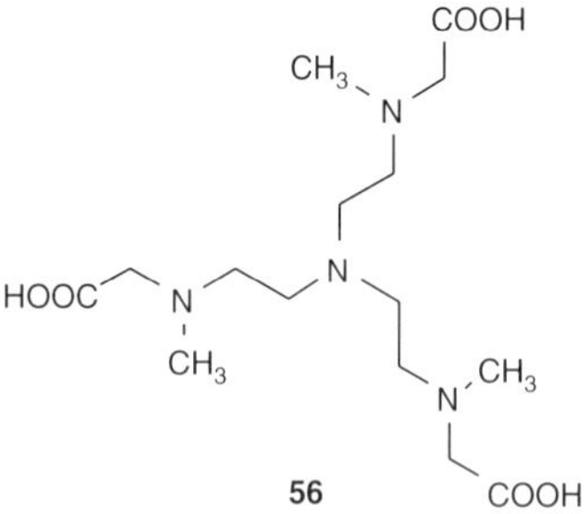

56

[Chart 3.4]

cheapest commercially available polyaminopolycarboxylic ligand. However, the tolerability of Gd-EDTA in animals proved poor and the compound was dropped in favor of the much more tolerable Gd-DTPA [2]. More recently, in a comparative study performed on a series of Gd(III) complexes of polyamino-polycarboxylic ligands it was found that release *in vivo* of gadolinium from Gd-EDTA is not negligible [35]. Therefore, it must be stressed that evaluation of the *in vivo* stability of the Gd(III) complexes can not be simply derived from the extent of the thermodynamic stability constant (not even the conditional stability constant). We refer the reader to more qualified sources for better insights into this matter [36].

As expected, replacement of one of the carboxymethyl side arms of EDTA with a weakly (e.g. **55a**) or non-coordinating residue (e.g. **55b**) (Chart 3.4) leads to a drop of two [37] and five [38] orders of magnitude in the stability of the corresponding Gd(III) complexes, respectively. Such a decreased stability is likely to prevent safe *in vivo* applications for gadolinium complexes of this kind. The hydroxyethyl derivative **55c** (Chart 3.4) has been also studied to some extent [39].

The basic structure of EDTA has also been selected for the preparation of Mn(II) complexes that could be used as MRI contrast agents. Indeed, relaxivity of such complexes is sufficiently large due to the presence of one molecule of water directly coordinated to the metal ion. This feature is also evident from the X-ray crystal structure of Mn-EDTA [40]. EDTA ligands in which one or two of the carboxymethyl residues bear a lipophilic substituent have been prepared developing novel methodologies [41]. However, due to the scanty importance in the MRI scenario of the Mn(II) complexes derived from these ligands, for their synthesis we refer the reader to the original patents [41].

A structure quite different from those of the polyaminopolycarboxylic ligands seen so far is represented by ligand **56** (Chart 3.4). This ligand formally contains seven (four amino groups and three carboxylates) potential binding sites. However, on complexation of the Gd(III) ion apparently only six sites (and not the pivotal tertiary amine) are involved in the coordination of the metal ion. This result was inferred on the basis of potentiometric studies which led to assess a very low thermodynamic stability for this complex [42].

2.3 DECADENTATE LIGANDS

Two decadentate polyaminopolycarboxylic ligands need to be mentioned now, namely **57** (TTHA) and **58** (TTAHA) (Chart 3.5). Both compounds are classically obtained by carboxymethylation of the corresponding tetraamine with either sodium cyanide and formaldehyde, followed by acidic hydrolysis [3, 43] or chloroacetic acid at controlled pH [37].

The Gd(III) complex of **57**, although not being of practical use as a contrast agent, is quite often used as a 'reference standard' for a complex in which the relaxivity derives from outer-sphere contributions only (i.e. no water molecules are allowed in the first coordination sphere of the gadolinium ion) [39]. This is clearly evident from (i) the X-ray crystal structure, which shows that all but one (i.e. a carboxylate group) of the binding sites of the ligand participate to the coordination of the gadolinium ion [44], and (ii) studies in solution on the corresponding Dy(III) complex [45]. Differently, in the case of the complex of the isomeric ligand **58**, the X-ray crystal structure shows that one iminodiacetic sub-unit is not involved in the coordination of the metal ion and hence two water molecules have access to the first coordination sphere of the gadolinium ion [44]. This has been confirmed by relaxometry and luminescence studies [44].

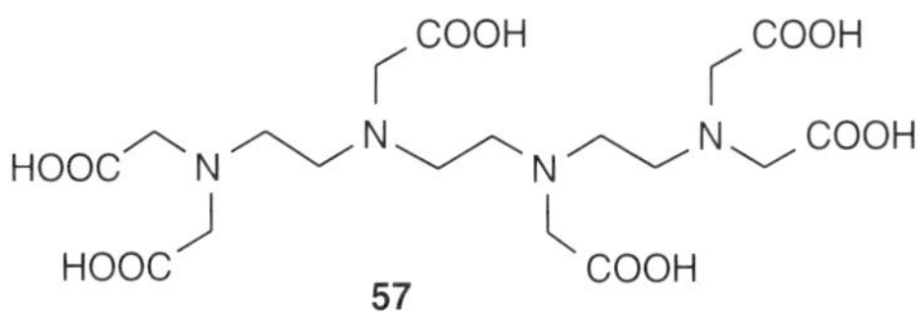

57 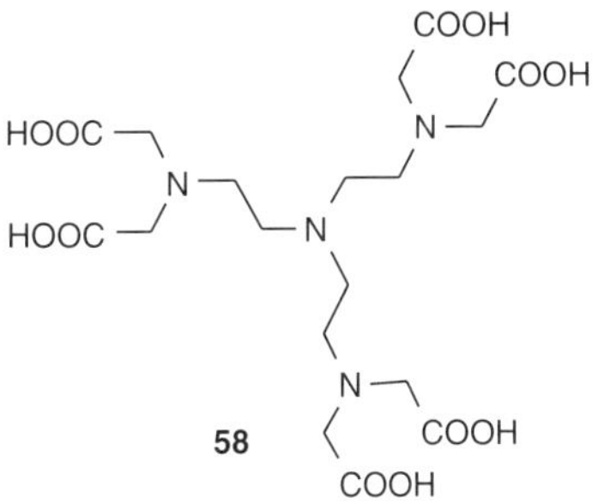 **58**

[Chart 3.5]

3 POLYAMINOPOLYCARBOXYLIC LIGANDS CONTAINING OTHER BINDING SITES

The ligands described in this section are polyaminopolycarboxylic ligands in which one or more amine or carboxylate binding sites have been replaced by coordinating residues of different nature.

3.1 OCTADENTATE LIGANDS (DTPA DIAMIDES AND MONOAMIDES)

Besides Gd-DTPA, the gadolinium complexes of DTPA diamides have probably been the most studied class of MRI contrast agents. The reason lies in the extremely easy synthetic access to these ligands, which are obtained by reaction of the DTPA bisanhydride **59** with an excess of the desired amine (Scheme 3.11). Bisanhydride **59** is a commercially available compound but, for clean formation of diamides with no or very little contamination by monoamide by-products, it is usually preferable to freshly prepare **59** from DTPA by reaction with acetic anhydride in pyridine [46, 47]. The diamidation reaction occurs smoothly by addition of the solid bisanhydride **59** to a solution of the amine. A variety of solvents which range from water (e.g. for volatile amines) [47–50] to alcohols (e.g. *i*-PrOH) [47] and dipolar aprotic solvents (e.g. DMF, DMSO) [47, 51–54] have been used. Direct addition of **59** to neat amine has also been employed in some cases [55, 56].

For the sake of conciseness, we report in Table 3.1 only a few of the diamide compounds which have been widely investigated. However, a plethora of

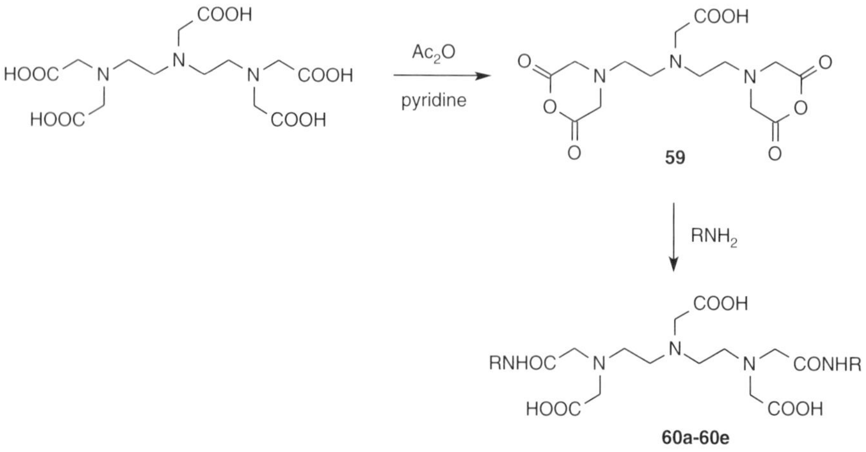

COOH
HOOC N N N COOH
HOOC COOH
Ac_2O
pyridine
COOH
O N N N O
O O
O O
59
RNH_2
COOH
RNHOC N N N CONHR
HOOC COOH
60a-60e

Scheme 3.11

Table 3.1 DTPA diamide ligands for complexation with gadolinium.

Diamide[a]	R	Reference
60a	CH_3	47, 48
60b	C_2H_5	47, 49
60c	C_3H_7	57
60d	C_4H_9	47, 51
60e	$CH_2CH_2OCH_3$	47

[a] General structure is shown in Scheme 3.11.

different amines have been employed for the preparation of DTPA diamides. Indeed, primary and secondary [58] aliphatic amines, very hydrophilic amines such as those derived from carbohydrates [50], amino acids [59], aryl [52–54] and heterocyclic [58] amines have been successfully used. Furthermore, poly-chelating polymeric backbones have been prepared by polycondensation of bisanhydride **59** with α, ω-diamines of different length [60].

Solid-state structures obtained by X-ray crystallographic analysis have shown that in Gd(III) complexes of DTPA diamide ligands the metal ion is nine-coordinate. Not only the three nitrogen atoms and the three carboxylate groups, but also the carbonyl oxygen atoms of the two carboxamide residues are involved in the complexation of the metal ion. One molecule of water occupies the ninth coordination site [49, 61]. NMR studies on complexes with Nd(III) have shown that, for these complexes, the same kind of coordination is observed in solution [62].

A ligand in which the two CH_2CONHR moieties are replaced by two methyl groups was also obtained and the corresponding Gd(III) complex prepared to demonstrate that the contribution of the two amide oxygens to the coordination of the metal ion is not negligible at all [63].

The stability of Gd(III) complexes of DTPA diamides (e.g. log K_{ML} 16.85 for the complex of 60a) [64] is a few orders of magnitude smaller than that of Gd-DTPA and similar complexes in terms of stability constants. However, if the conditional stability constants at physiological pH are considered, such a difference is noticeably smaller [64].

A couple of gadolinium complexes of DTPA diamides, namely **60a** (DTPA-BMA) and **60e** (DTPA-BMEA), have gone through the whole process of pharmaceutical development and are now available as contrast agents under the trade names of Omniscan® and Optimark®, respectively [65, 66]. In comparison with Gd-DTPA-like complexes, which are salified (e.g. with *N*-methyl-glucammonium or sodium), Gd(III) complexes of DTPA diamides are neutral complexes. Therefore, at the same concentration, the latter compounds give rise to solutions which feature a far smaller osmolality. This aspect is of great importance for the administration of multiple doses of contrast agent (up to 0.3 mmol kg^{-1} and more, instead of 0.1 mmol kg^{-1}, which is the most

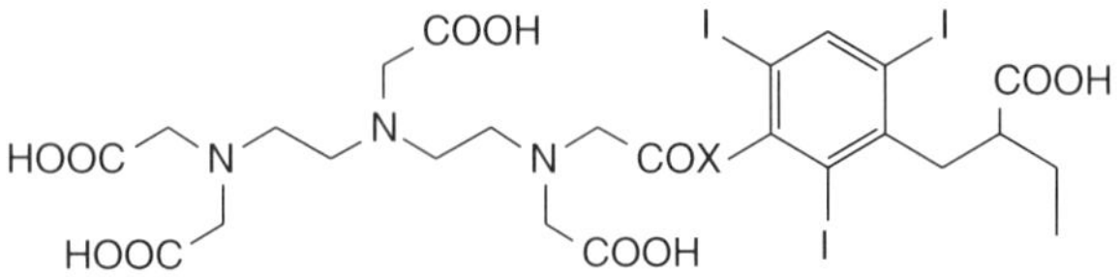

61a X = NH

61b X = $NH(CH_2)_5CONH$

Figure 3.2 Chemical structures of the ligands **61a** and **61b**.

commonly used dose level). In addition, the Dy(III) complex of **60a** has been developed to be used as a T_2 contrast enhancing agent [65].

A few derivatives of DTPA in which only one of the terminal carboxylic groups has been converted into a carboxamide have also been described. The most direct approach to these DTPA monoamides involves the reaction of bisanhydride **59** with an equimolar (or even smaller) amount of amine [67]. This procedure almost inevitably leads to reaction mixtures in which non negligible amounts of DTPA and DTPA diamide are formed along with the desired monoamide. The isolation of the monoamide from these mixtures is quite often lengthy and tedious. A stepwise alkylation of diethylenetriamine, which closely parallels that used for the preparation of **1**, has been exploited for the achievement of DTPA monoamides containing iodinated residues **61a** and **61b** (Figure 3.2) [68]. The Gd(III) complexes of such ligands have been studied as hepatospecific contrast agents.

However, a few alternative methodologies for the preparation of DTPA monoamides are worth mentioning here. First of all, the procedure of Krejcarek and Tucker to activate one of the carboxylic groups of DTPA has been employed [69]. Accordingly, DTPA is first converted into the corresponding pentakis(triethylammonium) salt and then reacted in acetonitrile with isobutyl chloroformate to give the monoactivated species which is coupled with the appropriate amine [52]. This methodology has also been used to covalently link several units of DTPA to polymeric backbones such as polylysines [70]. Monoamides of **1**, such as **62a–62c**, have been achieved by taking advantage of the different esterification rates of acetic and propanoic groups (Scheme 3.12). Indeed, ligand **1**, by reaction with thionyl chloride in MeOH, gives the monoacid tetraester **63**, in 50–60 % yield, after isolation by selective extraction. The carboxylic group of **63** is then activated with diethoxyphosphoryl cyanide and reaction with diamines afford monoamide tetraesters **64**, which after alkaline hydrolysis yield **62a–62c**. The use of a secondary amine in the amidation reaction under these conditions is a stringent requirement. Indeed, reaction of monoacid **63** with *N,N*-dimethylethylenediamine led almost exclusively to compound **65** (Figure 3.3) [71]. After complexation with Gd(III), ligands **62a–62c** give complexes that, owing to the presence of an additional

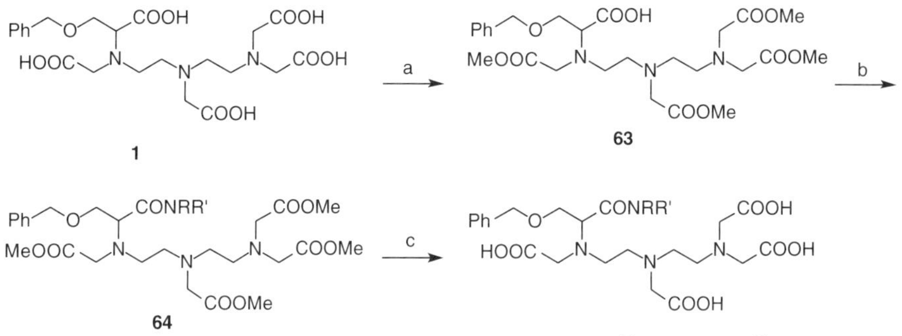

(a) $SOCl_2$, MeOH; (b) RR'NH, DEPC, DMF, 0–25 °C;

(c) NaOH, MeOH, pH 12

	R	R'
62a	CH_3	$(CH_2)_2N(CH_3)_2$
62b	CH_3	$(CH_2)_3N(CH_3)_2$
62c	$-CH_2CH_2N(CH_3)CH_2CH_2-$	

Scheme 3.12

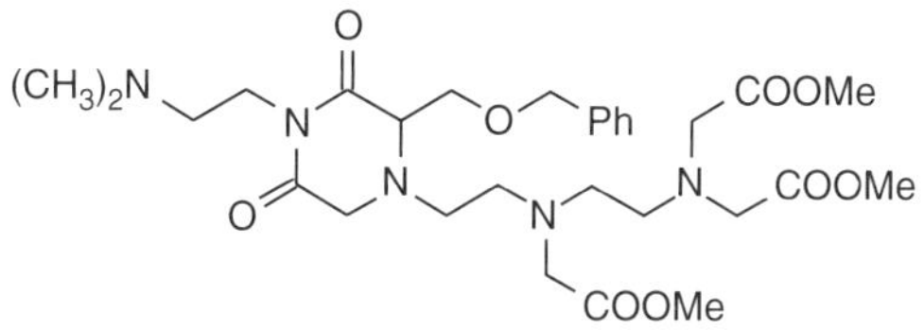

Figure 3.3 Chemical structure of the product obtained by reaction of the monoacid **63** with *N*,*N*-dimethylethylenediamine.

amino group, are inner salts. The aqueous solutions of these complexes have roughly the same osmolality of that of an aqueous solution of the gadolinium complex of the DTPA diamide **60a**. However, the poor tolerability of the former complexes prevented further investigations.

A useful synthon for the synthesis of DTPA monoamides is the monoanhydride **66** [72] (Chart 3.6), although its preparation is not straightforward. As a matter of fact, bisanhydride **59** is first converted with ethanol into the DTPA diethyl ester which is then hydrolyzed to give the DTPA monoethyl ester [73]. Finally, the DTPA monoethyl ester is converted into **66** with acetic anhydride in pyridine. After conjugation to an amino-functionalized synthon, hydrolysis of the ethyl ester group is required to obtain a stronger chelating sub-unit. Monoanhydride **66** has also been used to prepare DTPA monoamides bearing alkylphosphonic groups (e.g. **67**) and the corresponding Gd(III) complexes have been investigated as contrast agents which feature an affinity for calcified tissues [74].

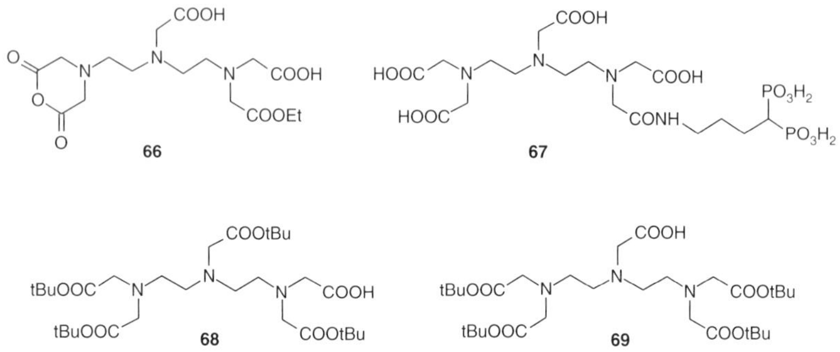

[**Chart 3.6**]

As an alternative to **66**, the monoacid tetraester **68** (Chart 3.6) can be used for the preparation of non-symmetric DTPA monoamides. Compound **68** has been independently prepared by different procedures developed by Platzek *et al.* [75] and Arano *et al.* [76].

DTPA derivatives in which the central acetic moiety is amidated can be obtained by amidation of the monoacid tetraester **69** (Chart 3.6) followed by hydrolysis of the *t*-butyl ester groups [77]. Compound **69** has been prepared by a Rapoport-like procedure from glycine [14] or by a completely different route involving a protection/deprotection sequence starting from diethylenetriamine [78].

Ligand **70a** (Figure 3.4), generally known as EGTA, is an octadentate ligand which is formally derived from DTPA by replacement of the central glycine unit with an ethylenebisoxy unit, so that the number of binding sites is the same in the two structures. EGTA is commercially available, but can be prepared by conventional carboxymethylation procedures on the corresponding diamine [79]. Not unexpectedly, this ligand forms a nine-coordinate complex with Gd(III) with one molecule of water in the first coordination sphere of the metal ion. Thorough studies on lanthanide complexes of EGTA have revealed a peculiar rigidity which could be exploited for the design of improved contrast agents [80]. In this respect it is worth mentioning that an EGTA derivative functionalized with a *p*-aminobenzyl residue (**70b**) (Figure 3.4) has been synthetised for different purposes [81].

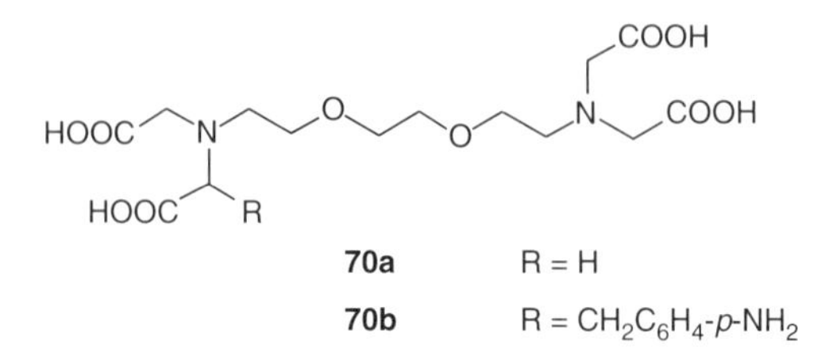

Figure 3.4 Chemical structures of the ligands **70a** and **70b**.

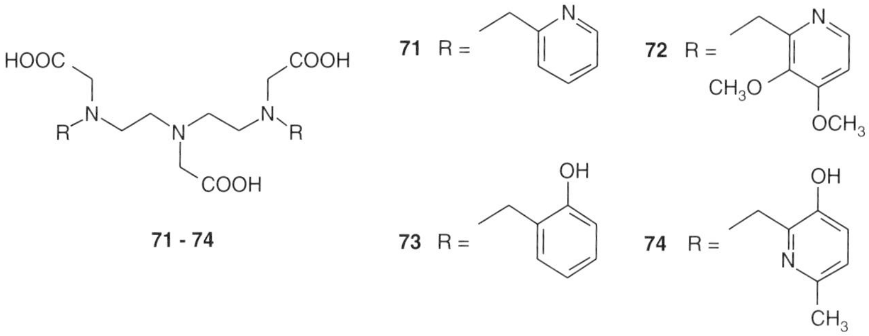

[Chart 3.7]

Other classes of octadentate ligands (e.g. **71–74**) (Chart 3.7) have been obtained by replacing, in the DTPA backbone, two terminal carboxylate groups with binding groups of different nature. Despite the fact that compounds **71** and **72** are true polyaminopolycarboxylic ligands, they have been dealt with in this section for homogeneity with other ligands which are likewise synthesized.

A common issue in the synthesis of ligands **71–74** is the introduction of the two terminal residues which occurs via formation of the double Schiff base of diethylenetriamine. As an example, the synthesis of ligand **71** is depicted in Scheme 3.13. The addition product of diethylenetriamine and 2 mol equivalents of pyridine-2-carboxaldehyde is generated and hydrogenated *in situ* to give compound **75**. This product is peralkylated with *t*-butyl bromoacetate and the resulting triester classically deprotected with trifluoroacetic acid to give ligand **71** [64]. The Gd(III) complex of **71** [64], as well as the Gd(III) and Mn(II) complexes of **72** [82], have been investigated as potential MRI contrast agents. Two slightly different synthetic pathways have been proposed for the synthesis of ligand **73** [83]. Ligand **74** deserves some comments: (i) the two phenol functions (and not the pyridine nitrogens) are involved in the coordination of the metal ion and this is favored by the presence of *meta*-pyridine nitrogens which increase the acidity of the phenol functions, and (ii) the two methyl

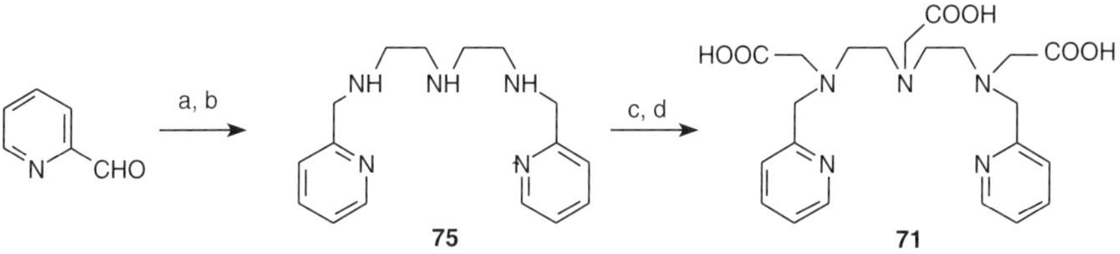

(a) diethylenetriamine; (b) H_2, 10 % Pd/C; (c) $BrCH_2COOtBu$, DIEA, CH_2Cl_2; (d) CF_3COOH, CH_2Cl_2

Scheme 3.13

groups *ortho* to the pyridine nitrogens were introduced in the structure to increase the steric hindrance around these positions. According to Martell and co-workers [84], this prevents, in the final alkylation step with *t*-butylbromoacetate, the concurrent quaternization of pyridine nitrogens, which is an annoying side-reaction in the synthesis of structurally similar ligands (see below). The Gd(III) complex of **74** has been investigated as a candidate MRI contrast agent [85].

3.2 HEXADENTATE LIGANDS

In this section, we will describe ligands which can be formally seen as EDTA-like derivatives in which two carboxylate moieties have been replaced by binding sites of different nature.

Ligand **76a**, universally known as EHPG, and the substituted derivatives **76b–76d** (Chart 3.8) are relevant from a historical point of view since the corresponding Fe(III) complexes were studied as prototypes of hepatobiliary MRI contrast agents [86]. Furthermore, the Fe(III) complexes of both diastereomeric forms of **76d** were used in a pioneering study aimed at investigating the interaction of such species with human serum albumin [87]. Later, studies on this class of iron complexes as contrast agents were dropped because these complexes do not have water molecules in the first coordination sphere of the metal ion and therefore their relaxivities are very poor. It must be stressed that the *in vivo* use of Fe(III) complexes with open sites for the direct coordination of water molecules to the metal ion appears very unlikely because such complexes would catalyze the formation of toxic hydroxyl radicals [86]. The basic ligand of this series, **76a**, is commercially available, while **76b–76d** can be prepared by reaction of the appropriate phenol with ethylenediamine and sodium glyoxylate or dichloroacetate [88].

Ligand **77** (Chart 3.8), which can be prepared by reaction of ethylenediamine-*N, N'*-diacetic acid with 2-acetoxybenzyl bromide, followed by alkaline hydrolysis [89], was also proposed for the preparation of Fe(III) complexes as MRI contrast agents [86].

Other hexadentate ligands, such as **78a** and **78b** (Chart 3.8), have had much better success in the arena of MRI contrast agents. Both ligands are synthesized [90–92] according to a pathway which is detailed in Scheme 3.14 for **78b** (DPDP). The condensation of ethylenediamine with two mol equivalents of pyridoxal 5-phosphate leads to the Schiff base **79** which, by catalytic hydrogenation, is converted into **80**. Carboxymethylation with bromoacetic acid at controlled pH of compound **80** leads to the ligand [92]. The interest in ligands **78a** and **78b** stems from their ability to give rise to relatively stable Mn(II) complexes. Initially, the complex derived from **78a** (PLED) was investigated [91], but later it was abandoned for the complex of **78b** which features a greater solubility in water and a slightly better thermodynamic stability. In both Mn(II)

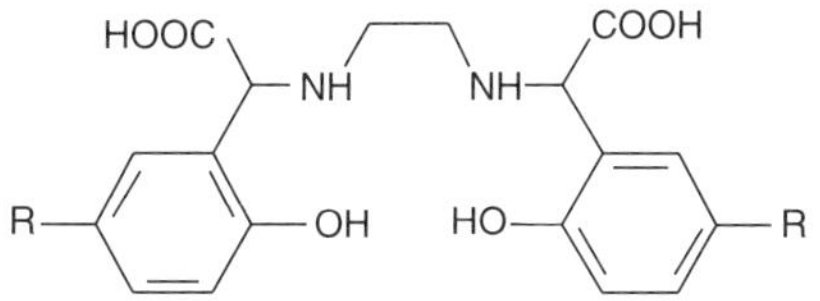

76a R = H
76b R = Me
76c R = Cl
76d R = Br

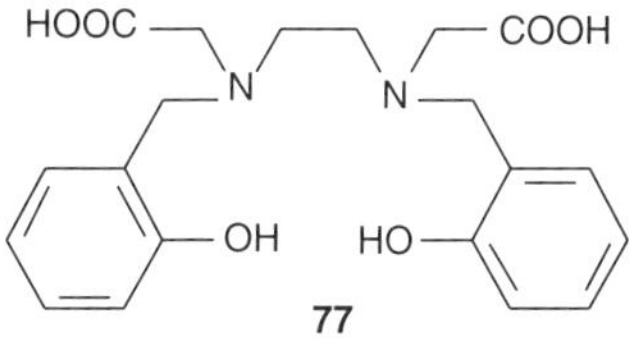

77

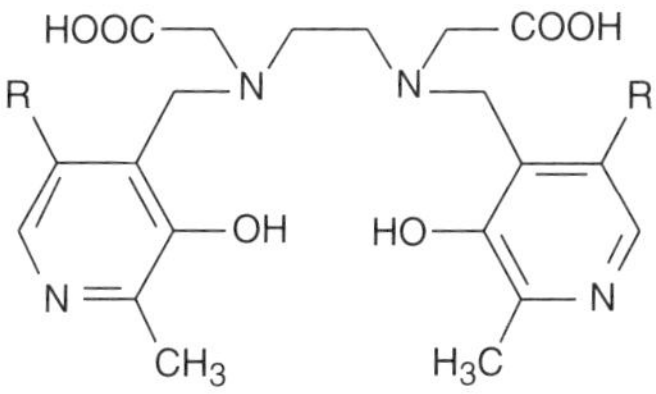

78a R = H
78b R = CH_2OPO_3H

[Chart 3.8]

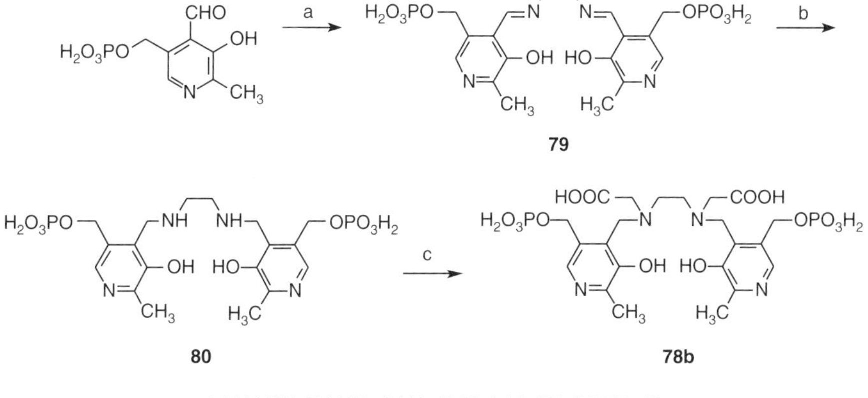

79

80

78b

(a) $H_2NCH_2CH_2NH_2$; (b) H_2, Pt/C; (c) $BrCH_2COOH$, pH11

[Scheme 3.14]

complexes, the metal ion is completely 'wrapped around' by the ligand. The geometry of these complexes is best represented by a distorted octahedron in which the two ethylenediamine nitrogens, two oxygen atoms of the carboxylates and the two phenoxide oxygens are involved in the coordination of the metal ion [91, 92]. Nowadays, the Mn(II) complex of **78b** (i.e. Teslascan®; mangafodipir trisodium) is available in clinics as a hepatospecific contrast agent.

In addition, Gd(III) and Mn(II) complexes of EDTA diamides have been proposed and investigated as potential contrast agents [93, 94]. However, the thermodynamic stability of such complexes appears quite poor. In close analogy to DTPA diamides, EDTA diamides are obtained by reaction of EDTA bisanhydride [46] with amines [93].

Hexadentate polyaminopolyphosphonic ligands, such as **81** and **82** (Chart 3.9), have been prepared by reaction of the corresponding diamine with H_3PO_3/HCl and formaldehyde (Moedritizer–Irani procedure). Because of the known affinity of aminophosphonates for calcified tissues, the Gd(III) complexes of **81** and **82** have been studied as potential 'bone-seeking' contrast agents [95].

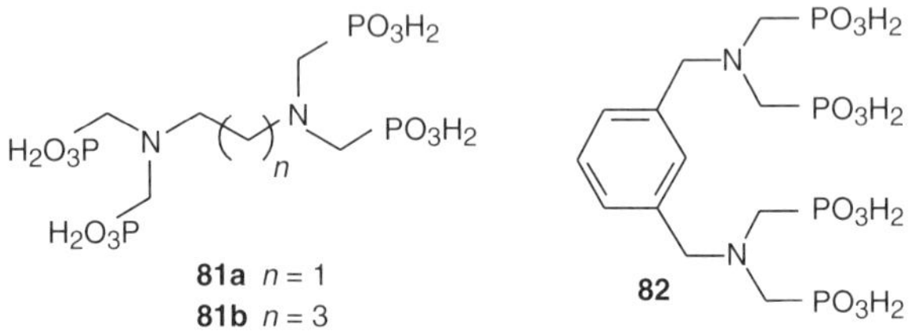

81a $n = 1$
81b $n = 3$

82

[**Chart 3.9**]

4 OTHER LIGANDS

A few acyclic ligands, whose structure is not of the polyaminopolycarboxylic type, have been developed to be used as MRI contrast agents after complexation with Gd(III).

Raymond and co-workers have synthesized a podand structure (**83**) in which three units of 3-hydroxy-2-pyridinone are held together by a unit of tris(2-aminoethyl)amine (tren) [96]. The synthesis of such a ligand is described in Scheme 3.15. Carboxylation of hydroxypyridinone **84** under Kolbe–Schmidt conditions gives the acid **85** which is protected by reaction with benzyl chloride followed by hydrolysis of the benzyl ester to restore the carboxylic group. Acid **86** is then activated and reacted with tren to afford **87**. Eventually, benzyl groups are removed by hydrogenolysis to yield ligand **83** which, after

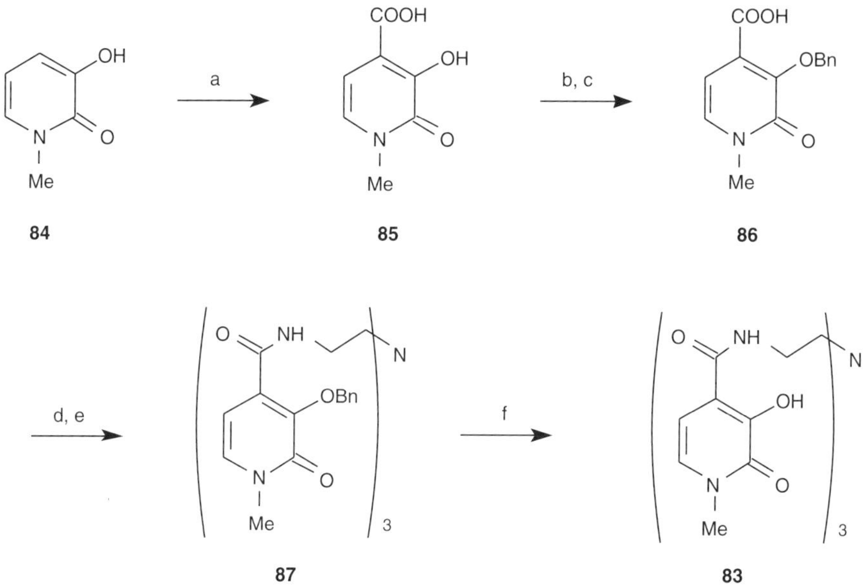

(a) CO_2, base; (b) BnCl, base; (c) hydrolysis;
(d) 1,3-thiazolidine-2-thione; (e) tren; (f) H_2, Pd/C

Scheme 3.15

complexation with Gd(III), produces a neutral complex. The X-ray crystal structure shows that the complex is unexpectedly eight-coordinate with two molecules of water in the first coordination sphere of the metal ion. Such a promising feature is in part obscured by the very poor solubility of the complex in water.

Chart 3.10 reports the structures of ligands which have been studied by other authors. Ligands **88** and **89** are obtained by condensation of the appropriate salicylaldehyde with tren and tetraethylenepentaamine, respectively [97]. Conventional reduction with $NaBH_4$ of the appropriate precursor affords the saturated ligands **90** and **91** [97]. It is quite remarkable that Gd(III) can act as a template to obtain in a 'one-pot' reaction the complexes corresponding to ligands **88b** and **89b** in over 90 % yield [97]. Gd(III) complexes derived from ligands **88** and **90** are neutral, whereas those obtained from ligands **89** and **91** are mono-cationic. On the basis of the X-ray crystal structure of the La(III) complex of **89a** [98], the corresponding Gd(III) complex is expected to have one molecule of water directly linked to the metal ion. Since the complexes obtained from **88a** and **89a** are poorly soluble in water, the derivatives bearing long polyoxyethylene chains were investigated. As a result, the solubilities in water of the corresponding complexes were noticeably improved. However, some

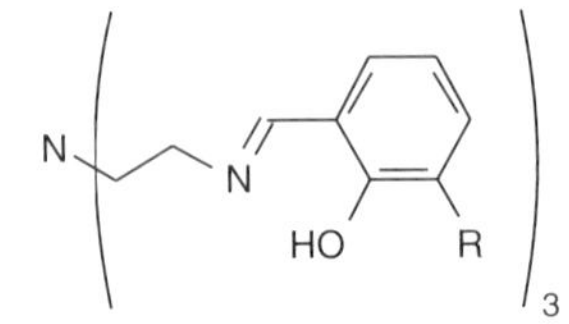

88a,b

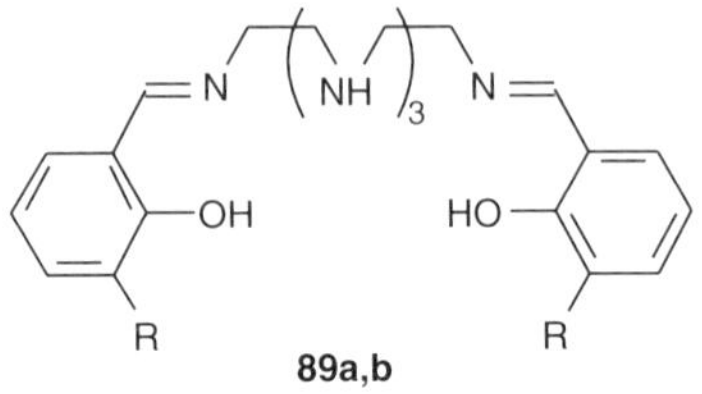

89a,b

(a) R = H; (b) R = polyoxyethylene chain

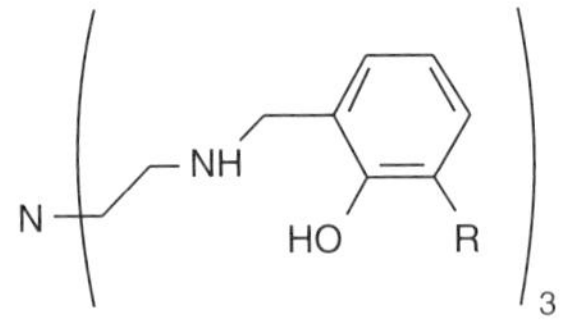

90

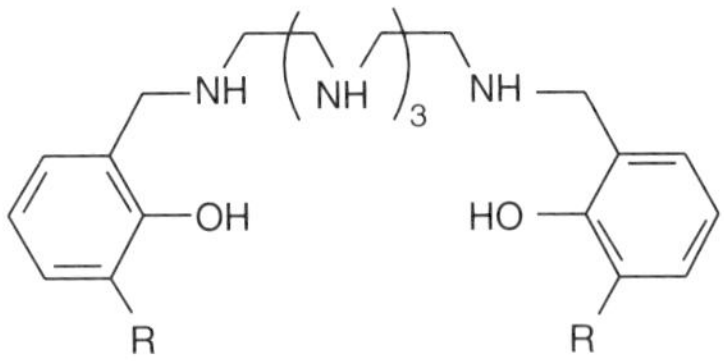

91

R = polyoxyethylene chain

[**Chart 3.10**]

concern about the stability in solution of these complexes was raised by the authors.

Podand-like ligands which were claimed to be effective for the preparation of Gd(III) complexes useful as MRI contrast agents have been reported in a few patents [99, 100]. In addition, Fe(III) complexes of siderophores such as desferrioxamine B and suitably modified derivatives have been investigated as MRI contrast agents [101].

5 FUNCTIONALIZED ACYCLIC LIGANDS FOR CONJUGATION

Functionalized ligands (or precursors of ligands) bearing a suitable reactive group are required for conjugation to (i) an appropriate carrier to achieve targeting, and (ii) macromolecular backbones containing multiple reactive groups (e.g. proteins, polyamino acids, dendrimers, etc.) to obtain contrast agents for specific applications (e.g. contrast agents for the cardiovascular compartment).

While describing the synthesis of the ligands reported in the previous sections, we have already touched upon some of these functionalized synthons. By and large, the most used reactive synthon for conjugation of DTPA moieties has been DTPA bisanhydride **59**. Indeed, coupling the bisanhydride **59** to proteins or polymers (e.g. polylysine) bearing multiple amino groups has been quite popular. However, the side-reaction which is observed when one single molecule of **59** reacts with two amine groups to give rise to a diamide can be a serious drawback. For instance, cross-linking to a certain degree has been observed in the reaction of **59** with proteins [102].

Monoreactive precursors of the DTPA moiety, such as monoanhydride **66**, have also been used. The activation procedure of Krejcarek and Tucker [69] has also been employed for linking DTPA moities to amino-functionalized macromolecules.

More recently, esters of DTPA bearing a single reactive function, such as **17** and **18** (Chart 3.1), and **68** and **69** (Chart 3.6), have been used. Remarkably, compounds **17** and **18** allow the preparation of conjugates in which all five carboxylate groups of DTPA are available for the complexation of the Gd(III) ion and therefore give rise to complexes which are characterized by a higher stability when compared to conjugates obtained by the bisanhydride route.

DTPA ligands containing an isothiocyanate group have been used to prepare conjugates via formation of a thiourea linkage. It is noteworthy that conjugation has been also achieved by reacting directly the isothiocyanate-functionalized Gd(III) complex (e.g. **9**) [11].

6 LIPOPHILIC LIGANDS

All complexes mentioned in the previous sections, which have been developed to a certain extent for clinical application, are water soluble. Their aqueous solutions are easily administered to the patient by intravenous injection. However, for some applications, even complexes which are almost insoluble in water have been studied.

Liposomes, which are spherical vesicles composed of lipid bilayers [103], are rapidly taken up *in vivo* by cells of the reticuloendothelial system and for this peculiarity they have been used as carriers for imaging agents with the main purpose of targeting liver and spleen [104]. For instance, Gd-DTPA was simply entrapped within the internal aqueous space of liposomes but the slow diffusion of water across the lipid bilayer critically limits the relaxivity [104a]. For better results, lipophilic chelates were synthesized and incorporated into the liposomal membrane.

The lipophilic character of a ligand can be increased by simply introducing aliphatic chains. In this respect, DTPA bis(octadecylester) **92** [105], bis(hexadecylamide) **93** [106] and bis(octadecylamide) **94** [105, 106] (Chart 3.11) are easily synthesized by reaction of bisanhydride **59** with the appropriate alcohol or amine.

The synthesis of diamide **95** (Chart 3.11) was also reported [107]. This unusual product was designed to build up polymerized liposomes which have an increased physical stability [108]. For this purpose, the Gd(III) complex of **95** was mixed with a phosphatidylcholine derivative containing the diyne moiety in the lipophilic chain. The mixture was sonicated to generate liposomes and then irradiated with UV light to induce the polymerization of the triple bonds

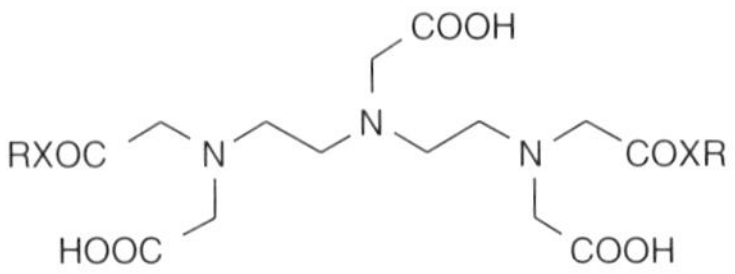

92 X = O R = $C_{18}H_{37}$
93 X = NH R = $C_{16}H_{33}$
94 X = NH R = $C_{18}H_{37}$

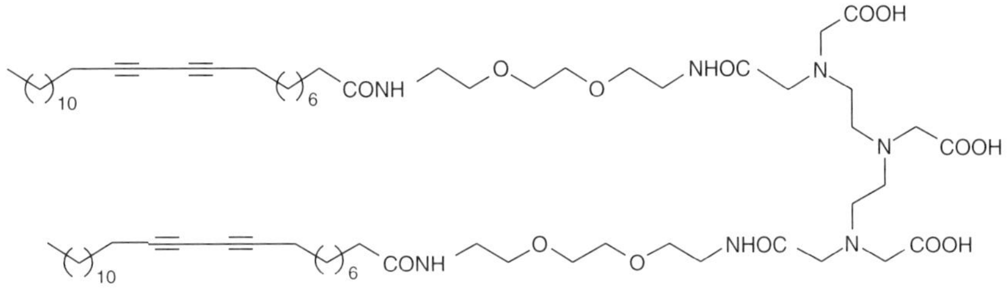

95

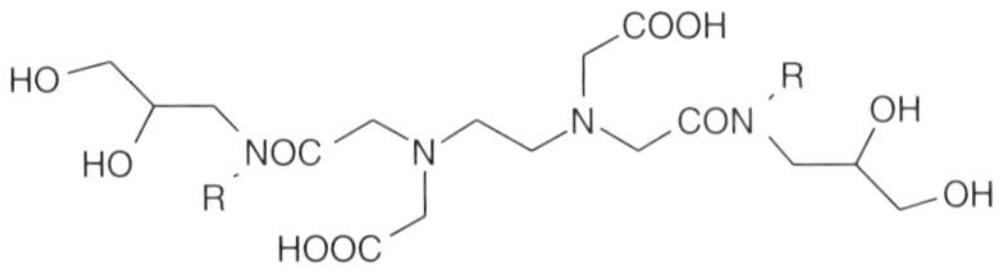

96 R = $C_{10}H_{21}$

[**Chart 3.11**]

[107]. The liposomes obtained in this way show an increased membrane rigidity, which is conferred by cross-linking. This feature may reduce the phospholipid exchange and the fusion with other liposomes or cell membranes, thus decreasing the rate of uptake by the reticuloendothelial system. Indeed, these paramagnetic polymerized liposomes have been studied as blood pool contrast agents because of their extended half-life in blood [109].

Ligand **96** (Chart 3.11) was obtained in the usual way by reaction of EDTA bisanhydride [46] and 3-decylamino-1,2-propandiol [110a], and the corresponding manganese complex has been obtained to study manganese-based liposomes [110].

In addition, lipophilic monoamides of DTPA and EDTA were synthesized for inglobation into liposomes. Reaction of dipalmitoylphosphatidylethanolamine with DTPA or EDTA bisanhydride gives compounds **97** [111, 112] and **98** [111] (Chart 3.12), respectively, while the synthesis of **99** takes advantage of the monoacid tetraester **69** (Chart 3.6), which is amidated with dioctadecylamine and EEDQ and then deprotected with trifluoroacetic acid [77b].

A series of DTPA diamides containing unsaturated aliphatic chains was synthesized by Cacheris *et al.* [113] by following the standard procedure. The Gd(III) complex of the most representative derivative, i.e. the bisoleylamide **100**

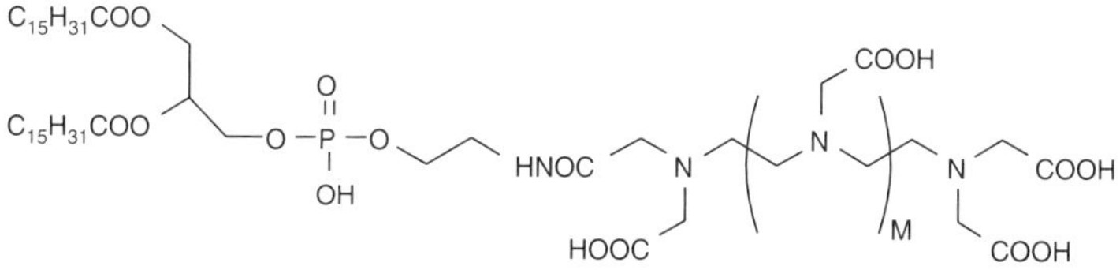

97 $n = 1$
98 $n = 0$

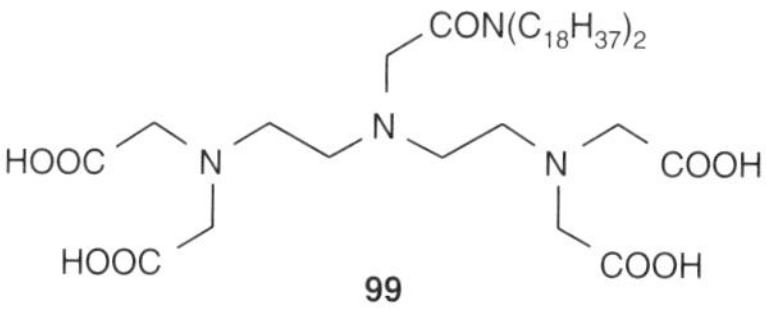

99

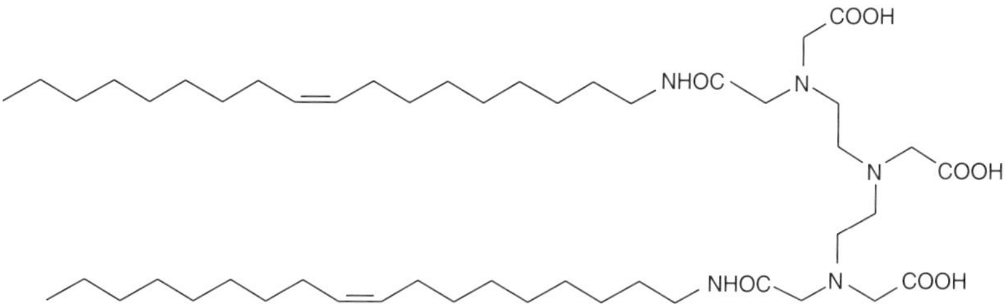

100

[Chart 3.12]

(Chart 3.12), was formulated as an extremely stable small particle emulsion and used as a blood pool MRI contrast agent [113].

Another application for Gd(III) complexes of compounds **92–94** is as components of mixed micelles [114]. Mixed micelles are formulated by using a lipophilic Gd(III) complex, an amphipatic compound (e.g. a phospholipid) and a non-ionic surfactant containing a polyoxyethylene chain (e.g. Synperonic® F 108). The presence of the polyoxyethylene chains prevents the recognition of the micelles by the reticuloendothelial system (Stealth® effect) [115] and results in a prolonged permanence of the micelles in blood. For this reason, several formulations of mixed micelles have been investigated for imaging the cardiovascular system [114].

7 PREPARATION OF THE COMPLEXES

A useful practical guide to the preparation of complexes of lanthanides on the laboratory scale appeared in the literature a few years ago [116]. The complexation issues for acyclic and cyclic ligands show several similarities and can be dealt with for both classes of ligands at once. Therefore, as sometimes it becomes a major problem for the latter compounds, it will be treated in depth in the following chapter of this book. Here we will only make some general comments.

Water-soluble Gd(III) complexes are usually prepared in water by using either $GdCl_3.6H_2O$ or Gd_2O_3 as the source of Gd(III) ion. Analogously, preparation of water-soluble Mn(II) complexes is performed by using salts such as $MnCl_2.4H_2O$ or $MnCO_3$ as the metal ion sources. When using a salt such as the chloride, the addition of a base (e.g. NaOH) is required to scavenge the hydrochloric acid which is produced by virtue of the complexation process. This means that the aqueous solution of the desired complex also contains a variable amount of a 'by-product' salt (e.g. NaCl) and desalting is needed. This apparently simple procedure is sometimes really tedious and time consuming. Use of classical ion-exchange resins is often prevented with ionic complexes and methodologies such as chromatography through non-functionalized resins, electrodialysis and nanofiltration have been developed to overcome this problem.

For the complexation of lipophilic ligands, which are insoluble in water, solvents such as pyridine [105], alcohols (e.g. methanol and ethanol) [105, 107] or even mixture of solvents (e.g. $CHCl_3/EtOH/H_2O$) [77b] have been used in association with the same Gd(III) ion sources previously described.

REFERENCES

1. Gries, H., Rosenberg, D. and Weinmann, H. J. *Ger. Offen. DE 3 129 906*, 1983; *Chem. Abstr.* 1983, **99**, 84766i.
2. Weinmann, H. J., Brasch, R. C., Press, W.-R. and Wesbey, G. E. *Am. J. Roentgenol.* 1984, **142**, 619.
3. Frost, A. E. *Nature* (*London*), 1956, **178**, 322.
4. Hart, J. R. In *Ullmann's Encyclopedia of Industrial Chemistry*, Vol. A 10, Gerhartz, W. (Ed.), VCH. Weinheim, 1987, pp. 95–100.
5. Samoilova, O. I. and Yashunskii, V. G. *USSR Patent 144 479*, 1962; *Chem. Abstr.* 1962, **57**, 12326d.
6. Moeller, T. and Thompson, L. C. *J. Inorg. Nucl. Chem.* 1962, **24**, 499.
7. Gries, H. and Miklautz, H. *Physiol. Chem. Phys. Med. NMR* 1984, **16**, 105.
8. Weinmann, H. J. in *Magnevist® Monograph*, Felix, R., Heshiki, A., Hosten, N. and Hricak, H. (Eds), Blackwell Scientific Publications, Oxford, 1994, pp. 5–14.
9. Uggeri, F., Aime, S., Anelli, P. L., Botta, M., Brocchetta, M., de Haën, C., Ermondi, G., Grandi, M. and Paoli, P. *Inorg. Chem.* 1995, **34**, 633.
10. Aime, S., Geninatti Crich S., Gianolio, E., Terreno E., Beltrami, A. and Uggeri, F. *Eur. J. Inorg. Chem.* 1998, 1283.
11. Keana, J. F. W. and Mann, J. S. *J. Org. Chem.* 1990, **55**, 2868.
12. Westerberg, D. A., Carney, P. L., Rogers, P. E., Kline, S. J. and Johnson, D. K. *J. Med. Chem.* 1989, **32**, 236.
13. Williams, M. A. and Rapoport, H. *J. Org. Chem.* 1993, **58**, 1151.

14. Anelli, P. L., Fedeli, F., Gazzotti, O., Lattuada, L., Lux, G. and Rebasti, F. *Bioconjugate Chem.* 1999, **10**, 137.
15. Wallace, R. A., Haar, J. P., Miller, D. B., Woulfe, S. R., Polta, J. A., Galen, K. P., Hynes, M. R. and Adzamli, K. *Magn. Reson. Med.* 1998, **40**, 733.
16. Anelli, P. L., Lolli, M., Fedeli, F. and Virtuani, M. *PCT Int. Appl. WO 98 05 626*, 1998; *Chem. Abstr.* 1998, **128**, 192928q.
17. (a) Maier, F.-K., Bauer, M., Krause, W., Speck, U., Schuhmann-Giampieri, G., Muhler, A., Balzer, T. and Press, W. R. PCT *Int. Appl. WO 96 16 929*, 1996; *Chem. Abstr.* 1996, **125**, 157019k; (b) Calabi, L., Maiocchi, A., Lolli, M. and Rebasti, F. *PCT Int. Appl. WO 98 05 625*, 1998; *Chem. Abstr.* 1998, **128**, 192927p; (c) Anelli, P. L., Beltrami, A., Uggeri, F. and Virtuani, M. *Eur. Pat. Appl. EP 822 180*, 1998; *Chem. Abstr.* 1998, **128**, 175395w.
18. Brechbiel, M. W., Gansow, O. A., Atcher, R. W., Schlom, J., Esteban, J., Simpson, D. E. and Colcher, D. *Inorg. Chem.* 1986, **25**, 2772.
19. (a) Schmitt-Willich, H., Platzek, J., Gries, H., Schuhmann-Giampieri, G., Vogler, H. and Weinmann, H. J. *Eur. Pat. Appl. EP 405 704*, 1991; *Chem. Abstr.* 1992, **116**, 194874v; (b) Schmitt-Willich, H., Brehm, M., Ewers, Ch. L. J., Michl, G., Müller-Fahrnow, A., Petrov, O., Platzek, J., Radüchel, B. and Sülzle, D. *Inorg. Chem.* 1999, **38**, 1134.
20. (a) Cummins, C. H., Rutter, E. W. and Fordyce, W. A. *Bioconjugate Chem.* 1991, **2**, 180; (b) Brechbiel, M. W. and Gansow, O. A. *J. Chem. Soc. Perkin Trans. 1* 1992, 1173.
21. Harre, M., Nickisch, K., Schulz, C. and Weinmann, H. *Tetrahedron Lett.* 1998, **39**, 2555.
22. Scott, D. M. and Lauffer, R. B. *PCT Int. Appl. WO 95 28 179*, 1995; *Chem. Abstr.* 1996, **124**, 111275s.
23. (a) Amedio, J. C., Bernard, P. J., Fountain, M. and Van Wagenen, G. *Synth. Commun.* 1999, **29**, 2377; (b) Sajiki, H., Ong, K. Y., Nadler, S. T., Wages, H. E. and McMurry, T. J. *Synth. Commun.* 1996, **26**, 2511.
24. Amedio, J. C., Bernard, P. J., Fountain, M. and Van Wagenen, G. *Synth. Commun.* 1998, **28**, 3895.
25. McMurry, T. J., Sajiki, H., Scott, D. M. and Lauffer, R. B. *PCT Int. Appl. WO 96 23 526*, 1996; *Chem. Abstr.* 1996, **125**, 230822 y.
26. Cox, J. P. L., Craig, A. S., Helps, I. M., Jankowski, K. J., Parker, D., Eaton, M. A. W., Millican, A. T., Millar, K., Beeley, N. R. A. and Boyce, B. A. *J. Chem. Soc. Perkin Trans. 1* 1990, 2567.
27. Sajiki, H. and Ong, K. Y. *Tetrahedron* 1996, **52**, 14507.
28. (a) Brechbiel, M. W. and Gansow, O. A. *Bioconjugate Chem.* 1991, **2**, 187; (b) Quadri, S. M. and Mohammadpour, H. *Bioorg. Med. Chem. Lett.* 1992, **2**, 1661; (c) Brechbiel, M. W., Gansow, O. A., Pippin, C. G., Rogers, R. D. and Planalp, R. P. *Inorg. Chem.* 1996, **35**, 6343.
29. (a) Williams, M. A. and Rapoport, H. *J. Org. Chem.* 1994, **59**, 3616; (b) Grote, C. W., Kim, D. J. and Rapoport, H. *J. Org. Chem.* 1995, **60**, 6987.
30. (a) Gries, H., Renneke, F.-J. and Weinmann, H.-J. *Ger. Offen. DE 3 621 025*, 1987; *Chem. Abstr.* 1989, **110**, 74830g; (b) Almen, T., Berg, A., Dugstad, H., Klaveness, J., Krautwurst, K. D. and Rongved, P. *PCT Int. Appl. WO 90 08 138*, 1990; *Chem. Abstr.* 1991, **114**, 101739a.
31. Caulfield, T. J., Guo, P., Illig, C. R., Kellar, K. E., Liversidge, E., Shen, J., Wellons, J., Ladd, D., Peltier, N. and Toner, J. L. *Bioorg. Med. Chem. Lett.* 1995, **5**, 1657.
32. Wang, Y.-M., Lee, C.-H., Liu, G.-C. and Sheu, R.-S. *J. Chem. Soc. Dalton Trans.* 1998, 4113.

33. Templeton, L. K., Templeton, D. H., Zalkin, A. and Ruben, H. W. *Acta Crystallogr. Sect. B* 1982, **38**, 2155.
34. Martell, A. E. and Smith, R. M. *Critical Stability Constants*, Vol. 1, Plenum, New York, 1974, p. 205.
35. Wedeking, P., Kumar, K. and Tweedle, M. F. *Magn. Reson. Imaging* 1992, **10**, 641.
36. (a) Brücher, E. and Sherry, D., Chapter 6 of this present book; (b) Caravan, P., Ellison, J. J., McMurry, T. J. and Lauffer, R. B. *Chem Rev.* 1999, **99**, 2293; (c) Bianchi, A., Valtancoli, B., Calabi, L., Losi, P., Maiocchi, A., Paleari, L., Corana, F. and Fontana, S. *Coord. Chem. Rev.* 2000, **204**, 309.
37. Ruloff, R., Arnold, K., Beyer, L., Dietze, F., Gründer, W., Wagner, M.and Hoyer, E. *Z. Anorg. Allg. Chem.* 1995, **621**, 807.
38. Miller, J. H. and Powell, J. E. *Inorg. Chem.* 1978, **17**, 774.
39. Chang, C. A., Brittain, H. G., Tesler, J. and Tweedle, M. F. *Inorg. Chem.* 1990, **29**, 4468.
40. Richards, S., Pedersen, B., Silverton, J. V. and Hoard, J. L. *Inorg. Chem.* 1964, **3**, 27.
41. (a) Brocchetta, M., Calabi, L., Palano, D., Paleari, L. and Uggeri, F. *PCT Int. Appl. WO 99 45 967*, 1999; *Chem. Abstr.* 1999, **131**, 237141h; (b) Brocchetta, M., Calabi, L., Fedeli, F., Manfredi, G., Paleari, L. and Uggeri, F. *PCT Int. Appl. WO 99 45 968*, 1999; *Chem. Abstr.* 1999, **131**, 225562h.
42. Geraldes, C. F. G. C., Brücher, E., Cortes, S., Koenig, S. H. and Sherry, A. D. *J. Chem. Soc. Dalton Trans.* 1992, 2517.
43. Hoyer, E., Wagler, D. and Anton, J. *Z. Phys. Chem.* 1969, **241**, 65.
44. Ruloff, R., Gelbrich, T., Sieler, J., Hoyer, E. and Beyer, L. *Z. Naturforsch. B.* 1997, **52**, 805.
45. Alpoim, M. C., Urbano, A. M., Geraldes, C. F. G. C. and Peters, J. A. *J. Chem. Soc. Dalton Trans.* 1992, 463, and references cited therein.
46. Eckelman, W. C., Karesh, S. M. and Reba, R. C. *J. Pharm. Sci.* 1975, **64**, 704.
47. Geraldes, C. F. G. C., Urbano, A. M., Alpoim, M. C., Sherry, A. D., Kuan, K.-T., Rajagopalan, R., Maton, F. and Muller, R. N. *Magn. Reson. Imaging* 1995, **13**, 401.
48. Quay, S. C. *PCT Int. Appl. WO 86 02 841*, 1986; *Chem. Abstr.* 1987, **106**, 46879d.
49. Konings, M. S., Dow, W. C., Love, D. B., Raymond, K. N., Quay, S. C. and Rocklage, S. M. *Inorg. Chem.* 1990, **29**, 1488.
50. Lammers, H., Maton, F., Pubanz, D., van Laren, M. W., van Bekkum, H., Merbach, A. E., Muller, R. N. and Peters, J. A. *Inorg. Chem.* 1997, **36**, 2527.
51. Geraldes, C. F. G. C., Delgado, R., Urbano, A. M., Costa, J., Jasanada, F. and Nepveu, F. *J. Chem. Soc. Dalton Trans.* 1995, 327.
52. Gèze, C., Mouro, C., Hindré, F., Le Plouzennec, M., Moinet, C., Rolland, R., Alderighi, L., Vacca, A. and Simonneaux, G. *Bull. Soc. Chim. Fr.* 1996, **133**, 267.
53. Aime, S., Botta, M., Dastrù, W., Fasano, M. and Panero, M. *Inorg. Chem.* 1993, **32**, 2068.
54. Aime, S., Fasano, M., Paoletti, S. and Terreno, E. *Gazz. Chim. Ital.* 1995, **125**, 125.
55. Wang, Y.-M., Cheng, T.-H., Liu, G.-C. and Sheu, R.-S. *J. Chem. Soc. Dalton Trans.* 1997, 833.
56. Wang, Y.-M., Lin, S.-T., Wang, Y.-J. and Sheu, R.-S. *Polyhedron* 1998, **17**, 2021.
57. Sherry, A. D., Cacheris, W. P. and Kuan, K.-T. *Magn. Reson. Med.* 1988, **8**, 180.
58. Periasamy, M., White, D., deLearie, L., Moore, D., Wallace, R., Lin, W., Dunn, J., Hirth, W., Cacheris, W., Pilcher, G., Galen, K., Hynes, M., Bosworth, M., Lin, H. and Adams, M. *Invest. Radiol.* 1991, **26** (Suppl. 1), S217.
59. (a) Best, T. A., Jin, J. Q., Kang, S.-I., Lin, L., Marinkovic, D., Shevlin, G. I., Tith, S. and Nunally, R. L. *Abstracts of Papers, Proceedings of the Society of Magnetic Resonance*, San Francisco, CA, Second Meeting, August 6–12, 1994, Vol. 1, 264; (b) Vexler, V. S. and White, D. L. *Abstracts of Papers, Proceedings of the Society of*

Magnetic Resonance, San Francisco, CA, Second Meeting, August 6–12, 1994, Vol. 2, 896.
60. Ladd, D. L., Hollister, R., Peng, X., Wei, D., Wu, G., Delecki, D., Snow, R. A., Toner, J. L., Kellar, K., Eck, J., Desai, V. C., Raymond, G., Kinter, L. B., Desser, T. S. and Rubin, D. L. *Bioconjugate Chem.* 1999, **10**, 361, and references cited therein.
61. Bligh, S. W. A., Chowdhury, H. M. S., McPartlin, M., Scowen, I. J. and Bulman, R. A. *Polyhedron* 1995, **14**, 567.
62. Geraldes, C. F. G. C., Urbano, A. M., Hoefnagel, M. A. and Peters, J. A. *Inorg. Chem.* 1993, **32**, 2426.
63. Paul-Roth, C. and Raymond, K. N. *Inorg. Chem.* 1995, **34**, 1408.
64. Cacheris, W. P., Quay, S. C. and Rocklage, S. M. *Magn. Reson. Imaging* 1990, **8**, 467.
65. Watson, A. D. *J. Alloys Compd.* 1994, **207/208**, 14.
66. Adzamli, K., Periasamy, M. P., Spiller, M. and Koenig, S. H. *Invest. Radiol.* 1999, **34**, 410, and references cited therein.
67. (a) Seri, S., Yamauchi, H., Azuma, M. and Arata, Y. *Eur. Pat. Appl. EP 451 824*, 1991; *Chem. Abstr.* 1992, **116**, 79512p; (b) Selvin, P. R., Jancarik, J., Li, M. and Hung, L.-W. *Inorg. Chem.* 1996, **35**, 700; (c) Anelli, P. L., Calabi, L., de Haën, C., Lattuada, L., Lorusso, V., Maiocchi, A., Morosini, P. and Uggeri, F. *Acta Radiol.* 1997, **38** (Suppl. 412), 125.
68. Anelli, P. L., Calabi, L., de Haën, C., Fedeli, F., Losi, P., Murru, M. and Uggeri, F. *Gazz. Chim. Ital.* 1996, **126**, 89.
69. Krejcarek, G. E. and Tucker, K. L. *Biochem. Biophys. Res. Commun.* 1977, **77**, 581.
70. Sieving, P. F., Watson, A. D. and Rocklage, S. M. *Bioconjugate Chem.* 1990, **1**, 65.
71. Anelli, P. L., Beltrami, A., Calabi, L., Lolli, M., Paleari, L., Uggeri, F. and Virtuani, M. *Abstracts of Papers, XXth International Symposium on Macrocyclic Chemistry*, Jerusalem, Israel, July 2–7, 1995.
72. Gries, H., Radüchel, B., Weinmann, H. J., Mützel, W. and Speck, U. *Ger. Offen. DE 3 633 245*, 1988; *Chem. Abstr.* 1989, **110**, 192267b.
73. Guilmette, R. A. G., Parks, J. E. and Lindenbaum, A. *J. Pharm. Sci.* 1979, **68**, 194.
74. Adzamli, I. K., Gries, H., Johnson, D. and Blau, M. *J. Med. Chem.* 1989, **32**, 139.
75. Platzek, J., Mareski, P., Niedballa, U. and Radüchel, B. *Ger. Offen. DE 19 601 060*, 1997; *Chem. Abstr.* 1997, **127**, 135556x.
76. Arano, Y., Uezono, T., Akizawa, H., Ono, M., Wakisaka, K., Nakayama, M., Sakahara, H., Konishi, J. and Yokoyama, A. *J. Med. Chem.* 1996, **39**, 3451.
77. (a) Anelli, P. L., de Haën, C., Lattuada, L., Morosini, P. and Uggeri, F. *PCT Int. Appl. WO 95 32 741*, 1995; *Chem. Abstr.* 1996, **124**, 192411h; (b) Platzek, J., Niedballa, U., Radüchel, B. Mareski, P., Weinmann, H.-J., Mühler, A. and Misselwitz, B. *PCT Int. Appl. WO 96 26 182*, 1996; *Chem. Abstr.* 1996, **125**, 275257j.
78. Platzek, J., Niedballa, U. and Radüchel, B. *US Patent 5 514 810*, 1996; *Chem. Abstr.* 1996, **125**, 87212s.
79. Geigy, J. R. *Br. Patent 695 346*, 1953; *Chem. Abstr.* 1954, **48**, 11486i.
80. (a) Aime, S., Botta, M., Nonnato, A., Terreno, E., Anelli, P. L. and Uggeri, F. *J. Alloys Compd.* 1995, **225**, 274; (b) Aime, S., Barge, A., Borel, A., Botta, M., Chemerisov, S., Merbach, A. E., Müller, U. and Pubanz, D. *Inorg. Chem.* 1997, **36**, 5104.
81. Kline, S. J., Betebenner, D. A. and Johnson, D. K. *Bioconjugate Chem.* 1991, **2**, 26.
82. (a) Kohl, B., Hummel, R. P., Lodemann, K. P., Beller, K. D. and Boss, H. *PCT Int. Appl. WO 94 01 406*, 1994; *Chem. Abstr.* 1994, **120**, 244695t; (b) Schneider, G., Urbschat, K., Sperber, H., Otto, J., Uder, M. and Kramann, B. *Abstracts of Papers, The European Society for Magnetic Resonance in Medicine and Biology*,

Vienna, Austria, April 20–24, 1994, Abstract 409; (c) Urbschat, K., Otto, J., Defreyne, L., Kramann, B. and Schneider, G. *Abstracts of Papers, The European Society for Magnetic Resonance in Medicine and Biology*, Vienna, Austria, April 20–24, 1994, Abstract 106.

83. (a) Ma, R., Murase, I. and Martell, A. E. *Inorg. Chim. Acta* 1994, **223**, 109; (b) Hancock, R. D., Cukrowski, I., Cukrowska, E., Hosken, G. D., Iccharam, V., Brechbiel, M. W. and Gansow, O. A. *J. Chem. Soc. Dalton Trans.* 1994, 2679.
84. Sun, Y., Martell, A. E., Reibenspies, J. H. and Welch, M. J. *Tetrahedron* 1991, **47**, 357.
85. Wiegers, C. B., Welch, M. J., Sharp, T. L., Brown, J. J., Perman, W. H., Sun, Y., Motekaitis, R. J. and Martell, A. E. *Magn. Reson. Imaging* 1992, **10**, 903.
86. Lauffer, R. B., Greif, W. L., Stark, D. D., Vincent, A. C., Saini, S., Wedeen, V. J. and Brady, T. J. *J. Comput. Assist. Tomogr.* 1985, **9**, 431.
87. Lauffer, R. B., Vincent, A. C., Padmanabhan, S. and Meade, T. J. *J. Am. Chem. Soc.* 1987, **109**, 2216.
88. Dexter, M. and Cranston, R. I. *US Patent 2 824 128*, 1958; *Chem. Abstr.* 1959, **53**, 6158c.
89. L'Eplattenier, F., Murase, I. and Martell, A. E. *J. Am. Chem. Soc.* 1967, **89**, 837.
90. Taliaferro, C. H., Motekaitis, R. J. and Martell, A. E. *Inorg. Chem.* 1984, **23**, 1188.
91. Rocklage, S. M., Sheffer, S. H., Cacheris, W. P., Quay, S. C., Hahn, F. E. and Raymond, K. N. *Inorg. Chem.* 1988, **27**, 3530.
92. Rocklage, S. M., Cacheris, W. P., Quay, S. C., Hahn, F. E. and Raymond, K. N. *Inorg. Chem.* 1989, **28**, 477.
93. Wang, Y.-M., Wang, Y.-J. and Wu, Y.-L. *Polyhedron* 1998, **18**, 109.
94. Hynes, M. R., Galen, K. P., Kuan, K.-T., Wallace, R. A. and Moore, D. A. *Abstracts of Papers, Proceedings of the Society of Magnetic Resonance*, San Francisco, CA, Second Meeting, August 6–12, 1994, Vol. 2, 921.
95. Bligh, S. W. A., Harding, C. T., McEwen, A. B., Sadler, P. J., Kelly, J. D. and Marriot, J. A. *Polyhedron* 1994, **13**, 1937.
96. Xu, J., Franklin, S. J., Whisenhunt, D. W. and Raymond, K. N. *J. Am. Chem. Soc.* 1995, **117**, 7245.
97. Kocian, O., Chiu, K. W., Demeure, R., Gallez, B., Jones, C. J. and Thornback, J. R. *J. Chem. Soc. Perkin Trans. 1* 1994, 527, and references cited therein.
98. Claire, P. P. K., Jones, C. J., Chiu, K. W., Thornback, J. R. and McPartlin, M. *Polyhedron* 1992, **11**, 499.
99. Dunn, T. J., Moore, D. A., Periasamy, M. P., Rogic, M. M., Wallace, R. A., White, D. H. and Woulfe, S. R. *PCT Int. Appl. WO 95 01 124*, 1995; *Chem. Abstr.* 1995, **123**, 137673z.
100. Peng, W.-J. and Aguilar, D. A. *PCT Int. Appl. WO 98 22 148*, 1998; *Chem. Abstr.* 1998, **129**, 35711s.
101. Muetterties, K. A., Hoener, B.-A., Engelstad, B. L., Tongol, J. M., Wikstrom, M. G., Wang, S.-C., Eason, R. G., Moseley, M. E. and White, D. L. *Magn. Reson. Med.* 1991, **22**, 88, and references cited therein.
102. Maisano, F., Gozzini, L. and de Haën, C. *Bioconjugate Chem.* 1992, **3**, 212.
103. Lasic, D. *Am. Sci.* 1992, **80**, 20.
104. (a) Tilcock, C., Unger, E., Cullins, P. and MacDougall P. *Radiology* 1989, **171**, 77; (b) Kabalka, G., Buonocore, E., Hubner, K., Moss, T., Norley, N. and Huang, L. *Radiology , 1987,* **163***, 255.*
105. Kabalka, G. W., Davis, M. A., Moss, T. H., Buonocore, E., Hubner, K., Holmberg, E., Maruyama, K. and Huang, L. *Magn. Reson. Med.* 1991, **19**, 406.
106. Jasanada, F. and Nepveu, F. *Tetrahedron Lett.* 1992, **33**, 5745.

107. Storrs, R. W., Tropper, F. D., Li, H. Y., Song, C. K., Kuniyoshi, J. K., Sipkins, D. A., Li, K. C. P. and Bednarski, M. D. *J. Am. Chem. Soc.* 1995, **117**, 7301.
108. Ringsdorf, H., Schlarb, B. and Venzmer, J. *Angew. Chem. Int. Ed. Engl.* 1988, **27**, 113.
109. Storrs, R. W., Tropper, F. D., Li, H. Y., Song, C. K., Sipkins, D. A., Kuniyoshi, J. K., Bednarski, M. D., Strauss, H. W. and Li, K. C. P. *J. Magn. Reson. Imaging* 1995, **5**, 719.
110. (a) Unger, E., Fritz, T., Shen, D. K. and Wu, G. *Invest. Radiol.* 1993, **28**, 933; (b) Unger, E., Fritz, T., Wu, G., Shen, D., Kulik, B., New, T., Crowell, M. and Wilke, N. *J. Liposome Res.* 1994, **4**, 811, and references cited therein.
111. Urizzi, P., Souchard, J.-P. and Nepveu, F. *Tetrahedron Lett.* 1996, **37**, 4685.
112. Grant, C. W. M., Karlik, S. and Florio, E. *Magn. Reson. Med.* 1989, **11**, 236.
113. Cacheris, W. P., Grabiak, R. C., Lee, A. C., Richard, T. J., Goodin, T. H. and Kaufman, R. J. *Abstracts of Papers, 4th Special Topic Seminar of the European Magnetic Resonance Forum*, Santiago de Compostela, Spain, September 28–30, 1994, 37.
114. Tournier, H., Lamy, B. and Hyacinthe, R. *PCT Int. Appl. WO 97 00 087*, 1997; *Chem. Abstr.* 1997, **126**, 168810h.
115. Lasic, D. D. *Angew. Chem. Int. Ed. Engl.* 1994, **33**, 1685.
116. Desreux, J. F. in *Lanthanide Probes in Life, Chemical and Earth Sciences. Theory and Practice*, Bünzli, J.-C. and Choppin, G. R. (Eds), Elsevier, Amsterdam, 1989, pp. 43–64.

4 Synthesis of MRI Contrast Agents II. Macrocyclic Ligands

VINCENT JACQUES and JEAN-FRANÇOIS DESREUX
University of Liège, Belgium

1 MACROCYCLIC LIGANDS: BENEFITS AND DRAWBACKS

Since the advent of Gd(III) chelates as MRI contrast agents, there has been a continuous search for improving their efficacy. As Gd(III) is characterized by a high toxicity, it is of the utmost importance to use Gd(III) complexes that are both thermodynamically stable and kinetically inert. Both of these properties are crucial but kinetic inertness is essential: the chelate should not decompose while in the body of the patient.

Macrocyclic ligands are particularly well suited to form such inert complexes as has been amply demonstrated by studies of the Gd(III) chelate of DOTA which is currently used as an MRI contrast agent under the brand name Dotarem (Guerbet, France). DOTA or 1,4,7,10-tetra(carboxymethyl)-1,4,7,10-tetraazacyclododecane, **1**, became the parent molecule of an ever-growing family of macrocyclic lanthanide(III) chelating agents. This ligand forms lanthanide complexes that are exceedingly stable and inert at physiological pH and in blood serum. This stability and inertness has been ascribed to the macrocyclic structure of the ligand. In the complex, the macrocyclic ring adopts the very stable [3.3.3.3] square conformation (Figure 4.1) in which all of

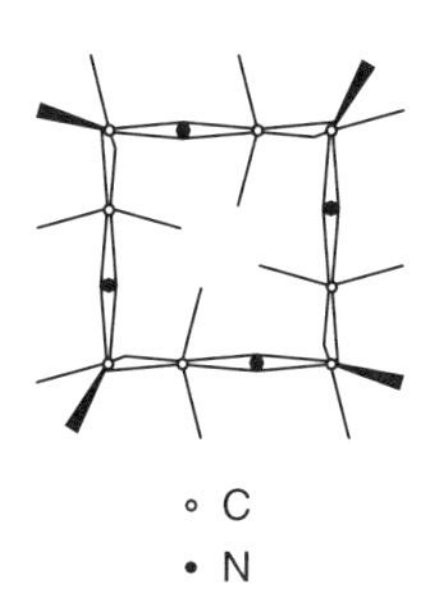

Figure 4.1 The [3.3.3.3] geometry of the 1,4,7,10-tetraazacyclododecane ring.

The Chemistry of Contrast Agents in Medical Magnetic Resonance Imaging
Edited by A. E. Merbach and É. Tóth.

the methylene groups are fully staggered and all of the nitrogen atoms have their lone electron pair pointing towards the metal ion. This arrangement hinders the entry of a proton and slows down the dissociation of the chelate [1, 2].

It thus comes as no surprise that much effort has been devoted in the last few years into improving the already promising characteristics of DOTA. Numerous new molecules have thus been synthesized by substituting chemical groups on the macrocyclic ring and/or on the pendant arms. Bifunctional chelating agents have also been designed in order to broaden the fields of application of the DOTA derivatives. All of these efforts have in common the development of new synthetic methodologies that would be adapted to macrocyclic polyamines. In this chapter, we will give an overview of the new organic strategies that have been developed and will also cover some of the old work which is pertinent to the synthesis of such molecules.

One should stress here that one of the biggest drawbacks of macrocyclic ligands lies in their synthesis. Even the preparation of the symmetrical simple 1,4,7,10-tetraazacyclododecane (or cyclen) requires several steps and has long been a very tedious process, involving numerous separations and crystallizations of intermediates. These synthetic problems have nonetheless been tackled recently and several new shorter syntheses of cyclen have appeared that could certainly be applied to other macrocyclic derivatives. Another major problem lies in the preorganization of the ligand. It is greatly desirable to have a highly preorganized molecule, as it will ultimately result in an increased stability of the chelate. However, preorganization is also a drawback when it comes to complexing a metal ion, as the kinetics of complexation can become exceedingly slow. This problem is not as crucial for MRI contrast agents as it is for radiopharmaceuticals where the half-life of the metal center is sometimes very short.

Today, DOTA and its derivatives are still the only family of ligands that can account for real-world applications: Dotarem (Guerbet, France), which is the gadolinium complex of DOTA, (**1**), ProHance (Bracco, Italy), the gadolinium chelate of HPDO3A, (**2**), and Gadobutrol (Schering, Germany), the gadolinium complex of DO3AB, (**3**), (Figure 4.2) are the main contenders. This chapter will

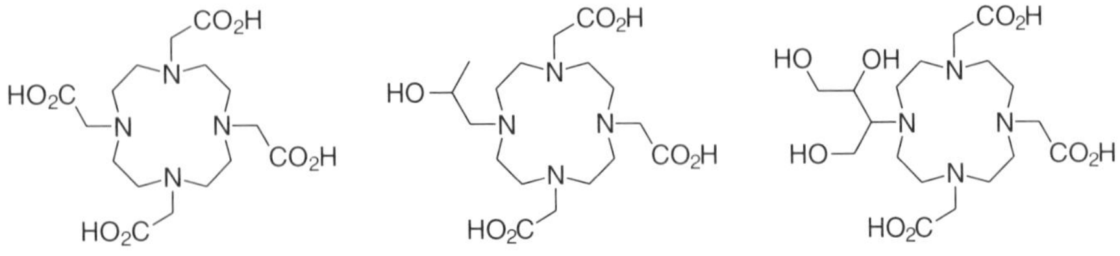

CO_2H HO_2C N N N N CO_2H HO_2C

DOTA **1**

CO_2H HO N N N N CO_2H HO_2C

HPDO3A **2**

CO_2H HO OH HO N N N N CO_2H HO_2C

DO3AB **3**

Figure 4.2 The structures of DOTA, HPDO3A and DO3AB (see text for details).

thus be mostly devoted to a description of the synthesis of DOTA and its many derivatives, although we will also provide some details of other promising macrocyclic compounds.

2 A CHALLENGE: AN EASY SYNTHESIS OF 1,4,7,10-TETRAAZACYCLODODECANE (OR CYCLEN)

1,4,7,10-Tetraazacyclododecane is by far the most important building block for the preparation of macrocyclic lanthanide (III) complexing agents for use in MRI. The development of a cost-effective, easy and efficient synthesis of this macrocyclic tetraamine is thus of the utmost importance. Several new synthetic methods have been developed over the years.

2.1 FIRST SYNTHESIS OF A MACROCYCLIC POLYAMINE

In 1957, Stetter and Marx [3] reported on the first general procedure for the synthesis of macrocyclic polyamines. The preparation of cyclen and other polyamines is based on the reaction of a sodium bis-tosylamide with a dihalogen derivative (Scheme 4.1) in highly diluted solutions of the reactants (0.001 mol/dm^3). This technique is thus not very useful for the synthesis of large quantities of macrocyclic polyamines.

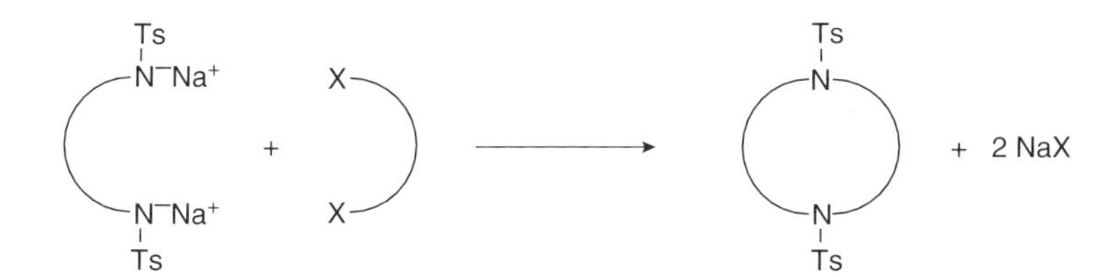

Scheme 4.1 General scheme for the Stetter and Marx (X = Br) [3] and Richman and Atkins (X = p-toluenesulfonyloxy) [4] macrocyclic polyamine synthesis.

2.2 RICHMAN AND ATKINS MACROCYCLIC POLYAMINE SYNTHESIS

In 1974, Richman and Atkins [4], building on the work of Stetter and Marx, proposed a new procedure for the specific synthesis of macrocyclic polyamines (see Scheme 4.1). This method has become, in actual fact, the procedure that has been used for the last twenty years or so for the preparation of 1,4,7,10-tetraazacyclododecane (cyclen).

The Richman and Atkins procedure is a very general method that does not require high-dilution conditions – hence its use for the industrial production of large amounts of cyclen. For example, when the sodium bis-tosylamide of 1,4,7-tris(*p*-toluenesulfonyl)-1,4,7-triazaheptane is reacted with 1,4,7-tris(*p*-toluene-sulfonyl)-4-aza-1,7-dioxaheptane in 0.07 mol/l concentration in *N,N*-dimethyl-formamide, 1,4,7,10-tetra(*p*-toluenesulfonyl)-1,4,7,10-tetraazacyclododecane is obtained in 90 % yield.

The yields strongly depend on the nature of the nucleofuge X : *p*-toluene-sulfonates and methanesulfonates give higher yields than halogen derivatives. Two effects can account for this observation: first, sulfonates are better nucleo-fuges than halides and thus favor cyclization over polymerization and secondly, the internal entropy loss is surprisingly low when the macrocyclization takes place [5]. This effect is particularly important when *p*-toluenesulfonates are used, as these bulky substituents hinder free rotation in the reactants and thus decrease their internal entropy while simultaneously promoting a favorable orientation of the two molecules for the cyclization. No template effect of the sodium ion has been observed. The yields of macrocycles are still high when tetraethylammonium bis-tosylamides are used [4].

2.3 MODIFICATION OF THE RICHMAN AND ATKINS PROCEDURE

A modification of the Richman and Atkins procedure, based on original work by Kellogg and co-workers [6], has been proposed by Chavez and Sherry [7] who reacted bis(*p*-toluenesulfonamides) with dibromides in *N,N*-dimethylform-amide in the presence of an alkali metal carbonate as a base. This method provides an interesting alternative to the original synthetic scheme because it does not require the preparation of sodium bis-(*p*-toluenesulfonamides) and may be used for the synthesis of macrocycles containing 9 to 15 atoms. Yields of up to 62 % in 1,4,7,10-tetra(*p*-toluenesulfonyl)cyclen, (**6**), may be obtained by reacting 1,4,7,10-tetra(*p*-toluenesulfonyl)-1,4,7,10-tetraazadecane, (**5**), and 1,2-dibromoethane, (**4**), in *N,N*-dimethylformamide in the presence of potassium carbonate at 30 °C for 24 h (Scheme 4.2).

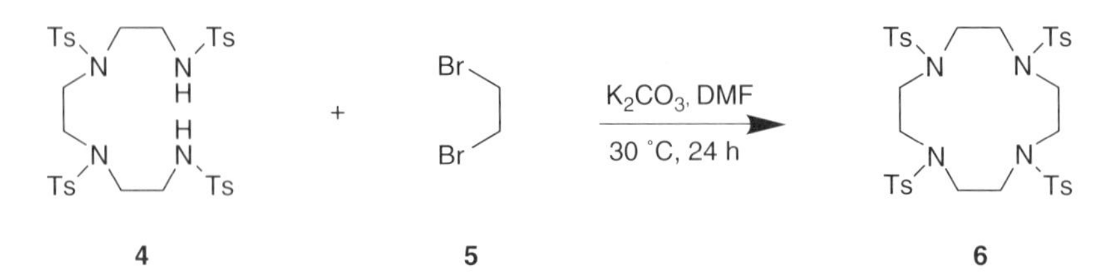

Scheme 4.2 Synthesis of tetra(*p*-toluenesulfonyl)cyclen according to Chavez and Sherry [7].

Cesium carbonate, which is an even better base for the deprotonation of the bis-(*p*-toluenesulfonyl)amide, may be substituted for potassium carbonate.

2.4 'ATOM-ECONOMIC' METHODS

The Richman and Atkins synthesis, as well as its variant present one particularly important disadvantage. It is not 'atom-economic' [8], i.e. it requires four *p*-toluenesulfonyl protecting groups which are to be removed in the final step of the synthesis. Furthermore, this procedure, while giving high yields of macrocyclic polyamine, is quite tedious and requires large amounts of dry solvent as well as long reaction times. Owing to the strategic importance of 1,4,7,10-tetraazacyclododecane, chemists have thus been looking for new and clever ways to synthesize this key compound, and several procedures have been reported in the last few years, as described in the following paragraphs.

2.4.1 The Bis-Imidazolidine Pathway [9]

As can be seen in Scheme 4.3, this synthesis involves the reaction of 1,4,7,10-tetraazadecane (**7**) with *N*,*N*-dimethylformamide dimethylacetal to form a bis-imidazolidine (**8**) which is then reacted with dibromoethane to give the tetracyclic ionic intermediate (**9**). This compound is refluxed in sodium hydroxide to give the pure macrocycle (**10**). The authors, Athey and Kiefer, claim that high yields of the pure macrocyclic tetraamine are obtained when the reaction conditions are right.

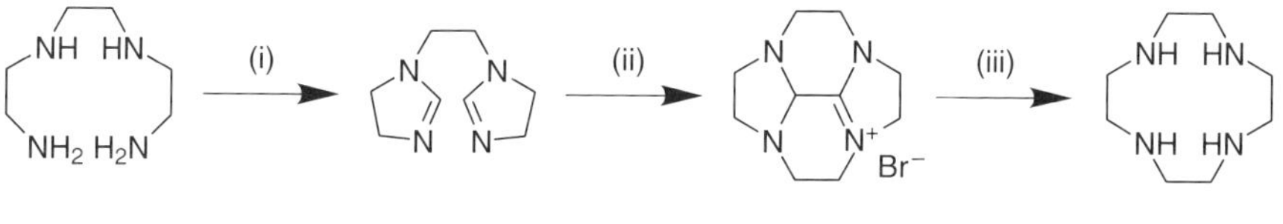

NH HN NH2 H2N (i) N N N N (ii) N N N N+ Br⁻ (iii) NH HN NH HN

7 **8** **9** **10**

Scheme 4.3 Synthesis of 1,4,7,10–tetraazacyclododecane: (i) dimethylformamide dimethylacetal, toluene, reflux; (ii) 1,2-dibromoethane, *N*,*N*-dimethylformamide, potassium carbonate, 100 °C, 30 min; (iii) aqueous sodium hydroxide, reflux [9].

The reaction between the bis-imidazolidine **8** and 1,2-dibromoethane has been tested in several solvents. The highest yields were obtained in *N*,*N*-dimethylformamide and acetonitrile. However, if the reaction mixture is left for too long at high temperature then the ionic tetracyclic intermediate **9** rearranges to form the neutral compound **11**, as shown in Scheme 4.4. This rearranged product is unfortunately less amenable to deprotection, i.e. if

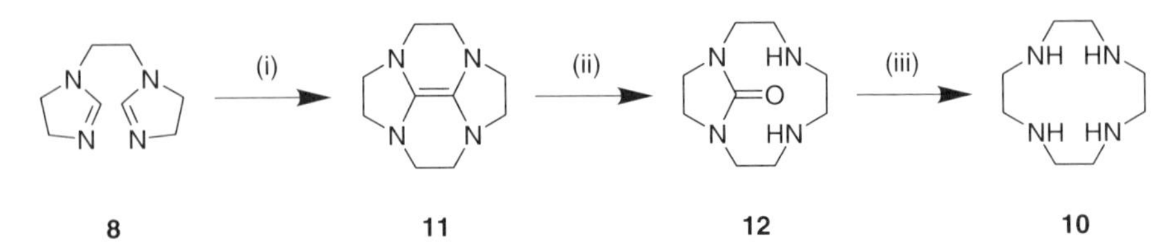

Scheme 4.4 The rearranged intermediate and its deprotection: (i) 1,2-dibromoethane, *N*,*N*-dimethylformamide, potassium carbonate, 100 °C, 12–14 h; (ii) aqueous sodium hydroxide, reflux; (iii), aqueous sodium hydroxide, 200 °C, Parr bomb [9].

refluxed in aqueous sodium hydroxide, it gives a very stable urea (**12**) that can only be deprotected by reaction with the same base in a Parr bomb at 200 °C (Scheme 4.4).

This method has also been used to produce a 2-hydroxymethyl substituted cyclen by using bromoepichlorhydrine instead of dibromoethane [9]. Furthermore, the bis-imidazolidine may be synthesized from imidazolidine and dibromoethane.

2.4.2 The Bridged Bis-Imidazolidine Synthesis [10, 11]

Weisman and Reed [10] developed an original and short (two-step) synthetic procedure for the production of cyclen. Dithiooxamide (**13**) and 1,4,7,10-tetraazadecane (**7**) are refluxed in ethanol to give 81 % of the tricyclic intermediate (**14**). Pure cyclen (**10**) is then obtained in 83 % yield after sublimation of the crude product resulting from a diisobutylaluminium hydride (DIBALH) reduction (Scheme 4.5). From the start, this technique has been designed for small quantities only (several grams at the most). It is particularly well suited for an academic laboratory as it gives good yields of a very pure cyclen. In our experience, the Athey and Kiefer method [9], while cheaper, does not give such an easy access to this key molecule when performed on small amounts of reactants.

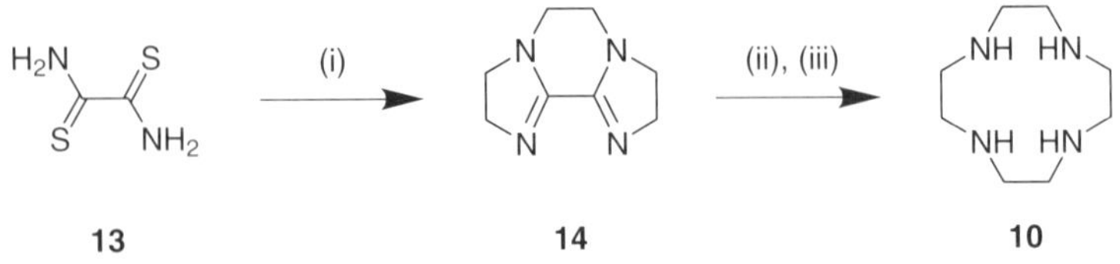

Scheme 4.5 The Weismann and Reed cyclen synthesis; (i) 1,4,7,10-tetraazadecane, ethanol, reflux, (ii) DIBALH, toluene; (iii) sodium fluoride, water [10, 11].

2.4.3 The Bis-Aminal Route [12–15]

Another short synthesis of cyclen (Scheme 4.6) has recently been proposed and carefully studied by several groups. The crucial step here consists in the preparation of the bis-aminal species **15** by condensation of 1,4,7,10-tetraazadecane (**7**) with either glyoxal [12, 13, 15] or butanedione [14]. The deprotection of the tetraazatctramacrocyclic intermediate **16** can be achieved with various reactants, i.e. hydroxylamine hydrochloride, hydrazine hydrate, bromine, potassium permanganate, etc., for the glyoxal adduct, and with hydrochloric acid for the butanedione derivative.

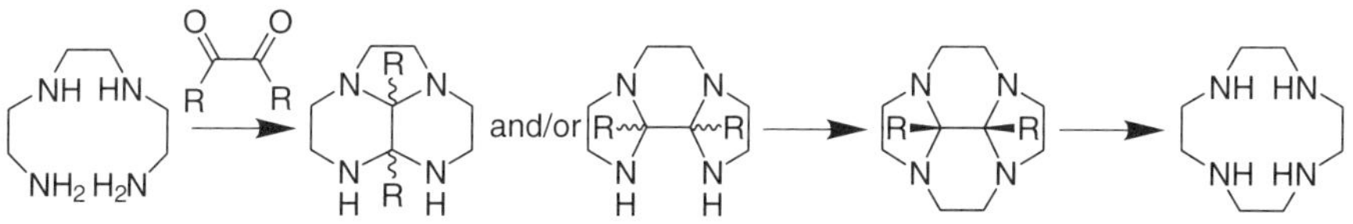

Scheme 4.6 Synthesis of cyclen by means of a bis-aminal intermediate (R = H or CH_3) [12–15].

Hervé *et al.* [14, 15] studied the formation and the deprotection of the bis-aminal intermediate. They were able to reach the following conclusions:

(i) The glyoxal-based bis-aminal species is obtained as a mixture of the four possible stereoisomers (*gem-cis*, *gem-trans*, *vic-cis* and *vic-trans*, see Figure 4.3). The relative proportions of these change with time and temperature and this isomerization is catalyzed by water. The *vic-trans* compound rearranges into the *gem-cis* isomer and the *vic-cis* compound into the *gem-trans* isomer, thus leading to products characterized by the maximum number of six-membered rings.

(ii) The bis-aminal species derived from butanedione, on the other hand, is isolated as the sole *gem-cis* isomer.

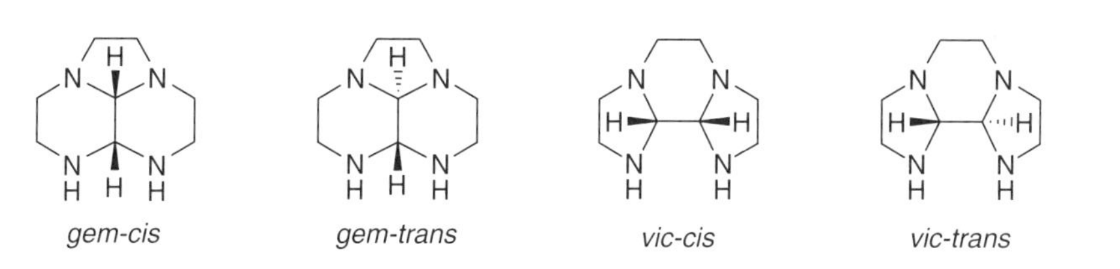

Figure 4.3 The four possible stereoisomers of the bis-aminal intermediate (R = H or CH_3)

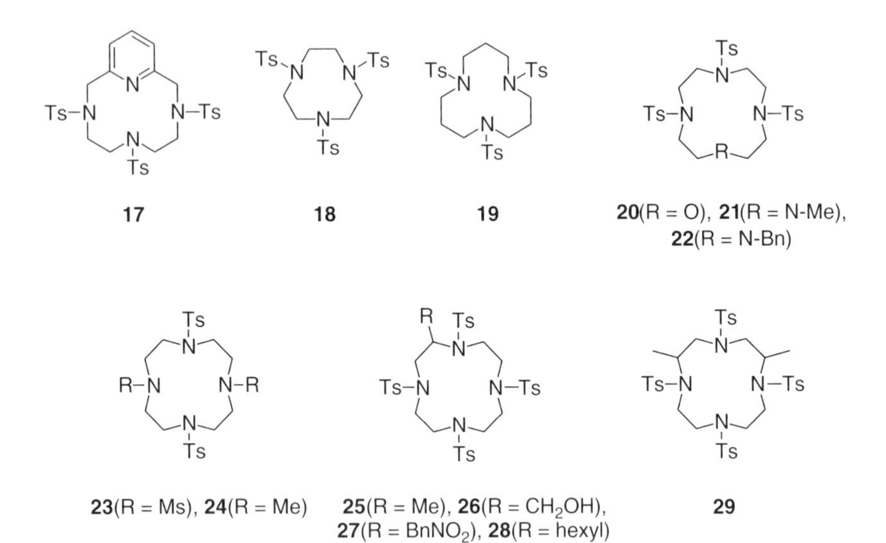

Figure 4.4 Polyaza macrocycles obtained by the Richman and Atkins method.

(iii) The *cis*-isomers are more favorable for cyclization. Indeed, dried recrystallized bis-aminals react in dry acetonitrile with 1,2-dibromoethane to give mostly polymers. An isomerization of the stereoisomers of the glyoxal adduct is thus necessary prior to its reaction with 1,2-dibromoethane. This can be achieved by prolonged heating of the bis-aminal intermediate in acetonitrile containing traces of water.

(iv) The deprotection of the final tetracyclic bis-aminal **16** is best achieved by boiling in neat hydrazine hydrate. The resulting polyazine precipitates in chloroform and is thus quite easily eliminated.

3 SYNTHESIS OF C-SUBSTITUTED MACROCYCLES

The introduction of a substituent on a carbon atom of the macrocyclic ring has been the subject of several publications. Indeed, such a modification of the DOTA structure could result in various new properties, e.g. an increased rigidity of the ring and thus an increase in the stability of its chelates or a change in lipophilicity, which would result in a different biodistribution. Furthermore, C-substituted macrocycles bearing a new functional group (bifunctional chelating agents) can be grafted to macromolecules or can be self-assembled around a central metal ion.

Several procedures have been developed. Some are carbon copies of the methods described in the preceding section, while others are entirely new.

3.1 RICHMAN AND ATKINS APPROACH

The Richman and Atkins cyclization has been one of the most successful methods in the production of large amounts of various macrocyclic polyaza rings, whether derivatives of cyclen or not (Figure 4.4).

The reaction between reactants of different sizes has been explored by Stetter *et al.* who reported the synthesis of several macrocyclic tetra-and pentaamines [16]. A pyridine-containing macrocycle, **17**, was also obtained. This compound has been reinvestigated by Aime *et al.* [17] (scheme 4.7), as well as by Takalo and Kankare [18]. Derivatives **18** and **19** were first synthesized by Sherry [19]. Tweedle and co-workers have reported the synthesis of several new polyaza rings such as **20**, **21** and **22** [5, 20, 21]. The latter (**22**) has also been prepared in a similar way by Gries *et al.* [22]. *N*,*N*″-disubstituted tetraazacyclododecanes have also been prepared: 1,7-bis(methanesulfonyl)-4,10-bis(*p*-toluenesulfonyl)-1,4,7,10-tetraazacyclododecane (**23**) by Dumont *et al.* [23] and the 1,7-dimethyl derivative (**24**) by Ciampolini and co-workers [24, 25]. Schaeffer *et al.* [26, 27] have used cesium carbonate as a base in *N*,*N*-dimethylformamide to synthesize **25** and **26** by condensation, respectively, of either *N*,*N*′-bis(*p*-toluenesulfonyl)-1-methyl-1,2-diaminoethane or *N*,*N*′-bis(*p*-toluenesulfonyl)-1-hydroxymethyl-1,2-diaminoethane with 1,4,7,10-tetra(*p*-toluenesulfonyl)-4,7-diaza-1,10-dioxadecane. The 2,6-dimethyl derivative (**29**) was synthesized by using the simple Richman and Atkins procedure, while the 2-hexyl macrocycle (**28**) required a two-phase mixture of toluene and aqueous sodium hydroxide with tetrabutylammonium hydrogensulfate as a phase-transfer agent.

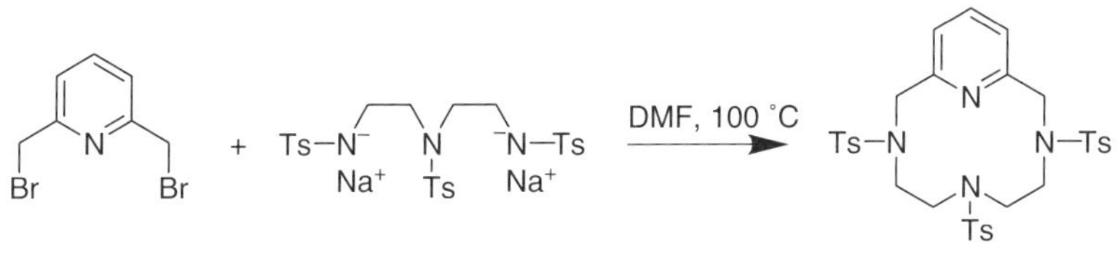

Scheme 4.7 Synthesis of a pyridine-substituted tetraazamacrocycle [16–18].

An interesting approach has been introduced by Meares and co-workers [28, 29] who reported the synthesis of the tetra (*p*-toluenesulfonamide) of 2-(*p*-nitrobenzyl)-1,4,7,10-tetraazacyclododecane, (**27**), by an intramolecular-ring-closure reaction akin to the Chavez and Sherry approach (Scheme 4.8a). This substituted macrocycle has been further transformed into the very promising bifunctional chelating agent bromoacetamidobenzyl-DOTA (BAD) (Figure 4.5).

A number of other synthetic procedures leading to this ligand have now been reported in the literature. It should be noted that inserting certain substituents

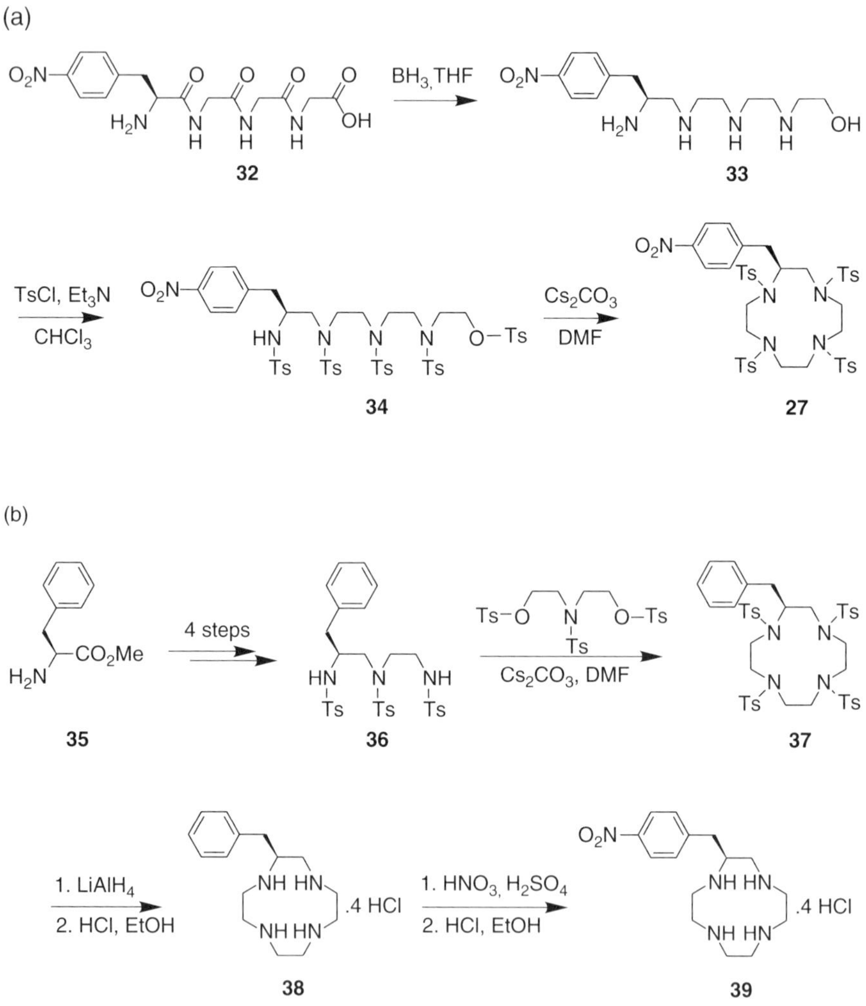

(a)

O_2N O O O O BH_3, THF O_2N
H_2N N H N H N H OH H_2N N H N H N H OH
32 **33**

TsCl, Et_3N / $CHCl_3$ O_2N HN Ts N Ts N Ts N Ts O−Ts Cs_2CO_3 / DMF O_2N Ts N N Ts N Ts N Ts
34 **27**

(b)

H_2N CO_2Me 4 steps HN Ts N Ts NH Ts Ts−O N Ts O−Ts / Cs_2CO_3, DMF Ts N N Ts N Ts N Ts
35 **36** **37**

1. $LiAlH_4$ / 2. HCl, EtOH NH HN NH HN .4 HCl 1. HNO_3, H_2SO_4 / 2. HCl, EtOH O_2N NH HN NH HN .4 HCl
38 **39**

Scheme 4.8 (a) Synthesis of 1,4,7,10-tetra(*p*-toluenesulfonyl)-2-(*p*-nitrobenzyl)-1,4,7,10-tetraazacyclododecane, according to Renn and Meares [29]. (b) Synthesis of the same compound via a Richman and Atkins macrocyclization, as proposed by Garrity *et al.* [31].

can hinder the cyclization reaction. For example, while Ansari *et al.* claimed the successful synthesis of this same compound by the Chavez and Sherry procedure [30], Garrity *et al.* reported difficulties in getting the same results. This led them to propose another method in which the cyclization was carried out with a benzyl-substituted reactant (**36**), with the nitration being performed in a later step [31] (Scheme 4.8b).

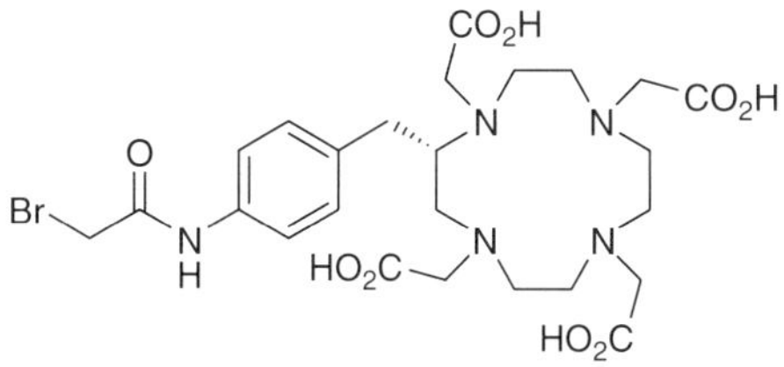

Figure 4.5 A promising bifunctional ligand, bromoacetamidobenzyl-DOTA (BAD).

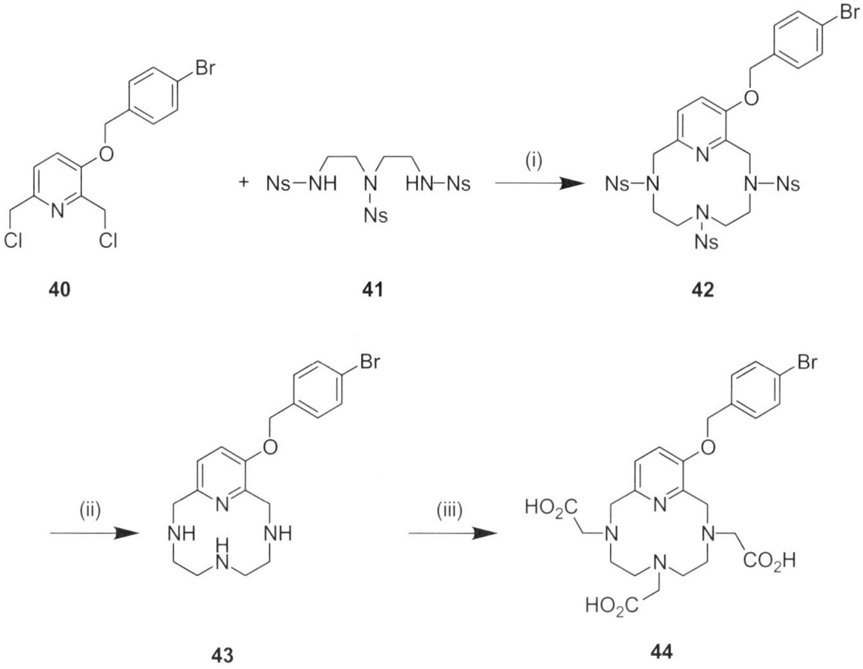

Scheme 4.9 Synthesis of a pyridine-substituted macrocycle: (i) K_2CO_3, acetonitrile, room temp; (ii) $HSCH_2COOH$, LiOH, DMF, room temp; (iii) chloroacetic acid, KOH, water, pH = 10, 70 7 °C (Ns = *p*-nitrobenzenesulfonyl) [17].

An overview of the Richman and Atkins syntheses of polyaza macrocycles would not be complete without mentioning the work that has been carried out in order to replace the *p*-toluenesulfonyl protecting groups. Indeed the *p*-toluenesulfonyl species is not what may be called a good protecting group, i.e. a substituent that is easy to attach and which protects the reactive group but which can also be easily removed. Regeneration of the macrocyclic polyamines after the cyclization requires harsh conditions that may not be compatible with the substituents, e.g. hydrolysis by concentrated sulfuric acid at 100 °C [29, 30] or at 180 °C [32], hydrolysis by 33 % hydrobromic acid in the presence of phenol and acetic acid [16, 23, 33], reduction with sodium or lithium in liquid ammonia [34], reduction with lithium aluminum hydride [31], reduction with a sodium amalgam in methanol [35, 36] or even cathodic reduction [37, 38]. Trifluoromethanesulfonyl protecting groups [39] have been proposed, but a Chavez-like cyclization did not work for the cyclen derivative. Furthermore deprotection is achieved by reduction with sodium in liquid ammonia. Aime *et al.* [17] used *p*-nitrobenzenesulfonyl (nosyl) protecting groups in the synthesis of a substituted pyridine containing tetraazacyclododecane (**42**) (Scheme 4.9). The nosyl group is removed in relatively mild conditions (lithium thioglycolate in DMF at room temperature).

3.2 SUBSTITUTED-AZIRIDINE OLIGOMERIZATION

Oligomerization of 1-benzylaziridine was reported several years ago by Hansen and Burg [40] and good yields of the twelve-membered macrocyclic tetramer **45** were isolated (Scheme 4.10). 1-Methyl and 1-phenylaziridine, on the other hand, gave only polymers. In 1970, this potentially useful synthetic reaction was re-examined by Tsuboyama *et al.* [41]. These authors were able to show that, depending on temperature, (*R*)-1-benzyl-2-ethylaziridine can react in the presence of boron trifluoride etherate to give a cyclic tetramer that was demonstrated to be the stereo chemically pure (R, R, R, R)1,4,7,10-tetrabenzyl-2,5,8,11-tetraethyl-1,4,7,10-tetraazacyclododecane (**46**). Later, Kossaï and Simonet [42] reported on the anodic tetramerization of *N*-benzylaziridine and proposed a chain reaction mechanism for the macrocyclization process by electrochemical means. The oligomerization of 1-benzylaziridine was then reassessed by Messerle *et al.* [43] who tested different catalysts and solvent conditions. The best yield of the cyclic tetramer **45** was obtained with *p*-toluenesulfonic acid in 95 % ethanol. The pentamer, 1,4,7,10,13-penta(benzhydryl)-1,4,7,10,13-pentaazacyclopentadecane (**47**), was obtained when 1-benzhydrylaziridine was used. However, the yield in cyclen is too low for this procedure to supersede the other methods. Ranganathan *et al.* [44] used this tetramerization procedure in order

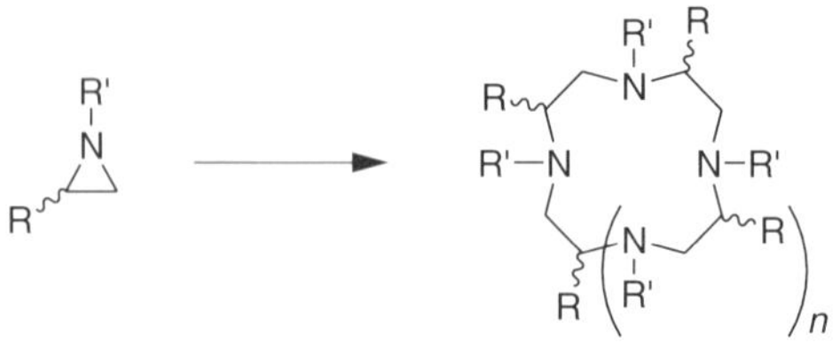

45 R = H, R' = Bn, $n = 1$
46 R = Et, R' = Bn, $n = 1$
47 R = H, R' = benzhydryl, $n = 2$
48 R = Me, R' = Bn, $n = 1$
49 R = CH_2OBn, R' = Bn, $n = 1$

Scheme 4.10 Aziridine oligomerization (Bn = benzyl; benzhydryl = diphenylmethyl) [40–45].

to gain access to an otherwise elusive macrocycle, namely [(2*S*)-(2α,5α,8α,11α)]-2,5,8,11-tetramethyl-1,4,7,10-tetraazacyclododecane (**48**). Using the same procedure, Uggeri and co-workers prepared a tetra(benzyloxymethyl) tetraazacyclododecane, (**49**) [45].

3.3 HIGH-DILUTION REACTIONS

A large number of strategies have been reported for the preparation of macrocycles [46]. The crucial step remains the macrocyclic ring closure. The procedures described above proved effective, but two other approaches should also be mentioned, i.e. the high-dilution method and the metal-templated synthesis. The latter will be described in the next section.

High-dilution techniques involve the slow and simultaneous addition of equal amounts of two bifunctional reagents to a vigorously stirred, large quantity of solvent [47]. Cyclization then competes favorably with polymerization and high yields of macrocyclic compound may be obtained provided that the right conditions are used. However, so many factors are involved that the procedure is still a highly empirical process and one may have to resort to trial-and-error in order to find the best choice of solvent, stirring, temperature and speed of addition. The high-dilution technique has nonetheless been applied with some success to the synthesis of several functionalized tetraaza macrocycles. Mono-protected *N*-benzylcyclen has been obtained after borane reduction of the macrocyclic bislactam **52**, while the latter was obtained from the high-dilution reaction of the methyl ester of *N*-benzyl iminodiacetic acid (**50**) with 1,4,7-triazaheptane (**51**) in dry methanol [5, 21] (Scheme 4.11a). Three different 2-substituted cyclens (**53**–**55**) have been synthesized by Takenouchi *et al.* who used the same method [48] (Scheme 4.11b). In each case, the yield of the high-dilution macrocyclization is fairly low (9–16 %).

Higher yields (40 %) were reported by McMurry *et al.* [49] who reacted the bis(succinimidyl) ester **56** with diamine **57** in dioxane at 90 °C (Scheme 4.12). As opposed to the high-dilution transamidification of a diester shown in Scheme 4.11, the cyclization proposed by McMurry and co-workers, and outlined in Scheme 4.12, is strongly favored over polymerization under high-dilution conditions because the reaction of an activated ester with an amine is substantially faster than the transamidification of an ester. However, this synthesis was reproduced by Kaspersen *et al.* [50] who only obtained a 10 % yield of the macrocyclic ring **58**.

Another method, closely related to the high-dilution technique, involves the slow reaction in the presence of a base in a large volume of boiling acetonitrile of a bis(bromo- or chloroacetamido) compound with a diamine, as originally proposed for the synthesis of cyclam (1,4,8,11-tetraazacyclotetradecane) and other polyaza-polyoxa macrocycles by Krakowiak, Bradshaw and Izatt who nicknamed the reaction 'crab-like cyclization' [51–53]. This reaction has been

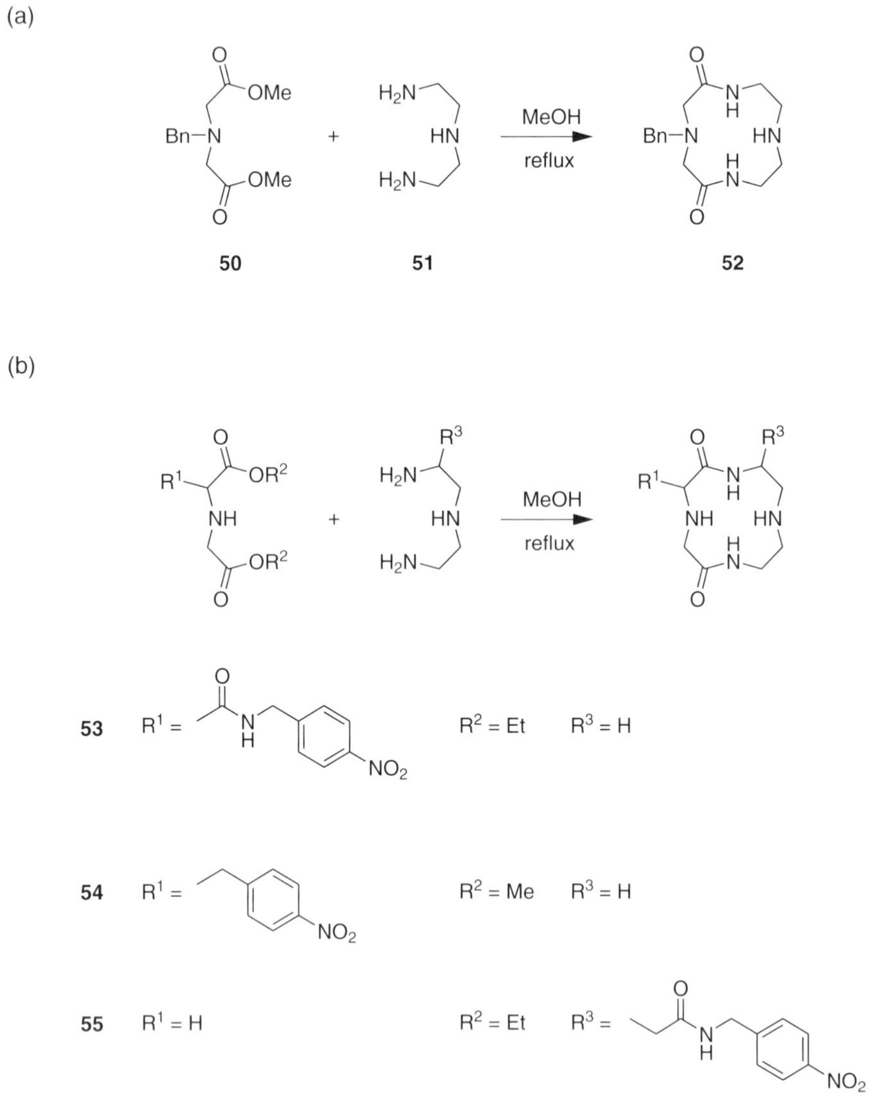

(a)

50 + 51 → (MeOH, reflux) → 52

(b)

53 R^1 = [$C(O)NHCH_2C_6H_4NO_2$] R^2 = Et R^3 = H

54 R^1 = [$CH_2C_6H_4NO_2$] R^2 = Me R^3 = H

55 R^1 = H R^2 = Et R^3 = [$CH_2C(O)NHCH_2C_6H_4NO_2$]

Scheme 4.11 High-dilution macrocyclization [5, 21, 48].

successfully applied for the synthesis of various substituted cyclens. Mishra *et al*. [54] proposed yet another synthesis of the 2-(*p*-nitrobenzyl)-substituted cyclen in which the macrocycle was obtained by a crab-like cyclization reaction. Desreux and co-workers [55, 56] made successful use of this macrocyclization technique to gain access to several rigidified tetraazacyclododecanes, as exemplified in Scheme 4.13 with the synthesis of the tetraline-substituted cyclen **62**.

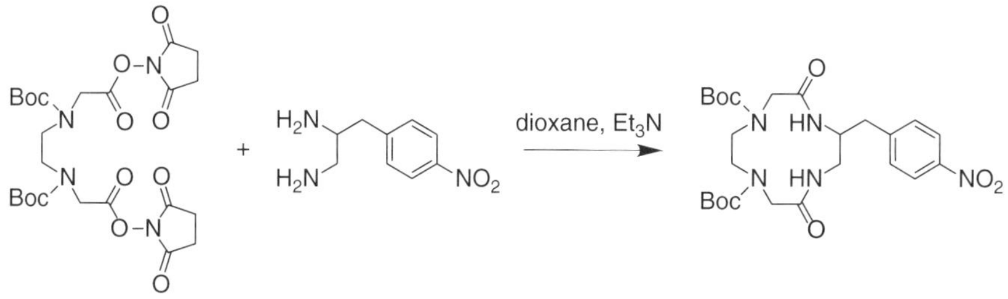

Scheme 4.12 High-dilution macrocyclization using a bis(succinimidyl) ester [49, 50].

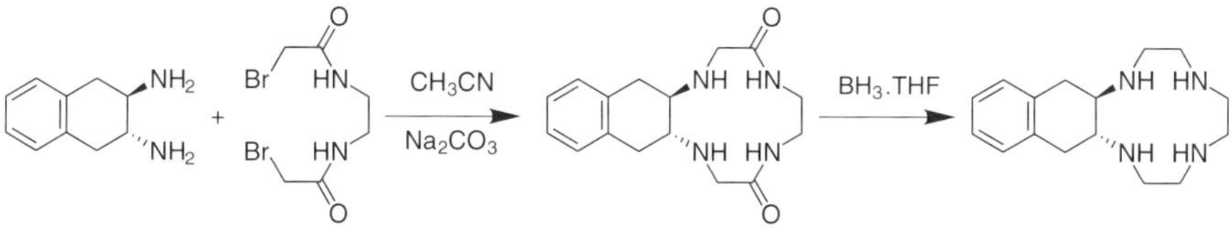

Scheme 4.13 Crab-like macrocyclization [55, 56].

3.4 TEMPLATE SYNTHESIS

Several polyaza macrocycles may be conveniently prepared by metal-templated synthesis, as exemplified by an efficient synthesis of 1,4,8,11-tetraazacyclotetradecane (**66**) by Barefield and co-workers [57, 58] (Scheme 4.14a). Takalo and Kankare [18] used a similar method for the preparation of **70** (R = Br). However the reduction of the diimine **69** was accompanied by a reduction of the aromatic bromo group and gave **70** (R = H) (Scheme 4.14b).

This simple and thus attractive reaction scheme has been applied to the synthesis of several 2-aryl substituted 1,4,7,10-tetraazacyclododecanes [59, 60] using iron(III) as the metal template (Scheme 4.15). Aryl-substituted tetraazamacrocycles (**73**) were isolated in reasonable to good yields (25–90 %). The *p*-nitrophenyl-substituted cyclen was found to be unstable due to the enhanced lability of the benzylic hydrogen leading to ring opening. These authors have attempted to extend their procedure to the synthesis of cyclen by using glyoxal instead of an arylglyoxal. Unfortunately, no detectable amount of the unsubstituted compound was found. They have thus postulated that the aryl group is mandatory for the reaction to proceed and that it could stabilize the intermediate diimine by conjugation. The lanthanide(III) chelates of the tetra(carboxymethyl) derivatives of these 2-arylcyclens have unfortunately been

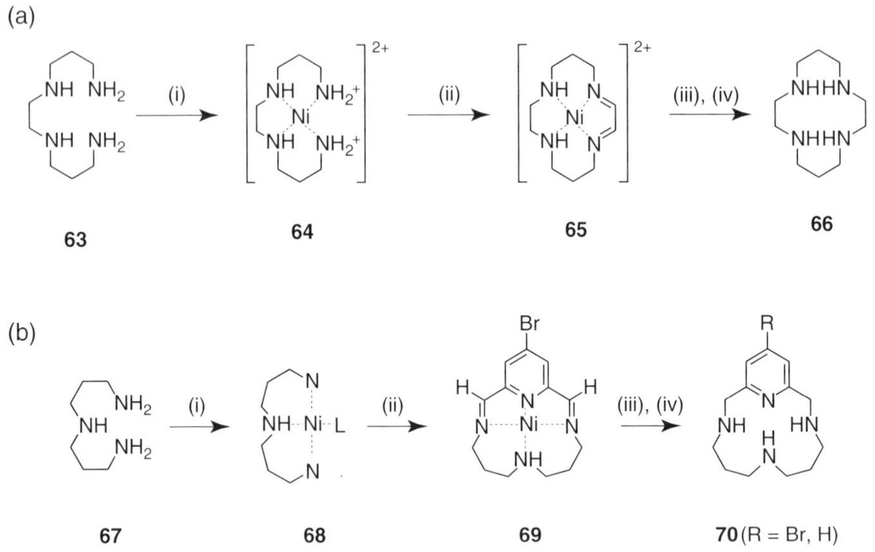

Scheme 4.14 Ni(II)-templated syntheses: (i) Ni(II) salt; (ii) glyoxal (a) or 4-bromo-2,6-diformyl-pyridine (b); (iii) sodium borohydride; (iv) sodium cyanide [18, 57, 58].

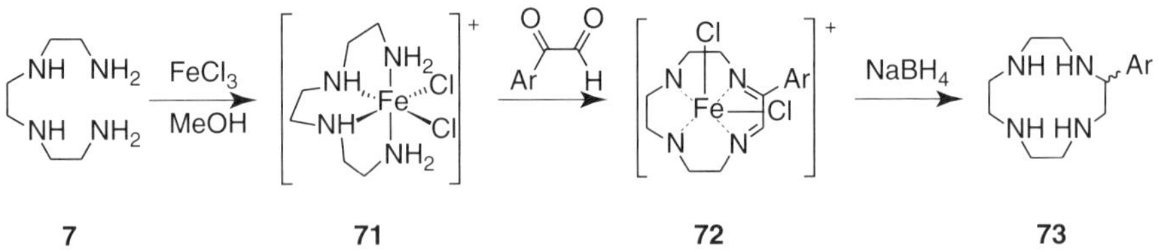

Scheme 4.15 Template synthesis of 2–arylcyclen [59, 60]

demonstrated to be exceedingly unstable, as the metal ion is easily removed with only a tenfold excess of EDTA [60].

4 FUNCTIONALIZATION

As has been already mentioned, substitution of the nitrogen atoms of cyclen by pendant arms containing donor groups is crucial in order to ensure a thermodynamically stable and kinetically inert Gd(III) chelate. Several functional groups have been used, such as carboxylates, phosphonates, phosphinates, amides or alcohols. Derivatives containing more than one type of pendant arm have also been synthesized. Selective substitution has thus become an important goal and this led to the design of new protecting groups.

4.1 ALKYLATION REACTIONS

A variety of electrophiles have been reacted with 1,4,7,10-tetraazacyclododecane. Often, several procedures have been implemented for the introduction of the same type of pendant arm. It can therefore prove quite difficult to choose the appropriate synthetic method. The choice should, however, be dictated by eventual cross-reactivity with other functional groups or by subsequent reactions, if any, in which the product could be involved.

Carboxymethyl pendant arms have been perused over the years, and hence a large number of methods have been reported. Potassium, sodium or lithium hydroxide is often added to an aqueous solution of the tetraamine and bromo- or chloroacetic acid in water at 70 °C, while maintaining the pH at about 10–11 [61, 62]. Refluxing a solution of potassium bromoacetate and polyamine in methanol in the presence of potassium carbonate has been proposed as an alternative when the polyamine is insoluble in water [63]. A two-phase system of toluene and potassium hydroxide was also used in such a case [64]. Several carboxymethyl esters have been prepared: methyl, ethyl, *t*-butyl and benzyl bromoacetate have been added to solutions of the macrocyclic tetraamine in *N*,*N*-dimethylformamide, dichloromethane, acetonitrile or methanol in the presence of a base such as diisopropylethylamine, triethylamine or sodium, potassium or cesium carbonate [21, 61, 65–67]. These esters may be purified as such by chromatography on silica or alumina. Precipitation of a crystalline hydrobromide has also been a great help in an easy synthesis [68, 69] of a very interesting intermediate, namely the tris(*t*-butyl) ester of 1,4,7-tris(carboxymethyl)-1,4,7,10-tetraazacyclododecane (**74**) (Figure 4.6). This compound has been selectively obtained by reacting cyclen at room temperature with three equivalents of *t*-butyl bromoacetate in *N*,*N*-dimethylacetamide in the presence of three equivalents of sodium acetate. The free carboxylic acids are obtained by base hydrolysis (methyl and ethyl esters), acid hydrolysis (*t*-butyl ester) or Pd/C catalytic hydrogenolysis (benzyl ester). Ion-exchange chromatography

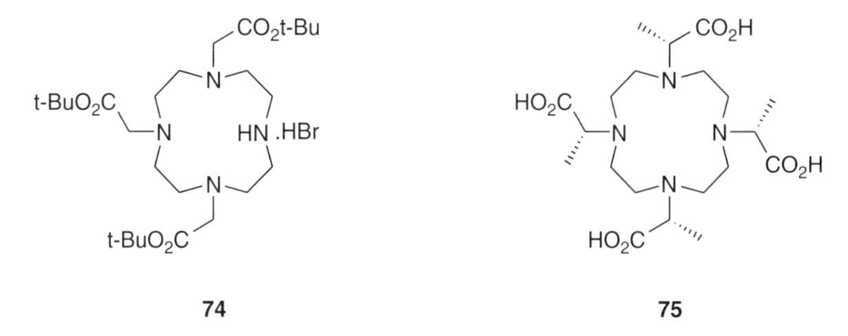

Figure 4.6 Structures of the tris(*t*-butyl) ester of 1,4,7-tris(carboxymethyl)-1,4,7,10-tetraazacyclododecane and of (*R*,*R*,*R*,*R*)-1,4,7,10-tetra(1-carboxyethyl)-1,4,7,10-tetraazacyclododecane.

and/or preparative HPLC techniques may then be implemented in order to get analytically pure macrocyclic ligands.

Van Westrenen and Sherry [70] have proposed a selective synthesis of DO2A, (**77**), that does not require protection of the macrocycle. Reaction of cyclen **10** with two equivalents of the sodium salt of formaldehyde bisulfite at neutral pH gives specifically the 1,7-bis(methylsulfonate) (**76**) in 95 % yield. Facile replacement of the sulfonate moiety is achieved by reaction with sodium cyanide, and hydrolysis with concentrated hydrochloric acid gives the title compound in 60 % overall yield (Scheme 4.16).

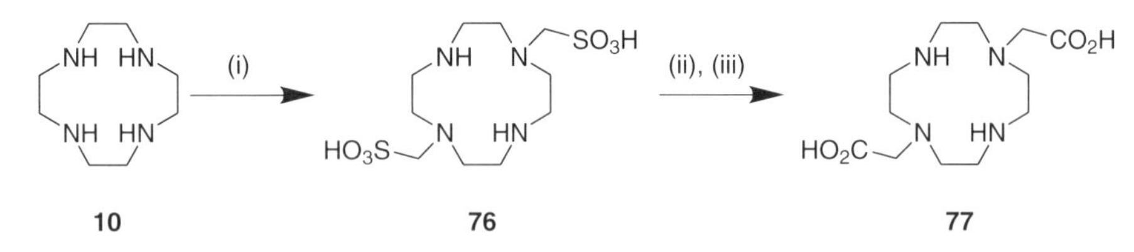

Scheme 4.16 Synthesis of 1,7-di(carboxymethyl)-1,4,7,10-tetraazacyclododecane: (i) $2 \times HOCH_2SO_3$ Na, pH = 7; (ii) 2 × sodium cyanide; (iii) conc. hydrochloric acid [70].

The reaction of cyclen with 2-chloropropionic acid in a basic aqueous medium has been reported to give 1,4,7,10-tetra(1-carboxyethyl)-1,4,7,10-tetraazacyclododecane (**75**) (Figure 4.6). This reaction proceeds via a stereospecific S_N2 pathway when optically active 2-chloropropionic acid is used [71]. Kang *et al.* [72] chose to synthesize the corresponding tetrabenzyl ester by reaction of cyclen with benzyl (*S*)-2-trifluoromethanesulfonyloxypropanoate. Indeed, racemization of an amino acid is very likely in basic medium but the triflate moiety is a much better leaving group than the chloride ion, and hence the S_N2 pathway and thus the stereoselectivity of the reaction are strongly favored. Other substituted carboxymethyl groups have been grafted to 1,4,7,10-tetraazacyclododecane via a base-promoted S_N2 reaction of the secondary amines with the corresponding 2-bromocarboxylic acid esters [73–76].

Functionalization of the tetraaza macrocycle is not limited to carboxymethyl and substituted carboxymethyl pendant arms. Electrophiles such as chloro- and bromoacetamides have been used to prepare *N*-carbamoylmethyl derivatives that can be substituted on the amide nitrogen and/or the methylene carbon atom [5, 21, 77–79]. Carboxymethyl pendant arms may also be converted to carbamoylmethyl groups by activation and reaction with an amine. The procedure usually involves the preparation of a mixed anhydride [80–82] by reaction of DOTA with one equivalent of *iso*-butyl chloroformate in the presence of a base (triethylamine and tetramethylguanidine have been proposed). Activated esters have also been used such as the *N*-hydroxysulfosuccinimidyl ester of DOTA [83]. The activated compound is then reacted with an amine and the product is purified by chromatography. Methylenephosphonate pendant arms have been added to 1,4,7,10-tetraazacyclododecane via a Mannich-type reaction of the tetraamine with formaldehyde and phosphorous acid in acidic medium [84–86]. Replacing phosphorous acid by trialkyl phosphite gives dialkyl phosphonates that may be partially hydrolyzed with sodium or potassium hydroxide [87–89]. *N*-methanesulfonates may be converted to *N*-methylphosphonates; albeit, this requires the oxidative hydrolysis of the aminomethanesulfonate to an aminomethanol with sodium triiodide prior to the reaction with phosphorous acid, the nucleophilicity of which is too low to displace sulfonate [70]. However, due to the extensive hydrolysis of the hemiaminal intermediate, the 1,7-bis(methanesulfonate) (**76**) gives only the mono(methylenephosphonate) (**78**) (Scheme 4.17). Carboxymethyl pendant arms are also amenable to conversion to methylenephosphonates by reaction with phosphorous acid in the presence of phosphorous trichloride in toluene [90].

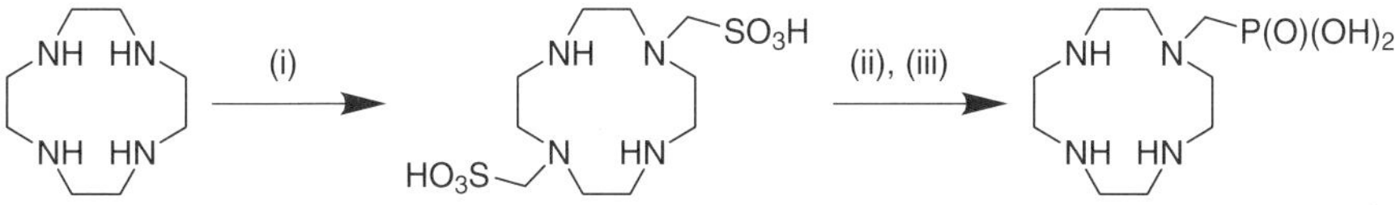

Scheme 4.17 Synthesis of 1,4,7,10-tetraazacyclododecane-1-methylenephosphonate: (i) 2 × $HOCH_2SO_3$ Na, pH = 7; (ii) sodium triiodide; (iii) molten phosphorous acid.

The Mannich reaction has also been applied to the synthesis of methane(alkyl or aryl)phosphinate-substituted polyaza macrocycles. Both phosphinic acids [88, 89, 91] and phosphinate esters [92–94] have been reacted with polyamines in the presence of formaldehyde. In the latter case, the ester is hydrolyzed by refluxing in an acidic medium or stirring after the addition of a base.

The opening of epoxides provides a very clean and simple way to introduce 2-hydroxyethyl and 2-substituted-2-hydroxyethyl functionalities [5, 21, 34, 95, 96].

Other substituents have been introduced on the tetraaza ring. Aime *et al.* [97] have attached a squaric ester moiety to the tris(*t*-butyl) ester of 1,4,7-tris(carboxymethyl)-1,4,7,10-tetraazacyclododecane (DO3A) (Scheme 4.18). The

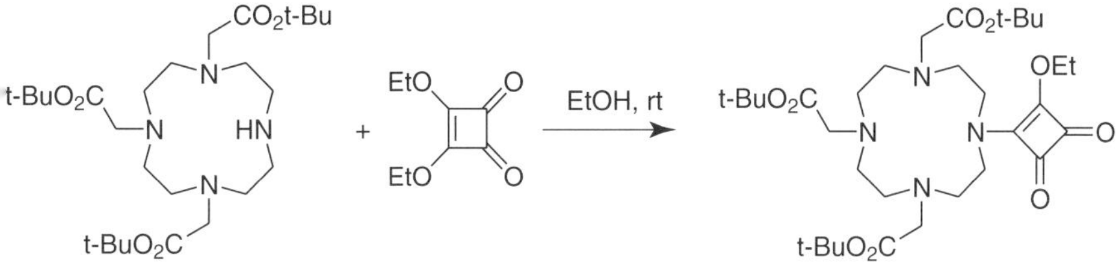

Scheme 4.18 Synthesis of a squarate ester derivative of DO3A [97]

squaric ester indeed reacts quite easily with amines, thus providing a potential linkage to a macromolecule. Heterocyclic moieties have also been appended to cyclen or its derivatives [50, 98].

4.2 MONOPROTECTION

Two different protecting groups have been widely used for the monoprotection of cyclen, namely the *N*-benzyl and *N*-formyl moieties. *N*-benzyl-protected cyclen has been obtained by a Richman and Atkins macrocyclization, followed by detosylation with a mixture of hydrobromic acid, acetic acid and phenol [23] or with sodium in liquid ammonia [5, 20, 21]. *N*-benzyl cyclen has also been synthesized, as already mentioned, by the reaction of dimethyl *N*-benzyl iminodiacetate with 1,4,7-triazaheptane in high-dilution conditions, followed by reduction of the macrocyclic bis(lactam) with borane. The benzyl protecting group can be removed by hydrogenation with Pd/C as a catalyst.

The synthesis of *N*-formyl cyclen is surprisingly very simple. Reacting cyclen **10** with *N*,*N*-dimethylformamide dimethylacetal gives a triprotected species (**81**) which can be hydrolyzed in a mixture of water and alcohol at low temperature to form quantitatively the monoprotected N-formyl cyclen **82** [5, 20, 21] (Scheme 4.19). The *N*-formyl protection is easily removed by acid hydrolysis (sulfuric acid and hydrochloric acid have been used [5, 20, 21, 72]).

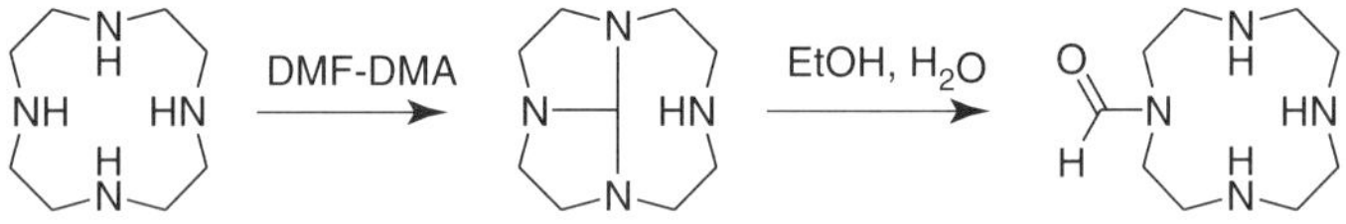

Scheme 4.19 Synthesis of *N*-formyl cyclen [5, 20, 21].

Protection is, however, not always required. Indeed, several authors have reported monoalkylation products when the alkylation is performed with a large excess of macrocycle (10:1), as could have been expected from simple statistics [99]. Kruper, Rudolf and Langhoff have also shown that the reaction between one equivalent of cyclen and one equivalent of a sterically hindered electrophile in a non-protic solvent such as dichloromethane or chloroform gives selectively the mono-*N*-alkylation product in a 70–80 % yield [75]. This selectivity has been explained by a decreased nucleophilicity of the remaining nitrogens in the monoalkylated compound. This procedure has been applied for example by Aime *et al.* [100] who easily isolated a monoalkylated compound in a 47 % yield when using a three to one mixture of electrophile to cyclen.

4.3 DIPROTECTION

Two regioisomers, i.e. 1,4- and 1,7-diprotected cyclens, are theoretically possible and both have been synthesized. Dumont *et al.* [24] have described the preparation of three different 1,7-diprotected cyclens (**83, 84** and **86**) (Figure 4.7). A Richman and Atkins macrocyclization allows the synthesis of 1,7-bis(methanesulfonyl)-4,10-bis(*p*-toluenesulfonyl) cyclen, which can be hydrolyzed in concentrated sulfuric acid to give the 1,7-bis(methanesulfonyl) derivative **83**. However, the removal of the methanesulfonyl protecting groups requires strong reducing agents such as Red-Al®, which may be incompatible with other substituents. The 1,7-bis(*p*-toluenesulfonyl) derivative **84** may be obtained specifically by reaction of two equivalents of *p*-toluenesulfonyl chloride with one equivalent of cyclen in pyridine. If the reaction is performed in chloroform with triethylamine as a base, the product is a mixture of mono-, di-and triprotected macrocycles that can nonetheless be easily separated owing to their different solubility properties. If *p*-toluenesulfonyl chloride is replaced by *p*-methoxybenzenesulfonyl chloride, the same synthetic method gives the 1,7-derivative (**85**) that could be of interest as *p*-methoxybenzenesulfonamides are generally more sensitive to acid hydrolysis than *p*-toluenesulfonamides [78]. Finally, the 1,7-bis(diethylphosporamidate) (**86**) is the only compound isolated when one equivalent of cyclen is reacted with two equivalents of diethylphosphite and carbon tetrachloride in a two-phase system with a transfer agent. The phosphoramidate-protecting group can be easily cleaved by using 6M hydrochloric acid.

Kovacs and Sherry [101] reacted cyclen with several chloroformates in acidic solution (pH = 2–3) and were able to isolate 1,7-diprotected cyclens (**87**) in good yields (88 % for benzyl to 98 % for ethyl chloroformate) (Figure 4.7). This regioselectivity was ascribed to the microprotonation sequence of the macrocycle which favors protonation of nitrogens that are as far apart as possible. The 1,7-bis(benzylcarbamate) derivative proved to be the most versatile protecting group as it is easily cleaved by an acid or by catalytic hydrogenation [102]. This compound was used to prepare several chelating agents based upon

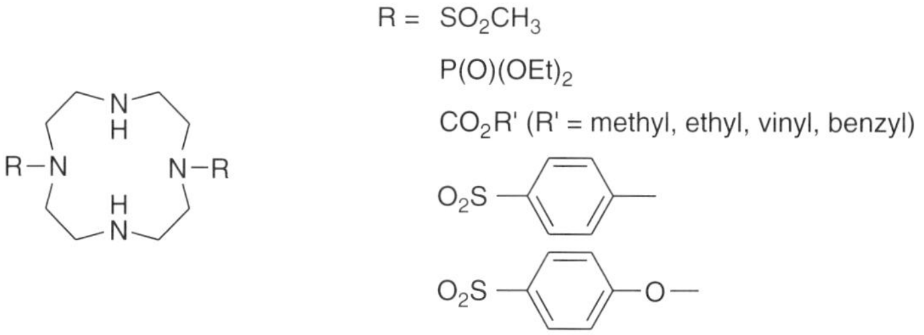

R = SO_2CH_3
$P(O)(OEt)_2$
CO_2R' (R' = methyl, ethyl, vinyl, benzyl)
O_2S–(p-tolyl)
O_2S–(p-methoxyphenyl)

Figure 4.7 Diprotected 1,4,7,10-tetraazacyclododecanes.

1,7-bis(carboxymethyl)-1,4,7,10-tetraazacyclododecane [88], as well as 1,7-phosphonate- and phosphinate-disubstituted cyclens [89].

Anelli *et al.* [103, 104] managed to place different protecting groups in positions 'one' and 'seven' of cyclen or 'one' single protecting group in position one and another substituent in position 'seven'. The monoprotected cyclen **88** (obtained by reaction of an adequate electrophile with 10 equivalents of cyclen) was reacted with *N,N*-dimethylformamide diethylacetal. The tricyclic intermediate **89** was then opened regioselectively in a mixture of ethanol and water to give the 1,7-disubstituted cyclen **90** that can be further derivatized before removal of the first and then the second protecting group (Scheme 4.20).

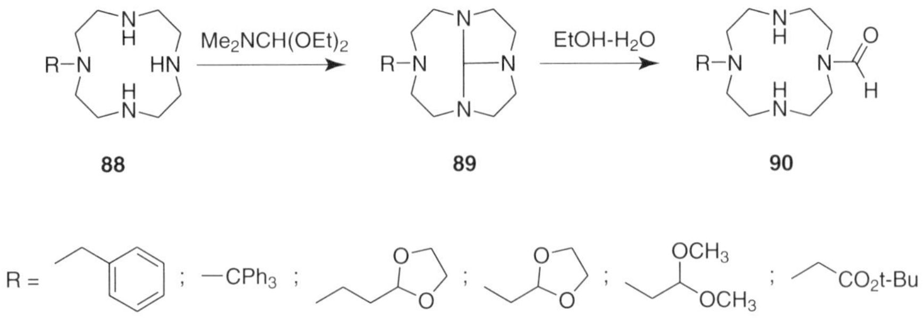

Scheme 4.20 Synthesis, of 1,7-diprotected cyclen, according to Anelli *et al.* [104].

An unusual 1,4-diprotection of cyclen has been reported by Li and Undheim [105]. These authors reacted the hydrochloride salt of cyclen with ethyl orthocarbonate and isolated a guanidinium salt (**91**) that was transformed into a tetracyclic neutral tetraamine (**92**) by reaction with a base. This compound, upon reaction with benzyl bromide, yields a substituted guanidinium salt (**93**) that can be reduced with Red-Al® to give the tricyclic derivative **94**, which is analogous to that reported by Anelli *et al.* [104]. When reacted with benzyl bromide, this intermediate gives a 1,4-dibenzyl amidinium salt (**95**) which is hydrolyzed by successive base and acid treatment to 1,4-dibenzyl-1,4,7,10-tetraazacyclododecane (**96**) (Scheme 4.21).

Another very interesting 1,4-diprotected cyclen is the 1,4,7,10-tetraazabicyclo[8.2.2]tetradecan-11-one (**97**) [106]. This bicyclic lactam (Scheme 4.22) is obtained by heating an aqueous solution of the addition compound of cyclen and glyoxal with piperazine at a pH between five and nine. Not withstanding its six-membered-ring lactam, this compound reacts surprisingly easily with a base to give **98**, a monosubstituted derivative of cyclen. In this case, the lactam acts more as a monoprotected than a diprotected cyclen. It can, however, be reacted with two equivalents of a haloacetate, thus yielding a disubstituted diprotected cyclen (**99**) that can be deprotected with base and further derivatized. This versatile compound can thus give access to a variety of macrocyclic ligands.

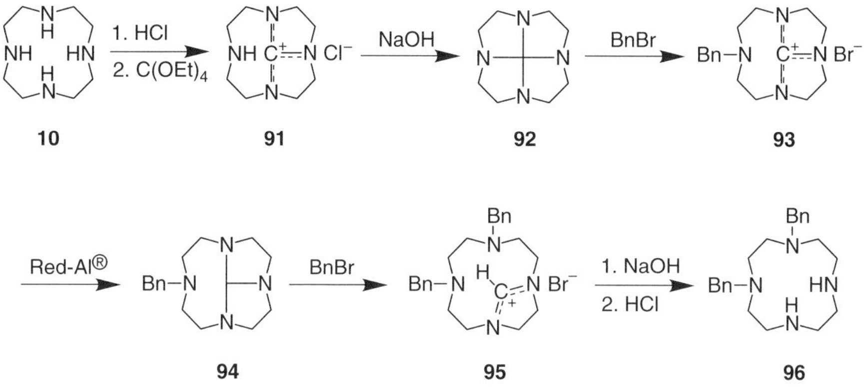

Scheme 4.21 Synthesis of 1,4-dibenzyl-1,4,7,10-tetraazacyclododecane [105].

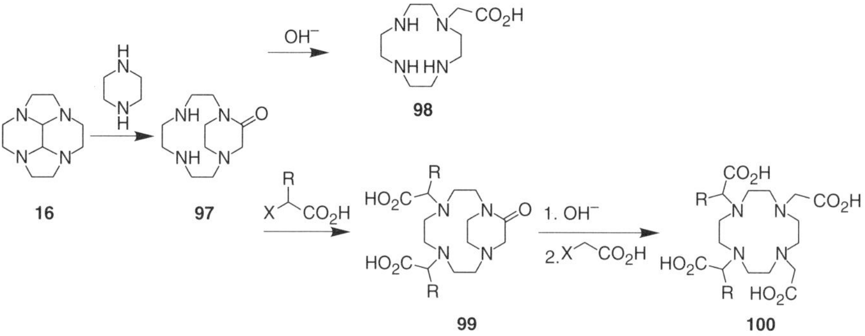

Scheme 4.22 Synthesis and reactions of 1,4,7,10-tetraazabicyclo[8.2.2]tetradecan-11-one [106].

The authors even reported the synthesis of a 1,4,7,10-tetraazacyclododecane substituted by four different pendant arms.

4.4 TRIPROTECTION

A variety of 1,4,7-triprotected 1,4,7,10-tetraazacyclododecane moieties have been reported in the literature and we have already mentioned some of these (Figure 4.8).

Reacting cyclen with *N*,*N*-dimethylformamide dimethylacetal in benzene yields the tricyclic triprotected derivative **81** [107] that can be functionalized [108, 109] provided that neutral anhydrous conditions are used because this intermediate is sensitive to water, acid and base solvolysis. This tricyclic intermediate can also be partially hydrolyzed to give the monoprotected *N*-formyl

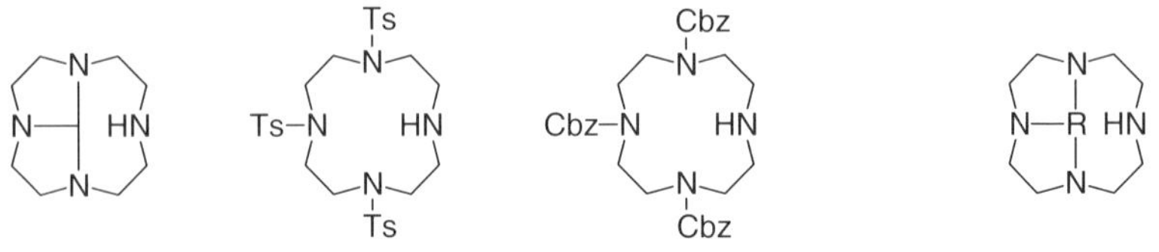

81 101 102 103

R = B, P(O), $Co(CO)_6$, $Mo(CO)_6$

Figure 4.8 Triprotected 1,4,7,10-tetraazacyclododecanes.

cyclen that has been described above in Section 4.2. The 1,4,7-tris(*p*-toluenesulfonyl)-1,4,7,10-tetraazacyclododecane (**101**) has been prepared by several groups [5, 24, 110]. However, the deprotection of functionalized derivatives requires harsh reaction conditions (concentrated sulfuric acid at 100 °C or sodium or lithium in liquid ammonia) that may be incompatible with many functional groups. Prasad *et al.* [62] have described the synthesis of the 1,4,7-tris(benzylcarbamate) derivative (**102**) of cyclen that was obtained in 49 % yield by reacting cyclen with 3.2 equivalents of benzylchloroformate in dichloromethane in the presence of triethylamine. This compound is stable to basic media and is thus a good candidate for a variety of electrophilic nitrogen substitution reactions that are usually carried out in the presence of a base. Furthermore, benzylcarbamates are easily deprotected by catalytic hydrogenation, conditions that are well tolerated by many functional groups.

The Handel group has synthesized boron (**103**) [111], phosphoryl (**103**) [112], chromium or molybdenum carbonyl (**103**) [113] and methylsilyl (**104**) [114] 1,4,7-triprotected 1,4,7,10-tetraazacyclododecane macrocycles. The latter compound is isolated as a hydrochloride salt and can give, after reaction with three equivalents of *n*-butyllithium, either the mono- or the 1,7-disubstituted-cyclen derivatives **105** and **106** in fair to good yields (60–85 %). However, these syntheses were limited to two electrophiles only, i.e. methyl iodide and benzyl bromide (Scheme 4.23).

The moisture-sensitive boron derivative **103** (R = B) (Figure 4.8) was prepared by reaction of cyclen with tris(*N*,*N*-dimethylamino)borane. It was deprotonated with *n*-butyllithium and the resulting monoamide was reacted with benzyl bromide. The product was treated with aqueous sodium hydroxide to give a 50 % yield of the 1-benzyl-1,4,7,10-tetraazacyclododecane. The phosphoryl-protected macrocycle **103** (R = PO) was obtained by a two-step procedure, first, by the reaction of cyclen with hexamethylphosphoric triamide to give a P(III)/P(V) tautomer **107**, which was subsequently oxidized to a phosphonium salt (**108**) with carbon tetrachloride. The latter can be hydrolyzed with sodium hydroxide to a phosphoryl-triprotected tetraaza macrocycle that is stable and can be purified by distillation. Reaction with benzylbromide and hydrolysis in acidic medium allow isolation of the mono(*N*-benzyl) cyclen (**110**)

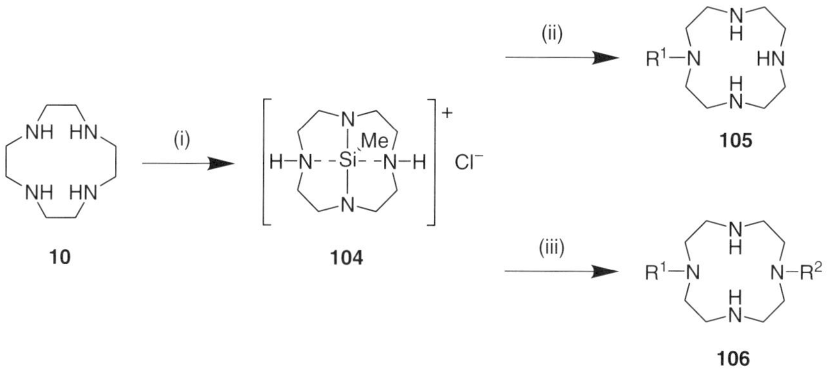

Scheme 4.23 Synthesis and reactions of methylsilyl-triprotected cyclen: (i) trichloromethylsilane, triethylamine; (ii) (a) 3 × *n*-butyllithium, (b) R^1 X, (c) H_3O^+; (iii) (a) 3 × butyllithium, (b) R^1 X, (c) R^2 X , (d) H_3O^+ [8, 114].

in 90 % yield. The phosphonium salt can also be reacted with an alcohol prior to base hydrolysis to get the same result but with a somewhat lower (80 %) overall yield (Scheme 4.24). The phosphoryl-protected cyclen (**109**) (R = H) can also be used for the preparation of 1,7-disubstituted cyclens by simple reactions with two equivalents of the alkylating agent [115].

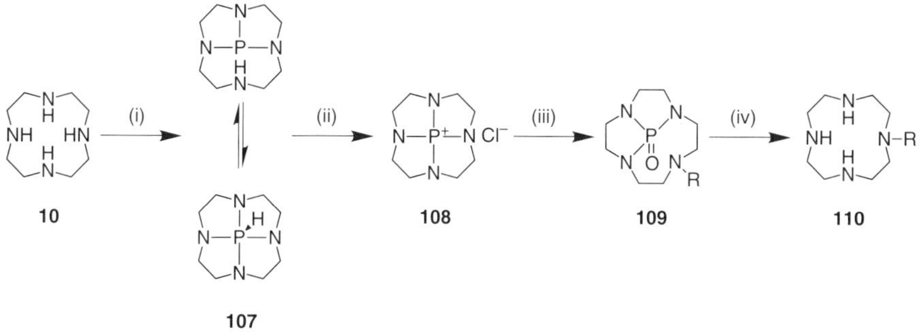

Scheme 4.24 Synthesis and substitution reaction of phosphoryl-protected cyclen: (i) $P(NMe_2)_3$; (ii) carbon tetrachloride; (iii) ROH then OH^-, or OH^-, then RX; (iv) H_3O^+ [112].

Finally, reacting 1,4,7,10-tetraazacyclododecane with hexacarbonylchromium or molybdenum gives the air-and moisture-sensitive tricarbonylchromium or molybdenum complex **103** (R = $Mo/Cr(CO)_3$) featuring one free nitrogen atom. Yaouanc *et al.* [113] have limited their investigation of alkylating agents to simple alkyl bromides, while *N*-alkyl and *N*,*N*-dialkyl bromoacetamides have been used by other groups [93, 94]. The demetallated products have been easily isolated by aerial oxidation in acidic medium. The tricarbonylchromium and molybdenum complexes (**103**) also give symmetrically 1,7-disubstituted cyclens when alkyl bromides are replaced by alkyl iodides [116]. An even more interesting feature of these triprotected cyclens is their ability to form unsymmetrically 1,7-disubstituted cyclens in a one-pot reaction. This selectivity has been ascribed to kinetic reasons as the first alkylation is a very fast reaction consuming all of the electrophile, while the second is much slower and only takes place when a second equivalent of the same or another alkyl halide is added [117].

5 POLYMACROCYCLIC LIGANDS

The efficacy of MRI contrast agents has been constantly improved over the years but there remains a need for target-specific agents that would highlight specific areas of the body. Such an ambitious goal obviously requires the synthesis of new molecules that would not only be highly specific but that would also display higher relaxivities. This has prompted researchers to synthesize new polymeric gadolinium chelates. Indeed, grafting several gadolinium complexes onto one backbone should result in an increased relaxivity because of slower tumbling rates. Furthermore, such large molecules should also prove useful for MR angiography, as their size would prevent any diffusion to the interstitial space, thus highlighting only the blood vessels. Several approaches have been developed in order to prepare polymeric materials.

Covalent bonding of a macrocycle to a polymer backbone has been achieved by reacting the mono-*N*-(2-isothiocyanatoethyl)amide of DOTA (**111**) or the mono-*N*-(2-bromoacetamidoethyl)amide of DOTA (**112**) (Figure 4.9) with poly-L-lysine [118]. A mixed anhydride of DOTA was also directly bound to poly-L-lysine. Gadolinium chloride is readily complexed by these polychelating agents. The resulting polymers were covalently attached to human serum albumin (HSA) via activation of the latter with 2-iminothiolane and of the polymer with succinymidyl-4-(*N*-maleimidomethyl)cyclohexane-1-carboxylate [118]. A covalent binding of polychelates to biologically active macromolecules such as antibodies can thus be envisaged as a means of achieving targeted imaging.

Polymers have been synthesized that incorporate a substituted DOTA in their backbone. The tris(*t*-butyl) ester of cyclen was reacted with methyl bromoacetate. The reaction of the resulting compound with tris(2-aminoethyl)amine gave the corresponding tris(*t*-butyl) ester monoamide (**113**) (Figure 4.9). Finally, the two free primary amino groups of **113** were allowed to react with a PEG-bis(activated ester). The resulting polyamide was deprotected and complexed with gadolinium. The final high-molecular-weight polymer has a long

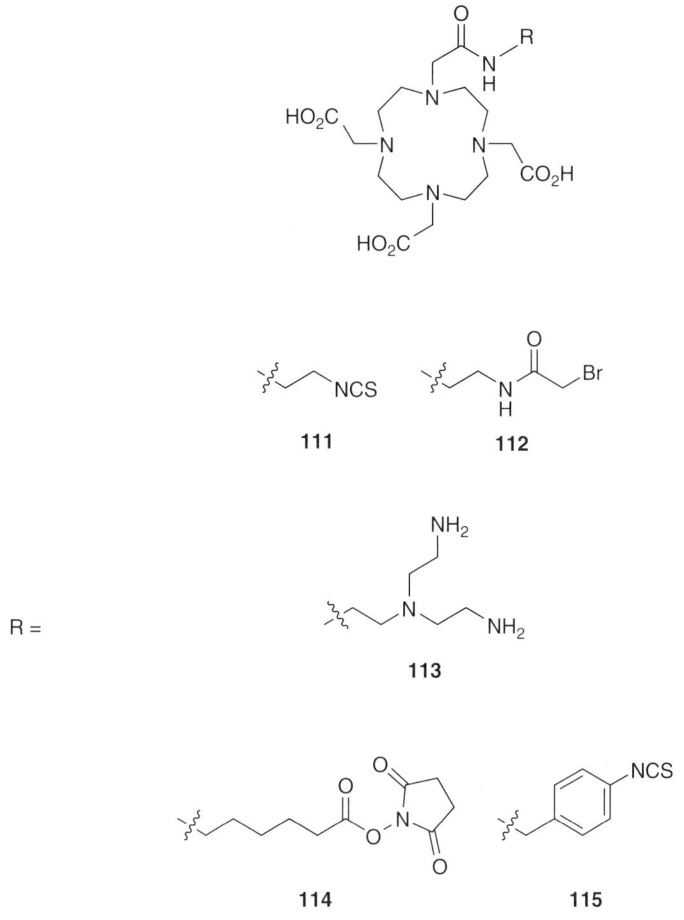

Figure 4.9 DOTA derivatives used for the synthesis of polymeric chelating agents.

blood retention time [119]. This makes it a very promising contrast agent for MR angiography.

Dendrimers have also been considered as backbones for the synthesis of high-molecular-weight MRI contrast agents. The simple polyamidoamine (PAMAM) dendrimer has thus been modified with substituted DOTA-like ligands: an active ester (**114**) [119] and an isothiocyanate (**115**) [120] have been used for that purpose (Figure 4.9). However, these compounds are characterized by shorter blood retention time and are thus less interesting in themselves.

Another class of interesting multimeric materials is obtained by self-assembly. In contrast to the preceding examples, the polychelates are generated by using intermolecular bonding. This approach has permitted the synthesis of a rigid trimer of a gadolinium complex, namely Gd(phenHDO3A) (**116**) [55]. The ligand is a bifunctional entity able to complex two different metal ions, as the substituted DO3A-like macrocycle shows a high affinity for gadolinium while the 5,6-dihydrophenanthroline moiety is a very good bidentate chelating agent of iron(II). The gadolinium complex Gd(phenHDO3A) has been shown to spontaneously self-assemble around one iron(II) to form a tris-complex (**117**) of high molecular weight that displays a higher relaxivity per gadolinium ion than the corresponding monomer (12.2 instead of 3.7 $mM^{-1}\ s^{-1}$) (Scheme 4.25).

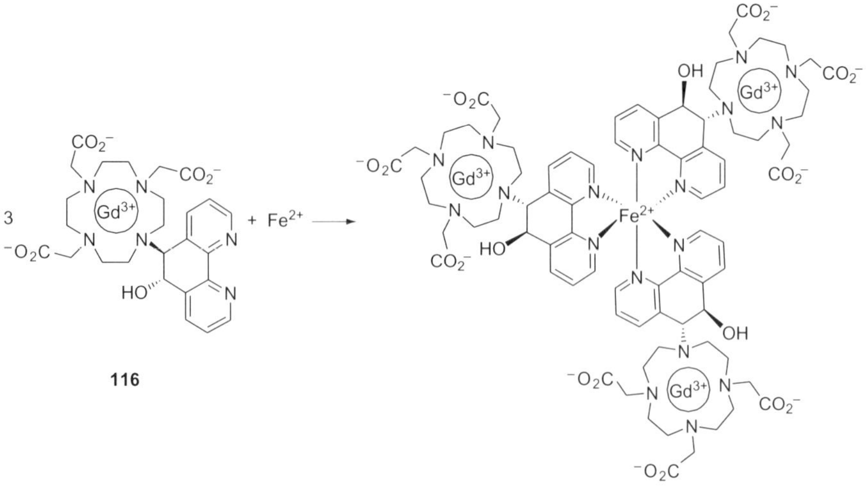

Scheme 4.25 Self-assembly of Gd(phenHDO3A) around Fe(II) [55].

6 NOT ALL MACROCYCLIC LIGANDS ARE DERIVED FROM TETRAAZA RINGS

The high affinity of macrocyclic polyaminopolycarboxylic ligands for gadolinium has been extensively studied, but this has not deterred the search for other efficient gadolinium-containing MRI contrast agents.

At least two different, yet somehow related, classes of interesting ligands have been explored, i.e. Schiff bases and texaphyrins (Figure 4.10). Gadolinium chelates of the macrocyclic Schiff base derivatives **118** are easily synthesized by a metal-templated reaction [121–123]. Six almost coplanar nitrogen atoms chelate the gadolinium ion and the complex is able to accept three water molecules in its first coordination sphere as has been demonstrated in the solid state and by measurement of the relaxivity of the chelate ($R = 11\ mM^{-1}\ s^{-1}$) [124]. However, these compounds are not kinetically stable. On the other

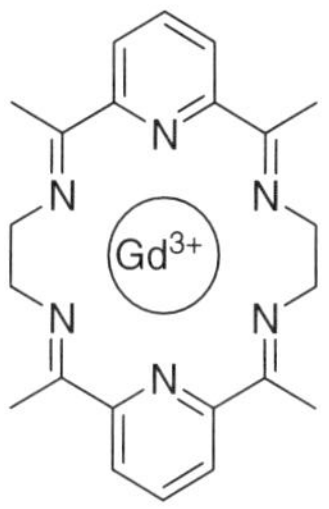

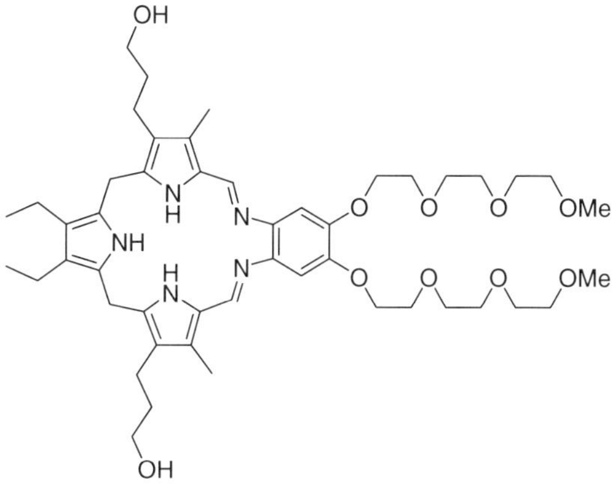

118 119

Figure 4.10 Structure of non-polyaza polycarboxylic macrocycles (see text for details).

hand, the synthesis of texaphyrins (**119**) and of their metal complexes has been described by Sessler *et al.* [125, 126]. The nitrogen atoms of the macrocycle are coordinated to the lanthanide ion in an almost coplanar arrangement, leaving enough space around the metal ion for several water molecules, hence the very high relaxivity of the gadolinium chelate ($R = 19\ mM^{-1}\ s^{-1}$). So far, the main potential application of gadolinium-containing texaphyrins is as radiation sensitizers.

7 PURITY AND ANALYSIS

The purity of the ligands and of their gadolinium complexes is of the utmost importance from a clinical point of view. As already mentioned in Section 4.4, esters of macrocyclic tetraaza tetracarboxylic ligands are isolated by column chromatography on silica gel or alumina. Polyacetic ligands can be purified by ion-exchange chromatography (see, for example, Prasad *et al.* [62]) or, in very difficult cases, by preparative reversed-phase high performance liquid chromatography (HPLC) [29]. Ion-exchange purification is a two-step process. An acidic solution of the ligand is added to a column of a sulfonated cation-exchange resin and the compound is eluted with a dilute ammonium hydroxide solution. The concentrated basic eluate is then loaded onto a column of a quaternary ammonium anion-exchange resin and eluted with formic acid solutions of increasing concentrations. The fractions containing the desired compound are collected and formic acid is evaporated as an azeotrope. Preparative HPLC is usually performed at a fixed pH and thus requires the use of additives that must be easy to dispose of. Trifluoroacetic acid and ammonium acetate are commonly used as they can be removed by lyophilization. DOTA, on the other hand, can be very easily purified by precipitation of its potassium chloride adduct. Whatever the isolation technique, the purity of the ligand is easily checked by HPLC using a UV-visible detection method. Direct detection is possible and should be used whenever a chromophor is available on the molecule; complexation with copper ions [127, 128] may be useful when the ligand does not absorb UV light. Absorption of the carbonyl moiety may be monitored but the low detection wavelength makes it very sensitive to solvent impurities.

Gadolinium complexes are easily obtained by reaction of the fully protonated ligand with an equimolar amount of a gadolinium salt. However, polyazapolycarboxylic macrocyclic ligands are usually isolated as hydrates and ligand solutions should always be titrated before complexation experiments are performed. Gadolinium oxide can be used as a standard after calcination at 1200 °C. It can be reacted directly with the ligand [5, 72] but this reaction is often very slow. Gadolinium chloride may be purchased or is readily prepared from the oxide. Constant monitoring of the pH, which is adjusted with a sodium hydroxide solution, ensures a fast reaction of this salt with the chelating agent [129] but this gives a solution of the complex that contains inorganic salts. If salts are to be avoided, gadolinium hydroxide may be reacted with the neutral ligand [130]. This hydroxide is obtained quantitatively by precipitation with sodium hydroxide from a solution of the metal chloride. It is filtered and thoroughly washed before being transferred without delay to the reaction flask. An excess of the hydroxide should always be avoided as ligands that contain four or more acidic protons may dissolve an excess of the metal ion. Gadolinium carbonate [131] and acetate [72] have also been used. In any case, the purity of the metal chelates has to be checked. HPLC on a reversed-phase support is the method of choice. The complexes may be detected by fluorescence of the metal ion [132] and, provided that the chelate bears a chromophoric group, by UV-visible spectroscopy. If needed, crystallization [130] or preparative HPLC [5], as a last resort, may be used to purify the complex. Large quantities of pure chelates have also been obtained by column chromatography on polymeric materials [72, 133].

ACKNOWLEDGEMENTS

The authors would like to gratefully acknowledge the financial support of the Fonds National de la Recherche Scientifique of Belgium. VJ is Chercheur Qualifié at this institution. We are also thankful for the support of the Institut Interuniversitaire des Sciences Nucléaires and the European Co-operation in the Field of Scientific and Technical Research, COST D8.

REFERENCES

1. Desreux, J. F. *Inorg. Chem.* 1980, **19**, 1319.
2. Wang, X. Y., Jin, T. Z., Comblin, V., Lopez-Mut, A., Merciny, E. and Desreux, J. F. *Inorg. Chem.* 1992, **31**, 1095.
3. Stetter, H. and Marx, J. *Chem. Brit.* 1957, **607**, 59.
4. Richman, J. E. and Atkins, T. J. *J. Am. Chem. Soc.* 1974, **96**, 2268.
5. Dischino, D. D., Delaney, E. J., Enswiler, J. E., Gaughan, G. T., Prasad, J. S., Srivastava, S. K. and Tweedle, M. F. *Inorg. Chem.* 1991, **30**, 1265.
6. Vriesema, B. K., Buter, J. and Kellogg, R. M. *J. Org. Chem.* 1984, **49**, 110.
7. Chavez, F. and Sherry, A. D. *J. Org. Chem.* 1989, **54**, 2990.
8. Trost, B. M. *Science* 1991, **254**, 1471.
9. Athey, P. S. and Kiefer, G. E. *Int. Pat. WO 95/14726*, 1996; *Chem. Abstr.* 1996, **124**, 8862.
10. Weisman, G. R. and Reed, D. P. *J. Org. Chem.* 1996, **61**, 5186.
11. Weisman, G. R. and Reed, D. P. *J. Org. Chem.* 1997, **61**, 4548.
12. Argese, M., Ripa, G., Scala, A. and and Valle, V. *Int. Pat. WO 97/49691*, 1997; *Chem. Abstr.* 1998, **128**, 114973.
13. Argese, M., Ripa, G., Scala, A., Valle, V. *Int. Pat. WO 98/45296*, 1998; *Chem. Abstr.* 1998, **129**, 316210.
14. Hervé, G., Bernard, H., Le Bris, N., Yaouanc, J.-J., Handel, H. and Toupet, L. *Tetrahedron Lett.* 1998, **39**, 6861.
15. Hervé, G., Bernard, H., Le Bris, N., Le Baccon, M., Yaouane, J.-J. and Handel, H. *Tetrahedron Lett.* 1999, **40**, 2517.
16. Stetter, H., Frank, W. and Mertens, R. *Tetrahedron* 1981, **37**, 767.
17. Aime, S., Botta, M., Frullano, L., Geninatti Crich, S., Giovenzana, G. B., Pagliaro, R., Palmisano, G. and Sisti, M. *Chem. Eur. J.* 1999, **5**, 1253.
18. Takalo, H. and Kankare, J. *J. Heterocycl. Chem.* 1990, **27**, 167.
19. Sherry, A. D. *Int. Pat. WO 86/02352*, 1986; *Chem. Abstr.* 1987, **106**, 64034.
20. Tweedle, M. F., Gaughan, G. T., Hagan, J. J. *Eur. Pat. EP 232751*, 1987; *Chem. Abstr.* 1988, **108**, 56130.
21. Tweedle, M. F., Gaughan, G. T., Hagan, J. J. *Eur Pat. EP 292689*, 1988; *Chem. Abstr.* 1989, **110**, 173270.
22. Gries, H., Raduchel, B. E., Speck, U. and Weinmann, H.-J. *Eur. Pat. EP 255471*, 1988; *Chem. Abstr.* 1988, **198**, 6552.
23. Dumont, A., Jacques, V., Qixiu, P. and Desreux, J. F. *Tetrahedron Lett.* 1994, **35**, 3707.
24. Bencini, A., Bianchi, A., Borselli, A., Chimichi, S., Ciampolini, M., Dapporto, P., Micheloni, M., Nardi, N., Paoli, P. and Valtancoli, B. *Inorg. Chem.* 1990, **29**, 3282.
25. Ciampolini, M., Micheloni, M., Nardi, N., Paoletti, P., Dapporto, P. and Zanobini, F. *J. Chem. Soc. Dalton Trans.* 1984, 1357.
26. Schaeffer, M., Meyer, D., Beaute, S. and Doucet, D. *Magn. Reson. Med.* 1991, **22**, 238.
27. Schaeffer, M., Doucet, D., Bonnemain, B. and Meyer, D. *Eur. Pat. EP 287465*, 1998; chem. Abstr. 1989 **111**, 69850.
28. Moi, M. K. and Meares, C. F. and DeNardo, S. J. *J. Am. Chem. Soc.* 1988, **110**, 6266.
29. Renn, O. and Meares, C. F. *Bioconjugate Chem.* 1992, **3**, 563.
30. Ansari, M. H., Ahmad, M. and Dicke, K. A. *Bioorgan. Med. Chem. Lett.* 1993, **3**, 1067.
31. Garrity, M. L., Brown, G. M., Elbert, J. E. and Sachleben, R. A. *Tetrahedron Lett.* 1993, **34**, 5531.

32. Lázár, I. *Synth. Commun.* 1995, **25**, 3181.
33. Chen, D., Squattrito, P. J., Martell, A. E. and Clearfield, A. *Inorg. Chem.* 1990, **29**, 4366.
34. Platzek, J., Gries, H., Weinmann, H. J., Schuhmann-Giampieri, G.and Press, W. R. *Eur. Pat. EP 448191*, 1991; *Chem. Abstr.* 1992, **116**, 21084.
35. Sink, R. M., Buster, D. C. and Sherry, A. D. *Inorg. Chem.* 1990, **29**, 3645.
36. Chappell, L. L., Voss, D. A. Jr, Horrocks, W. D. J. and Morrow, J. R. *Inorg. Chem.* 1998, **37**,1.
37. Lebouc, A., Martigny, P., Carlier, R., and Simonet, J. *Tetrahedron* 1985, **41**, 1251.
38. Kossai, R., Simonet, J. and Jeminet, G. *Tetrahedron Lett.* 1979, 1059.
39. Panetta, V., Yaouanc, J. J. and Handel, H. *Tetrahedron Lett.* 1992, **33**, 5505.
40. Hansen, G. R. and Burg, T. E. *J. Heterocycl. Chem.* 1968, **5**, 305.
41. Tsuboyama, S., Tsuboyama, K., Higashi, I., Yanagita, M. *Tetrahedron Lett.* 1970, 1367.
42. Kossaï, R. and Simonet, J. *Tetrahedron Lett.* 1980, **21**, 3575.
43. Messerle, L., Amarasinghe, G., Fellman, J. D. and Garrity, M. *Int. Pat. WO 96/28420*, 1996; *Chem. Abstr.* 1996, **125**, 300809.
44. Ranganathan, R. S., Pillai, R., Ratsep, P. C., Shukla, R., Tweedle, M. F. and Zhang, X. *Int. Pat. WO 95/314447*, 1995; *Chem. Abstr.* 1996, **124**, 2324990.
45. Uggeri, F., Fedeli, F., Maiocchi, A., Franzini, M. and Virtuani, M. *Eur. Pat. EP 872479*, 1998; *Chem. Abstr.* 1998, **129**, 325347.
46. Dietrich, B., Viout, P. and Lehn, J. M. *Aspects de la Chimie des Composés Macrocycliques*, Interéditions, Paris, 1991.
47. Knops, P., Sendhoff, N., Mekelburger, H. B. and Vögtle, F. *Top. Curr. Chem.* 1992, **161**, 1.
48. Takenouchi, K., Tabe, M., Watanabe, K., Hazato, A., Kato, Y., Shionoya, M., Koike, T. and Kimura, E. *J. Org. Chem.* 1993, **58**, 6895.
49. McMurry, T. J., Brechbiel, M. W., Kumar, K. and Gansow, O. A. *Bioconjugate Chem.* 1992, **3**, 108.
50. Kaspersen, F. M., Reinhoudt, D. N., Verboom, W. and Staveren, C. J. *Int. Pat. WO 95/01346*, 1995; *Chem. Abstr.* 1995, **125**, 254760.
51. Bradshaw, J. S., Krakowiak, K. E., Izatt, R. M. and Zamecka-Krakowiak, D. J. *Tetrahedron Lett.* 1990, **31**, 1077.
52. Krakowiak, K. E., Bradshaw, J. S. and Izatt, R. M. *J. Heterocycl. Chem.* 1990, **27**, 1585.
53. Bradshaw, J. S., Krakowiak, K. E. and Izatt, R. M. *J. Heterocycl. Chem.* 1989, **26**, 1431.
54. Mishra, A. K., Gestin, J. F., Benoist, E., Faivre-Chauvet, A. and Chatal, J. F. *New J. Chem.* 1996, **20**, 585.
55. Comblin, V., Gilsoul, D., Hermann, M., Humblet, V., Jacques, V., Mesbahi, M., Sauvage, C. and Desreux, J. F. *Coord. Chem. Rev.* 1999, **185–186**, 451.
56. Desreux, J. F., Tweedle, M. F., Ratsep, P. C., Wagler, T. R. and Marinelli, E. R. *US Pat. US 5358704*, 1994; *Chem. Abstr.* 1995, **122**, 75611.
57. Barefield, E. K., Wagner, F., Herlinger, A. W., Dahl, A. R. and Holt, S. *Inorg. Synth.* 1976, **16**, 220.
58. Barefield, K. E. *Inorg. Chem.* 1972, **11**, 2273.
59. Edlin, C. D., Faulkner, S., Parker, D. and Wilkinson, M. P. *J. Chem. Soc. Chem. Commun.* 1996, 1249.
60. Edlin, C. D., Faulkner, S., Parker, D., Wilkinson, M. P., Woods, M., Lin, J., Lasri, E., Neth, F. and Port, M. *New J. Chem.* 1998, **22**, 1359.
61. Cox, J. P. L., Craig, A. S., Helps, I. M., Jankowski, K. J., Parker, D., Eaton, M. A. W., Millican, A. T., Millar, K., Beeley, N. R. A. and Boyce, B. A. *J. Chem. Soc. Perkin Trans.* 2 1990, 2567.

62. Prasad, J. S., Okuniewicz, F. J., Delaney, E. J. and Dischino, D. D. *J. Chem. Soc. Perkin Trans. 2* 1991, 3329.
63. Jacques, V., Mesbahi, M., Boskovic, V. and Desreux, J. F. *Synthesis* 1995, 1019.
64. Brechbiel, M. W., Gansow, O. A., Atcher, R. W., Schlom, J., Esteban, J., Simpson, D. E. and Colcher, D. *Inorg. Chem.* 1986, **25**, 2772.
65. Parker, D. and Millican, T. A. *Int. Pat. WO 89/01476*, 1989; *Chem. Abstr.* 1989, **111**, 174141.
66. Brücher, E., Cortes, S., Chavez, F. and Sherry, A. D. *Inorg. Chem.* 1991, **30**, 2092.
67. Delgado, R., Sun, Y. Z., Motekaitis, R. J. and Martell, A. E. *Inorg. Chem.* 1993, **32**, 3320.
68. Berg, A., Almen, T., Klaveness, J., Rongved, P. and Thomassen, T. *US Pat. US 5419893*, 1993; *Chem. Abstr.* 1989, **111**, 233674.
69. Schultze, L. M. and Bulls, A. R. *US Pat. US 5631368*, 1997; *Chem. Abstr.* 1996, **125**, 315328.
70. van Westrenen, J. and Sherry, A. D. *Bioconjugate Chem.* 1992, **3**, 524.
71. Brittain, H. G. and Desreux, J. F. *Inorg. Chem.* 1984, **23**, 4459.
72. Kang, S. I., Ranganathan, R. S., Emswiler, J. E., Kumar, K., Gougoutas, J. Z., Malley, M. F. and Tweedle, M. F. *Inorg. Chem.* 1993, **32**, 2912.
73. Kline, S. J., Betebenner, D. A. and Johnson, D. K. *Bioconjugate Chem.* 1991, **2**, 26.
74. Aime, S., Botta, M., Ermondi, G., Fedeli, F. and Uggeri, F. *Inorg. Chem.* 1992, **31**, 1100.
75. Kruper, W. J., Rudolf, P. R. and Langhoff, C. A. *J. Org. Chem.* 1993, **58**, 3869.
76. Aime, S., Botta, M., Ermondi, G., Terreno, E., Anelli, P. L., Fedeli, F. and Uggeri, F. *Inorg. Chem.* 1996, **35**, 2726.
77. Forsberg, J. H., Delaney, R. M., Zhao, Q., Harakas, G. and Chandran, R. *Inorg. Chem.* 1995, **34**, 3705.
78. Beeby, A., Parker, D. and Williams, J. A. G. *J. Chem. Soc., Perkin Trans.* 2 1996, 1565.
79. Raganathan, R. S., Marinelli, E. R., Pillai, R. and Tweedle, M. F. *US Pat. US 5573752*, 1996; *Chem. Abstr.* 1996, **124**, 146216.
80. Sherry, A. D., Brown, R. D., III; Geraldes, C. F. G. C., Koenig, S. H., Kuan, K.-T. and Spiller, M. *Inorg. Chem.* 1989, **28**, 620.
81. Sieving, P. F., Watson, A. D., Quay, S. C. and Rocklage, S. M. *Int. Pat. WO 90/12050*, 1990; *Chem. Abstr.* 1991, **115**, 15665.
82. Li, M. and Meares, C. F. *Bioconjugate Chem.* 1993, **4**, 275.
83. Lewis, M. R. and Shively, J. E. *Bioconjugate Chem.* 1998, **9**, 72.
84. Geraldes, C. F. G. C., Sherry, A. D. and Cacheris, W. P. *Inorg. Chem.* 1989, **28**, 3336.
85. Swinkels, D. W., Vanduynhoven, J. P. M., Hilbers, C. W. and Tesser, G. I. *Rec. Trav. Chim. Pays-Bas* 1991, **110**, 124.
86. Lázár, I., Hrncir, D. C., Kim, W. D., Kiefer, G. E. and Sherry, A. D. *Inorg. Chem.* 1992, **31**, 4422.
87. Kim, W. D., Kiefer, G. E., Huskens, J. and Sherry, A. D. *Inorg. Chem.* 1997, **36**, 4128.
88. Huskens, J., Torres, D. A., Kovacs, Z., André, J. P., Geraldes, C. F. G. C. and Sherry, A. D. *Inorg. Chem.* 1997, **36**, 1495.
89. Burai, L., Ren, J., Kovacs, Z., Brücher, E. and Sherry, A. D. *Inorg. Chem.* 1998, **37**, 69.
90. Loussouarn, A., Duflos, M., Benoist, E., Chatal, J. F., Le Baut, G. and Gestin, J. F. *J. Chem. Soc., Perkin Trans. 1* 1998, 237.
91. Lázár, I., Sherry, A. D., Ramasamy, R., Brücher, E. and Kiraly, R. *Inorg. Chem.* 1991, **30**, 5016.

92. Broan, C. J., Cole, E., Jankowski, K. J., Parker, D., Pulukkody, K., Boyce, B. A., Beeley, N. R. A., Millar, K. and Millican, A. *Synthesis* 1992, 63.
93. Pulukkody, K. P., Norman, T. J., Parker, D., Royle, L. and Broan, C. J. *J. Chem. Soc., Perkin Trans. 2* 1993, 605.
94. Aime, S., Botta, M., Dickins, R. S., Maupin, C. L., Parker, D., Riehl, J. P. and Williams, J. A. G. *J. Chem. Soc., Dalton Trans.* 1998, 881.
95. Schmitt-Willich, H., Platzek, J., Gries, H., Schuhmann-Giampieri, G. and Frenzel, T. *Eur. Pat. EP 512661*, 1992; *Chem. Abstr.* 1993, **118**, 109826.
96. Tilstam, U., Borner, H., Nickisch, K., Gries, H. and Platzek, J. *Ger. Pat. DE 418744*, 1994; *Chem. Abstr.* 1994, **121**, 57541.
97. Aime, S., Botta, M., Geninatti Crich, S., Giovenzana, G., Palmisano, G. and Sisti, M. *Bioconjugate Chem.* 1999, **10**, 192.
98. Wainwright, K. P. *Coord. Chem. Rev.* 1997, **166**, 35.
99. Anelli, P. L., Calabi, L., De Haen, C., Fedeli, F., Losi, P., Murru, M. and Uggeri, F. *Gazz. Chim. Ital.* 1996, **126**, 89.
100. Aime, S., Anelli, P. L., Botta, M., Fedeli, F., Grandi, M., Paoli, P. and Uggeri, F. *Inorg. Chem.* 1992, **31**, 2422.
101. Kovacs, Z. and Sherry, A. D. *J. Chem. Soc. Chem. Commun.* 1995, 185.
102. Kovacs, Z. and Sherry, A. D. *Synthesis* 1997, 759.
103. Anelli, P. L., Murru, M., Uggeri, F. and Virtuani, M. *J. Chem. Soc. Chem. Commun.* 1991, 1317.
104. Anelli, P. L., Calabi, L., Dapporto, P., Murru, M., Paleari, L., Paoli, P., Uggeri, F., Verona, S. and Virtuani, M. *J. Chem. Soc. Perkin Trans. 2* 1995, 2995.
105. Li, Z. and Undheim, K. *Acta Chem. Scand.* 1998, **52**, 1247.
106. Argese, M. and Ripa, G. *Int. Pat. WO 99/05145*, 1999; *Chem. Abstr.* 1999, **130**, 153679.
107. Atkins, T. J. *J. Am. Chem. Soc.* 1980, **102**, 6364.
108. Tóth, E., Kiraly, R., Platzek, J., Raduchel, B. E. and Brücher, E. *Inorg. Chim. Acta* 1996, **249**, 191.
109. Platzek, J., Blaszkiewicz, P., Gries, H., Luger, P., Michl, G., MülleFahrnow, A., Radüchel, B. and Sülzle, D. *Inorg. Chem.* 1997, **36**, 6086.
110. Broan, C. J., Cox, J. P. L., Craig, A. S., Kataky, R., Parker, D., Harrison, A., Randall, A. M. and Ferguson, G. *J. Chem. Soc. Perkin Trans. 2* 1991, 87.
111. Bernard, H., Yaouanc, J. J., Clement, J. C., Abbayes, H. D. and Handel, H. *Tetrahedron Lett.* 1991, **32**, 639.
112. Filali, A., Yaouanc, J. J. and Handel, H. *Angew. Chem., Int. Ed. Engl.* 1991, **30**, 560.
113. Yaouanc, J. J., Le Bris, N., Le Gall, G., Clement, J. C., Handel, H. and Abbayes, H. D. *J. Chem. Soc. Chem. Commun.* 1991, 206.
114. Roignant, A., Gardinier, I., Bernard, H., Yaouanc, J. J. and Handel, H. *J. Chem. Soc. Chem. Comm.* 1995, 1233.
115. Gardinier, I., Bernard, H., Chuburu, F., Roignant, A., Yaouanc, J. J. and Handel, H. *J. Chem. Soc. Chem. Commun.* 1996, 2157.
116. Patinec, V., Yaouanc, J. J., Handel, H., Clément, J. C. and des Abbayes, H. *Inorg. Chim. Acta* 1994, **220**, 347.
117. Patinec, V., Gardinier, I., Yaouanc, J. J., Clément, J. C., Handel, H. and des Abbayes, H. *Inorg. Chim. Acta* 1996, **244**, 105.
118. Sieving, P. F., Watson, A. D., Quay, S. C. and Rocklage, S. M. *US Pat. US 5364613*, 1994; *Chem. Abstr.* 1995, **122**, 75613.
119. Ladd, D. L., Hollister, R., Peng, X., Wei, D., Wu, G., Delecki, D., Snow, R. A., Toner, J. L., Kellar, K., Vinay, J. E., Desay, C., Raymond, G., Kinter, L. B., Desser, T. S. and Rubi, D. L. *Bioconjugate Chem.* 1999, **10**, 361.

120. Tóth, E., Pubanz, D., Vauthey, S., Helm, L. and Merbach, A. E. *Chem. Eur. J.* 1996, **2**, 1607.
121. Bombieri, G., Benetollo, F., Polo, A., De Cola, L., Smailes, D. L. and Vallarino, L. M. *Inorg. Chem.* 1986, **25**, 1127.
122. Tsubomura, T., Yasaku, K., Sato, T. and Morita, M. *Inorg. Chem.* 1992, **31**, 447.
123. Alexander, V. *Chem. Rev.* 1995, **95**, 273.
124. Bligh, S. W. A., Choi, N., Evagorou, E. G., McPartlin, M., Cummins, W. J. and Kelly, J. D. *Polyhedron* 1992, **11**, 2571.
125. Sessler, J. L., Mody, T. D., Hemmi, G. W., Lynch, V., Young, S. W. and Miller, R. A. *J. Am. Chem. Soc.* 1993, **115**, 10368.
126. Sessler, J. L., Mody, T. D., Hemmi, G. W. and Lynch, V. *Inorg. Chem.* 1993, **32**, 3175.
127. Chinnick, C. C. T. *Analyst* 1981, **106**, 1203.
128. Grushka, E., Levin, S. and Gilon, C. *J. Chromatogr.* 1982, **235**, 401.
129. Seri, S., Azuma, M. and Iwai, K. *Eur. Pat. EP 481420*, 1991; *Chem. Abstr.* 1992, **117**, 43666.
130. Kumar, K., Chang, C. A., Francesconi, L. C., Dischino, D. D., Malley, M. F., Gougoutas, J. Z. and Tweedle, M. F. *Inorg. Chem.* 1994, **33**, 3567.
131. Inoue, M. B., Inoue, M., Munoz, I. C., Bruck, M. A. and Fernando, Q. *Inorg. Chim. Acta* 1993, **209**, 29.
132. Hagan, J. J., Taylor, S. C. and Tweedle, M. F. *Anal. Chem.* 1988, **60**, 514.
133. Dischino, D. D. and Swigor, J. E. *J. Labelled Compd. Radiopharm.* 1992, **31**, 455.

5 Protein-Bound Metal Chelates

SILVIO AIME, MAURO FASANO, ENZO TERRENO
Università degli Studi di Torino, Torino, Italy

and

MAURO BOTTA
Università del Piemonte Orientale 'Amedeo Avogadro', Alessandria, Italy

1 INTRODUCTION

The study of the interaction of metallic chelates with proteins had been extensively addressed well before the development of MRI applications. In fact, modification of proteins by introducing chelating sites can provide useful probes for various physical studies of protein structure (through EPR, NMR, X-ray analysis and spectrophotometric investigations). Moreover, the chelation of radioactive metal ions has been exploited in nuclear medicine and physiology where a protein can be used to measure a biological function. As far as the use of paramagnetic chelate/protein adducts (either covalent or non-covalent) as contrast agents for MRI is concerned, the driving force has been primarily determined by the expectation that paramagnetic macromolecular complexes would show a marked enhancement of the relaxation rate of tissue protons [1]. In fact, the large molecular size of such adducts results in a slowing down of the molecular motion, thus allowing τ_R to reach values in the range of tenths of nanoseconds. In principle, such an elongation of τ_R with respect to the small-sized paramagnetic units (τ_R of $[Gd\text{-}DTPA(H_2O)]^{2-}$ and related complexes is of the order of 10^{-10} s) should cause a dramatic increase of the observed relaxivity for each metallic center, provided that the values of the other parameters involved in the paramagnetic relaxation process fall within an optimal range (see Chapter 2) [2]. Thus, there is the expectation that macromolecular Gd(III) or Mn(II) chelates endowed with enhanced relaxivity can yield exceptionally dose-effective agents. In particular, the attainment of high relaxivities will allow

The Chemistry of Contrast Agents in Medical Magnetic Resonance Imaging
Edited by A. E. Merbach and É. Tóth.

the development of specific targeting applications. It has been estimated that visualization of receptors on the surface of cells requires 10^2-10^3 paramagnetic chelates to bring about sufficient reduction in the T_1 relaxation time [3]. It is obvious that the availability of systems endowed with higher relaxivities than those of the currently available contrast agents will allow a drastic reduction of such a number.

Up to now, most of the applications involving protein/Gd(III) chelate conjugates have exploited their prolonged intravascular retention as opposed to relatively small molecules, such as $[Gd\text{-}DTPA(H_2O)]^{2-}$, which rapidly equilibrates between the plasma and the extravascular interstitial space. Thus, these systems have been primarily considered for the enhancement of the blood pool and of the sites of abnormal endothelial permeability [4]. Their use permits contrast-enhanced imaging of multiple body regions without repeated dosing of contrast agent. This is possible because their plasma concentration remains relatively stable for the time of the examination as the (renal) elimination first requires their degradation to the individual units of the contrast agent. The observed enhancement in the MR images of the various tissues is primarily a function of blood volume. Thus, highly vascular tissues, such as kidneys, liver, lung and myocardium, show the greatest enhancement, whereas brain and skeletal muscle, having lower blood volumes, display a lesser enhancement.

The formation of macromolecular adducts between a protein and a metal complex can be pursued through the establishment of either covalent or non-covalent linkages. Both approaches have been widely investigated and several interesting systems have been developed. Basically, one of the major drawbacks of the covalently bound conjugates is represented by their metabolic fate. On the other hand, these multimeric derivatives appear to be the only available systems for delivering a high number of paramagnetic complexes at a given site of interest.

The non-covalent approach is based on the use of complexes containing on their surfaces suitable moieties which are able to recognize specific targeting proteins. When the targeting protein is confined in the blood, the paramagnetic adduct between human serum albumin (HSA) and the functionalized complex may display analogous properties to the covalent conjugates in providing detailed images of the vascular system. However, these systems would still maintain excretory pathways typical of small complexes.

2 COVALENTLY BOUND CONJUGATES

The background to this field is provided by a number of interesting systems developed for various applications in medicine (either as diagnostic or therapeutic agents) and in biophysical experiments. In fact, these applications have led to the synthesis of several 'bifunctional chelating agents', which have chelating groups on one end and a chemically reactive functional group on the other. The latter's functionality is then exploited to develop novel classes of biological probes containing metal complexes with particular properties, which are covalently attached to biological macromolecules [5].

2.1 SYNTHESIS

The chelating moiety commonly used is represented by a polyamino-polycarboxylic acid. Several routes are available for its conjugation to a protein:

(i) Use of cyclic DTPA bis-anhydride as the acylating agent at the ε-amino groups of lysine residues [6–8]. The reaction is carried out by adding aliquots of DTPA bis-anhydride (dissolved in DMSO) to a diluted solution of the protein (HEPES buffer, pH 8.8) (Scheme 5.1).

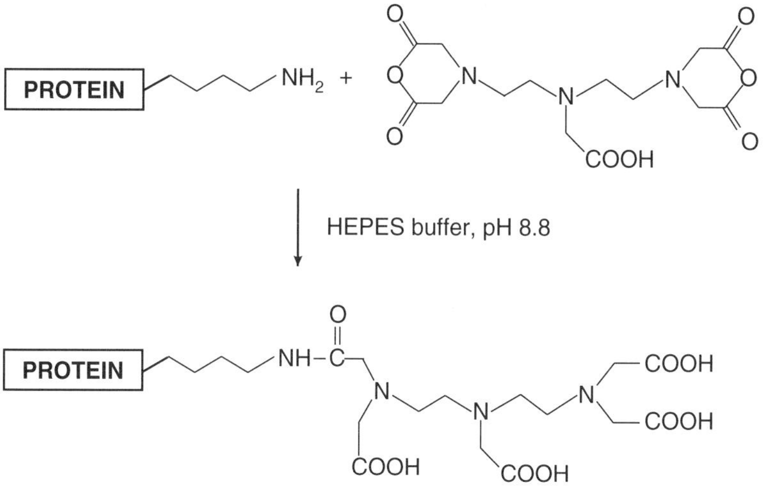

PROTEIN–NH_2 + DTPA bis-anhydride

HEPES buffer, pH 8.8

PROTEIN–NH–C(=O)–DTPA (COOH, COOH, COOH, COOH)

Scheme 5.1

This is the most straightforward procedure to attain conjugation, but it has the main drawback of causing extensive intra- and intermolecular cross-linking. This might not only result in a wide distribution of molecular weights of the synthetized products, but also a decreased affinity towards Gd(III) ions because two of the carboxylic acid groups of DTPA ligand have been transformed into peptide linkages [9].

To limit this undesired reactivity, it has been suggested that work should be carried out at low protein concentrations, although the hydrolysis reaction of the DTPA bis-anhydride in water leads to extensive (and uncontrolled) formation of DTPA. Another source of free DTPA can arise from other undesirable side-reactions of DTPA bis-anhydride that may attack other functionalities on the protein such as the hydroxy group of tyrosine, the mercapto group of cysteine and the imidazole group of hystidine. These reactions yield very unstable acylated products which convert readily to the starting amino acids with the concomitant formation of free DTPA.

(ii) Use of DTPA mixed anhydride [10–11]. As an example, a mixed anhydride can be synthetised by reacting DTPA (in acetonitrile and in the presence of NEt_3 at low temperature) with a stoichiometric amount of isobutylchloroformate (IBCF). The coupling to the protein is carried out as described in (i) (Scheme 5.2).

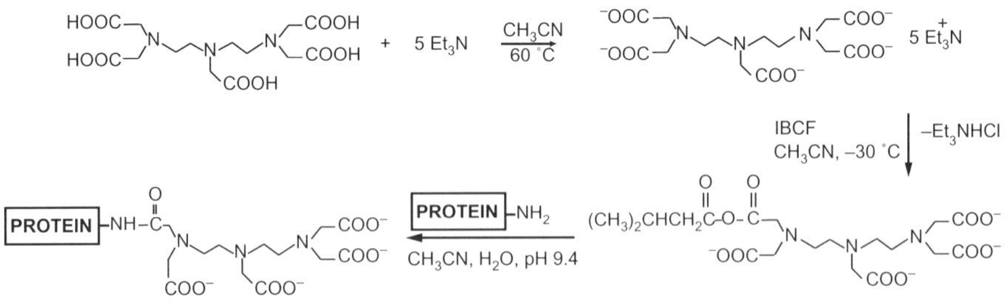

HOOC HOOC N N N COOH COOH COOH + 5 Et_3N $\xrightarrow[60\ ^{\circ}C]{CH_3CN}$ ^-OOC ^-OOC N N N COO^- COO^- COO^- 5 $Et_3\overset{+}{N}$

IBCF, CH_3CN, –30 °C; –Et_3NHCl

$(CH_3)_2CHCH_2C(=O)$-O-C(=O)… N N N ^-OOC COO^- COO^- COO^-

PROTEIN–NH_2; CH_3CN, H_2O, pH 9.4

PROTEIN–NH–C(=O)… N N N COO^- COO^- COO^- COO^-

Scheme 5.2

(iii) Use of squaric acid (SQ) esters as a linker. It has recently been reported that this moiety reacts with the amino groups of proteins under mild conditions, leading to very high yields of the conjugation product [12]. We have applied this procedure by forming a DO3A derivative containing the squaric moiety bound at the unsubstituted nitrogen of the tetraazamacrocycle (Scheme 5.3) [13].

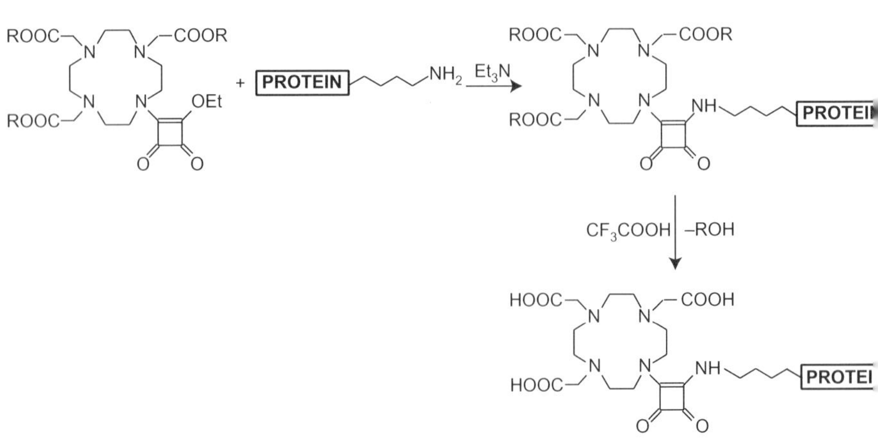

ROOC ROOC COOR N N N N OEt O O + PROTEIN NH_2 $\xrightarrow{Et_3N}$ ROOC ROOC COOR N N N N NH PROTEIN O O

CF_3COOH | –ROH

HOOC HOOC COOH N N N N NH PROTEI O O

Scheme 5.3

It has been found that all of the added chelate binds almost quantitatively. The main drawback appears to be a marked decrease of the affinity of the functionalized macrocycle towards Gd(III) ions as compared to the parent DO3A ligand.

(iv) Use of *N*-hydroxysuccinic ester [14]. As in the case of squaric acid, functionalization of the chelating molecule is activated only at one end and so the possibility of cross-linking is ruled out. Again, this method has been reported to give highly reproducible conjugation yields (Scheme 5.4).

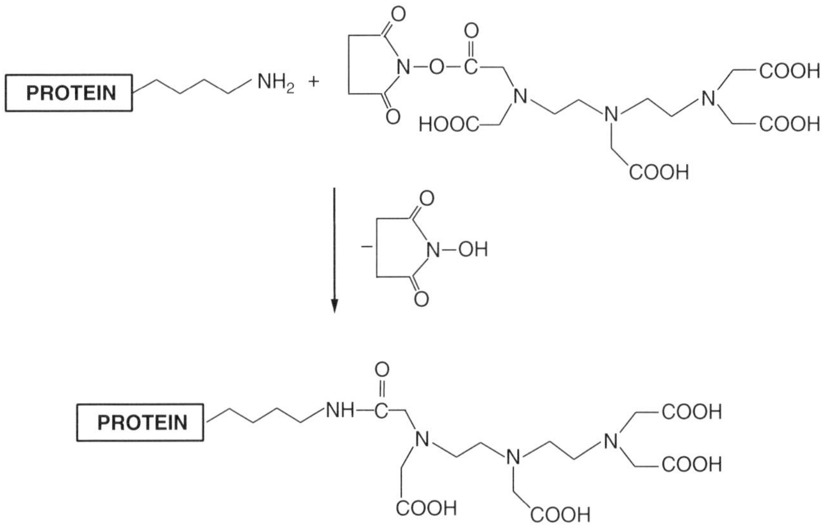

Scheme 5.4

(v) Use of isothiocyanatobenzyl groups as the linking agents [15]. Since the ligating moiety is bound at the ethylenic carbons of the ligand, its coordination denticity is not modified by the introduction of the functionality. This procedure has been applied both to DTPA and TRITA, a polyaminocarboxylate ligand which differs from DOTA in the fact that it is based on a thirteen-membered macrocycle (Chart 5.1)

In general, whatever synthetic approach is used, the macromolecular system is separated by the free ligand by exhaustive dialysis. It has been suggested that the percentage conjugation efficiency can be determined by treating the reaction mixture with radioactive $^{111}InCl_3$ in the presence of nitrilotriacetic acid (NTA). The ^{111}In bound to the protein and the unbound ^{111}In are separated by means of size-exclusion high performance liquid chromatography and quantified by using a flow-through radiometer detector [8].

The complexation of Gd(III) ions to the chelating moieties of the macromolecular system is usually carried out at low temperature (i.e. 4 °C) by stirring

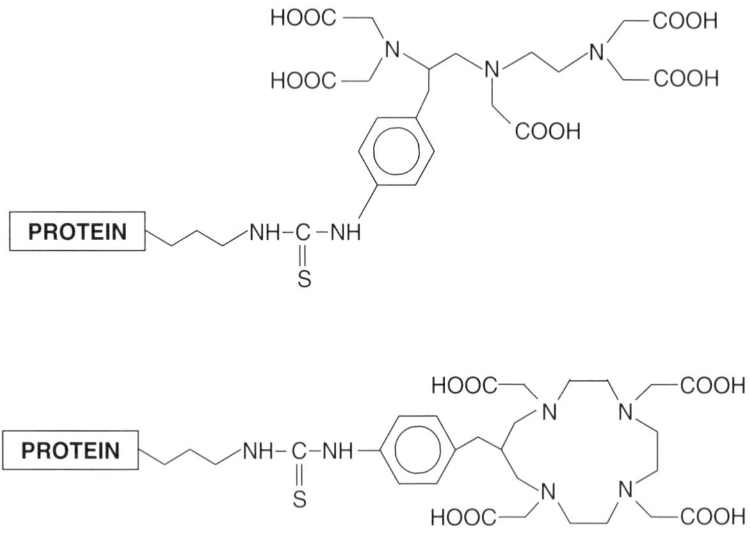

PROTEIN NH-C-NH S HOOC HOOC N N N COOH COOH COOH

PROTEIN NH-C-NH S HOOC HOOC N N N N COOH COOH

[**Chart 5.1**]

a solution of the functionalized protein with $Gd(NTA)_2$. The ternary protein/Gd(III) chelate is purified by exhaustive dialysis with several changes of citrate buffer (pH 6.5) and finally with distilled water.

2.2 RELAXOMETRIC STUDIES

The most widely studied system is represented by albumin labeled with Gd-DTPA, which has been prepared with the aim of providing an agent that would distribute primarily in the intravascular space, thus giving the physician direct information on the status of the blood volume in various tissues. HSA contains 58 lysine residues and DTPA has been coupled to it by using one of the five routes described above.

The degree of Gd-DTPA substitution on the protein carrier varied among different preparations and the relaxivity (r_1 per Gd(III) ion) was reported to fall in the range of about 10.0 to 20.0 mM^{-1} s^{-1}, depending on the experimental conditions [4, 14, 16, 17].

The issue concerning the relationship between the longitudinal relaxivity and the number of Gd-DTPA units on albumin has been specifically addressed by Spanoghe *et al.* [14], who reported that there is a significant increase of the relaxation rate upon increasing the DTPA/BSA ratios. This effect was most pronounced up to a DTPA/BSA ratio of 8, whereas at higher ratios the relaxation enhancement reached a plateau (note: these measurements were carried out at 2.4 T).

We have investigated in detail the frequency and temperature dependence of proton and oxygen-17 relaxivities of a specimen of HSA containing 43 units of Gd-DTPA, prepared according to route (ii) by L. Gozzini and co-workers at Bracco s.p.a, Milan (Figures 5.1, 5.2 and 5.3). At 25 °C and 0.47 T, r_1 of this system is 14.9 mM^{-1} s^{-1}, which increases to 18.8 mM^{-1} s^{-1} as the temperature is raised to 39 °C. The observed behavior is typical of systems whose relaxivity is limited by the occurrence of a long exchange lifetime of the coordinated water molecule, i.e. $\tau_M > T_{1M}$ (see Chapter 2). By measuring $^{17}O - R_{2p}$ as a function

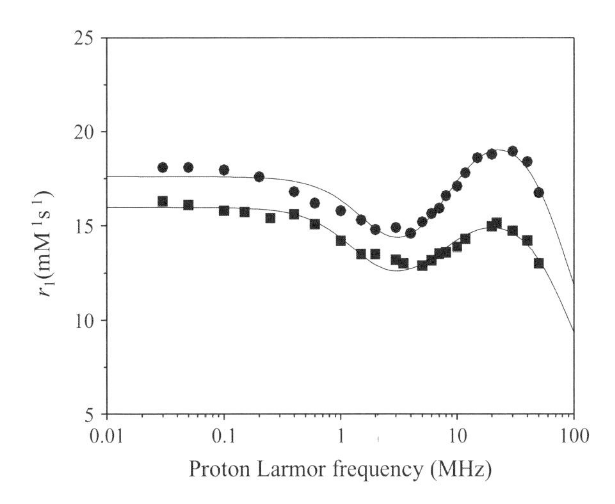

Figure 5.1 ^{1}H NMRD profiles of (Gd-DTPA)$_{45}$–HSA recorded at 25 °C (■) and 39 °C (•) (pH 7, 50 mM borate buffer, 0.1 M NaCl).

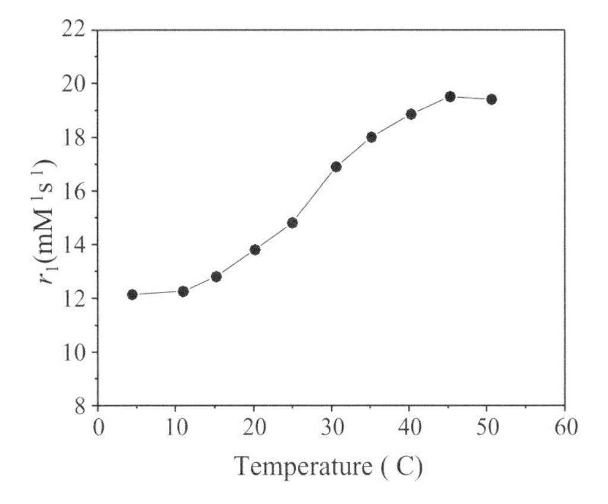

Figure 5.2 Temperature dependence of r_1 for (Gd-DTPA)$_{45}$–HSA at 0.47 T; the relaxivity is limited by a long water-exchange lifetime (pH 7, 50 mM borate buffer, 0.1 M NaCl).

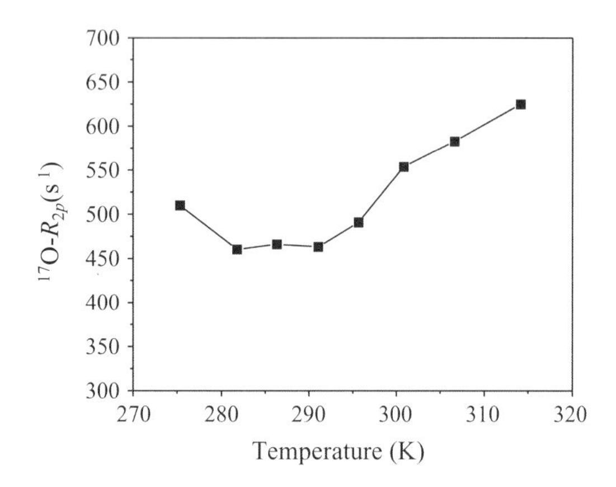

Figure 5.3 Temperature dependence of water $^{17}O\text{-}R_{2p}$ for $(Gd\text{-}DTPA)_{45}$–HSA at 2.1 T. The profile suggests that Gd(III) chelates with different water-exchange lifetimes are present in solution ([Gd] = 40 mM, pH 7, 50 mM borate buffer, 0.1 M NaCl).

of temperature, we gained further support to this view, although the resulting profile does not display the well-defined bell-shaped pattern as is commonly found for the small complexes. It is likely that the different environment of each Gd-DTPA unit can affect the exchange properties of the coordinated water, thus resulting in a much flattened distribution of the R_{2p} values with non-negligible contributions from moieties endowed with high exchange rates.

The 'quenching' effect of a relatively long exchange lifetime of the coordinated water can be drawn also from the work of Niemi and co-workers [15], who found that BSA (ITCB-Gd-DTPA)$_n$ and (ITCB-Gd-TRITA)$_n$ show an opposite behavior with temperature, i.e. the nuclear magnetic resonance dispersion (NMRD) profile of the latter at low temperature is much higher than that at 39 °C, as expected for systems where $\tau_M < T_{1M}$, whereas the NMRD profile of the DTPA-containing derivative is higher at 39 °C than at 5 °C.

In order to enhance markedly the relaxation enhancement per albumin molecule, Sieving *et al.* [10] developed an interesting approach based on the anchoring to HSA of polylysine chains containing up to ninety Gd(III) chelates. They first synthetised a poly-(L-lysine) (degree of polymerization ca 100) containing 60 to 90 chelating groups (DTPA or DOTA) per molecule. Once complexed to Gd(III) ions, the paramagnetic chains were conjugated to HSA (Scheme 5.5) as a prototype for systems which may be used for specific targeting applications. The polychelate approach described in this paper allows the attachment of large numbers of chelates to a protein through a single covalent bond, with a minimum amount of functional modification of the protein.

The relaxivities of the conjugates were essentially the same as those of the polychelates from which they were prepared (ca 10.0 mM^{-1} s^{-1} for Gd-PL-DTPA-HSA and ca 13.0 mM^{-1} s^{-1} for Gd-PL-DOTA-HSA).

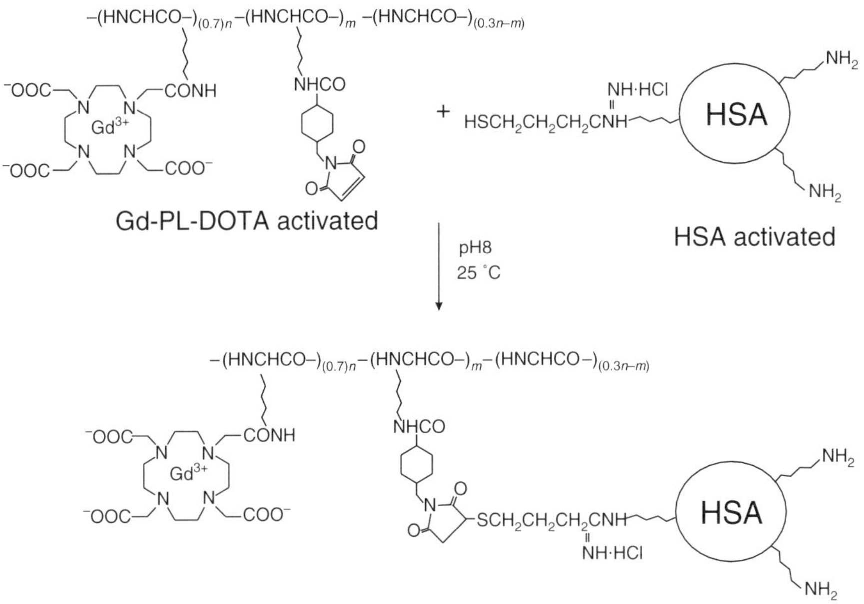

Scheme 5.5

The relatively small relaxation enhancement shown by Gd(III) chelates when bound to poly(L-lysine) is mostly accounted for by the high internal mobility of the paramagnetic moiety, allowed by the free rotation around the four-carbons chain of each Lys unit at which the substitution takes place [18]. An analogous result has been observed by using the squaric acid unit as a linker between the chelating cage and the macromolecule (route (iii) above). In fact, in the case of a conjugate with BSA containing ca 50 Gd-DO3A-SQ moieties, an r_1 value of 15.5 $mM^{-1}\ s^{-1}$ was determined at 0.47 T and 25 °C [17]. Interestingly, in the case of the related polyornithine-Gd-DO3A-SQ (ca 114 ornithine residues and 30 paramagnetic chelates) it has been found that the water proton relaxation rate is pH dependent [19]. The observed behavior is significantly higher than for the BSA derivative ($r_1 = 23.0\ mM^{-1}\ s^{-1}$ at pH < 4.5 and $r_1 = 32.0\ mM^{-1}\ s^{-1}$ at pH > 8) and is related to structural changes occurring in the polypeptide upon protonation of the side-chain NH_2 groups. In fact, at acidic pH values, the repulsion among positively charged amino groups induces the occurrence of a highly flexible structure, whereas at basic pH values the formation of intra-chain hydrogen bonds yields the formation of an α-helix, thus resulting in an overall rigidity of the macromolecular system.

The methodologies developed for pursuing conjugation to albumin (and polylysine) have been exploited to link Gd-chelates to other proteins. For instance, Niemi and co-workers reported the binding of stable paramagnetic

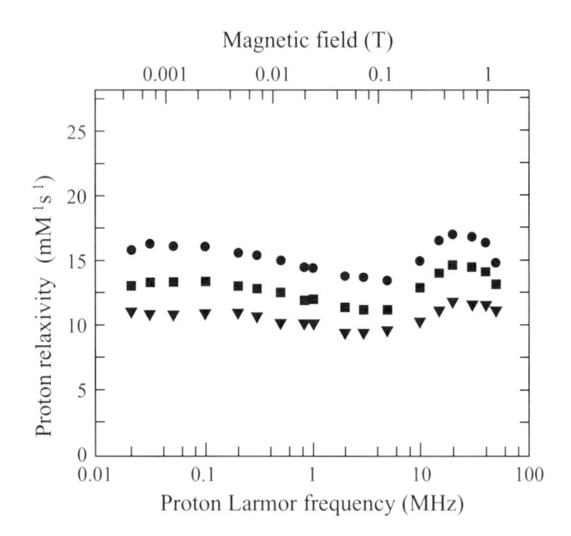

Figure 5.4 ^{1}H NMRD curves of (Gd-DTPA)$_n$–FIB recorded at 35 °C: n = 45 (●); n = 98 (■); n = 134 (▼). The increase of the number of paramagnetic chelates covalently bound to the fibrinogen decreases the relaxivity per Gd(III) ion. Adapted from [15].

chelates to porcine serum fibrinogen (FIB) [15, 16]. They found that the loading of the protein with progressively more Gd(III) chelates causes the relaxivity to decrease (Figure 5.4).

Finally, labeling of immunoglobulins and monoclonal antibodies with Gd(III) chelates has been considered for targeting tumors. To this purpose, Curtet *et al.* [20] developed systems based on Polylysine chains containing a high number (n = 24–28) of paramagnetic Gd(III) chelates (with DTPA or DOTA ligands) grafted to anti-carcinoembryonic antigen (CCA)F$(ab')_2$ fragments. They found that, 24 h after intravenous injection of the Gd-labeled immunoconjugate to mice bearing human colorectal carcinoma, there was a detectable relaxation enhancement in the tumor [20]. Further improvements are expected upon increasing the number of paramagnetic metal ions at the targeting sites. Wu *et al.* [21] proposed a route based on the use of a monoclonal antibody bearing polyamidoamine dendrimers chemically modified by reaction with DOTA and DTPA bifunctional chelators.

The possibility of visualizing receptors overexpressed on the surface of tumor cells is a very important (and challenging) direction for further work in this field. In order to pursue this goal, one needs to conjugate high-relaxivity chelates to synthons endowed with high-binding affinity for the targeting site. In this context, an important issue (which is not so demanding for nuclear medicine applications) appears to be the biological problem of the accessibility of such large-sized molecules to the receptor sites [3].

3 NON-COVALENT CONJUGATES

The toxicological problems associated with the use of covalently bound conjugates have prompted the search for paramagnetic metal complexes which are able to form non-covalent adducts with endogenous proteins. Most of the work has been focused on serum albumins, whose concentration in blood is by far the highest [22]. Moreover, their main physiological role deals with the transport of a number of substrates and detailed investigations of the binding interaction have been carried out for several drugs and metabolites. Over the years, on the basis of extensive competitive assays, it has been possible to gain a detailed picture of the various domains at which the binding of a given substrate takes place. Finally, in the last few years, the solid-state X-ray structures of HSA have become available (PDB codes: 1AO6 and 1BM0[23]; 1BJ5 and 1BKE[24]; 1UOR[25]) and a better definition of the structural properties of the binding interactions between the various substrates and the protein now appears possible. Thus, the information gained from investigations on the binding interactions between albumin and a variety of substrates provided a good basis for the design of contrast agents able to form tightly bound adducts with HSA. Of course, a high binding affinity is an important requisite because the relaxation enhancement promoted by these systems in blood serum is dependent not only on the relaxivity of the free and bound complex, but also on the molar fraction of contrast agent bound to the protein. As anticipated for the paramagnetic macromolecular conjugates discussed above, such contrast agents are expected to have a prolonged lifetime in the vascular system and then to act as reporters of the blood volume in the region of interest.

3.1 EXPERIMENTAL METHODS FOR DETERMINING THE PROTEIN BINDING OF METAL CHELATES

In spite of the large number of methodologies now available for investigating the binding of a small molecule to a protein, the data published so far concerning the protein binding of metal chelates of interest in the field of contrast agents for MRI applications have been collected mainly by using three techniques, i.e. equilibrium dialysis, ultrafiltration and Proton Relaxation Enhancement.

The first two methods are the most widely used for following the drug-protein binding in pharmacological studies [26], whereas the latter technique is the method of choice when the substrate interacting with the protein is a paramagnetic species [1].

Equilibrium dialysis and ultrafiltration are separative methods, in which a membrane or a filter of suitable size are employed to allow the separation and the subsequent quantitative analysis of the unbound substrate.

In a typical equilibrium dialysis experiment involving serum albumin, a membrane with a molecular weight cut-off of ca 10 000 Da is used. The two compartments of the dialysis cell are filled with equal volumes of a solution containing the substrate at a given concentration, whereas the protein (usually at physiological concentration) is added only to one compartment. Then, the system is allowed to reach the equilibrium condition, and this operation, owing to the very slow diffusion processes, usually requires very long times (> 10 h). At equilibrium, the concentration of the substrate in the cell without the protein (the unbound substrate) is determined by means of conventional analytical methods. Finally, the experiment is completed by repeating the same procedure with different starting concentrations of substrate and the same amount of protein.

The main drawbacks of this technique are represented by the long time necessary to perform the analysis and by some concerns about the possible interference of the membrane on the binding equilibrium [27].

The influence of the separative device on the determination of the binding parameters also represents the main limitation for the ultrafiltration technique, but in this case the time required for the analysis is significantly reduced.

In this method, the solution containing both the protein and the substrate is put in a vial containing the filter and the vial is centrifuged until a minimum amount of filtrate (ca 5 % with respect the initial volume) is obtained. In this way, the consequent increase of the protein concentration during the filtration process may be neglected.

In analogy to the case of equilibrium dialysis, the concentration of the unbound species is determined by analyzing the filtrate obtained and this procedure is repeated by using different initial concentrations of substrate.

In separative techniques, the data are usually analyzed according to the Scatchard equation [28] which, in the presence of multi-classes (j) of multi-binding sites (i) on the protein, assumes the following form:

$$r = \sum_{i=1}^{J} \frac{n_i C_F K_{Ai}}{1 + C_F K_{Ai}} \tag{5.1}$$

where r indicates the number of substrate molecules bound per protein molecule, as follows:

$$r = \frac{C_B}{P_{r_T}} \tag{5.2}$$

In the above, n_i refers to the number of equivalent binding sites for each class, K_{Ai} is the association constant of the n_i sites, C_F and C_B are the molar concentration of the free and bound substrate, respectively, and P_{r_T} is the total protein concentration.

The determination of the binding parameters may be successfully performed through a non-linear least-square regression analysis of Equation (5.1), with the

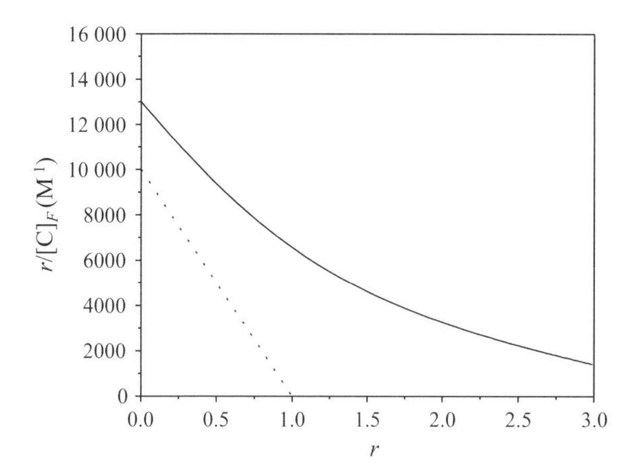

Figure 5.5 Simulated Scatchard plots showing a comparison between binding to a single class of binding sites with $n = 1$ and $K_A = 1 \times 10^4\ \mathrm{M}^{-1}$ (dotted line) or to two classes of binding sites with $n_1 = 1, K_{A1} = 1 \times 10^4\ \mathrm{M}^{-1}$and $n_2 = 3, K_{A2} = 1 \times 10^3\ \mathrm{M}^{-1}$ (continuous line).

data often being reported in the form of a Scatchard plot (r/C_F vs. r) (Figure 5.5).

This diagram is linear in the presence of a single class of binding sites (and in this case, the number of the equivalent sites n can be easily obtained from the intercept to the x-axis), but often the Scatchard plot is non-linear and this may be accounted for by the presence of additional classes of weaker binding sites available to the substrate. A correct analysis of the Scatchard plot requires, therefore, the choice, a priori, of the binding model which best fits the experimental data.

The Proton Relaxation Enhancement (PRE) method is a non-separative technique in which the binding parameters can be obtained by exploiting the differences in the NMR water solvent relaxation rates between the bound and the unbound substrates [1]. Since the relaxation rate can be markedly increased in the presence of a paramagnetic substrate interacting with a protein, this method is perfectly tailored to investigate the binding of a paramagnetic metal chelate. Besides providing information about the association constant and the number of the interaction sites on the protein, this technique also allows the assessment of the relaxivity of the paramagnetic complex bound to the protein (r_1^b) which is a parameter strictly related to the efficacy of the contrast agent.

Like most of the spectroscopic methods used for determining an equilibrium constant, in the PRE technique the concentration of the species involved in the binding equilibrium can not be determined directly. This is the main limitation of this method, thus making the analysis more difficult in the case of multi-site binding.

Given that in the field of contrast agents for MRI, most of the data reported on the non-covalent interactions between a contrast agent and a protein has been obtained by the PRE technique, it is useful to discuss here in more detail the basic principles of this method.

In the case of a binding equilibrium between a paramagnetic substrate (S) and a protein (Pr):

$$\mathrm{S} + \mathrm{Pr} \rightleftharpoons \mathrm{S} - \mathrm{Pr} \tag{5.3}$$

when involving a single class of equivalent binding site, the association constant may be expressed as follows:

$$K_A = \frac{[\mathrm{S} - \mathrm{Pr}]}{[\mathrm{S}]\ [n\mathrm{Pr}]} \tag{5.4}$$

in which the term $[n\mathrm{Pr}]$ indicates the concentration of the equivalent and independent binding sites on the protein.

In an aqueous solution containing the two interacting species, the measured longitudinal water proton relaxation rate (R_{1obs}) is given by the sum of the contributions arising from the unbound and the bound species, as well as the diamagnetic contribution of the protein (R_{1Pr}) itself:

$$R_{1obs} = \left(r_1[\mathrm{S}] + r_1^b[\mathrm{S} - \mathrm{Pr}]\right)1000 + R_{1\mathrm{Pr}} \tag{5.5}$$

where r_1 and r_1^b are the millimolar relaxivities of the unbound and bound substrate, respectively.

Combination of Equations (5.4) and (5.5) allows us to correlate the measured $R_{1\mathrm{obs}}$ to the binding parameters, K_A and n, as follows:

$$R_{1obs} = \frac{(K_A S_T + nK_A Pr_T + 1) - \sqrt{(K_A S_T + nK_A Pr_T + 1)^2 - 4K_A^2 S_T n Pr_T}}{2K_A} - (r_1^b - r_1 + r_1 S_T)1000 + R_{1Pr} \tag{5.6}$$

where S_T and Pr_T are the total molar concentrations of the substrate and the protein, respectively.

The experimental procedure consists of carrying out two distinct titrations, called 'E'- and 'M-', in accordance to the nomenclature proposed by Dwek [1].

In the E-titration, a fixed concentration of substrate (usually an S_T value lower than 0.2 mM is used) is titrated with the protein and the observed relaxation rates increase according to the K_A value (Figure 5.6). One may note that the curves reported in the figure are not asymptotic, as would be expected when the substrate is fully bound to the protein, owing to the diamagnetic contribution of the added protein to the overall relaxation rate.

By using a low S_T value, the formation of the macromolecular adduct is favored, and furthermore the presence of additional weaker binding sites on the protein has a smaller effect on the measured relaxation rate.

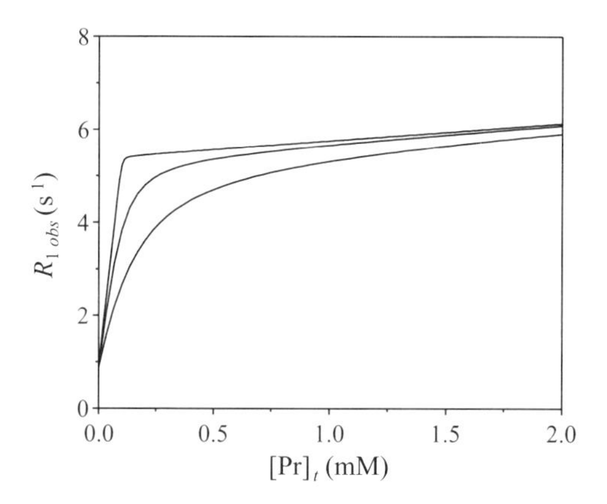

Figure 5.6 Calculated PRE E-titrations reporting the effect of increasing binding strength on the observed relaxation rate: $K_A = 5 \times 10^6\,M^{-1}$ (top curve); $5 \times 10^4\,M^{-1}$ (intermediate curve); $1 \times 10^4\,M^{-1}$ (bottom curve). The other parameters are $n = 1, r_1 = 5\,mM^{-1}\,s^{-1}, r_1^b = 50\,mM^{-1}\,s^{-1}$, $[Gd] = 0.1$ mM (the calculated diamagnetic contribution refers to HSA).

On this basis, the quantitative analysis of an E-titration provides an accurate estimation of the relaxivity of the substrate (r_1^b) bound to the site with higher affinity, whereas the determination of K_A and n is not as straightforward because this experiment is not particularly sensitive to n, particularly when K_A is small. For this reason, the protein affinity of a given substrate, as determined by the E-titration, may be more conveniently expressed in terms of the nK_A term.

A route to obtain an independent evaluation of K_A and n may be pursued through analysis of the data obtained from an M-titration in which a fixed concentration of protein (usually the physiological concentration is used for HSA) is titrated with the paramagnetic substrate.

In the presence of a single class of binding sites with relatively high affinity, the increase of R_{1obs} upon addition of the substrate is almost linear until the interaction sites are fully saturated (Figure 5.7). After this point, the change in the slope of the curves reflects the difference in the relaxivity between the bound and the unbound form of the substrate. It follows that the S_T/Pr_T ratio at the inflection point roughly corresponds to n.

On this basis, the determination of K_A is carried out through the fitting of the experimental data collected at S_T values lower than the inflection point and by fixing n and r_1^b (taken from the E-titration).

Unfortunately, in real cases and particularly when HSA is used, the inflection point may be rather difficult to identify, mainly because of the presence of weaker binding sites which cause an increase of R_{1obs} after the higher-affinity sites have been saturated. The same difficulty is encountered when the substrate binds weakly to the protein (see the bottom curve of Figure 5.7).

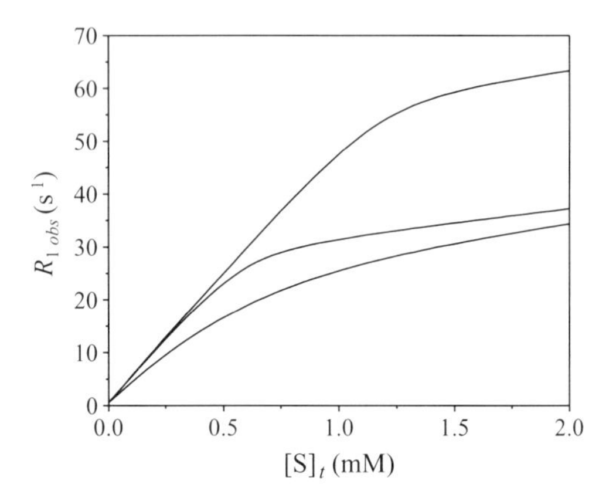

Figure 5.7 Simulated PRE M-titrations showing the effect of different binding strengths and number of equivalent binding sites on the observed relaxation rate: $K_A = 5 \times 10^4\,M^{-1}, n = 2$ (top curve); $K_A = 5 \times 10^4\,M^{-1}, n = 1$ (intermediate curve); $5 \times 10^3\,M^{-1}, n = 1$ (bottom curve). The other parameters are $r_1 = 5\,mM^{-1}\,s^{-1}$, $r_1^b = 50\,mM^{-1}\,s^{-1}$ (the calculated diamagnetic contribution refers to 0.6 mM HSA).

In these cases, it may be useful to represent the experimental data as a Scatchard plot, even if much care has to be exercised in interpreting the results.

For this purpose, it is convenient to re-arrange Equation (5.5) for the concentration of the bound substrate [S-Pr]:

$$[\mathrm{S-Pr}] = \frac{r_1 S_T 1000 - R_{1obs} + R_{1Pr}}{(r_1^b)1000} \tag{5.7}$$

from which the values of r and [S] may be easily calculated.

In principle, unlike the separative methods, the Scatchard plot resulting from a PRE titration may provide useful information only for the binding sites whose r_1^b values are known. Usually, since this method allows us to determine accurately only the r_1^b values for the equivalent binding sites with higher affinity, the presence of weaker interactions is not considered in the analysis.

Furthermore, it has to be kept in mind that the error associated with the indirect determination of the concentration of the unbound substrate is larger than in the separative methods. Thus, an increased data scattering in the Scatchard plot from a PRE measurement is expected, in particular when the unbound concentration is very small, i.e at low r values and for substrates with high affinities towards the protein.

The only possibility for emphasizing the presence of non-equivalent binding sites through PRE measurements occurs when the relaxivity of the weaker site is significantly higher than that of the stronger one. Jenkins *et al.* [29] observed such behavior by investigating the binding properties to HSA of a series of Fe(III) chelates.

In the literature, few papers report a comparison between the results obtained on the same systems by using equilibrium dialysis and PRE technique [30, 31]. In all cases, the agreement was widespread and, furthermore, the PRE method allowed the estimation of weak association constants not detectable with the equilibrium dialysis approach.

3.2 BINDING STUDIES

As it has been already pointed out, most of the data regarding the non-covalent interactions between a metal chelate of interest in the field of contrast agents for MRI applications and a protein has been obtained by using serum albumin.

It is well established that the presence of hydrophobic residues as well as negative electric charges are the basic requirements for binding a substrate to serum albumin [22] and, therefore, it is expected that the same features exist for the metal chelates as well.

The effect of the hydrophobicity of the paramagnetic complex on the affinity towards serum albumin has been observed by comparing the K_A values for a series of Gd(III) complexes bearing an increasing number of lipophilic residues (Chart 5.2) [32].

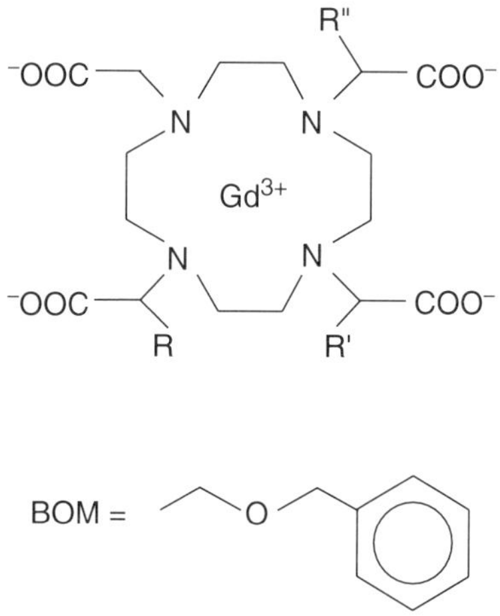

BOM =

Gd-DOTA(BOM): R = R'' = H; R' = BOM

cis-Gd-DOTA(BOM)$_2$: R'' = H; R' = R = BOM

trans-Gd-DOTA(BOM)$_2$: R' = H; R'' = R = BOM

Gd-DOTA(BOM)$_3$: R = R' = R'' = BOM

[Chart 5.2]

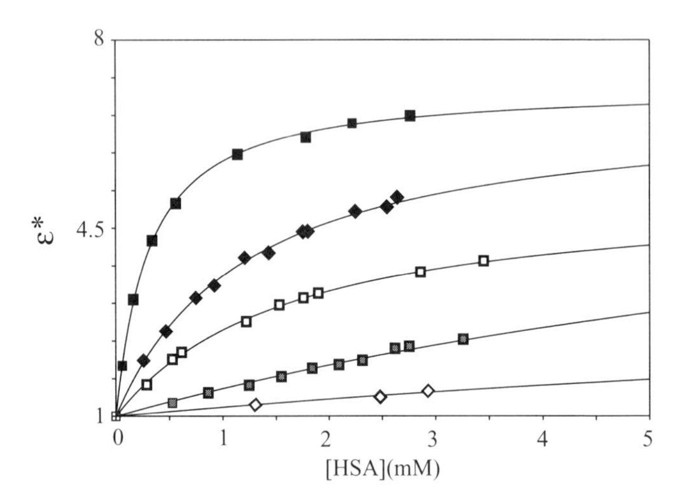

Figure 5.8 PRE E-titration data obtained for HSA with 0.1 mM solutions of [Gd-DOTA(H_2O)]$^-$ (□), [Gd-DOTA(BOM)(H_2O)]$^-$ (◆), *cis*-[Gd-DOTA$(BOM)_2$(H_2O)]$^-$ (□), *trans*-[Gd-DOTA$(BOM)_2$(H_2O)]$^-$ (■) and [Gd-DOTA$(BOM)_3$(H_2O)]$^-$ (■) (0.47 T, 25 °C, 50 mM phosphate buffer, pH 7.4).

In Figure 5.8, the PRE data obtained from E-titrations for such complexes in the presence of HSA are reported and compared with the behavior of the parent [Gd-DOTA(H_2O)]$^-$ species. The results are expressed in terms of the enhancement factor ε^*, which is related to the measured R_{1obs}, according to the following relationship:

$$\varepsilon^* = \frac{R_{1obs} - R_{1HSA}}{R_{1GdL} - R_{1water}} \tag{5.8}$$

where R_{1HSA}, R_{1GdL} and R_{1water} refer to the measured relaxation rates of solutions containing only the protein, the paramagnetic complex and pure water, respectively.

The [Gd-DOTA(H_2O)]$^-$ complex, which possesses only one negative electric charge, showed a small increase of the relaxation rate in the presence of HSA. Although this result may be interpreted on the basis of a very weak interaction to the protein, it has been pointed out that the observed relaxation enhancement may also be a consequence of the reduced molar concentration of water protons in the presence of a large amount of protein [33].

All of the substituted Gd-DOTA complexes increase their relaxation rates in the presence of HSA, owing to the formation of the macromolecular adduct, but the enhancement appears more steeply for the three-substituted [GdDOTA$(BOM)_3$(H_2O)]$^-$ species, thus suggesting a higher affinity for this complex. In fact, the K_A value shown by this complex is almost one order of magnitude higher than those obtained for the two isomers (*cis*- and *trans*-) of [GdDOTA$(BOM)_2$(H_2O)]$^-$, whereas for the [GdDOTA(BOM)(H_2O)]$^-$ chelate only a weak interaction is detectable (Table 5.1).

Table 5.1 Effect of the number of hydrophobic residues on the binding strength to HSA for a series of Gd-DOTA derivatives (see Chart 5.2) ($n = 2$, 25 °C, pH 7.4, 50 mM phosphate buffer).

Metal complex	$K_A (M^{-1})$
$[Gd\text{-}DOTA(BOM)(H_2O)]^-$	$< 1 \times 10^2$
cis-$[Gd\text{-}DOTA(BOM)_2(H_2O)]^-$	3.2×10^2
trans-$[Gd\text{-}DOTA(BOM)_2(H_2O)]^-$	3.6×10^2
$[Gd\text{-}DOTA(BOM)_3(H_2O)]^-$	1.7×10^3

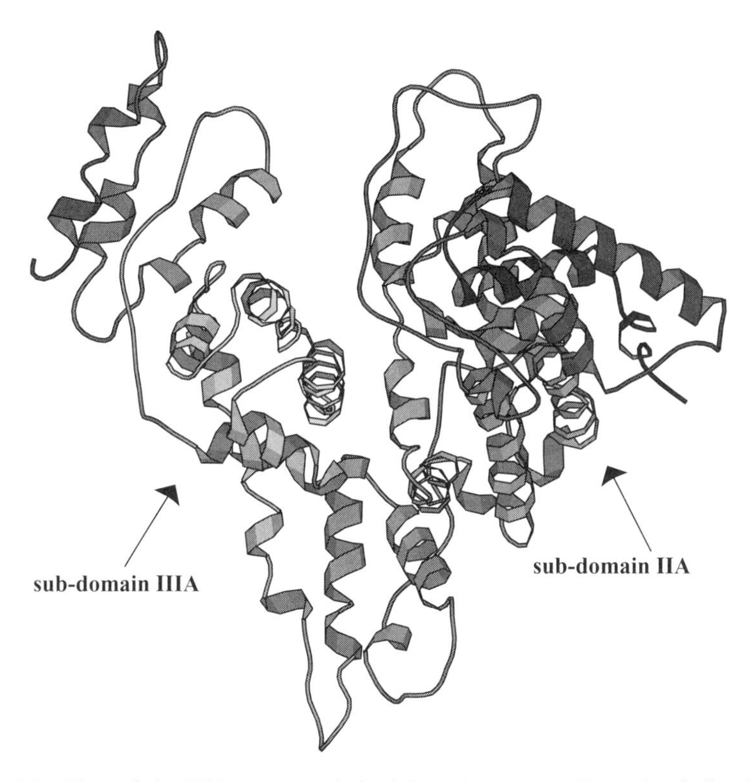

Figure 5.9 View of the HSA structure derived from X-ray crystallography, indicating the location of the sub-domains IIA and IIIA [25].

A more detailed study to assess the number and the location of the binding regions occupied by these complexes on the protein was performed through analysis of the M-titrations in the presence of competitor probes such as ibuprofen and warfarin, which are known to strongly bind at regions located in the sub-domains IIA and IIIA of the protein, respectively (Figure 5.9) [22].

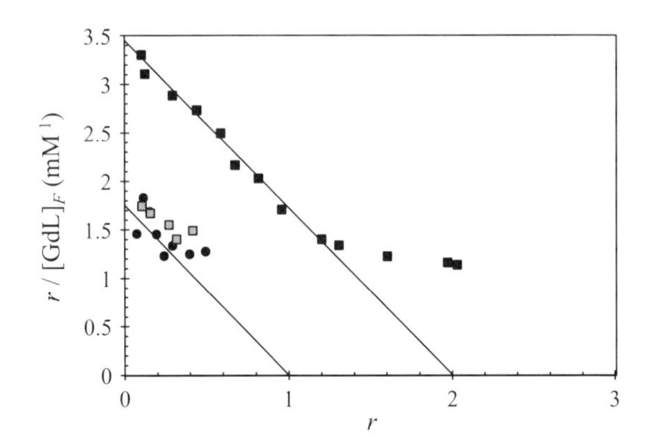

Figure 5.10 Scatchard plots reporting the binding to HSA of $[Gd\text{-}DOTA(BOM)_3(H_2O)]^-$ (■), suggesting the presence of two equivalent binding sites. The deviation from the straight line at $r > 1.2$ may be ascribed to the interaction with weaker binding sites. The metal complex is displaced in the presence of ibuprofen (□) and warfarin (•), at proportions of 1:1 to HSA (0.47 T, 25 °C, 50 mM phosphate buffer, pH 7.4, [HSA] 0.6 mM).

The results of such measurements (Figure 5.10) clearly indicate that two nearly equivalent binding regions, which nicely correspond to those used by the two competitor drugs warfarin and ibuprofen, are available on HSA for $[GdDOTA(BOM)_3(H_2O)]^-$.

The fundamental role of the hydrophobic substituent to drive the binding to the HSA is confirmed by evaluating the temperature dependence of the association constant. From this analysis, the interaction turns out to be entropically driven, as it is expected when the binding is dominated by hydrophobic forces.

In addition, as mentioned above, the number of negative electric charges on the complex is an important factor for the enhancement of the affinity towards the protein. This effect is evident by analyzing the behavior of a series of Gd-DTPA derivatives functionalized with the same hydrophobic residue, but with an additional negative charge on the coordination cage of the complex (Chart 5.3).

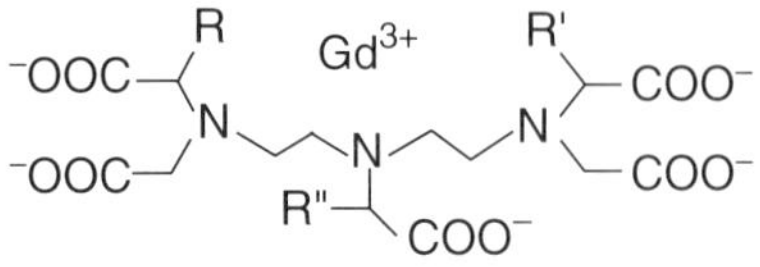

Gd-BOPTA: R' = R" = H; R = BOM

Gd-DTPA$(BOM)_2$: R" = H; R = R' = BOM

Gd-DTPA$(BOM)_3$: R = R' = R" = BOM

[Chart 5.3]

Table 5.2 Effect of the number of hydrophobic groups on the binding to HSA for a series of Gd-DTPA derivatives (see Chart 5.3) (25 °C, pH 7.4 50 mM phosphate buffer).

Metal complex	$nK_A(M^{-1})$
$[Gd\text{-}BOPTA\ (H_2O)]^{2-}$	4.0×10^2
$[Gd\text{-}DTPA(BOM)_2(H_2O)]^{2-}$	3.6×10^3
$[Gd\text{-}DTPA(BOM)_3(H_2O)]^{2-}$	4.0×10^4

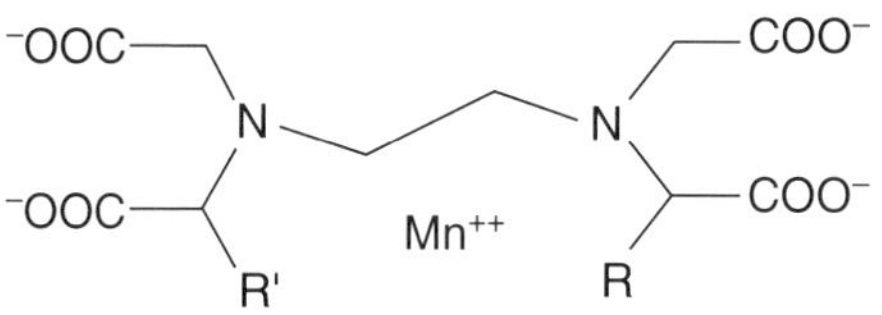

$^{-}$OOC, N, N, COO^{-}, $^{-}$OOC, COO^{-}, Mn^{++}, R', R

Mn-EDTA(BOM): R = H; R' = BOM

Mn-EDTA(BOM)$_2$: R = R' = BOM

[Chart 5.4]

Table 5.3 Association constants to HSA for two Mn(II)-EDTA chelates (see Chart 5.4) bearing BOM substituents (25 °C, pH 7.4, 50 mM phosphate buffer).

Metal complex	$nK_A(M^{-1})$
$[MnEDTA(BOM)(H_2O)]^{2-}$	1.5×10^3
$[MnEDTA(BOM)_2(H_2O)]^{2-}$	1.9×10^4

The bis-anionic DTPA-complexes bind tightly to HSA and show affinities of almost one order of magnitude higher than the corresponding macrocyclic derivatives with the same number of hydrophobic synthons (Table 5.2) [17]. Furthermore, keeping the same type of hydrophobic synthon and the same number of negative charges, it can be shown that the affinity to HSA is dependent on the overall size of the metal chelate [34].

In Table 5.3, the K_A values for two Mn(II)EDTA complexes (Chart 5.4) bearing one and two BOM substituents, respectively, are reported.

Interestingly, the affinity towards HSA of the Mn(II)EDTA chelates is almost one order of magnitude higher than the structurally related Gd(III)DTPA complexes with the same number of hydrophobic groups. It is likely that this effect may be accounted for by the reduced size of the Mn(II) complexes which might facilitate the adhesion of the BOM groups to the hydrophobic pocket of the binding site of the protein.

Although the results showed so far suggest that the presence of a hydrophobic residue is fundamental for the interaction with HSA, in some cases only the presence of negative charges may be sufficient to encourage a fairly strong binding to the albumin.

Typical examples are represented by macrocyclic Gd(III) complexes having methylenephosphonate groups instead of acetate as coordinating arms. Both $[GdPCTP\text{-}[13](H_2O)]^{3-}$ and $[Gd - DOTP]^{4-}$ complexes (Chart 5.5) show good affinity towards HSA (nK_A of $6.0 \times 10^2\ M^{-1}$ and $3.1 \times 10^3\ M^{-1}$ at pH 7 and 25 °C, respectively) which is directly related to the number (3 and 4 at physiological pH) of negative charges on the complexes [35].

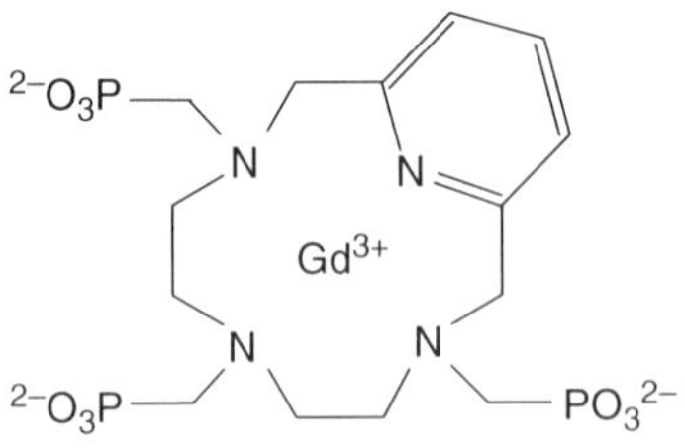

Gd-PCTP[13]

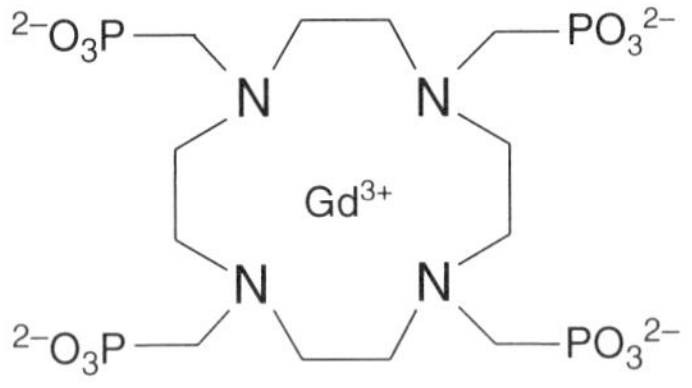

Gd-DOTP

[Chart 5.5]

On the other hand, some neutral metal chelates endowed with hydrophobic residues may also show a significative binding to HSA. In particular, it is only when the hydrophobic residue is represented by an aliphatic chain that the binding to the serum protein is not negligible, and relatively high affinities have been observed in the case of Gd-DTPA bisamide derivatives bearing *n*-heptyl and 2-ethylhexyl groups [36, 37].

In order to promote the binding to HSA, Gd(III) chelates, endowed with hydrophobic synthons whose high affinity to the serum protein was already known, have been synthesized. For example, the binding properties of a series of Gd(III) complexes, (DOTA-and DTPA-like) bearing iopanoic acid and 2, 4, 6-triiodobenzoic acid residues linked through an amidic bond (Chart 5.6) have been investigated (Table 5.4) [38].

Table 5.4 Affinity to HSA for a series of Gd(III) chelates (see Chart 5.6) containing iodinated residues ($n = 2$, 25 °C, pH 7.4, 50 mM phosphate buffer).

Metal complex	K_A (M^{-1})
[Gd-DOTA-IOP(H_2O)]$^-$	6.2×10^2
[Gd-DOTA-IOPsp(H_2O)]$^-$	2.9×10^3
[Gd-DOTA-TIBsp(H_2O)]$^-$	5.3×10^2
[Gd-DTPA-IOP(H_2O)]$^{2-}$	3.8×10^2
[Gd-DTPA-IOPsp(H_2O)]$^{2-}$	4.8×10^3

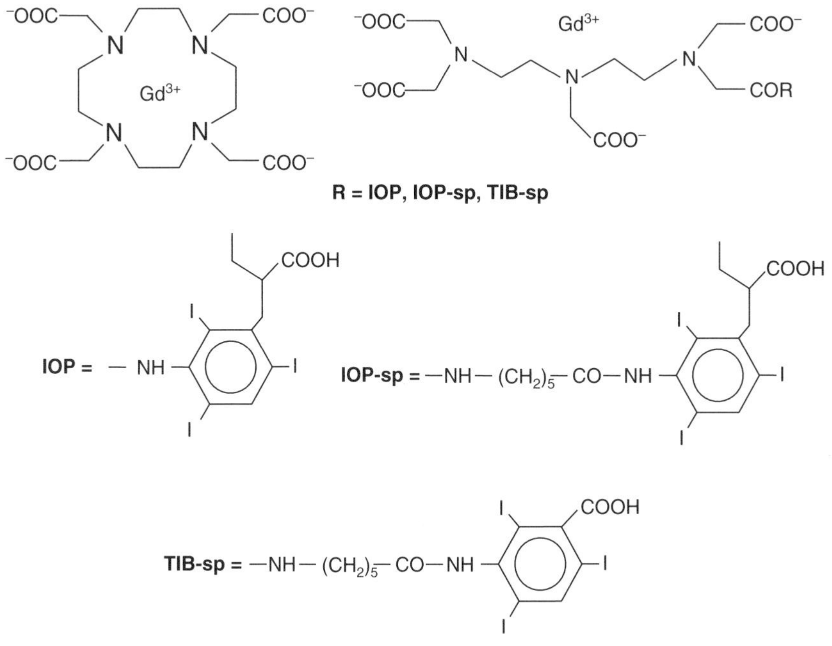

R = IOP, IOP-sp, TIB-sp

[Chart 5.6]

The two iodinated synthons bind quite strongly to two nearly equivalent binding regions located in the sub-domain IIA and IIIA, as observed in the X-ray structure [25, 39]. The affinity of iopanoic acid is slightly higher (average K_A for the two sites of 2.4×10^6 M^{-1}) than triiodobenzoic acid (2.2×10^5 M^{-1}) and, interestingly, this difference is kept in the corresponding Gd(III) complexes, [Gd-DOTA-TIBsp(H_2O)]$^-$ and [Gd-DOTA-IOPsp(H_2O)]$^-$, even if the presence of the metal chelate reduces the binding by almost two orders of magnitude.

The interference effect on the binding induced by the presence of the metal chelate is confirmed by comparing the K_A values for the complexes with the same synthon but differing in the length of the aliphatic spacer. Furthermore, the presence of a recognition synthon containing both a hydrophobic moiety and a negative charge seems to significantly reduce the role of the metal chelate structure. In fact, unlike the case of the Gd(III) complexes substituted with BOM residues, the affinities of the iodinated complexes are very similar for DOTA- and DTPA-like chelates.

Further support to this view has been obtained by comparing the affinities of two Gd(III) complexes with a different chelating unit, (DOTA- and DTPA-like, respectively), and the same recognition synthon represented by a taurocholic acid (Chart 5.7), which is known to be transported in plasma by serum albumin [22].

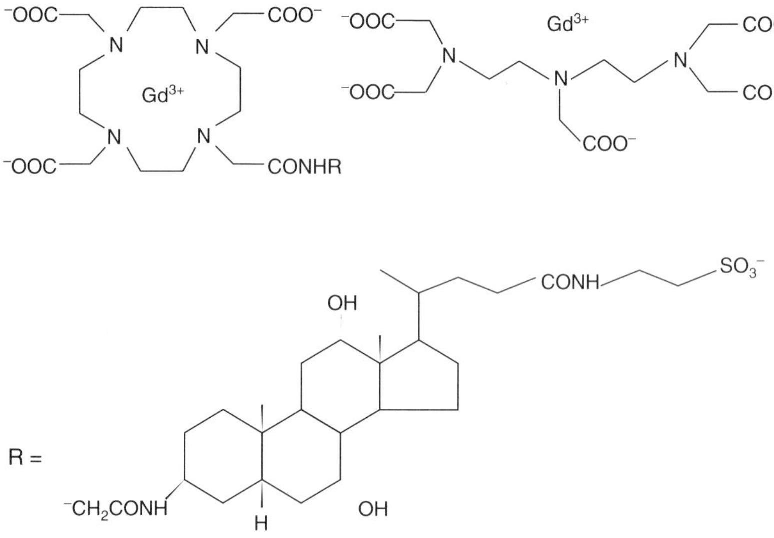

[Chart 5.7]

In addition, in this case the structure of the chelate does not affect the affinities of the two complexes, which are very similar (nK_A of 5.2×10^3 M^{-1} and 5.7×10^3 M^{-1} at 25 °C for the linear and the macrocyclic complexes, respectively) [17].

A wide array of hydrophobic residues have been considered in order to pursue a strong binding of a metal chelate to the serum albumin. For instance, a lipophilic tail bearing a bicyclic ring was linked to a DTPA ligand through the intercalation of a dipeptide and the resulting Gd(III) complex (MP2269) shows an appreciable binding to BSA (K_A of 9.1×10^3 M^{-1} and $n = 2$) [40].

A stronger binding to HSA has been observed by using a bulky hydrophobic residue consisting of two phenyl rings attached to a cyclohexyl moiety linked, through a negatively charged phosphate group, to one of the two ethylenediamine sides of the DTPA backbone. The binding properties of the resulting Gd(III) complex, MS-325, have been extensively investigated, either by means of PRE techniques or ultrafiltration analysis. [41–43]. The results are in good agreement and indicate that this complex interacts with the serum protein mainly to a single strong binding site (K_A of 3.0×10^4 M^{-1} at 25 °C [41] and 6.1×10^3 M^{-1} at 37 °C [42] from PRE measurements, and of the order of 10^4 M^{-1} from ultrafiltration analysis [43]).

The presence of a single binding site is particularly evident from the analysis of the M-titration with the PRE technique (Figure 5.11), in which the inflection point shows that the saturation of the stronger binding sites occurs at a [MS-325]/[HSA] ratio of ca 1 [41].

At a higher metal complex concentration, other weaker binding regions are available on the protein and, as inferred from the ultrafiltration analysis, up to 20 molecules of metal complex can bind a single protein molecule when a large excess (80:1) of MS-325 is used [43].

Obviously, with the doses of contrast agents employed in clinical applications, the interaction with the stronger site is the more relevant one.

In order to disclose its location, competition assays were carried out by using several competitor probes able to bind in different regions of the protein. Surprisingly, all of the competitors tested (ibuprofen, warfarin, bilirubin, linolenic acid and 1,3,5-triiodobenzoic acid) have an effect on the binding interaction of MS-325 by causing a slight decrease of its affinity constant (Figure 5.12). The observed behavior has been interpreted as an effect of the occurrence of conformational changes in the protein structure induced by the binding of the competitors [41].

More information about the characteristics of the binding site has been gained by exploring the pH dependence of the association constant between HSA and MS-325 [41].

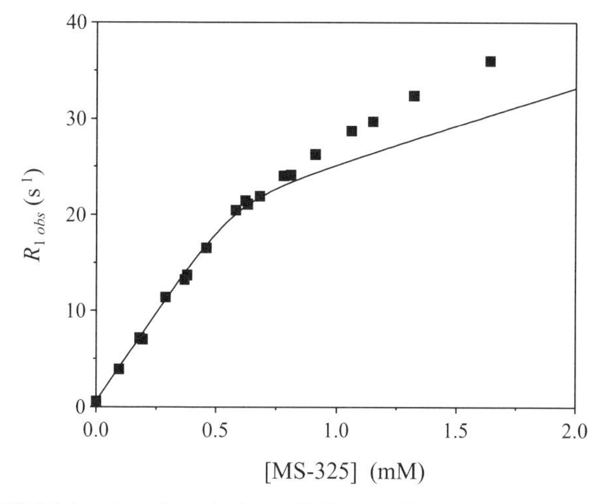

Figure 5.11 PRE M-titration of a solution of 0.6 mM HSA upon addition of MS-325. Note the inflection point at the [MS-325]/[HSA] ratio of 1:1 which indicates the presence of a single strong binding site (pH 7.5, 50 mM HEPES buffer, 25 °C, 0.47 T).

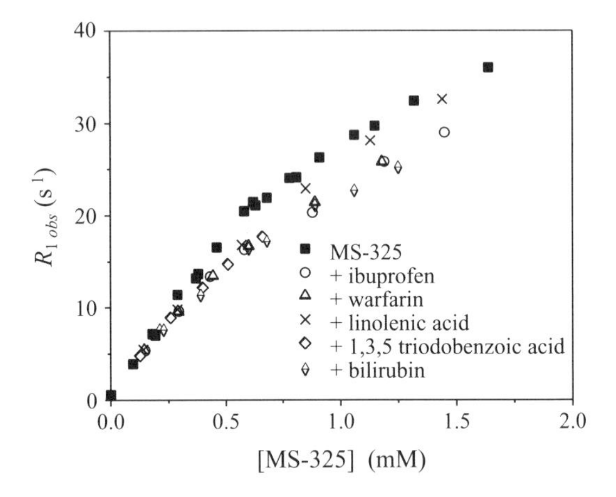

Figure 5.12 PRE M-titration of MS-325 in the presence of several competitor substrates in a 1:1 ratio to HSA (except for 1, 3, 5–triiodobenzoic acid, used in a 2:1 ratio). Each substrate appears to displace, to a similar small extent, the metal complex from the protein (pH 7.5, HEPES buffer, 25 °C, 0.47 T).

Starting from pH 7, the K_A values increase at lower pH (Figure 5.13) up to pH 5.5. Since no conformational changes are reported for HSA in this pH range, such a result may be interpreted in terms of an increase in the electrostatic component of the binding caused by the protonation of some basic groups (likely histidine) on the protein surface.

On the other hand, at pH > 7, K_A decreases as a probable consequence of the N–B conformational change in the tertiary protein structure occurring as the pH of the solution moves from the neutral to the basic side [22].

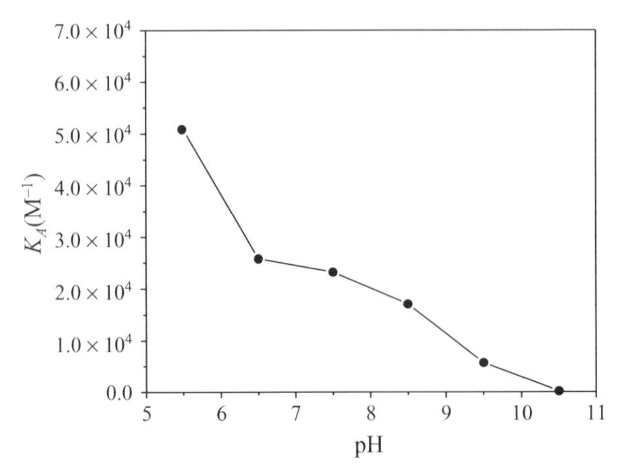

Figure 5.13 The pH dependence of the binding affinity of MS-325 to HSA (25 °C, 50 mM HEPES buffer, 0.47 T).

As it is known that such a phenomenon promotes the binding of substrates interacting in the sub-domain IIA of the protein, it is likely that MS-325 does not bind in this region.

The good binding to HSA, along with its favorable relaxometric and pharmacological properties, make MS-325 (brand name AngioMARK®) the first Gd(III)-based contrast agents for angiographic applications to proceed to human trials [43].

It is known that the binding to serum albumin may be stereospecific and, therefore, it is expected that different stereoisomers of a metal complex can display a different affinity towards the protein [22].

In fact, a different binding affinity has been found for the two enantiomers of $[Gd\text{-}EOBDTPA(H_2O)]^{2-}$, a contrast agent developed for the hepatobiliary tract [31].

This complex possesses two negative charges and a hydrophobic residue linked to the diethylenetriamine side of the DTPA backbone. As expected, its affinity to HSA is not particularly high, but, neverthless, the *S*-enantiomer binds the protein about three times more strongly than the corresponding *R*-form (nK_A of $6.8 \times 10^2\ M^{-1}$ and $2.0 \times 10^2\ M^{-1}$, respectively, at 37 °C) [44].

Another very interesting example concerning the stereoselectivity of the binding to HSA has been reported by Lauffer and co-workers, who investigated the HSA binding of two diastereisomers (*rac*- and *meso*-) of $[Fe-EHPG]^-$ (Chart 5.8) [30].

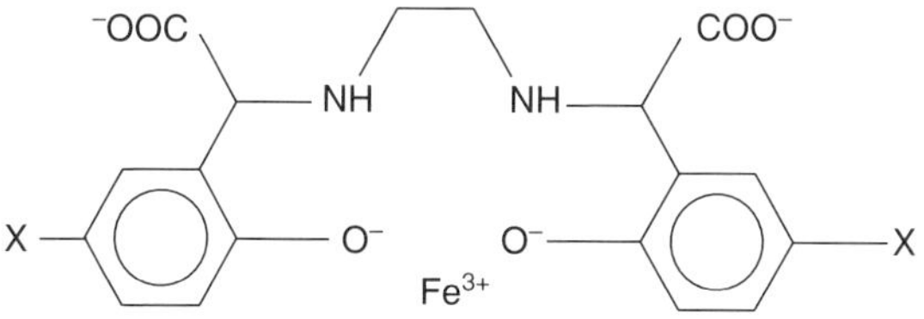

Fe-EHPG: X = H

Fe-5-Br-EHPG: X = Br

[Chart 5.8]

The affinity of the racemic form is, either in the unsubstituted or in the 5-Br-substituted compound, higher than the *meso*-form, but, interestingly, the interaction for the 5-Br derivatives involves at least two non-equivalent binding sites which are not the same for the two diastereoisomers. Competition tests, performed at 5 °C by means of both equilibrium dialysis and PRE measurements, helped to determine that the stronger site of the *meso*-form corresponds to the bilirubin binding region, whereas the same site is the secondary site for the racemic form.

Even more surprinsingly, at 37 °C the behavior is reversed and the bilirubin site becomes the stronger site for the racemic isomer and the secondary site for the *meso*-form. The different orientation of the two phenolate rings in the two diastereoisomers is likely to be responsible for this peculiar behavior.

Finally, a very interesting modality of the non-covalent interaction between a metal chelate and a protein is represented by the formation of ternary adducts, in which donor groups on the protein enter into the coordination sphere of the metal ion. Such macromolecular species have been observed in the case of the interaction with HSA of some Gd(III) complexes with heptadentate ligands such as the Gd-DO3A derivative shown in Chart 5.9 [45].

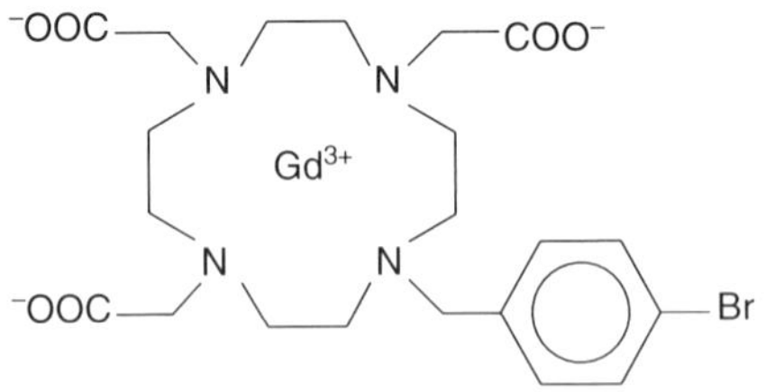

[Chart 5.9]

On the basis of the structural requirements necessary to have a strong binding to HSA, this chelate should not display a significative interaction because it is neutral and, furthermore, its hydrophobic synthon is quite close to the chelating unit. Nevertheless, this compound shows a relatively high association constant (nK_A of 5.4×10^3 M^{-1} at 25 °C). Relaxometric and luminescence measurements clearly indicate that the hydration number of the metal is reduced upon interaction with the serum albumin, thus suggesting that some donor groups (most probably carboxylates) of the protein may coordinate the metal ion by replacing its inner-sphere water molecules. Thus, the formation of the ternary adduct may account for the good binding shown by this chelate.

3.3 RELAXOMETRIC RESULTS

In addition to profoundly affecting the excretion pathway and increasing the confinement into the vascular system of the paramagnetic agent, the conjugation with plasma proteins results in a strong enhancement of the longitudinal relaxation rate of the water protons. Furthermore, the non-covalent interaction mode preserves the structural integrity of the complexes and thus the values of the corresponding relaxation parameters (q, r, a), whereas the rotational dynamics is markedly reduced with large effects on the τ_R and D parameters. At values of the magnetic field strengths typical of routine MRI, the relaxivity of small metal chelates is largely controlled by τ_R, which is one or two orders of magnitude shorter than τ_M and τ_S, and then the reduced mobility

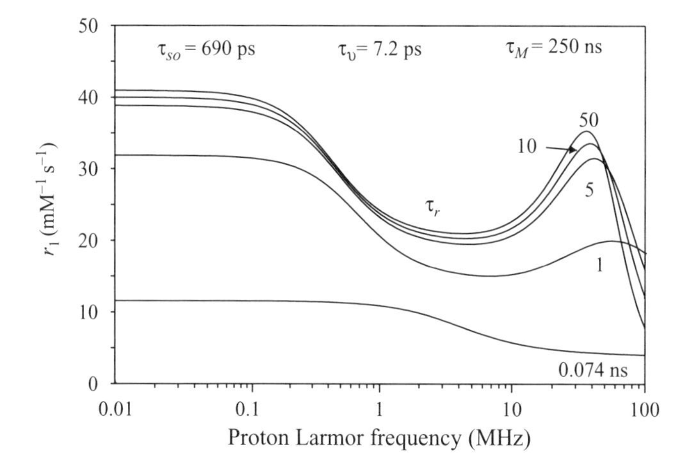

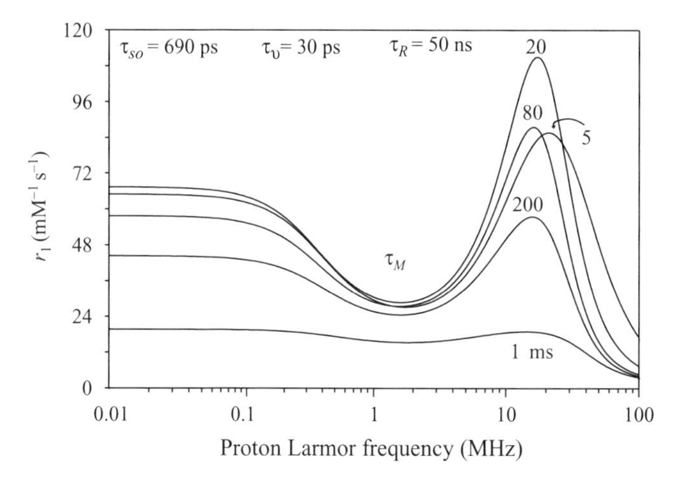

Figure 5.14 Simulated ^{1}H NMRD profiles showing the effect of (a) τ_R and (b) τ_M on the relaxivity of a Gd(III) chelate.

of the Gd complex in the macromolecular adduct represents the primary cause of the relaxivity gain (Figure 5.14). In the bound form, the Gd(III) chelate can be assumed, to a first approximation, to tumble coherently with the protein and then, if the other parameters remain unchanged, the contribution of τ_R to τ_C reduces significantly.

Conversely, only small perturbations of the electronic relaxation times and of the diffusion coefficent occur upon interaction and consequently only a small relaxivity enhancement is expected in the case of pure outer-sphere complexes.

3.3.1 Outer-Sphere Complexes ($q = 0$)

It was established early on that the relaxivities of the anionic coordinatively saturated complexes, $[Fe(5\text{-}Br\text{-}EHPG)]^-$ (X = Br) and $[Fe(HBED)^-]$, substantially increase when non-covalently linked to HSA, passing from about $1.0\,mM^{-1}\,s^{-1}$ in the free complexes to $2\text{–}4\,mM^{-1}\,s^{-1}$ in the macromolecular adducts [30]. The increase in relaxivity for these outer-sphere complexes is much higher than what could be expected by variations in the parameters τ_D and τ_S upon binding. Analogous results were found for an outer-sphere Gd(III) complex bearing hydrophobic substituents ($[Gd\text{-}DOTPMB]^-$, Chart 5.10), whose binding to HSA ($K_A = 9.3 \times 10^2\ M^{-1}$, $n = 1.4$) promotes a significant increase in the relaxivity of the complex, which passes from $2.8\,mM^{-1}\,s^{-1}$ for the free chelate to $13.4\,mM^{-1}\,s^{-1}$ for the macromolecular adduct (0.94 T, 25 °C) [46].

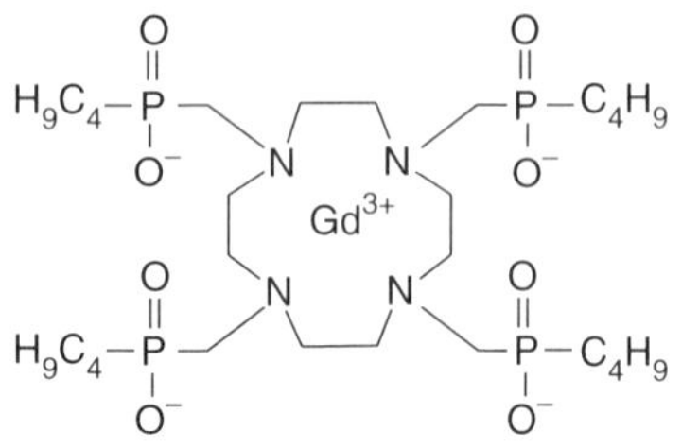

[Chart 5.10]

Such a marked relaxivity enhancement (for a $q = 0$ Gd(III) complex) has also been found for the complex $[Gd\text{-}BzDOTP]^-$ (Chart 5.11) upon formation of an adduct with BSA ($nK_A = 3.6 \times 10^3\ M^{-1}$) (Figure 5.15) [47]. Moreover, a remarkably high performance of the complex in liver and bile has been observed in MRI trials. In all of these cases, a major contribution to the relaxivity enhancement is likely due to the exchange of the mobile protons of the protein which are dipolarly relaxed by the proximity to the paramagnetic center. Another possible contribution could arise from the high structural organization and the consequent reduced mobility of the solvent molecules in the hydration sphere of the protein, near the binding site of the complex, which

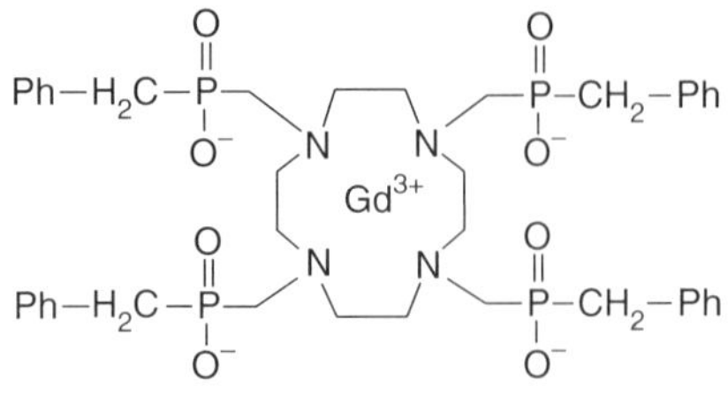

[Chart 5.11]

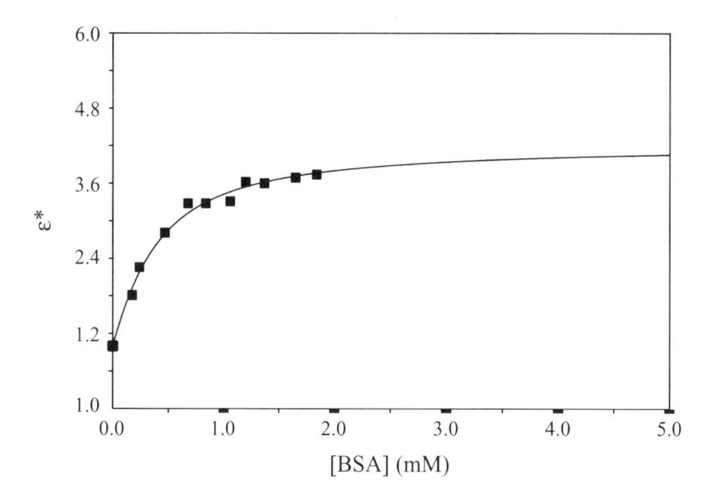

Figure 5.15 PRE E-titration for $[Gd\text{-}BzDOTP]^-$ in the presence of BSA (0.47 T, 25 °C, pH 7, [Gd] 0.2 mM).

allows the observation of second-shell interactions. In fact, the NMRD profile of the bound complex shows the relaxivity peak at around 20 MHz (the same behavior was observed for $[Gd\text{-}DOTPMB]^-$ [46]), typical of a slowly tumbling system and thus of a dependence of the relaxivity upon the rotational dynamics of the metal chelate. The aryl-methoxy-substituted complex showed both a stronger interaction with the protein ($K_A = 9.1 \times 10^3\ M^{-1}$) and, in particular, a significantly higher ($\sim 20\ \%$) enhancement of the relaxivity (Figure 5.16) [48]. Since a change of the coordination number of the metal ion upon interaction of the complex with albumin can be excluded, this relaxivity increase can be

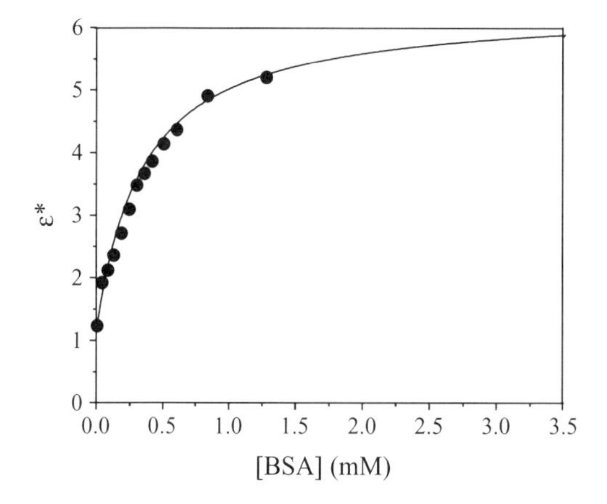

Figure 5.16 PRE E-titration with BSA for the aryl-methoxy derivative of $[Gd\text{-}BzDOTP]^-$ (0.47 T, 25 °C, pH 7, [Gd] 0.2 mM).

attributed to the ability of the methoxy groups to promote the formation of a network of hydrogen-bonded water molecules in the second coordination sphere of the Gd(III), near the surface of the protein. This effect is not detected in the aqueous solution of the free complex since it is likely that the mean residence lifetime of the hydrogen-bonded water molecules approaches the solvent translational diffusion correlation time. On the other hand, the high structural organization and the consequent reduced mobility of the water molecules of the hydration sphere of the protein allows the observation of the 'second-sphere' interactions, which result in an enhancement of the overall relaxivity. As expected, the enhancement that can be obtained depends strongly on the site of the interaction. A markedly higher relaxivity gain has been observed in the case of the adduct of $[Gd\text{-}DOTP]^{4-}$ with BSA [32], and the NMRD profile (Figure 5.17) is rather similar both in shape and amplitude to those measured for $q = 1$ complexes (see below). In this case, the interaction is governed only by electrostatic forces and the binding sites are localized on the positively charged domains of the proteins where a higher number of well-organized water molecules can be expected.

Clearly, the mechanism governing the relaxation enhancement of Gd(III) complexes with $q = 0$ is poorly understood and relies on several features of the structure and dynamics of the hydration layers of the protein and on the proton-exchange processes between the mobile protons on the protein and water molecules. The results so far obtained indicate that this contribution is always present and represents the largest component of the relaxivity of the bound

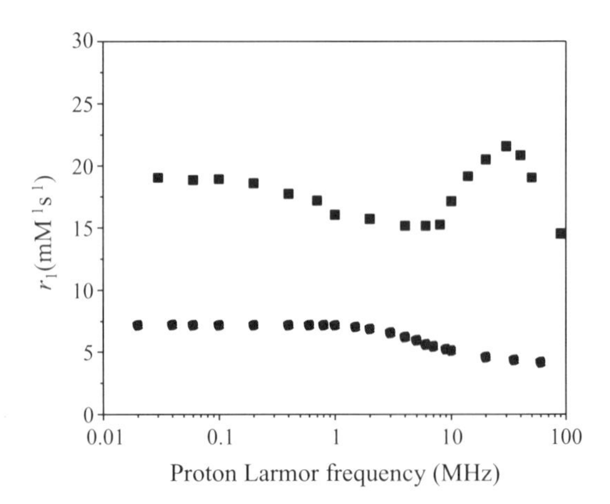

Figure 5.17 ^{1}H NMRD profiles for $[Gd\text{-}DOTP]^{4-}$ chelate, free (•) and fully bound to BSA (■). Since the complex has no water directly coordinated to the metal ion, the relaxivity increase shown upon binding to HSA has been accounted for by the exchange of water and mobile protons on the surface of the protein in close proximity to the paramagnetic center (25 °C, pH 7).

complexes. At present, this contribution can not be modeled and prevents the quantitative evaluation of their NMRD profiles.

3.3.2 Inner-Sphere Gd(III) Complexes ($q \geq 1$)

When one or more water molecules are directly bound to the metal center, the main component of the relaxivity of the free complex is attributable to the inner-sphere relaxation mechanism. This contribution determines to a large extent the ability of small Gd(III) chelates to enhance the longitudinal water proton relaxation rate and, with the magnetic field strengths currently employed in MRI (0.5–1.5 T, corresponding to proton Larmor frequencies of 20–60 MHz), it is mainly dependent on the value of the molecular reorientational time, τ_R. Therefore, it was recognized early on that higher water proton relaxation rates may be achieved through an elongation of this parameter, since the increase of the number of metal-bound water molecules (q), which would lead to a similar result, is likely to be accompanied by a decrease in the stability of the complex. The gain in attainable relaxivity and the changes in shape and amplitude of the NMRD profiles are well illustrated by the adducts of [Gd-DOTA(H_2O)]$^{-}$ and [Gd-DTPA(H_2O)]$^{2-}$ complexes incorporating one or more benzyloxymethylenic residues and model substrates such as β-cyclodextrin and cationic micelles (Figure 5.18) [49, 50]. The common feature is the appearance of a relaxivity peak at high fields showing the field-dependence of the electronic relaxation time that can be expressed when τ_R is lengthened. In the case of the

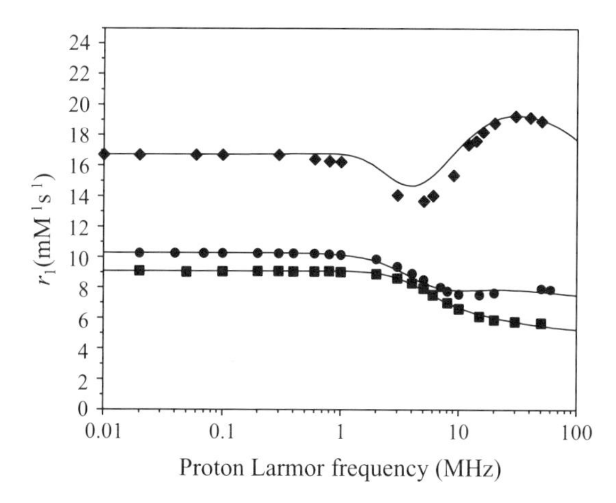

Figure 5.18 ^{1}H NMRD curves for Gd-BOPTA, free (■) and non-covalently fully bound to β-cyclodextrin (●) and cationic micelles of CTA$^+$ (◆). The elongation of the reorientational time τ_R promotes a relaxivity enhancement which is particularly evident at frequencies higher than 10 MHz (25 °C, pH 7)

non-covalent adduct with albumin, the decrease of the rotational mobility of the complex is such that a much higher relaxivity enhancement can in principle be obtained. However, the relaxivity peaks in the NMRD profiles of the bound Gd(III) complexes investigated until now are much lower than expected. Whereas the experimental r_1^b values, at 20 MHz and 25 °C, are comprised in the range 12–55 $mM^{-1} s^{-1}$, the calculated values considering an immobilized complex are much higher ($\geq$ 100 $mM^{-1} s^{-1}$) [2, 33]. This largely depends upon the fact that the advantage associated with the lengthening of τ_R can only be fully exploited if the water exchange lifetime and the electronic relaxation time have optimal values. These are parameters which depend on the properties of the metal complex, which need to be finely tuned with a proper choice of the ligand, rather than on the structure of the binding site on the protein. In particular, the residence lifetime of the bound water molecule(s) appears to be the limiting factor that prevents the currently investigated Gd(III) chelates, based on DOTA and DTPA derivatives, to attain higher relaxivity values. The dominant role of τ_M has been clearly demonstrated by studying the temperature dependence of the relaxation rate of a solution of [Gd-DOTA(BOM)$_3$(H$_2$O)]$^-$ (0.1 mM) with HSA (2.9 mM), in a condition where most of the metal chelate is bound to the protein. [32]. A careful dissection of the observed relaxivity into its different components demonstrated that r_1^b increases with temperature in the range 5–45 °C (Figure 5.19). This implies that the system is in the slow-exchange regime ($\tau_M > T_{1M}$) and its relaxivity follows the temperature dependence of the water exchange rate.

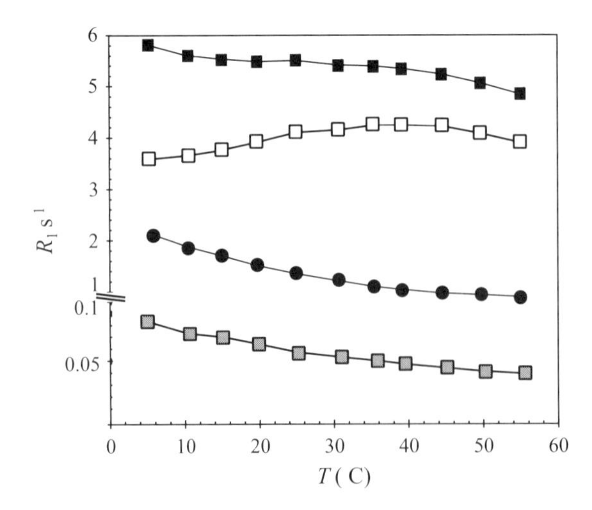

Figure 5.19 Temperature dependence of the longitudinal water proton relaxation rate for a 0.1 mM solution of [Gd-DOTA(BOM)$_3$(H$_2$O)]$^-$ in the presence of 2.9 mM HSA (0.47 T, 50 mM phosphate buffer; pH 7.4): observed values (■); diamagnetic contributions from the protein (●); paramagnetic contributions for the free (▩) and the bound (□) complexes.

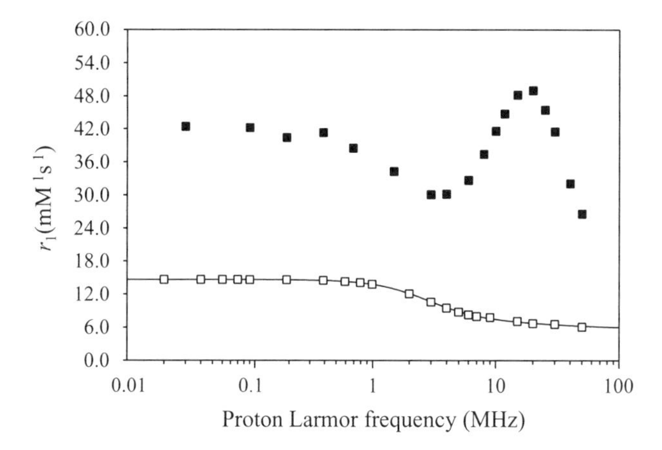

Figure 5.20 ^{1}H NMRD profiles for the $[Gd\text{-}DOTA(BOM)_3(H_2O)]^-$ chelate, free (□) and bound to BSA (■) (25 °C, pH 7).

The NMRD profiles of the free and protein-bound $GdDOTA(BOM)_3$ are shown in Figure 5.20. The relaxivity gain due to binding to HSA is remarkable, but the complexity of the system and the number of different contributions to the relaxivity prevents a reliable quantitative evaluation [2]. The main limitation is the need of an accurate estimation of the outer- and second-sphere components, which, as shown by the profile of $[Gd\text{-}DOTP]^{4-}$, make sizeable contributions to the observed relaxivity.

In most cases, the long values of the water residence lifetimes appear to be characteristic of the complex, rather than a consequence of the binding to the protein. In the case of the $[Gd\text{-}MP2269(H_2O)]^{4-}$ complex, this has been directly assessed by an ^{17}O NMR study which indicated that the water exchange rate on the Gd(III) center does not decrease appreciably when bound to BSA [40]. However, this could be due to the existence of a relatively large distance between the hydrophobic binding site of the protein and the metal site on the complex, confirmed by the high degree of internal flexibility around the long side-chain. Support of this conclusion can be found by plotting r_1^b (at 0.47 T) for a series of Gd(III) complexes versus the τ_M values as measured for the free complexes at 25 °C (Table 5.5). A straight line is obtained, which decreases with the increasing τ_M (Figure 5.21), whose values fall in the range 80–300 ns for DOTA and DTPA derivatives incorporating one or more hydrophobic side-chains. Thus, either the water exchange rate does not change appreciably upon binding, or the decrease is proportionally similar for all of the complexes. A slightly longer value of τ_M, which might be related to the interaction at the surface of the protein, has been calculated by fitting the NMRD profile of the (GdEOB-DTPA)–HSA adduct [31]. A much more remarkable decrease of the water exchange rate upon binding to HSA was calculated in the case of the macrocyclic phosphonate complex, $[Gd\text{-}PCTP\text{-}[13](H_2O)]^{3-}$ [35]. In this case,

Table 5.5 Relaxivity values (0.47 T, 25 °C) of HSA/Gd(III) chelate adducts versus the τ_M value (25 °C) of the free complex.

Metal complex	$r_1^b(s^{-1}mM^{-1})$	$\tau_M(ns)$
$[Gd\text{-}DOTA(BOM)_3(H_2O)]^-$	53.2	80
trans-$[Gd\text{-}DOTA(BOM)_2(H_2O)]^-$	44.2	130
cis-$[Gd\text{-}DOTA(BOM)_2(H_2O)]^-$	35.2	175
$[Gd\text{-}DTPA(BOM)_3(H_2O)]^{2-}$	44.0	180
MS-325	35.0	250
$[Gd\text{-}DTPA(BOM)_2(H_2O)]^{2-}$	28.2	260
$[Gd\text{-}BOPTA\ (H_2O)]^{2-}$	33.0	280
$[Gd\text{-}DOTA\text{-}IOPsp(H_2O)]^-$	20.8	550
$[Gd\text{-}DTPA\text{-}IOPsp(H_2O)]^{2-}$	19.9	630
$[Gd\text{-}DOTA\text{-}IOP(H_2O)]^-$	24.1	730
$[Gd\text{-}DTPA\text{-}IOP(H_2O)]^{2-}$	16.1	860

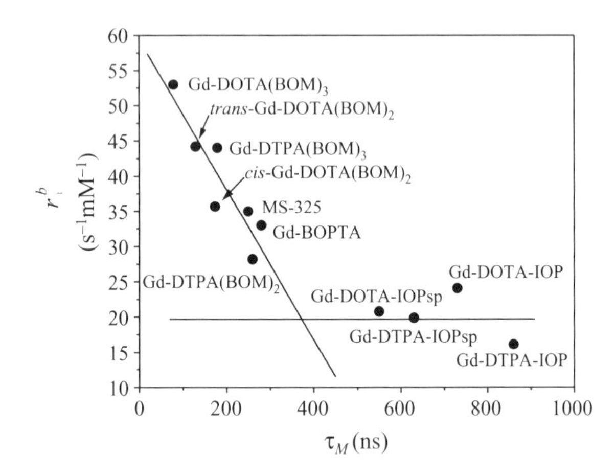

Figure 5.21 Correlation between the relaxivities (0.47 T, 25 °C) of some Gd(III) complexes fully bound to HSA (r_1^b) and their τ_M values determined for the free chelates at the same temperature.

the effect might have been enhanced by the high negative charge of the complex, which promotes strong electrostatic interactions with the positively charged groups on the surface of the protein eventually involving the coordinated water molecule. In spite of this limitation, the Gd-PCTP-[13]-HSA adduct presents a very high relaxivity, ca $60\,mM^{-1}\,s^{-1}$ at 30 MHz, which is attributed to a large contribution from the water molecules outside of the first coordination sphere of the metal ion. This is due to the ability of phosphonate groups to involve a number of solvent molecules in a hydrogen-bond network, in analogy with the behavior observed for $[Gd\text{-}DOTP]^{4-}$. This contribution depends on the characteristic features of the binding interaction and the charge density on the surface of the complex, and decreases with increasing temperature.

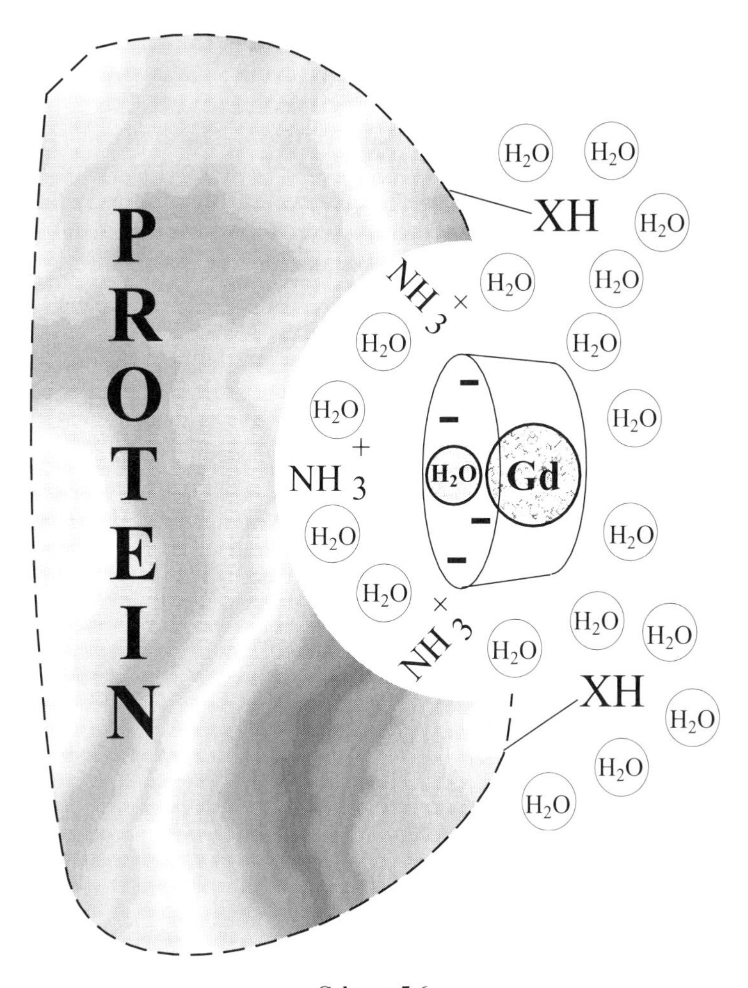

Scheme 5.6

A schematic representation of the non-covalent interaction between a negatively charged complex and the serum protein is shown in Scheme 5.6.

It may be noted that the derivatives with iodinated synthons (see Chart 5.6) present very similar r_1^b values, centered at $20\,mM^{-1}\,s^{-1}$, which indicates that the water exchange rate is so slow that it completely cancels any differential effect on the relaxivity enhancement due to a different internal mobility associated with a different length and rigidity of the side-chain. Such an effect was observed in the case of MS-325, where a relaxivity peak of $50\,mM^{-1}\,s^{-1}$ was measured for the protein-bound adduct, lower than the value ($59\,mM^{-1}\,s^{-1}$) allowed by the rate of water exchange, as a consequence of a certain degree of mobility of the bound complex [42]. Interestingly, macrocyclic Gd(III) complexes with heptadentate ligands ($q = 2$) bearing a side chain which promotes interaction with the serum proteins, do not show the expected relaxation enhancement upon binding. Data obtained from ^{17}O NMR studies have shown that in the presence of the protein the complexes do not possess metal-bound water molecules and thus the observed (small) relaxivity enhancement is entirely due to the contributions of the water molecules of the hydration layers of HSA and possibly of the dipolarly relaxed exchanging protons in the vicinity of the interaction site [45]. Clearly, the two inner-sphere water molecules in these complexes are replaced by coordinating atoms (likely to be carboxylate groups) on the surface of the protein with formation of ternary adducts. Support of this conclusion was gained by observing a relaxivity decrease upon the addition of malonate to an aqueous solution of the complexes. Thus, the strategy of using complexes with $q > 1$ in order to attain higher relaxivities has to be optimized, and this might require the design of a suitable rigid spacer between the recognition synthon and the coordination polyhedron in order to avoid close contact with the surface of the protein and prevent water displacement.

A relaxivity limited by τ_M was also observed in the HSA adduct of the cationic Gd(III) texaphyrin complex (Chart 5.12) [51]. The analysis of the NMRD profile of the free chelate gave an estimated q value of 3.5, leading to a relaxivity of ca 15.0 $mM^{-1}\,s^{-1}$ at 0.47 T and 37 °C. The relaxivity of this chelate decreases by increasing the temperature, thus suggesting that $T_{1M} > \tau_M$. In the presence of 5 % HSA, an increase of relaxivity was observed (from 15.0 to 27.0 $mM^{-1}s^{-1}$ at 0.47 T and 37 °C) and the typical relaxivity peak centered at

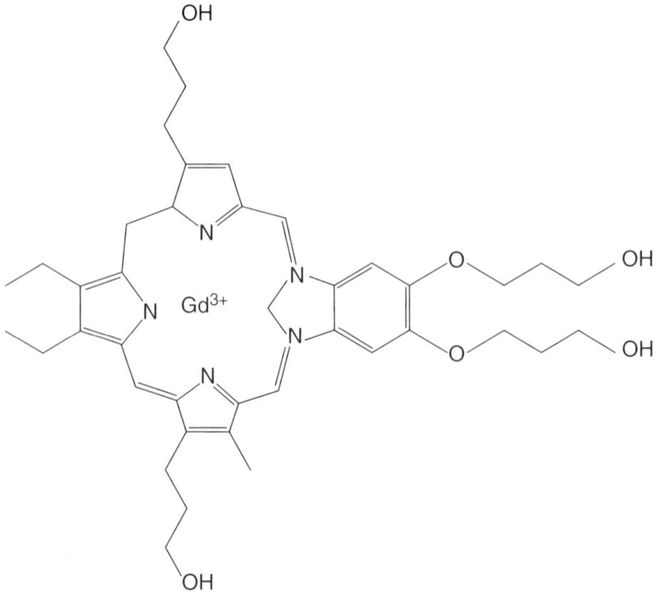

[Chart 5.12]

20 MHz appeared in the NMRD profile. The relaxivity of the complex in the presence of HSA was not affected by temperature in the range 5–37 °C. This behavior is a clear indication that the relaxivity of the macromolecular adduct is limited by τ_M. It was suggested that, with respect to the free complex, there is a change in the acessibility of the water molecules to the bound chelate which is likely to be buried in the HSA structure or there may be a change in the hydration state of the lanthanide(III) ion upon formation of a ternary adduct between the complex and donor groups of the protein.

The limiting effect on the relaxivity of a Gd(III)-based macromolecular adduct induced by a long water exchange lifetime may be overcome by using Mn(II) complexes [52].

In fact, it was reported early on that Mn(II) chelates with $q = 1$ display very short τ_M values [53].

For this reason, the relaxometric behavior of two Mn-EDTA derivatives bearing one and two BOM substituents (see Chart 5.4) have been investigated in the presence of HSA [34].

In these complexes, in analogy with the parent $[\text{Mn-EDTA}(H_2O)]^{2-}$ [54, 55], the metal ion is heptacoordinated and a water molecule is present in its inner coodination sphere. From ^{17}O NMR studies, τ_M values of 11.0 and 8.0 ns for $[\text{Mn-EDTA(BOM)}(H_2O)]^{2-}$ and $[\text{Mn-EDTA(BOM)}_2(H_2O)]^{2-}$, respectively, were obtained, i.e. more than one order of magnitude smaller than those found for similar polyaminopolycarboxylic Gd(III) complexes with $q = 1$. Obviously, the relaxivities of these complexes are lower than those of the analogous Gd(III) chelates, but, interestingly, the adducts with HSA show higher r_1^b values (at 0.47 T) (Figure 5.22).

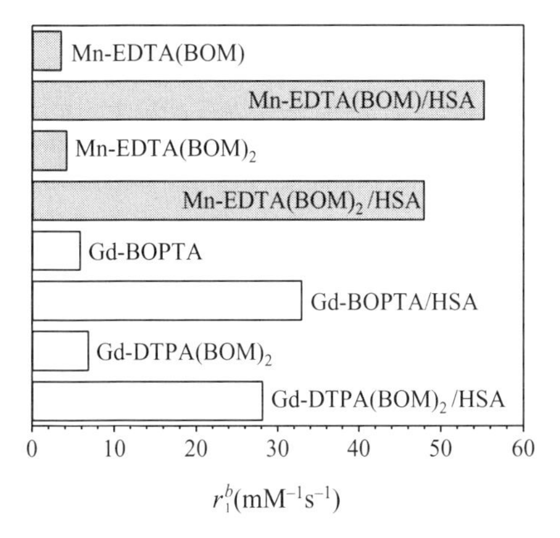

Figure 5.22 Unlike free chelates, the relaxivities of the Mn(II) chelates fully bound to HSA are significantly higher than those of the Gd(III) complexes with the same kind and number of hydrophobic synthons (0.47 T, 25 °C, pH 7.4, 50 mM phosphate buffer).

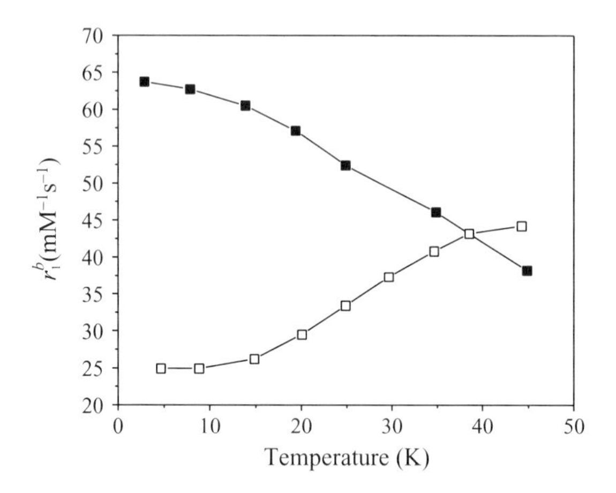

Figure 5.23 The different temperature dependences of r_1^b for Mn-EDTA(BOM)/HSA (■) and Gd-BOPTA/HSA (□) can be accounted for by shorter τ_M values for the former adduct (0.47 T, pH 7.4, 50 mM phosphate buffer).

It is important to note that the r_1^b values for these Mn(II) chelates are in the range of the highest values observed until now.

The comparison between the temperature dependence of r_1^b for MnEDTA (BOM)–HSA and GdBOPTA–HSA adducts (Figure 5.23) supports the view that the higher relaxivity shown by the Mn(II) chelate is primarily due to a shorter τ_M.

In fact, τ_M values of 22.0 and 38.0 ns have been estimated from ^{17}O NMR measurements for MnEDTA(BOM)–HSA and Mn-EDTA$(BOM)_2$–HSA adducts, respectively. These values are slightly longer than those previously determined for the free chelates, but still lie in the optimal order of magnitude to fully exploit the increase of τ_R. As corollary information, one may note that, analogously to the Gd(III) chelates which bear hydrophobic substituents, a slight elongation of the water exchange lifetime occurs upon interaction with HSA.

4 FURTHER APPLICATIONS

This section deals with studies concerning the formation of paramagnetic adducts between Gd(III) chelates and a few selected targeting proteins. The resulting relaxation enhancement acts as a reporter of a specific property of the targeting protein. In principle, this approach may provide the route for the development of novel *in vitro* assays for the quantitation of specific analytes based on the measurement of water proton relaxation rates. Moreover, this kind of investigation may provide useful insights for the design of a new generation of 'smart' contrast agents whose relaxivities act as probes of their physico-chemical environments.

4.1 INTERACTION OF A Gd(III)-LABELED PHOSPHATE WITH HEMOGLOBIN

The oxygen binding and spectroscopic properties of human adult and fetal hemoglobin (HbA and HbF, respectively) are modulated physiologically by heterotropic effectors, such as 2,3-D-glycerate bisphosphate (BPG), inorganic phosphate, chloride ions, carbon dioxide and protons. A similar effect has been reported for synthetic compounds displaying pharmaceutical activity, such as the antihyperlipidemic drugs bezafibrate (BZF) and clofibric acid (CFA). Heterotropic effectors bind to Hb not only at the BPG pocket (i.e. at the dyad axis, in between the β-chains), but also to different multiple functionally linked interaction sites [56]. The effect of heterotropic interactions results in the stabilization of the low-oxygen-affinity conformational state (T-state), thus allowing the transport of oxygen from HbA to myoglobin (Mb) in the muscle cells. Oxygen binding to HbA in the lung reverts the conformational equilibrium towards the oxygenated (R) state.

We thought that a negatively charged metal complex, such as $[Gd\text{-}DOTP]^{4-}$ (see Chart 5.5), could set up an electrostatic heterotropic interaction with human hemoglobin in a similar fashion to that shown by BPG. It is known that the latter molecule forms salt bridges with the positively charged *N*-terminal nitrogens of Val1 residues and with side chains of His2, Lys82 and His143 of opposite β chains.

The binding isotherms of $[Gd\text{-}DOTP]^{4-}$ with oxygenated and deoxygenated HbA (R- and T-states, respectively) were measured by using the PRE methodology (Figure 5.24), and the dissociation constants of the adducts were determined [57]. It is worth noting that ferric hemoglobin (metHbA) binds $[Gd\text{-}DOTP]^{4-}$ with a similar affinity to that of oxygenated HbA. Actually, metHbA does not take part in the allosteric equilibrium, since it is not able to bind oxygen. However its conformation (as determined by X-ray crystallography) highly resembles that of oxygenated HbA (both have hexacoordinated heme iron).

A detailed investigation of the effect of $[Gd\text{-}DOTP]^{4-}$ and its diamagnetic La(III) complex analog on oxygen binding and on the spectroscopic properties of HbA and HbF showed that the allosteric effect of these complexes is less pronounced in the latter protein [58]. Notably, this is analogous to the behavior shown by the natural allosteric effector (BPG) and has been accounted for by the occurrence of a point-mutation His143Ser in the β chains (identified as γ chains in HbF) [59].

Although the three-dimensional structure of Hb in the presence of Ln(III) complexes has not been solved, there is enough evidence to support the view

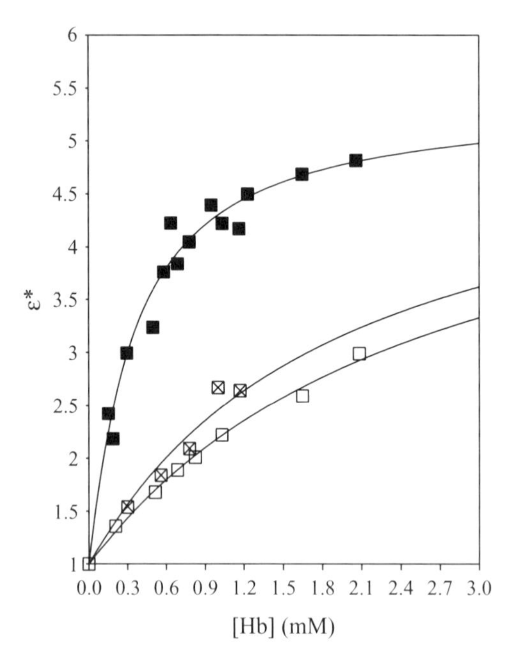

Figure 5.24 PRE E-titrations of Gd-DOTP with human deoxy-Hb (■), HbO_2 (□) and met-Hb (⊠). The paramagnetic chelate shows the highest affinity for the de-oxygenated form of the protein (0.47 T, [Gd] 0.1 mM, pH 7.2, 25 °C).

that $[Gd\text{-}DOTP]^{4-}$ and the corresponding La(III) complex associate to the tetramer at the BPG binding site.

In principle, the availability of a paramagnetic complex, able to bind with different affinity to the T- and R-states of Hb, could suggest their use as contrast agents for functional MRI. In fact, the Hb/HbO_2 molar ratio could be monitored from changes in the local contrast in magnetic resonance images. More generally, the peculiar stereochemical properties of Ln(III) complexes may be exploited to develop a novel class of NMR-detectable allosteric effectors of Hb.

4.2 INTERACTION OF A Gd(III)-LABELED BORONATE COMPLEX WITH GLYCATED ALBUMIN

Glycated human serum albumin (gHSA) appears to be an important marker in the clinical control of diabetes mellitus, in addition to the more popular glycated hemoglobin (HbA_{1c}). Diabetes is a very common metabolic disease, which is diagnosed solely on the basis of glucose levels in the blood or other body fluids. Monitoring of the glucose level is necessary to determine the posology of the administered drugs, although it gives limited information on the recent history of the disease. To this purpose, it has been assessed that HbA_{1c} is more informative. Indeed, hemoglobin is exposed to the electrophilic attack of glucose for a time equal to the lifetime of the red blood cells, and thus a peak in hematic glucose concentration will cause an increased HbA_{1c} concentration for the following three to four months. Furthermore, it has been suggested that the determination of gHSA can represent a useful complementary test to monitor the course of the disease on a mid-term scale. Indeed, the HSA lifetime is 27 days, and gHSA appears to be degraded at a rate which is 33 % higher than HSA. The determination of gHSA is thus a useful marker of changes in the glucose metabolism in a three to four week period.

In order to set up an easy and accurate determination of gHSA in diabetic patients, a relaxometric *in vitro* method based on the interaction of a Gd(III)-labeled boronate with gHSA has been proposed [60]. Phenylboronates are known to interact quantitatively with *syn*-diols, and the covalent, reversible interaction of the Gd(III)-boronate probe with gHSA offers a slowly tumbling paramagnetic adduct, which causes a relaxation enhancement of solvent water protons.

Binding isotherms are measured by means of the PRE approach, and the results are shown in terms of a linear relationship versus the relaxation rate the amount of gHSA, with the latter being determined by a colorimetric method (Figure 5.25).

The limitation of this probe molecule is the long lifetime of the coordinated water molecule in this [Gd-bis(*m*-boroxyphenylamide)DTPA(H_2O)] complex (Chart 5.13); a τ_M value of a few μs limits r_1 to about 9.0 $mM^{-1}\,s^{-1}$, and consequently the enhancement factor ε_b to about 1.8. The design of alternative probe molecules endowed with shorter τ_M values is currently under investigation, in order to overcome this limitation and enhance the

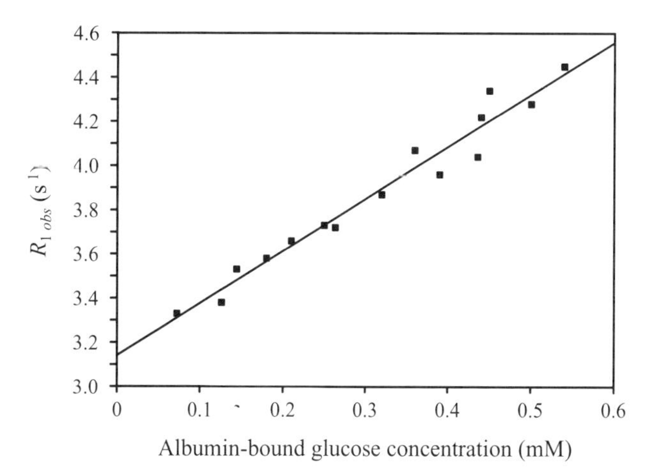

Figure 5.25 Correlation between the water proton relaxation rate of a 0.56 mM solution of Gd-bis(*m*-boroxyphenylamide)DTPA and the concentration of glycidic residues on HSA as evaluated by the fructosamine colorimetric method (0.47 T, 25 °C, pH 7).

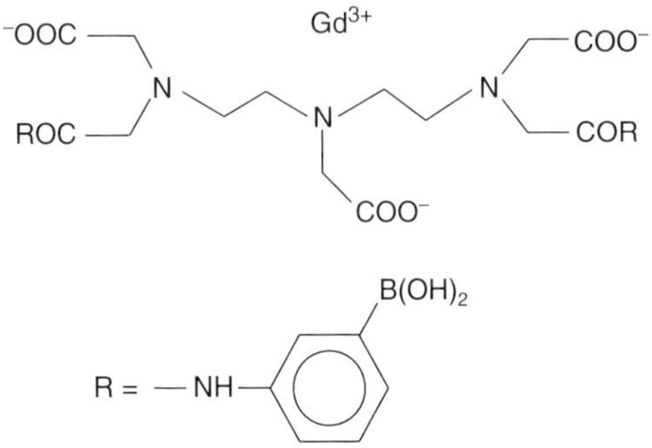

[Chart 5.13]

slope (and consequently the accuracy) of the linear relationship between r_1 and the gHSA concentration.

4.3 INTERACTION OF A Gd(III)-LABELED BORONATE WITH SERINE PROTEINASES

Many physiological and pathological processes are mediated by serine proteinases and by the products of their action (such as enzyme-generated peptides). However, the uncontrolled serine proteinase activity can be deleterious and dangerous to cells, tissues, organs and, eventually, to the entire organism. Thus, the ability to monitor the protease concentration in different body areas might be of valuable help in the early diagnosis of disease states which are accompanied by an overproduction of proteases, namely infarction, inflammation, and tumour invasion [61].

Boric acid and boronic derivatives are reversible and slow-binding inhibitors of serine proteinases. Boronic peptides bind either to catalytic His57 and Ser195 side-chains to give tetrahedral adducts or to form a diadduct involving both histidyl and seryl residues. The Gd(III) bis(*m*-boroxyphenylamide)DTPA complex described above was investigated with the PRE methodology in order to measure its binding isotherms to chymotrypsin, chymotrypsinogen, trypsin and trypsinogen, between pH 6.0 and 8.5, at 25.0 °C (Figure 5.26) [62]. The results obtained have been compared with those found in the study of enzymes assessed spectrophotometrically by measuring the hydrolysis of suitable chromogenic peptides. Values of K_i for the competitive inhibition of serine proteinases by the probe molecule are in accordance with the values of K_d obtained by the relaxometric approach, thus suggesting that the substrate and the paramagnetic complex bind to the same region. Moreover, the PRE methodology allows us to determine values of K_d for binding to chymotrypsinogen and trypsinogen, which are both devoid of catalytic activity. The increase of the water proton relaxation rate upon binding to serine (pro)enzymes may be useful

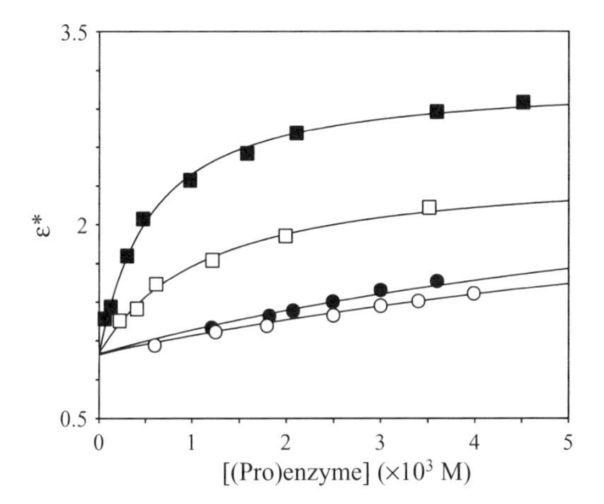

Figure 5.26 PRE E-titrations of a solution of Gd-bis(*m*-boroxyphenylamide)DTPA with chymotrypsin (■), chymotripsinogen (□), trypsin (●) and trypsinogen (○) (0.47 T, [Gd] 0.1 mM, 25 °C, pH 7.3).

in the design of novel functional contrast agents for magnetic resonance imaging which are able to report on the protein levels in specific organs and tissues.

Another approach to develop a reporter of the enzymatic activity has been published by McMurry and co-workers [63]. They demonstrated that the activity of alkaline phosphatase can be assessed by exploiting the changes in binding affinity to HSA of a paramagnetic substrate which is dephosphorylated by the enzyme. In fact, by introducing on the surface of a Gd-DTPA chelate a substituent such as $-CH_2-O-C_6H_4-C_6H_4-O-PO_3^{2-}$ which transforms to $-CH_2-O-C_6H_4-C_6H_4-OH$ upon the hydrolysis brought about by the enzyme, an increase of 70 % in $1/T_1$ in the presence of 4.5 % of HSA has been measured. This is clearly a consequence of a much higher binding affinity to HSA of the dephosphorylated form with respect to the phosphorylated one.

5 FINAL REMARKS

On the basis of the body of results obtained up until now, it is easy to foresee a number of potential applications based on the use of the relaxation enhancement promoted by the covalent or non-covalent adducts formed between a protein and a paramagnetic metal chelate. The relaxivity values found for these macromolecular systems are remarkable and represent important steps towards the attainment of contrast agents of improved efficiency (Figure 5.27). The work of the last few years on the elucidation of the relationship between the structural properties of the chelate moiety and the exchange lifetime of the coordinated water has been highly relevant in this context. More has to be

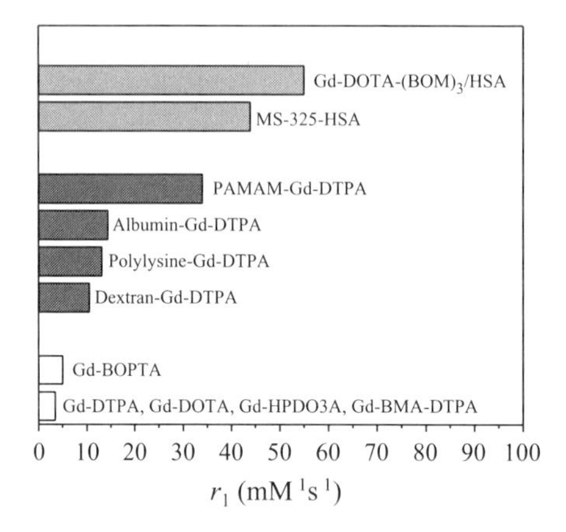

Figure 5.27 Relaxivity values (0.47 T, 25 °C) for the different classes of Gd(III) chelates: free (white); covalently bound to macromolecules (gray); and non-covalently bound to HSA (light gray). The values for dextran-Gd-DTPA and for the dendrimeric PAMAM-Gd-DTPA are taken from references [4] and [64], respectively.

done in order to get a better understanding of the determinants of the electronic relaxation time. Moreover, the effect of the internal rotation of the water molecule along its coordination axis on the overall reorientational time of the metal–proton vector has to be assessed. Furthermore, it is important to highlight the contribution to the observed relaxivities arising from water molecules and mobile protons at the surface of the protein in the proximity of the interacting site. It might be of interest to carry out more in-depth investigations on this property, with the expectation that the extent of this contribution may depend on the region involved in the binding, thus providing a route for its modulation through a proper design of the interaction scheme in the ternary complex.

Besides the goal of pursuing the highest relaxivities, the targeting procedures described in this chapter may lead to a number of applications in different areas of clinical diagnosis by designing systems whose relaxation enhancement is associated with the recognition of a specific protein. This approach is expected to provide the physician with an array of contrast agents endowed with enhanced specificities for assessing various pathological states. A feedback from these investigations would deal with the development of novel *in vitro* assays based on relaxometric determinations of the analytes of interest. Moreover, one may envisage further advantages associated with the formation of such paramagnetic macromolecular adducts through instrumental developments which will allow a precise positioning of the signal detection at the frequency at which the relaxivity peak occurs.

REFERENCES

1. Dwek, R. A. *Nuclear Magnetic Resonance in Biochemistry, Applications to Enzyme Systems*, Clarendon Press, Oxford, 1973, pp. 174–283.
2. Aime, S. A., Botta, M., Fasano, M. and Terreno, E. *Chem. Soc. Rev*. 1998, **27**, 19–29.
3. Nunn, A. D., Linder K. E. and Tweedle M. F. *Q. J. Nucl. Med*. 1997, **41**, 155–162.
4. Brasch, R. C. *Magn. Reson. Med*. 1991, **22**, 282–287.
5. Meares, C. F. and Wensel, T. G. *Acc. Chem. Res*. 1984, **17**, 202–209.
6. Hnatowich, D. J., Layne, W. W., Childs, R. L., Laintegne, D., Davis, M. A., Griffin, T. W. and Doherty, P. W. *Science* 1983, **220**, 613–615.
7. Lauffer, R. B. and Brady, T. J. *Magn. Res. Imag*. 1985, **3**, 11–16.
8. Ogan, M. D., Schmiedl, U., Moseley, M., Grodd, W., Paajanen, H. and Brasch, R. C. *Invest. Radiol*. 1987, **22**, 665–671.
9. Sherry, A. D., Cacheris, W. P. and Kuan, K. -T. *Magn. Res. Med*. 1988, **8**, 180–190.
10. Sieving, P. F., Watson, A. D. and Rocklage, S. M. *Bioconjugate Chem*. 1990, **1**, 65–71.
11. Maisano, F., Gozzini, L. and de Haën, C. *Bioconjugate Chem*. 1992, **3**, 212–217.
12. Tietze, L. F., Schroter, C., Gabius, S, Brinck, U., Goerlach-Graw, A. and Gabius, H. J. *Bioconjugate Chem*. 1991, **2**, 48–53.
13. Aime, S., Botta, M., Geninatti Crich, S., Giovenzana, G. and Palmisano, G., Sisti, M. *Bioconjugate Chem*. 1999, **10**, 192–199.
14. Spanoghe, M., Lanens, D., Dommisse, R., Van der Linden, A. and Alderweireldt, F. *Magn. Res. Imaging* 1992, **10**, 913–917.
15. Niemi, P., Reisto, T., Hemmilä, I. and Kormano, M. *Invest. Radiol*. 1991, **26**, 820–824.
16. Paajanen, H., Reisto, T., Hemmilä, I., Komu, M., Niemi, P. and Kormano, M. *Magn. Res. Med*. 1990, **13**, 38–43.
17. Aime, S., unpublished results.
18. Berthezène, Y., Vexler, V., Price, D. C., Wisner-Dupon, J., Moseley, M. E., Aicher, K. P. and Brasch, R. C. *Invest. Radiol*. 1992, **27**, 346–351.
19. Aime, S., Botta, M., Geninatti Crich, S., Giovenzana, G., Palmisano, G. and Sisti, M. *J. Chem. Soc. Chem. Commun*. 1999, 1577–1578.
20. Curtet, C., Maton, F., Havet, T., Slinkin, M., Mishra, A., Chatal, J. F. and Muller, R. N. *Invest. Radiol*. 1998, **33**, 752–761.
21. Wu, C., Brechbiel, W., Kozak, R. W. and Gansow, O. A. *Bioorg. Med. Chem. Lett*. 1994, **4**, 449–454.
22. Peters, T. J. *All About Albumin: Biochemistry, Genetics and Medical Applications*, Academic Press, San Diego, CA, 1996.
23. Sugio, S., Kashima, A., Mochizuki, S., Noda, M. and Kobayashi, K. *Protein Eng*. 1999, **12**, 439–446.
24. Curry, S., Mandelkow, H., Brick, P. and Franks, N. *Nat. Struct. Biol*. 1998, **5**, 827–835.
25. He, X. M. and Carter, D. C. *Nature* (*London*), 1992, **358**, 209–215.
26. Wright, J. D., Boudinot, F. D. and Ujhelyi, M. R. *Clin. Pharmacokinet*. 1996, **30**, 445–462.
27. Sebille, B., Zini, R., Madjar, C. V., Thuaud, N. and Tillement, J. P. *J. Chromatogr*. 1990, **531**, 51–77.
28. Scatchard, G. *Ann. NY Acad. Sci*. 1949, **51**, 660–672.
29. Jenkins, B. G., Armstrong, E. and Lauffer, R. B. *Magn. Res. Med*. 1991, **17**, 164–178.

30. Larsen, S. K., Jenkins, B. G., Memon, N. G. and Lauffer, R. B. *Inorg. Chem.* 1990, **29**, 1147–1152.
31. Vander Elst, L., Maton, F., Laurent, S., Seghi, F., Chapelle, F. and Muller, R. N. *Magn. Res. Med.* 1997, **38**, 604–614.
32. Aime, S., Botta, M., Fasano, M., Geninatti Crich, S. and Terreno, E. *J. Biol. Inorg. Chem.* 1996, **1**, 312–319.
33. Caravan, P., Ellison, J. J., McMurry, T. J. and Lauffer, R. B. *Chem. Rev.* 1999, **99**, 2293–2352.
34. Aime, S., Canton, S., Terreno, E., Fedeli, F., Manfredi, G., Paleari, L., Calabi, L. and Uggeri, F. paper in preparation.
35. Aime, S., Botta, M., Geninatti Crich S., Giovenzana, G. B., Pagliarin, R., Piccinini, M., Sisti, M. and Terreno, E. *J. Biol. Inorg. Chem.* 1997, **2**, 470–479.
36. Bligh, S. W. A., Chowdhury, A. H. M. S., Kennedy, D., Luchinat, C. and Parigi, G. *Magn. Res. Med.* 1999, **41**, 767–773.
37. Geraldes, C. F. G. C., Urbano, A. M., Alpoim, M. C., Sherry, A. D., Kuan, K. T., Rajagopalan, R., Maton, F. and Muller, R. N. *Magn. Res. Imaging* 1995, **13**, 401–420.
38. Aime, S., Anelli, P. L., Fedeli, F., Geninatti Crich, S., Murru, M. and Terreno, E. *Third International Society for Magnetic Resonance in Medicine Meeting*, Nice, France, 1995, Abstracts, p. 1096.
39. Ho, J. X., Holowachuk, E. W., Norton, E. J., Twigg, P. D. and Carter, D. C. *Eur. J. Biochem.* 1993, **215**, 205–212.
40. Tóth, E., Connac, F., Helm, L., Adzamli, K. and Merbach, A. E. *J. Biol. Inorg. Chem.* 1998, **3**, 606–613.
41. Aime, S., Chiaussa, M., Digilio, G., Gianolio, E. and Terreno, E. *J. Biol. Inorg. Chem.* 1999, **4**, 766–774.
42. Muller, R. N., Radüchel, B., Laurent, S., Platzek, J., Pièrart, C., Mareski, P. and Vander Elst, L. *Eur. J. Inorg. Chem.* 1999, 1949–1955.
43. Lauffer, R. B., Parmalee, D. J., Dunham, S. U., Ouellet, H. S., Dolan, R. P., Witte, S., McMurry, T. J. and Walowitch, R. C. *Radiology* 1998, **207**, 529–538.
44. Muller, R. N. personal communication.
45. Aime, S., Gianolio, E., Terreno, E., Giovenzana, G. B., Pagliarin, R., Sisti, M., Palmisano, G. and Botta, M. *Fifth FGIPS Meeting in Inorganic Chemistry*, Toulouse, France, 1999, Abstracts, OP F3.
46. Geraldes, C. F. G. C., Sherry, A. D., Lázár, I., Miseta, A., Bogner, P., Berenyi, B., Sumegi, B., Kiefer, G. E., McMillan, K., Maton, F. and Muller, R. N. *Magn. Res. Med.* 1993, **30**, 696–703.
47. Aime, S., Batsanov, A. S., Botta, M., Howard, J. A. K., Parker, D., Senanayake, K. and Williams G. *Inorg. Chem.* 1994, **33**, 4696–4706.
48. Aime, S., Batsanov, A. S., Botta, M., Dickins, R. S., Faulkner, S., Foster, C. E., Harrison, A., Howard, J. A. K., Moloney, J. M., Norman, T. J., Parker, D., Royle, L. and Williams J. A. G. *J. Chem. Soc. Dalton Trans.* 1997, 3623.
49. Aime, S., Botta, M., Panero, M., Grandi, M. and Uggeri, F. *Magn. Reson. Chem.* 1991, **29**, 923–927.
50. Aime, S., Barbero, L. and Botta, M. *Magn. Reson. Imaging* 1991, **9**, 843–847.
51. Geraldes, C. F. G. C., Sherry, A. D., Vallet, P., Maton, F., Muller, R. N., Mody, T. D., Hemmi G. and Sessler, J. L. *J. Magn. Res. Imaging* 1995, **5**, 725–729.
52. Aime, S., Botta, M., Fasano, M., Geninatti Crich, S. and Terreno, E. *Coord. Chem. Rev.* 1999, **185–186**, 321–333.
53. Zetter, M. S., Grant, M. W., Wood, E. J., Dodgen, H. W. and Hunt, J. P. *Inorg. Chem.* 1972, **11**, 2701–2706.

54. Koenig, S. H., Baglin, C., Brown R. D. III and Brewer, C. F. *Magn. Res. Med.* 1984, **1**, 496–501.
55. Geraldes, C. F. G. C., Sherry, A. D., Brown R. D. III and Koenig, S. H. *Magn. Res. Med.* 1986, **3**, 242–250.
56. Perutz, M. F. *Annu. Rev. Physiol.* 1990, **52**, 1–10.
57. Aime, S., Ascenzi, P., Comoglio, E., Fasano, M. and Paoletti, S. *J. Am. Chem. Soc.* 1995, **117**, 9365–9366.
58. Aime, S., Bellelli, A., Coletta, P., Fasano, M., Paoletti, S. and Ascenzi, P. *J. Inorg. Biochem.* 1998, **71**, 37–43.
59. Wyman, J. *Adv. Protein Chem.* 1964, **19**, 223–286.
60. Aime, S., Botta, M., Dastrù, W., Fasano, M., Panero, M. and Arnelli, A. *Inorg. Chem.* 1993, **32**, 2068–2071.
61. Powers, J. C. and Harper J. W. in *Proteinase Inhibitors*, Barrett, A. J. and Salvesen G. S. (Eds), Elsevier, Amsterdam, 1986, pp. 55–152.
62. Aime, S., Fasano, M., Paoletti, S., Viola, F., Tarricone, C. and Ascenzi, P., *Biochem. Mol. Biol. Int.* 1996, **39**, 741–746.
63. Lauffer, R. B., McMurry, T. J., Dunham S. O., Scott D. M., Parmalee, D. J. and Dumas, S. *PCT Int. Appl. WO 9736619*, 1997; *Chem. Abstr.* 1997, **127**, 316334.
64. Wiener, E. C., Brechbiel, M. W., Brothers, H., Magin, R. L., Gansow, O. A., Tomalia, D. A. and Lauterbur, P. C. *Magn. Res. Med.* 1994, **31**, 1–8.

6 Stability and Toxicity of Contrast Agents

ERNÖ BRÜCHER
Lajos Kossuth University, Debrecen, Hungary

and

A. DEAN SHERRY
University of Texas at Dallas, Richardson, TX, USA

1 PHYSICO-CHEMICAL PROPERTIES OF MRI CONTRAST AGENTS

Metal chelates used as contrast enhancement agents in MRI are administered intravenously but will distribute through all of the extracellular and intravascular spaces. Since the amount of metal chelate required for a significant increase in image contrast is relatively high (for low-molecular-weight Gd^{3+} complexes, a typical dose is 0.1–0.3 mmol/kg total body weight), the concentration of the injected solution must also be high. A typical concentration of a pharmaceutical constrast agent (CA) is 0.5 M, so the metal chelate by necessity must have a high water solubility. In order to achieve this, some ligands that form non-ionic complexes with Gd^{3+} contain one or more alcoholic OH groups as side-chains or functional groups (e.g. HP-DO3A and DO3A-butrol). The presence of an alkyl or aryl group in the ligand decreases the solubility of a complex. For example, the solubility of Gd(NP-DO3A) in butanol is greater than that of Gd(HP-DO3A) which contains an alcoholic OH group [1]. The requirement of high concentration also means that the osmolality of the drug is higher than that of the blood, which may result in adverse effects for patients. The osmolality is particularly high for ionic metal chelate complexes. Pharmaceutical preparations of the anionic complexes, $GdDTPA^{2-}$ and $GdDOTA^{-}$, contain *N*-methylglucamine (NMG) as the counter-cation (NMG, also known as meglumine (MEG), is protonated at neutral pH). Even so, the osmotic pressures of these ionic solutions are lower than expected for typical 2:1 or 1:1 electrolytes, presumably due to significant association of the ions at these high concentrations [2, 3].

The Chemistry of Contrast Agents in Medical Magnetic Resonance Imaging
Edited by A. E. Merbach and É. Tóth.

Table 6.1 Osmolalities and butanol/water partition coefficients (log *P*) of some Gd^{3+} complexes (the osmolalities were measured in 0.5 M GdL solutions at 37 °C; the osmolality of blood plasma is 0.3 osmol kg^{-1}).

Ligand[a]	EDTA	DTPA	DTPA-BMA	EOB-DTPA	BOPTA	DOTA	DO3A	HP-DO3A	DO3A butrol
osmol (kg^{-1})	—	1.96	0.65	0.89[b]	1.97	1.35	—	0.63	0.57
log *P*	−2.79	−3.16	−2.13	−2.11	−2.23	−2.87	−2.15	−1.98	−2.00

[a] Structures are shown in Chart 6.1.
[b] Concentration of Solution, 0.25 M GdL.

Some of the more important physico-chemical properties of a number of Gd^{3+} complexes are presented in Table 6.1, with the structures of the various ligands being shown in Chart 6.1. Although the osmolality of 0.5 M $(NMG)_2GdDTPA$ is significantly higher than that of blood, the osmolalities of the non-ionic CAs are much lower than the ionic agents and quite close to the osmolality of blood. Largely due to concerns about the hyper-osmolality of the ionic agents, most of the second-generation and newer agents are neutral. These include GdHP-DO3A (approved), GdDTPA-BMA (approved), GdDTPA-BMEA (in trials) and GdDO3A-butrol (in trials). A second property of importance is compound hydrophilicity. $GdDTPA^{2-}$, $GdDOTA^{-}$, GdHP-DO3A, GdDTPA-BMA, GdDTPA-BMEA and GdDO3A-butrol are all relatively hydrophilic and these complexes uniformly distribute into all of the extracellular space before being excreted, predominantly through the renal system. The half-time for excretion of $GdDTPA^{2-}$ (typical of hydrophilic CAs) is ~ 1.6 h in man [4].

The complexes $GdEOB\text{-}DTPA^{2-}$ and $GdBOPTA^{2-}$ possess lipophilic groups and consequently are partially excreted through the hepatobiliary system. This makes them particularly useful as imaging agents for the liver [5, 6]. These comparisons indicate that the biodistribution of a CA largely depends on its relative lipophilic and/or hydrophilic properties. The property is commonly measured by a partition coefficient (*P*), determined by the distribution ratio of a drug between equal volumes of an organic solvent (e.g. butanol, octanol, etc.) and water. Log *P* values are typically lowest for hydrophilic ionic agents ($GdEDTA^{-}$, $GdDTPA^{2-}$ and $GdDOTA^{-}$) and about one order of magnitude higher for hydrophilic neutral agents (GdDTPA-BMA, GdDO3A, GdHP-DO3A and GdDO3A-butrol). Interestingly, those agents designed to target the liver are also charged ($GdEOB\text{-}DTPA^{2-}$, $GdBOPTA^{2-}$) yet have log *P* values similar to the non-ionics (see Table 6.1). The viscosities of these pharmaceutical preparations (typically 0.5 M solutions) do not differ dramatically and in general this property does not preclude rapid administration of the drug.

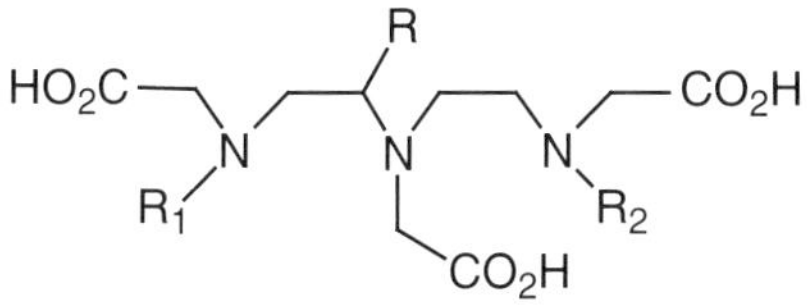

HO_2C R CO_2H N N N R_1 R_2 CO_2H

R = H, $R_1 = R_2 = CH_2COOH$	DTPA
R = H, $R_1 = R_2 = CH_2CONHCH_3$	DTPA-BMA
R = (4-ethoxybenzyl; OEt), $R_1 = R_2 = CH_2COOH$	EOB-DTPA
R = H, $R_1 = CH_2COOH$, R_2 = (benzyloxy-methyl fragment with COOH)	BOPTA

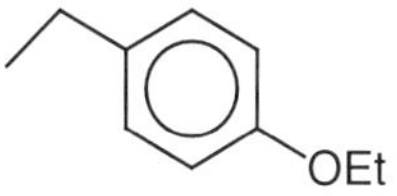

R_4 N N R_1 N N R_3 R_2

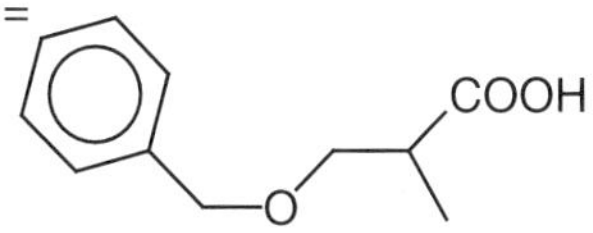

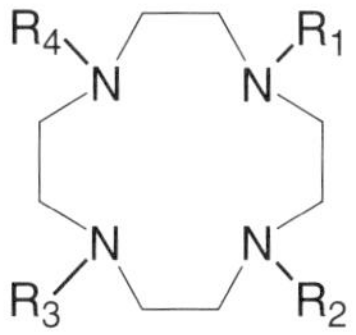

$R_1 = R_3 = H$, $R_2 = R_4 = CH_2COOH$	DO2A
$R_1 = H$, $R_2 = R_3 = R_4 = CH_2COOH$	DO3A
$R_1 = CH_2CH(OH)CH_3$, $R_2 = R_4 = CH_2COOH$	HP-DO3A
R_1 = (fragment with OH, OH, OH), $R_2 = R_3 = R_4 = CH_2COOH$	DO3A-butrol
$R_1 = R_2 = R_3 = R_4 = CH_2COOH$	DOTA
$R_1 = R_2 = R_3 = R_4 = CH_2PO_3H_2$	DOTP

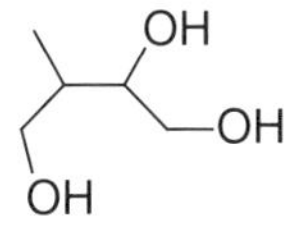

[**Chart 6.1**]

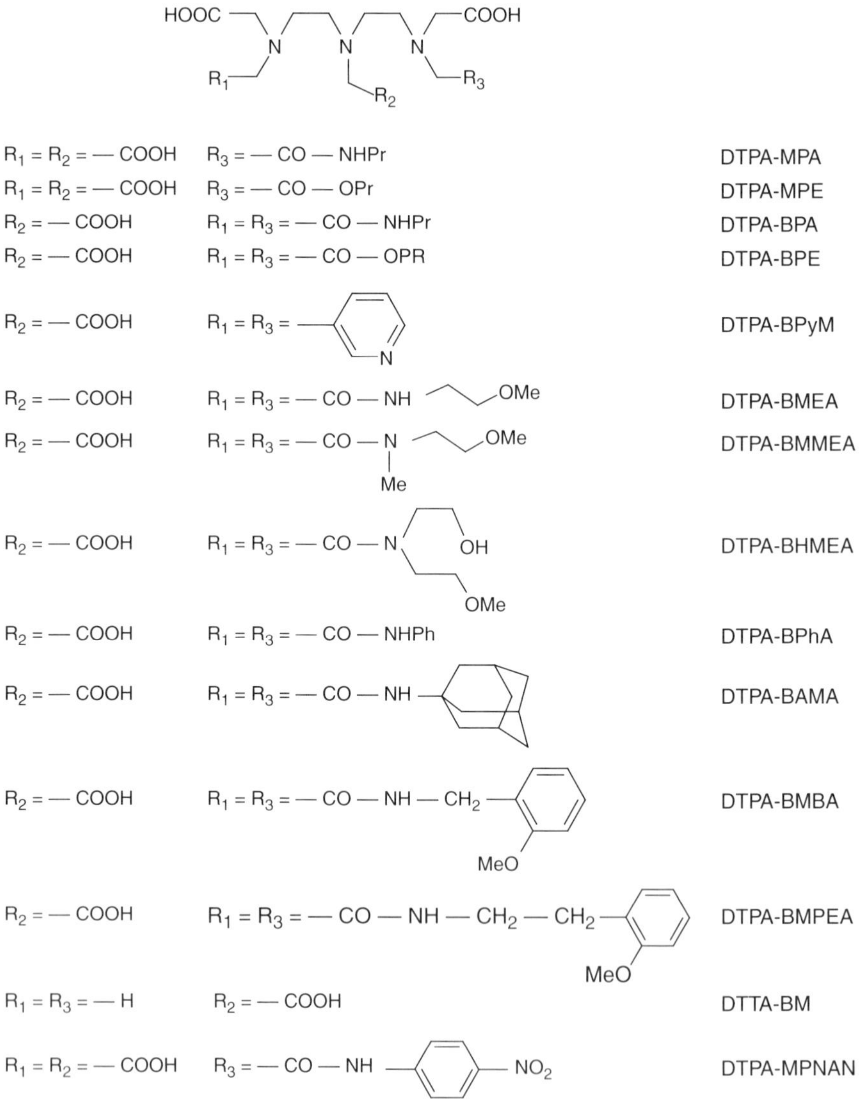

HOOC, COOH, N, N, N, R_1, R_2, R_3

$R_1 = R_2 =$ — COOH	$R_3 =$ — CO — NHPr	DTPA-MPA
$R_1 = R_2 =$ — COOH	$R_3 =$ — CO — OPr	DTPA-MPE
$R_2 =$ — COOH	$R_1 = R_3 =$ — CO — NHPr	DTPA-BPA
$R_2 =$ — COOH	$R_1 = R_3 =$ — CO — OPR	DTPA-BPE
$R_2 =$ — COOH	$R_1 = R_3 =$ — (pyridyl, N)	DTPA-BPyM
$R_2 =$ — COOH	$R_1 = R_3 =$ — CO — NH — (CH₂CH₂)OMe	DTPA-BMEA
$R_2 =$ — COOH	$R_1 = R_3 =$ — CO — N(Me) — (CH₂CH₂)OMe	DTPA-BMMEA
$R_2 =$ — COOH	$R_1 = R_3 =$ — CO — N(CH₂CH₂OH)(CH₂CH₂OMe)	DTPA-BHMEA
$R_2 =$ — COOH	$R_1 = R_3 =$ — CO — NHPh	DTPA-BPhA
$R_2 =$ — COOH	$R_1 = R_3 =$ — CO — NH — (adamantyl)	DTPA-BAMA
$R_2 =$ — COOH	$R_1 = R_3 =$ — CO — NH — CH_2 — (2-MeO-phenyl)	DTPA-BMBA
$R_2 =$ — COOH	$R_1 = R_3 =$ — CO — NH — CH_2 — CH_2 — (2-MeO-phenyl)	DTPA-BMPEA
$R_1 = R_3 =$ — H	$R_2 =$ — COOH	DTTA-BM
$R_1 = R_2 =$ — COOH	$R_3 =$ — CO — NH — (phenyl) — NO_2	DTPA-MPNAN

[Chart 6.2]

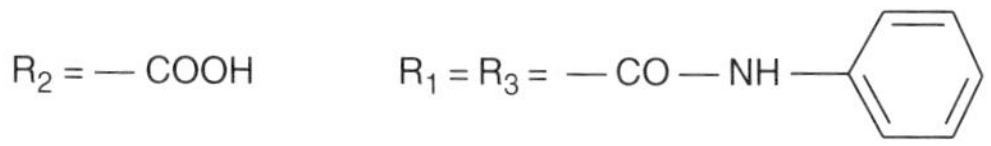

$R_2 = —COOH$ $R_1 = R_3 = —CO—NH—C_6H_5$ DTPA-BAN

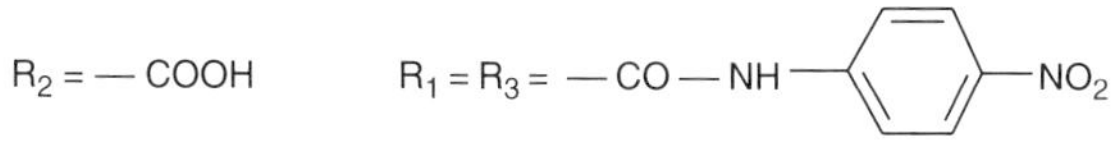

$R_2 = —COOH$ $R_1 = R_3 = —CO—NH—C_6H_4—NO_2$ DTPA-BPNAN

$R_2 = —COOH$ $R_1 = R_3 =$ (3-hydroxy-6-methylpyridin-2-yl; HO, N, Me) DTTA-HP

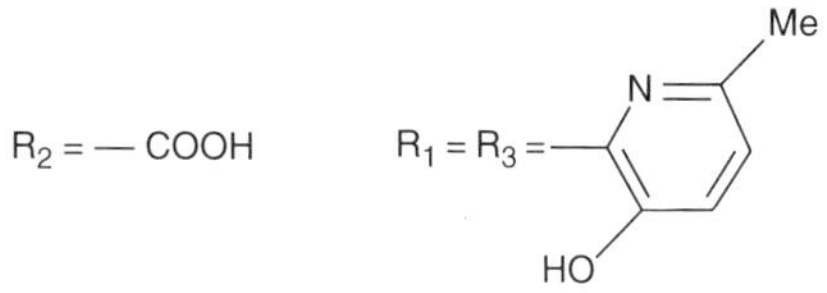

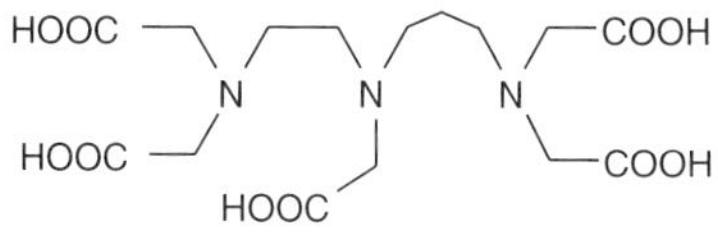

HOOC, HOOC, HOOC, N, N, N, COOH, COOH AEPDPA

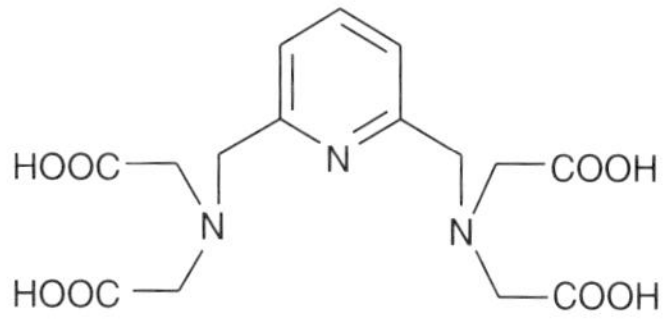

HOOC, HOOC, N, N, N, COOH, COOH PBMTA

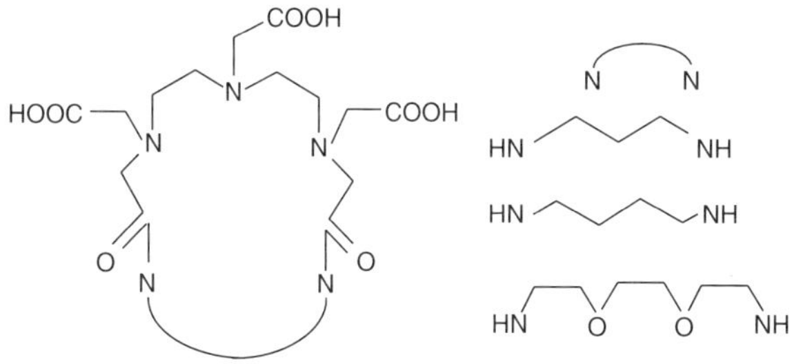

COOH, HOOC, N, N, N, COOH, O, O, N, N; N⌒N = HN—(CH₂)₃—NH DTPA-PAM; HN—(CH₂)₄—NH DTPA-BAM; HN—O—O—NH DTPA-OAM

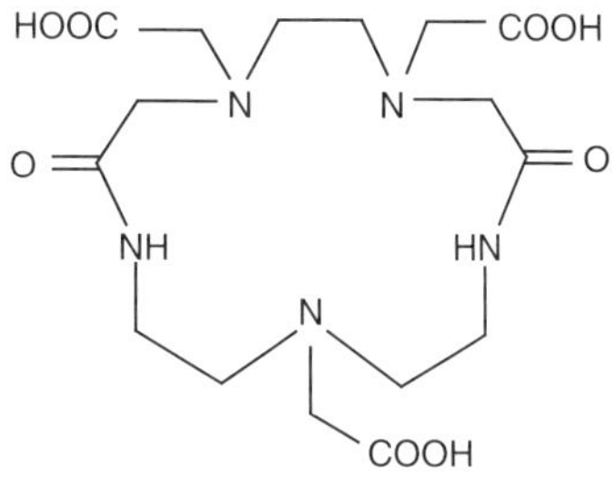

HOOC, COOH, N, N, O, O, NH, HN, N, COOH EDTA-DAM

[**Chart 6.2** *cont.*]

2 METAL–LIGAND STABILITY CONSTANTS

The metal chelates used as contrast agents in MRI investigations must meet several requirements. In addition to high water solubility and low osmolality, both important for tolerance reasons, the complexes must also not dissociate into the more toxic free metal ion and ligand species once injected into the body. Thus, other important criteria for safe *in vivo* use include high thermodynamic stability and kinetic inertness. The thermodynamic stability of a metal–ligand (ML) complex is expressed by the equilibrium constant shown in Equation (6.1). Here, K_{ML} refers to a specific equilibrium constant called a stability constant:

$$K_{ML} = \frac{[ML]}{[M][L]} \tag{6.1}$$

where [M], [L] and [ML] are the equilibrium concentrations of the metal ion, deprotonated ligand, and complex, respectively (the charges of the ions are omitted for simplicity). Complexes formed with multidentate ligands are often protonated at low pH values, and hence the following equilibrium must also be taken into account:

$$ML + iH \rightleftharpoons MH_iL \tag{6.2}$$

The constant(s) used to express the extent of protonation of the complex are defined as follows:

$$K_{MH_iL} = \frac{[M(H_iL)]}{[M(H_{i-1}L)][H^+]} \tag{6.3}$$

where $i = 1, 2, \ldots, n$, and $[H^+]$, $[M(H_iL)]$ and $[M(H_{i-1}L)]$ are the concentrations of the H^+ ions and protonated complexes, respectively.

In order to calculate a stability constant, the protonation constants of the ligand are also required. For convenience, these are defined by Equation (6.4) and reported as log K_i values (the inverse of pK_a).

$$K_i = \frac{[H_iL]}{[H_{i-1}L][H^+]} \tag{6.4}$$

where $i = 1, 2, 3 \ldots, n$.

The classical method for determining ML stability constants is pH-potentiometric titration of the ligand in the presence and absence of M [7]. This works extremely well for systems that reach equilibrium quickly (a few minutes) after addition of acid or base. Numerous programs have been written to treat such data and provide estimates of the uncertainties in the L protonation constants and ML stability constants. Of course all programs require the experimentalist to choose an appropriate model (L, ML, ML_2, M_2L, MLH, MLH_2, etc.) to fit the data and this can become problematical whenever independent evidence for the existence of such species is not available. Without independent data from NMR or other spectroscopic data collected as a function of pH, one normally fits the pH-potentiometric data to the simplest model possible (usually the model containing the fewest species).

From the early stages in the research and development of contrast agents, considerable effort has gone into correlating the stability constant of a Gd^{3+} complex with its *in vivo* toxicity. Although such a correlation has never been found, it has been shown that a conditional stability constant (K^{c}_{ML}, a stability constant for some specific condition such as pH) may be related to the selectivity of a ligand for Gd^{3+} over endogenous metals, such as Zn^{2+} and Cu^{2+} [8]. Hence, conditional stability constants are now widely used to predict the complexation behavior of ligands at physiological pH. In essence, a conditional stability constant reflects competition between the metal ion and protons for the ligand. The parameter K^{c}_{ML} is defined by the total free ligand concentration ($[L]_t = [L] + [HL] + [H_2L] + \ldots + [H_nL]$) instead of the free ligand concentration, [L]:

$$K^{c}_{ML} = \frac{[ML]}{[M][L]_t} = \frac{[ML]}{[M][L]\,\alpha_H} \tag{6.5}$$

where $\alpha_H = 1 + K_1[H^+] + K_1K_2[H^+]^2 + \ldots + K_1K_2\ldots K_n[H^+]^n$. A comparison of Equations (6.1) and (6.5) shows that K_{ML} and K^{c}_{ML} are proportionally related by α_H, i.e. $K_{ML} = K^{c}_{ML}\alpha_H$.

The pH-dependent K^{c}_{ML} conditional stability constants are widely used in analytical chemistry [9], but in complex biological systems such as body fluids, many other reactions can take place between the free ligand, the free metal ion, and the complex. By taking into account the most important side-reactions, a more general conditional stability constant (K*) was introduced by Ringbom [9]. The free ligand L can be protonated but it can also interact with endogenous metal ions such as Mg^{2+}, Ca^{2+}, Zn^{2+} and Cu^{2+}, plus perhaps others. Similarly, free Gd^{3+} can react with a number of important biological ligands such as citrate, phosphate, bicarbonate, transferrin, oxalic acid, and others. Gd^{3+} complexes can also be protonated or form ternary complexes with other small ligands such as carbonate, phosphate and dicarboxylic acids. If one includes all of the most significant side-reactions that are possible for all three species, then a new conditional stability constant, K^*, can be defined as follows:

$$K^* = \frac{[ML^*]}{[M^*][L^*]} \tag{6.6}$$

where the concentrations of $[M^*]$, $[L^*]$ and $[ML^*]$ are expessed as follows:

$$[M^*] = [M] + [MA] + [MB] + \ldots, \text{ etc.} \tag{6.7}$$

$$[L^*] = [L] + [HL] + [H_2L] + \ldots + [H_nL] + [M'L] + [M''L] + \ldots, \text{ etc.} \tag{6.8}$$

$$[ML^*] = [ML] + [MHL] + [MLX] + [MLY] + \ldots, \text{ etc.} \tag{6.9}$$

By expressing the concentrations of the metal and proton complexes using the stability and protonation constants and introducing the functions α_M, α_L and α_{ML}, the conditional stability constant can be simplified to the following:

$$K^* = \frac{[ML]}{[M][L]}\frac{\alpha_{ML}}{\alpha_M \alpha_L} \tag{6.10}$$

where:

$$\alpha_M = 1 + K_{MA}\,[A] + K_{MB}\,[B] + \ldots, \text{ etc.}$$
$$\alpha_L = 1 + K_1[H^+] + K_1K_2[H^+]^2 + \ldots + K_1K_2\ldots K_n[H^+]^n + K_{M'L}[M'] + K_{M''L}[M''] + \ldots, \text{ etc.}$$
$$\alpha_{ML} = 1 + K_{MHL}[H^+] + K_{MLX}[X] + \ldots, \text{etc.}$$

Thus, the conditional stability constant (K^*) is related to the thermodynamic stability constant (K_{ML}) by the following:

$$K^* = K_{ML}\frac{\alpha_{ML}}{\alpha_M \alpha_L} \tag{6.11}$$

Given either known or measured protonation and stability constants, K^* can be calculated and the concentration of 'free' paramagnetic metal ion, $[M^*]$, evaluated. The value of $[M^*]$ is considered important because it is generally assumed that the toxicity of a contrast agent is related to the amount of metal ion (e.g. Gd^{3+}) released from a complex into the body.

Due to their high magnetic moments and relaxation efficiencies, complexes of Gd^{3+}, Mn^{2+} and Fe^{3+} have received the most attention in the research and development of MRI contrast agents. However, in practice only Gd^{3+} complexes are widely used, while Mn^{2+} and Fe^{3+} containing agents have so far played only a marginal role. As a result, the equilibrium studies of the complexes relevant for MRI contrast agents are limited mainly to complexes of Gd^{3+}.

The first, clinically used contrast agent was $GdDTPA^{2-}$, but soon after its introduction the kinetically more inert $GdDOTA^-$ material was also approved. On the basis of these early successes, many other ligands, predominantly derived from either DTPA or DOTA, have been designed and studied. Since the properties of the open-chain DTPA and the macrocyclic DOTA derivatives differ in some respects, the equilibrium behavior of their complexes will be discussed separately. In body fluids, a few endogenous metal ions can compete with Gd^{3+} for the ligand so that the stability constants of the complexes formed with Ca^{2+}, Zn^{2+} and Cu^{2+}, which are present in relatively large concentrations in body fluids, will be compared with those of the Gd^{3+} complexes whenever data are available.

2.1 ML STABILITY CONSTANTS WITH LIGANDS DERIVED FROM LINEAR AMINES

DTPA is an octadentate ligand that coordinates to Gd^{3+} by using all three nitrogen and five carboxylate oxygen donor atoms. The ligands BOPTA and EOB-DTPA possess the same number and type of donor atoms and the resulting Gd^{3+} stability constants are quite similar (Table 6.2). Although conversion of one carboxylate group into an amide or an ester still yields octadentate ligands (DTPA-MPA, DTPA-N-MA, DTPA-N′-MA and DTPA-MPE; see structures in Chart 6.2), this simple removal of one anionic donor atom (carboxylate) and replacement by a non-ionic functional group (amide or ester) results in a decrease in stability of the resulting Gd^{3+} complexes by about three orders of magnitude. This shows that the charge on the ligand side-chain and

Table 6.2 Stability constants (log K_{ML}) of complexes formed with DTPA-derivative ligands.

Ligand	Gd^{3+}	Ca^{2+}	Zn^{2+}	Cu^{2+}	log (K_{GdL}/K_{ZnL})	Reference
DTPA	22.46	10.75	18.29	21.38	4.17	7
EOB-DTPA	23.46	11.74	—	—	—	6
BOPTA	22.59	—	—	—	—	5
DTPA-MPA	19.68	—	—	—	—	11
DTPA-MPE	18.91	—	—	—	—	11
DTPA-N-MA	19.37	—	16.00	18.71	3.37	107
DTPA-N'-MA	19.94	—	16.82	18.50	2.09	107
DTPA-BMA	16.85	7.17	12.04	13.03	4.81	8
DTPA-BPyM	16.83	7.97	14.02	17.50	2.81	8
DTPA-BPA	16.23	—	—	—	—	11
DTPA-BMEA	16.84	—	—	—	—	101
DTPA-BMMEA	17.68	—	—	—	—	101
DTPA-BHMEA	17.49	—	—	—	—	101
DTPA-BPhA	16.79	—	—	—	—	102
DTPA-BAMA	16.85	7.49	11.90	12.86	4.95	103
DTPA-BMBA	16.82	7.74	12.48	13.42	4.34	103
DTPA-BMPEA	16.91	7.54	12.79	13.28	4.12	103
DTTA-BM	13.12	7.25	—	—	—	16
DTTA-HP	23.65	—	—	—	—	105
DTPA-BPE	16.30	—	—	—	—	11
PBMTA	18.10	9.20	15.30	—	2.80	106
DTPA-MPNAN	17.63	—	—	—	—	111
DTPA-BPNAN	14.90	—	—	—	—	111
DTPA-BAN	15.16	—	—	—	—	111
AEPDPA	22.77	14.45	18.59	19.31	4.18	104
DTPA-PAM	14.49	—	11.74	—	2.75	19
DTPA-BAM	15.39	—	11.56	—	3.83	19
DTPA-OAM	17.44	—	12.19	—	5.25	19
EDTA-DAM	15.15	—	9.03	—	6.12	19

the basicity of that donor group (related properties) are important determinants of complex stability.

Bis(amide) derivatives of DTPA have proven to be very popular, partly due to ease of synthesis. A large number of bis(amides) have been synthesized and their equilibrium properties examined [10–12]. It was shown by ^{1}H and ^{13}C NMR spectroscopy that the amide groups are coordinated to the complexed lanthanide cation via the neutral carbonyl oxygen, thereby yielding octadentate structures similar to the parent DTPA complexes [13]. Of the various non-ionic functional group choices (amides, esters, hydroxyalkyls, etc.), it is the amide carbonyl oxygen that results in the largest contribution to complex stability [8, 14, 15]. This likely reflects an important resonance contribution characteristic of amides whereby the carbonyl oxygen maintains a partial negative charge. The importance of amide coordination in the stability constants of Gd^{3+} complexes was demonstrated by Paul-Roth and Raymond [16]. They synthesized bis(methyl)-diethylenetriamine triacetate (DTTA-BM, a hexadentate ligand; see structure in Chart 6.2) and reported that the stability constant of Gd(DTTA-BM) is 3.73 log K units lower than that of Gd(DTPA-BMA). This comparison does not just reflect the contribution made by the two amide groups since the basicities of the nitrogen donors in DTTA-BM are much higher than the respective nitrogen donors in DTPA-BMA [16].

Even though amide groups do participate in complexation in these systems, the stability constants of GdDTPA-mono(amide) and GdDTPA-bis(amide) complexes are about three and six orders of magnitude lower, respectively, than that of $GdDTPA^{2-}$ (see Table 6.2). This significant drop in the stability does not result in a decreased tolerance for the GdDTPA-bis(amide) complexes because the stability constants of DTPA-bis(amide) derivatives with endogenous Ca^{2+}, Zn^{2+} and Cu^{2+} decrease even more dramatically [8]. Furthermore, unlike DTPA, DTPA-bis(amide) ligands do not form dinuclear M_2L complexes with Cu^{2+} and Zn^{2+}. Since the DTPA is octadentate while typical coordination numbers for Zn^{2+} and Cu^{2+} are four to six, relatively stable dinuclear Cu_2DTPA^- and Zn_2DTPA^- complexes are formed (log $K_{Cu_2L} = 5.54$ and log $K_{Zn_2L} = 4.48$ for the equilibrium $ML + M \rightleftharpoons M_2L$) [17]. Conversely, Gd^{3+}, preferring a much larger coordination number, forms only a weak binuclear complex with DTPA, namely $Gd_2(DTPA)^+$ (log $K_{Gd_2L} \approx 1.0$) [18]. The greater stabilities of the $Cu_2(DTPA)^-$ and $Zn_2(DTPA)^-$ complexes contribute to a greater displacement of Gd^{3+} from $GdDTPA^{2-}$ by Zn^{2+} and Cu^{2+} [18]. Thus, even though the bis(amide) derivatives of DTPA do indeed have a lower affinity for Gd^{3+}, they also have a much less tendency to form dinuclear complexes with Cu^{2+} and Zn^{2+} and this turns out to be an unanticipated, and significant advantage for the amide systems in increasing the binding selectivity for Gd^{3+} over Cu^{2+} and Zn^{2+}.

The interesting properties of the DTPA-bis(amide) ligands prompted the synthesis of other cyclic derivatives. Condensation of DTPA-dianhydride or EDTA-dianhydride with diamines led to several new 15-, 16-, 17-, 18-and 21-

membered-ring DTPA-bis(amides) and a 15–membered-ring EDTA-bis(amide) derivative [19]. The stability constants of the resulting Gd^{3+}-ligand complexes increase with ring size, while the stabilities of the corresponding Zn^{2+}-ligand complexes are practically independent of ring size. This results in a very high selectivity for Gd^{3+} over $Zn^{2+}$for the EDTA-DAM ligand (see structure in Chart 6.2) [19]. Condensation of two DTPA-dianhydrides with a single 1, 4–diaminobutane yielded a 34-membered-ring cyclic derivative that forms both mononuclear and dinuclear complexes with Gd^{3+} [20]. The stability constants of this system (log $K_{GdL} = 16.46$ and log $K_{Gd_2L} = 30.36$) were found to be similar to the log K_{GdL} values of other GdDTPA-bis(amide) complexes.

2.2 ML STABILITY CONSTANTS WITH MACROCYCLIC LIGANDS

As noted above, pH-potentiometry is the classic method for determining ML stability constants for systems that reach equilibrium quickly. Unfortunately, this is simply not the case for most Ln^{3+}-macrocyclic ligand systems. The sluggishness of Ln^{3+}-macrocyclic ligand complex formation, as noted in the first report of $LnDOTA^-$ complexes [21], is at least partially responsible for the wide range of thermodynamic stability constants reported for many of the macrocyclic sytems, shown in Table 6.3 (Structures of the ligands are shown in Chart 6.1). However, since macrocyclic polyamines tend to be more basic than linear polyamines, a second factor in the wider variation in log K_{ML} values reported for macrocyclic systems is the uncertainty in the highest protonation constant(s) of the ligand itself [22]. There are several interesting trends in the log K_{ML} values listed in Table 6.3. As with linear polyamine systems, the log K_{ML} values increase substantially along the Ln^{3+} series. This reflects the largely electrostatic character of the bonding interactions between the highly charged trivalent Ln^{3+} cations and these polyaza-polyacetate/polyphosphonate ligands. Unlike the more conformationally flexible linear polyaza systems, the log K_{ML} values for most macrocyclic ligands do not increase in a smooth progression with decreasing Ln^{3+} size (increasing charge density). This indicates that macrocyclic ring size and hydration effects play a role in complex stability. This is especially evident for the $LnDOTA^-$ and $LnDOTP^{5-}$ complexes, where complex stability makes a noticable stepped increase near the center of the Ln^{3+} series for both ligands. Obviously, ligand charge also plays an important role in determining complex stability. For cyclen-based ligands, the log K_{ML} values for any given Ln^{3+} increase with increasing negative charge on the ligand. This is seen in the log K_{ML} values where DO2A < DO3A < DOTA < DOTP. The most stable Ln^{3+} complexes are those formed with the tetra-phosphonate ligand, DOTP. This is due to the combined effects of high ligand charge, the greater basicity of amines with pendant methylene phosphonate groups, and the smaller number of inner-sphere water molecules in the $LnDOTP^{5-}$ complexes (the

Table 6.3 Thermodynamic stability constants (log K_{ML}) for some Ln^{3+}-macrocyclic complexes; structures of the ligands are shown in Chart 6.1.

Ln^{3+}	NOTA	DO2A	DO3A	DOTA	HP-DO3A	DOTP	TETA	DO3A-butrol
La	13.5[a]	10.94[c]	—	22.9[a]	—	27.6[h]	11.60[b]	—
		16.6[f]		21.7[b]			12.74[i]	
Ce	13.2[a]	11.31[c]	19.7[d]	23.4[a]	21.2[d]	27.7[h]	13.12[i]	19.7[j]
				23.0[d]				
				24.6[k]				
Pr	13.3[a]	12.00[c]	—	23.0[a]	—	27.4[h]	—	—
Nd	13.1[a]	12.56[c]	—	23.0[a]	—	27.3[h]	13.76[i]	20.1[j]
Sm	13.4[a]	12.93[c]	—	23.0[a]	—	28.1[h]	14.47[i]	—
Eu	13.9[a]	12.99[c]	20.69[e]	23.5[a]	—	28.1[h]	14.66[i]	21.2[j]
Gd	14.3[a]	13.06[c]	21.0[d]	24.7[a]	23.8[d]	28.8[h]	13.77[b]	20.8[j]
		19.1[f]	22.02[g]	24.0[b]			14.73[i]	
				25.3[d]				
				24.67[g]				
Tb	14.5[a]	12.93[c]	—	24.2[a]	—	28.9[h]	14.81[i]	—
Dy	15.1[a]	13.13[c]	—	24.8[a]	—	—	—	21.2[j]
Ho	15.2[a]	13.00[c]	—	24.5[a]	—	29.2[h]	—	21.0[j]
Er	15.2[a]	13.31[c]	—	24.4[a]	—	29.6[h]	—	—
Tm	15.4[a]	13.19[c]	—	24.4[a]	—	29.5[h]	15.15[i]	21.1[j]
Yb	15.4[a]	13.26[c]	—	25.0[a]	—	29.5[h]	—	—
				26.4[k]				
Lu	15.6[a]	13.16[c]	23.0[d]	25.4[a]	—	29.6[h]	14.77[i]	20.9[j]
		20.6[f]						

[a] Batch method (≥ 18 h at 60 °C, followed by 12 h at 25 °C); spectrophotometry using arsenazo III as competing ligand [29].
[b] Batch method (10–16 days); pH-potentiometry [108].
[c] Batch method (at least 12 h); capillary electrophoresis [30].
[d] Batch method (≥ 18 h at 60 °C, followed by 12 h at 25 °C); spectrophotometry using arsenazo III as competing ligand [26].
[e] Batch method (70 °C for 2 days, followed by 4 days at 25 °C); laser-excited Eu^{3+} luminescence using EDTA as competing ligand [109].
[f] Batch method (up to 3 weeks); ^{1}H NMR using EDTA as competing ligand [28].
[g] Batch method (3 weeks); pH-potentiometry; no competing ligand [112].
[h] Batch method (several days); spectrophotometry using arsenazo III as competing ligand [33].
[i] Direct pH-potentiometric titration [110].
[j] Batch method (20 days); spectrophotometry using Eu^{3+} as reference cation [34].
[k] Batch method (4 weeks); pH-potentiometry; high concentration of DOTA and Ln^{3+} [22].

crystal structure of $TmDOTP^{5-}$ shows no inner-sphere waters; H.-P. Juretschke, personal communication).

Linear relationships between experimentally measured log K_{ML} values and ligand protonation constants have been noted in the literature for nearly 50 years [23]. Irving and Rossotti [24] discussed the theoretical basis for such relationships and pointed out reasons why log K_{ML} versus ligand pK_a relationships can deviate from linearity even for some simple mono-and bidentate ligands. Choppin [25] extended this to linear polyamino-polycarboxylate systems and demonstrated a single linear correlation between log K_{ML} and $\sum$ pK_a

for monodentate and polydentate ligands that form five-membered chelate rings with Ln^{3+} cations. He also showed that the ΔS of complexation for a given Ln^{3+} cation is proportional to the number of carboxyl groups in a linear polyamino-polycarboxylate ligand. Deviations from linearity were observed for polydentate ligands that formed six or seven-membered chelate rings, again reflecting the importance of entropy in determining complex stability [25]. More recently, such relationships have been extended to macrocyclic polyamino-polycarboxylates [26–28]. As ligands have become more diverse, one uncertainty in forming such relationships is the number of ligand protonation steps that should be included in the calculation of $\sum pK_a$. Kumar *et al.* [27] advocated including only those protonation constants that result in a neutral ligand, i.e., three for NTA, four for EDTA and DOTA, five for DTPA, etc. By using this concept, the empirical relationship, $\log K_{GdL} = 0.85 \pm 0.05 \sum pK_a$, for a wide range of linear and macrocyclic polyamino-polycarboxylate ligands was reported [27].

A plot of $\log K_{GdL}$ versus $\sum pK_a$ for a series of ligands derived from both linear and macrocyclic polyamines with carboxylate, phosphinate, and amide side-chain coordinating groups is reproduced in Figure 6.1 (the pK_a rule introduced by Kumar *et al.* [27] was used in deriving these data). It is interesting that this simple relationship at least qualitatively holds for a wide variety of linear and macrocyclic amines with differing side-chain coordinating functionalities. This relationship, however, does not predict accurate stability constants for a number of ligands, most notably NOTA, DOTEP, DTPA-BMA, and DO3MA. Of these, three are macrocyclic ligands (NOTA, DOTEP, and DOTMA) while one (DTPA-BMA) is derived from a linear amine. This simple basicity relationship predicts a $\log K_{ML}$ for GdNOTA of about 16, while an experimental value of 13.7 was found [29]. This difference likely reflects the fact that the triazacyclononane ring is too small for Gd^{3+}, so that the nitrogen donor atoms of the triazacyclononane ring are not optimally positioned to allow maximum electrostatic interactions between the nitrogen lone-pair electrons and the large Gd^{3+} ion. Interestingly, the triacetate analog of the larger triazacyclododecane macrocycle (DOTRA) should be flexible enough to allow maximum overlap, yet LnDOTRA complexes have not been isolated or even detected in solution. This may be a case where the free DOTRA ligand has little or no pre-organized structure in solution so formation of a complex may be limited by extremely slow kinetics (see below). The reasons why the $LnDOTEP^-$ complexes are less stable than expected, while the LnDTPA-BMA and $LnDOTMA^-$ complexes are more stable than expected, are less clear. The $LnDOTEP^-$ complexes may be destabilized somewhat by steric interactions between the ethyl groups on neighboring phosphinates (only a single complex stereoisomer is present in solution), while the $LnDOTMA^-$ complexes may gain stability by the extra rigidity imposed by the methyl acetate side-arms. In the case of the LnDTPA-BMA complexes, only three pK_as were included in the $\sum pK_a$ term for this ligand, two of which were nitrogen pK_as and a third was by necessity the highest carboxylate pK_a. One might anticipate that the carbonyl oxygen atoms of bis-amides in DTPA-BMA, once coordinated to Gd^{3+}, are more basic than predicted by solution nK_a's due to resonance stabilization with the dimethylamide groups. The deviations seen by these three, structurally unrelated ligands, illustrate that many factors other than simple donor basicity contribute to complex stability and that empirical relationships such as log K_{GdL} versus $\sum pK_a$, although qualitatively useful, should not be over-interpreted.

A second interesting trend is found in the comparison of the stability constants for the different macrocyclic ring sizes, where stability follows the order, NOTA $<$ TETA $<$ DOTA. The 9-membered macrocyclic ring of NOTA is clearly too small to adequately accommodate a rather sizeable Ln^{3+} cation and, consequently, the stabilities of the LnNOTA complexes are consistently about 10^{10} smaller than the corresponding $LnDOTA^-$ complexes. Interestingly, an increase in ring size from the 12–membered DOTA to the 14–membered TETA has nearly the same profound effect. In this case, the stabilities of the $LnTETA^-$ complexes are more than 10^{10} smaller than the corresponding $LnDOTA^-$ complexes. This comparison highlights the near optimal size of macrocyclic ligands derived from the 12–membered ring, cyclen, for the trivalent lanthanide cations.

A comparison of Ln^{3+} stability constants with DO2A, DO3A, DOTA, HP-DO3A, and DOTA-butrol (all derived from the 12–membered ring, cyclen) also illustrates the importance of the number and identity of side-chain chelating groups. Using DOTA as the reference, the Ln^{3+} stability constants drop $\sim 10^3$ after removal of a single acetate group in the LnDO3A complexes and by another factor of $\sim 10^8$ after removing a second acetate group to form the $LnDO2A^+$ complexes [30, 31]. Thus, a minimum of three chelating side-groups seems necessary to form reasonably stable complexes (the same conclusion would be reached by comparing $LnDO2P^-$ and $LnDOTP^{5-}$ complexes) [32, 33]. Addition of a sterically uncrowded hydoxypropyl group to DO3A (to form HP-DO3A) results in stabilization of the Ln^{3+} complexes by $\sim 10^2$, while addition of the sterically crowded 2, 3–dihydroxy-(1–hydroxymethyl)-propyl group (to form DO3A-butrol) results in a slight destabilization [34]. The latter is especially evident for the smaller Ln^{3+} cations, suggesting that the bulky 2, 3–dihydroxy-(1–hydroxymethyl)-propyl group actually destabilizes binding interactions between the Ln^{3+} and each of the acetate side-arms. Tóth *et al.* [34] have demonstrated that the bound hydroxy group of the LnDO3A-butrol complexes is more acidic than the bound hydroxy group of the LnHP-DO3A complexes (by factors of 10^2–$10^{4.4}$, depending upon Ln^{3+} cation size) and it is this factor that leads to less stable LnDO3A-butrol complexes. This indicates that the basicity of side-chain coordinating groups can significantly affect complex stability and that empirical relationships like that described above are indeed an over-simplification because they often do not consider the basicity of side-chain coordinating groups.

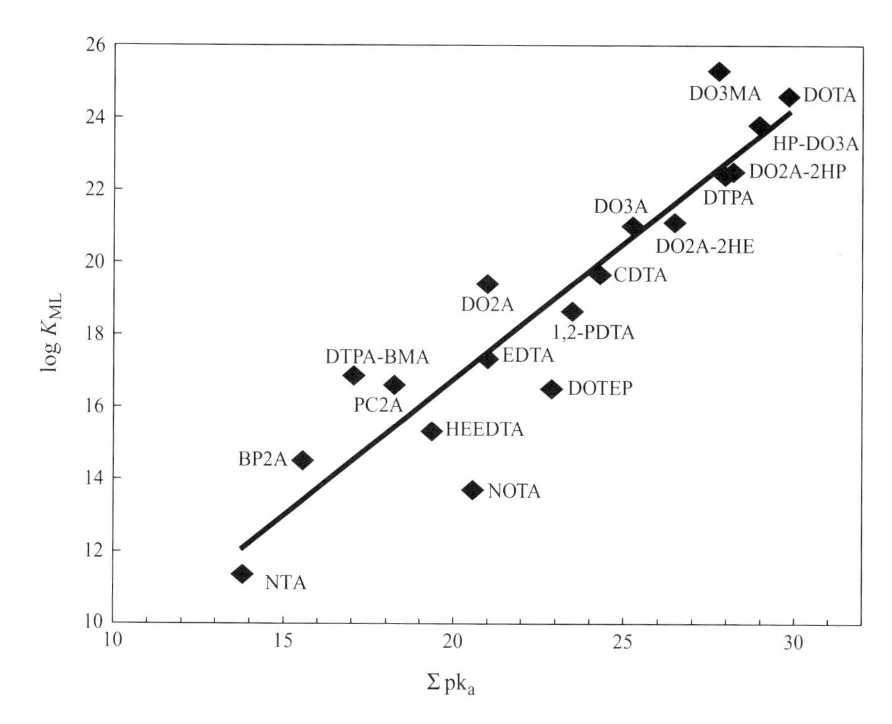

Figure 6.1 Plot of log K_{ML} versus $\sum pK_a$ for a series of Gd^{3+} complexes. The number of pK_a's used in the summation corresponds to the number of protonation steps required to give a neutral ligand (see text for details).

One might note that $GdDOTP^{5-}$ was not included in the plot of Figure 6.1, for two reasons. First, the highest pK_a of DOTP is relatively uncertain [33, 35] and second, if one uses the eight highest protonation constants to estimate $\sum pK_a$ for this ligand, one obtains an unreasonably high value of 58 and a predicted log K_{GdL} of 49! The much lower experimentally determined log $K_{ML} = 28.8$ likely reflects a much larger ΔS of dehydration of the phosphonate groups in DOTP in comparison to carboxylate groups found in the majority of ligands that adhere to this simple linear relationship. Interestingly, the greater basicity of the two most basic amines in DOTP in comparison to DOTA ($\Delta \sum pK_a \sim 4.4$) is alone sufficient to explain the differences in stability reported in Table 6.3 for these respective complexes.

3 FORMATION OF TERNARY COMPLEXES AND THE INFLUENCE ON WATER RELAXATION

The coordination number of Gd^{3+} in all complexes currently used for clinical MRI studies is nine, with the ligand contributing eight donor atoms and the ninth position being occupied by an H_2O molecule. Thus, the correct formulas of the complexes currently in clinical use are as follows: $GdDTPA(H_2O)^{2-}$, $GdDOTA(H_2O)^-$, $GdHP\text{-}DO3A(H_2O)$, and $GdDTPA\text{-}BMA(H_2O)$. The existence of a single inner-sphere H_2O molecule in these systems has been demonstrated by X-ray diffraction of single crystals [2, 15, 36], by luminescence spectroscopy [37–39] and by ^{17}O NMR spectroscopy [40]. This inner-sphere H_2O molecule is critical in transferring magnetic information from the highly paramagnetic Gd^{3+} to bulk water molecules via rapid water exchange. However, in a biological milleu there are a large number of organic and inorganic ligands that may, in principle, displace the H_2O molecule and occupy the ninth coordination position. This effect can be quite detrimental because removal of this water coordination site can either eliminate the inner-sphere relaxation effect entirely or, at least, make the complex substantially less efficient at producing MRI contrast. Displacement of inner-sphere waters by a bidentate or tridentate ligand from complexes such as $Ln(EDTA)(H_2O)_x^-$ ($x = 2$ or 3) is well known [41–45]. Ternary $Ln(EDTA)_X$ complexes have been reported for iminodiacetate, nitrilotriacetate, 8–hydroxyquinoline-5-sulfonate, oxalate, diglicolate, tartrate, fluoride and various amino acids [41–45].

Initially, formation of ternary complexes in the octadentate Gd^{3+}-ligand systems used as CAs did not seem particularly troublesome because it was thought that only a single water coordination site was available to form ternary complexes. Formation of a ternary complex between $LaDTPA^{2-}$ and F^- had been noted [43] but the stability constant of this complex ($K_{LaLF} = 3 \pm 1$) was too low to be considered significant. In 1988, however, Aime and Nano invoked ternary complex formation between $GdDTPA^{2-}$ and undefined components in blood plasma to explain the anamolous water relaxivity of this complex in blood [46]. Nevertheless it has not been necessary to include ternary complex formation in models to explain the species distributions of $GdEDTA^-$, $GdDTPA^{2-}$ or GdDTPA-BMA in blood plasma [8, 47]. Some evidence does exist for ternary complex formation for other complexes with higher water coordination numbers. For example, the ^{31}P relaxation rate of inorganic phosphate in the presence of $GdDO3A(H_2O)_2$ is most consistent with formation of a ternary $GdDO3A(P_i)$ complex [48] and a solid-state ternary complex, $[GdDO3A]_3Na_2CO_3 \cdot 17H_2O$, has been reported by Chang *et al.* [49]. The X-ray diffraction study showed that Gd^{3+} did not have an inner-sphere water molecule and that a single bridging CO_3^{2-} anion was coordinated to three GdDO3A complexes. These findings indicate that one should at least consider the possibility of ternary complex formation between CAs and species such as PO_4^{3-} or CO_3^{2-} in aqueous solution.

Burai *et al.* [50] measured the binding of carbonate, phosphate, and citrate to $GdDTPA^{2-}$, $GdDOTA^-$, GdDTPA-BMA, GdDO3A, and $GdEDTA^-$ by using pH-potentiometric methods. The stability constant of a ternary complex, K_{GdLX}, is defined in the following equation:

$$K_{GdLX} = \frac{[Gd(L)X]}{[Gd(L)][X]} \tag{6.12}$$

Here, [X] is the concentration of the CO_3^{2-}, PO_4^{3-} or Cit^{3-}, and GdL is the CA under study. The log K_{GdLX} value for GdDO3A was calculated from water relaxivity data [51]. The stability constants determined in this series of experiments are summarized in Table 6.4. As anticipated, the stability constants of the ternary complexes formed with $GdDTPA^{2-}$, $GdDOTA^{-}$ and GdDTPA-BMA are all quite low. There are two interesting features in these data. First, no evidence could be found by pH-potentiometry for ternary complex formation between citrate (a potential tridentate ligand) and any of these Gd^{3+} complexes and, secondly, CO_3^{2-} and PO_4^{3-} form stronger ternary complexes with GdDO3A than with $GdEDTA^{-}$, even though the latter complex has more inner-sphere water positions which can be replaced by the anionic ternary ligands. This suggests that the negative charge on $GdEDTA^{-}$ may be an important factor in reducing the tendency to form ternary complexes. These findings do indicate that complexes containing even one water coordination site ($GdDTPA^{2-}$, $GdDOTA^{-}$ and GdDTPA-BMA) can form weak ternary complexes with the small-size, highly negatively charged (CO_3^{2-} and PO_4^{3-}) ligands. The larger citrate can not displace the only H_2O molecule in the inner sphere of the Gd^{3+} complexes, probably for steric reasons. The absence of ternary complexes with citrate also indicates that the secondary ligand can not displace the carboxylate or amide oxygen donors from the inner-sphere of the Gd^{3+}.

There are many low-molecular-weight ligands present in blood plasma (citrate, lactate, dicarboxylic acids, amino acids, etc.) which are capable of forming complexes of varying stability with Ln^{3+} ions. Some ligands, such as carbonate, oxalate, phosphate and fluoride, form insoluble precipitates with the free Ln^{3+} aquo ions, but it has been shown both experimentally and by calculation that Gd^{3+} complexes used as contrast agents do not form precipitates with either carbonate or phosphate ions [8, 47]. Of the smaller ligands found in blood plasma, the concentrations of HCO_3^-/CO_3^{2-} (2.45×10^{-2} M), $H_2PO_4^-/HPO_4^{2-}$ (3.8×10^{-4} M) and citrate (1.1×10^{-4} M) are comparable to the concentration of a clinical dose of CA after distribution throughout all of the extracellular space ($1–5 \times 10^{-4}$ M) [4, 52]. Under these conditions, ternary

Table 6.4 Stability constants (log K_{GdLX}) of ternary complexes (25 °C, 0.1 M NaCl).

Ligand	$GdDTPA^{2-}$	$GdDOTA^{-}$	Gd(DTPA-BMA)	GdDO3A	$GdEDTA^{-}$
PO_4^{3-}	2.0	2.2	2.0	4.8	2.8
CO_3^{2-}	1.4	1.9	1.3	4.8	2.6
Cit^{3-}	—[a]	—[a]	—[a]	—[b]	3.6

[a] No complexation.
[b] Water relaxation rates indicate complexation.

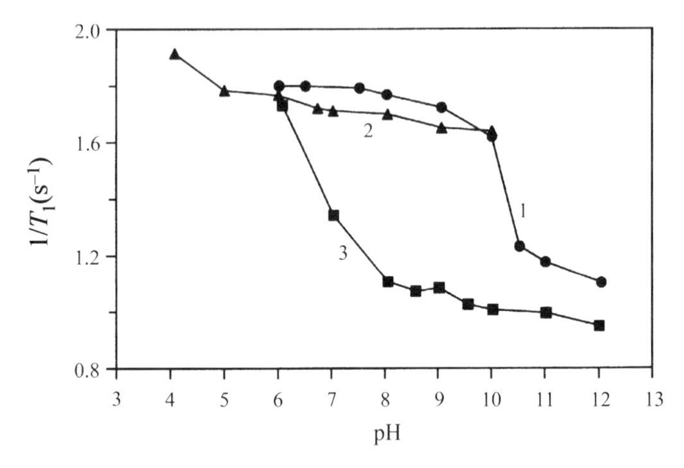

Figure 6.2 Relaxation rates of water protons at 9 MHz in the presence of 0.2 mM GdDO3A and (1) 0.5 mM phosphate, (2) 0.2 mM citrate, and (3) 25 mM carbonate, in 0.1 M NaCl at 25 °C.

complexes are indeed a possibility but, in practice, this appears not to be an important factor. Early studies indicated that Gd(DO3A) had several favorable features considered advantageous in CA design (two inner-sphere water molecules, high water relaxivity, and high kinetic stability), but under physiological conditions many endogenous ligands can displace the two water molecules and decrease the effective relaxivity of the complex. This is illustrated by the relaxivity data shown in Figure 6.2. In this experiment, the T_1 values for 0.2 mM solutions of $GdDTPA^{2-}$, $GdDOTA^{-}$ and GdDTPA-BMA in the presence and absence of CO_3^{2-} (25 mM) or PO_4^{3-} (0.5 mM) were independent of pH over the range 5–10 [50]. Conversely, the T_1 values of solutions containing GdDO3A and $GdEDTA^{-}$ in the presence of carbonate, phosphate or citrate showed that ternary complexes were formed after displacement of inner-sphere water. This results in an decrease in relaxation rate ($1/T_1$) at higher pH values and illustrates that ternary complex formation must be considered in the design of ligands to be used as CAs. This appears to be a most important consideration for complexes that have more than one inner-sphere water molecule positioned in a *cis*-arrangement. One can predict with some certainty that such structures will form ternary complexes in a biological milieu (unless perhaps the complex is highly negatively charged) and erase any relaxivity advantage that might be observed *in vitro*.

4 EQUILIBRIUM CALCULATIONS: LIGAND SELECTIVITY AND THE CONCENTRATION OF 'FREE' Gd^{3+}

Blood plasma is a relatively complex milieu in which a large number of ligands and metal ions can compete for binding. Computer simulations of such complex equilibria have been reported. In one early model, May *et al.* [52] included seven metal ions (Mg^{2+}, Ca^{2+}, Mn^{2+}, Fe^{3+}, Cu^{2+}, Zn^{2+} and Pb^{2+}) and forty low-molecular-weight ligands (amino acids, dicarboxylic acids, carbonate, phosphate, citrate, lactate, salicylate, etc.) in a complex equilibria involving ca 5000 binary and ternary complexes. Given known concentrations of all exchangeable metal ions and ligands and previously reported stability constants of the complexes, the program ECCLES [53] was used to estimate the concentrations of all species at equlibria. Jackson *et al.* [47] applied this plasma model to estimate the species distribution of all Gd^{3+} species in the presence and absence of DTPA. In the absence of DTPA, the species distribution strongly depends on the total concentration of Gd^{3+}. At Gd^{3+} concentrations below 10^{-4} M, the Gd^{3+} is bound largely to citrate, with small amounts of salicylate, glutaminate, threoninate and oxalate complexes also being present. In situations where concentration of transferrin (3.1×10^{-5} M) exceeds the available Fe^{3+}, then formation of Gd(transferrin) and Gd_2(transferrin) should also be considered. In this case, at total Gd^{3+} concentrations below 10^{-4} M, Gd^{3+} was predicted to be bound almost exclusively to transferrin. In the presence of $GdDTPA^{2-}$, the concentration of the free Gd^{3+} was also strongly influenced by the inclusion of transferrin as a binding component in the calculations. With transferrin included and [$GdDTPA^{2-}$] below 10^{-5} M, the ligand DTPA has little effect on the free metal concentration. At a CA concentration of 10^{-4} M, the concentrations of free Gd^{3+} in the presence and absence of transferrin were $\sim 10^{-13}$ M and 10^{-11} M, respectively [47]. Unfortunately, the authors reported only the concentration of free Gd^{3+} and did not include the extent of dissociation of $GdDTPA^{2-}$ (any dissociated Gd^{3+} would be present in the form of other complexes, such as Gd^{3+}-citrate, etc.). In this model, the effect of transferrin on the species distribution of Gd^{3+} was likely over-estimated because other metal ions present in the plasma (e. g. Zn^{2+} and Al^{3+}) also compete effectively for binding sites on transferrin [54].

Cacheris *et al.* [8] also used a plasma model to estimate the amount of Gd^{3+} released from $GdDTPA^{2-}$, GdDTPA-BMA, GdDTPA-BP, and $GdEDTA^{-}$. In this simplified model, competing equilibria between Gd^{3+}, H^{+}, Ca^{2+}, Cu^{2+} and Zn^{2+} and the ligand were assumed, as well as competitive binding of citrate, amino acids and albumin for all of the ions. The results of these equlibrium calculations indicated, not unexpectedly, that the amount of Gd^{3+} released from the complexes increases with increasing concentration of CA. At a typical dose of 0.1 mmol/kg, the Gd^{3+} released from GdDTPA-BMA, $GdDTPA^{2-}$, GdDTPA-BP, and $GdEDTA^{-}$ was estimated at $\sim 1.2 \times 10^{-6}$ M, 2.4×10^{-6} M, 3.4×10^{-6} M, and 1×10^{-5} M, respectively. The released Gd^{3+} was predicted to be largely in the form of Gd^{3+}-citrate, with insignificant complexation by amino acids or albumin [8]. The approximately twofold lower amount of Gd^{3+} released from GdDTPA-BMA compared to $GdDTPA^{2-}$ was largely due to the presence of Zn^{2+} and the binding selectivity of DTPA-BMA for Gd^{3+} over Zn^{2+}.

In order to make this a generally useful concept for predicting relationships between thermodynamic stability and toxicity, the concept of a selectivity binding constant, K_{sel}, was envoked. By definition, a selectivity constant can be expressed as:

$$K_{\text{sel}} = K_{\text{GdL}}(\alpha_{\text{H}} + \alpha_{\text{Ca}} + \alpha_{\text{Zn}} + \alpha_{\text{Cu}}) \tag{6.13}$$

where α is defined as in Equation (6.10). The parameter K_{sel} defined by Equation (6.13) is actually a conditional stability constant which takes into account competition not only between Gd^{3+} and protons for the ligand, but also the competition between Gd^{3+} and other endogenous metals. The selectivity constants of the CAs considered in that study fell in the order, GdDTPA-BMA > $GdDTPA^{2-}$ > GdDTPA-BP > $GdEDTA^{-}$, which is exactly the same order as observed for the experimental LD_{50} values of these complexes in mice [8]. These findings were the first to illustrate the importance of the selectivity of the ligand for Gd^{3+} over Zn^{2+} (and other endogenous metals) in determining the toxicity of a Gd^{3+} complex. The selectivity over the Zn^{2+} was deamed most important because the *in vivo* concentration of free Zn^{2+} is the largest among the endogenous metal ions.

In calculating the species distribution for the DTPA complexes, formation of the dinuclear $Zn_2(DTPA)^-$ and $Cu_2(DTPA)^-$ complexes were neglected in both the Jackson *et al.* [47] and Cacheris *et al.* [8] models. In a simpler equlibrium model where competition between Gd^{3+} (0.1 mM) and Zn^{2+} (0.05 mM) and the ligands DTPA (0.1 mM), lactate (2.0 mM), citrate (0.1 mM), succinate (0.04 mM), Gly (1.0 mM), Asp (0.07 mM), His (0.08 mM) and Cys (0.04 M) was considered, inclusion of a dinuclear Zn_2DTPA^- equilibrium resulted in a significant change in species distribution [18]. At the concentrations indicated above, this equilibrium calculation indicated that $GdDTPA^{2-}$ would dissociate to an extent of about 9.3 % and most of the released Gd^{3+} would be present as Gd^{3+}-citrate. The major Zn^{2+}-containing species included $Zn_2(DTPA)^-$ (19.1 %), Zn(His) (17.8 %), Zn(Cys) (45.3 %) and Zn(citrate) (12.4 %). This calculation suggests that Zn^{2+} may have a larger impact on displacement of Gd^{3+} from DTPA than predicted by the earlier models. Nevertheless, another important factor that can not be easily included in such estimates is the impact of rapid excretion of $GdDTPA^{2-}$ from the body ($t_{1/2} = 1.6$ h) [4]. Since both acid-catalyzed and metal-ion-assisted dissociation of $GdDTPA^{2-}$ is relatively slow, the excreted solution is far from equlbrium and the concentration of the Gd^{3+} released from the complex is much lower than that estimated by assuming equilbrium. Measurable differences in transmetallation kinetics with Zn^{2+} has also been reported for $GdDTPA^{2-}$, GdDTPA-BMA, and GdDTPA-BIGA (another bis-amide ligand) [55], so this is yet another kinetic factor that could substantially change the amount of free Gd^{3+} predicted by models based upon thermodynamic equilibria.

Early studies carried out in phosphate buffers of high concentration indicated that $GdPO_4$ slowly precipitates from solution [56, 57]. The stability of

Gd^{3+}-complexes in plasma with respect to phosphate precipitation has been explained by slow dissociation of the complex and the slow formation of the precipitate [47, 57]. For characterizing the stability of solutions, a solubilization constant, K_{sol} is used, which describes the amount of the complexed metal in the presence of the ligand, L ($K_{sol} = [ML]/[L]_t$). If the competition of the various metals and the proton for the ligand is taken into account, the K_{sol} value can be expressed as follows [8]:

$$K_{sol} = K_{ML}[M](\alpha_H + \alpha_{Ca} + \alpha_{Zn} + \alpha_{Cu} + \ldots, \text{ etc.})^{-1} \tag{6.14}$$

A positive value of log K_{sol} predicts that the ligand will be able to solubilize $GdPO_4$, while a negative value indicates the formation of precipitate. The calculations carried out using Equation (6.14) in the pH range 5–8 resulted in the positive K_{sol} values for the $GdDTPA^{2-}$, GdDTPA-BMA, and even for $GdEDTA^{-}$ [8]. Similar calculations were carried out for urine, where the concentration of the Gd^{3+} complex is significantly higher (ca 30 mM), but all of the complexes were found to be stable towards phosphate precipitation [8].

5 KINETICS OF ML COMPLEX FORMATION

Macrocyclic ligands tend to form complexes much more slowly with the trivalent lanthanide ions than with ligands derived from linear polyamines. Numerous reports have appeared on the complexation kinetics between the macrocyclic ligand, DOTA, and the trivalent lanthanide cations, beginning with the early report by Kasprzyk and Wilkins [58]. This was followed by reports by Brücher *et al.* [59], Wang *et al.* [60], Kumar and Tweedle [61], Tóth *et al.* [62] Wu and Horrocks [63] and Burai *et al.* [22]. Each of these investigators came to similar conclusions regarding the mechanism of complex formation (differing only in minor details). The free ligand has all four acetate side-arms extended above the plane of the four nitrogens so it is reasonable to suggest that the acetates provide an electrostatic charge potential to attract the trivalent cation toward the ligand cavity. It is clear that a weak complex is initially formed between the negatively charged DOTA molecule and a trivalent lanthanide cation, with kinetic rate constants similar to those measured for linear polyamine systems such as DTPA. This is then followed by a slow, pH-dependent rearrangement of this weak complex to form a final complex that is orders of magnitude more stable. In the solid state, the macrocyclic ring of DOTA [64, 65] is pre-organized into the familiar [3, 3, 3, 3] dodecane-like structure, virtually identical to that found in all crystal structures of the $LnDOTA^{-}$ complexes.

A minor difference between these early kinetic studies was disagreement about the number of protons attached to the macrocyclic nitrogens and whether a nitrogen atom is coordinated in the initially formed weak ML complex. Tóth *et al.* [62] showed by spectrophotometry and direct potentiometric titration that the kinetically stable intermediate has two protons, i.e. LnH_2DOTA^+, presumably with the H^+s attached to two macrocyclic nitrogen atoms in the *trans*-positions (as found in the crystal structure of the diprotonated ligand) and the Ln^{3+} coordinated only to the extended acetates. In a subsequent report by Wu and Horrocks [63], separate excitation bands ($^7F_0 \rightarrow ^5D_0$ transitions) were observed for the rapidly formed initial out-of-cage $Eu(H_2O)_qLH_x$ (579.2 nm) complex and the more slowly formed final in-cage $Eu(H_2O)L^-$ (579.77 nm) complex. The intensity of the peak representing the $Eu(H_2O)_qLH_x$ species was measured immediately after mixing at different pH values and this titration indicated that two protonations take place between pH 6 and 3, again presumably reflecting protonation of two macrocyclic nitrogen atoms. From these data, the thermodynamic stability constant of $Eu(H_2O)_qLH_2^+$ was estimated as $7.6 \times 10^5\ M^{-1}$. The beauty of the fluorescence technique is that the difference in luminescence lifetimes measured in H_2O versus D_2O provides a direct read-out of the number of water molecules (q) coordinated to the Eu^{3+}. The lifetime data indicated that $q = 4.5 \pm 0.5$. Combining this information with a molecular mechanics estimate of waters of hydration, assuming that four carboxylates are bound in this species, led to the conclusion that $q = 5$. Thus, the rapidly formed, out-of-cage species appears to be a nine-coordinate Eu^{3+} species with four carboxylate and five water ligand donors, i.e. $Eu(H_2O)_5LH_2^+$. As anticipated, based upon prior luminescence determinations [66], the final equilibrium in-cage species is also a nine-coordinate species with four carboxylate and four nitrogen ligand donors and a single inner-sphere water molecule, i.e. $Eu(H_2O)L^-$.

The rate of conversion of the out-of-cage complex, $Eu(H_2O)_5LH_2{}^+$, to the in-cage $Eu(H_2O)L^-$ complex is highly pH-dependent. The mechanism proposed by Wu and Horrocks [63] included as a first step the OH^--catalyzed removal of a single proton from $Eu(H_2O)_5LH_2^+$ to form the neutral species, $Eu(H_2O)_5LH$. Burai *et al.* [22] tested this proposal by using excess ligand and measuring the kinetics of complex formation at higher pH values where the $Ln(H_2O)_5LH$ species predominates. They found no need to invoke OH^--catalyzed removal of the first proton but found clear evidence for OH^--catalyzed removal of the second proton. The protonation constants for formation of $Ce(H_2O)_5LH_2$ and $Yb(H_2O)_5LH_2$ were found to be $4.4 \pm 0.5 \times 10^8 M^{-1}$ ($\log K_{CeHL} = 8.64$) and $2.5 \pm 1.4 \times 10^8 M^{-1}$ ($\log K_{YbHL} = 8.40$), respectively. These values are considerably lower than the second protonation constant of DOTA ($\log K_{HL} = 9.67$) [22], consistent with the expectation of electrostatic repulsion between the Ln^{3+} and nitrogen-bound H^+ in the out-of-cage $Ln(H_2O)_5LH$ species. It has been proposed that the rate determining step in formation of in-cage LnDOTA-complexes is deprotonation of the monoprotonated intermediate, Ln $(H_2O)_5LH$. Deprotonation is assisted by any Bronsted base, such as H_2O or OH^-, or a deprotonated buffer species. The mechanistic details of how these protons are catalytically removed from the $Ln(H_2O)_5LH_2^+$ and $Ln(H_2O)_5LH$ species to form the stable, Ln-encapsulated, $Ln(H_2O)L^-$ species has not been delineated, but it is now clear that the rate determining step in formation of this stable chelate is deprotonation of the monoprotonated species [22].

The 1,7-diacetate derivative, DO2A, forms complexes with Ln^{3+} cations particularly slowly [31] and this may be reflected in the wide range of log K_{ML} values reported for this system [28, 30, 31] (see Table 6.3). If one assumes a similar mechanism to that delineated for formation of the DOTA complexes, then the rate determining step in formation of the in-cage $LnDO2A^+$ complexes could also be removal of a proton from a $LnHDO2A^{2+}$ species in which the out-of-cage Ln^{3+} is coordinated to only two acetates and perhaps one or two nitrogen atoms. In the case of DO2A, it has been shown that the most basic nitrogen is one of the secondary amines[28] and so out-of-cage $LnHDO2A^{2+}$ species would also likely have a proton attached to one of the secondary nitrogen atoms. Regardless of the exact mechanism, formation of the stable, in-cage $LnDO2A^+$ complexes appears to occur at extraordinarily slow rates and this could account for the dramatic differences in stability constants reported for DO2A [28, 30, 31].

The phosphonate analog, DO2P, has provided some clues about the structures of the intermediates that are possible with macrocyclic systems [32]. The first two protonation constants of DO2P are considerably higher than those of DO2A and, unlike in DO2A, protonation takes place at the two ternary nitrogen atoms [32]. ^{31}P NMR has revealed that DO2P forms a variety of out-of-cage and in-cage species (depending upon pH) [32]. Although a thorough kinetic comparison of the two ligands has not been made, DO2P at least qualitatively, forms in-cage complexes with the Ln^{3+} cations more quickly than does DO2A. The same qualitative conclusion is reached if one compares the rates of $LnDOTP^{5-}$ versus $LnDOTA^-$ complex formation. All four ligands, DOTA, DO2A, DOTP and DO2P, quickly form out-of-cage complexes that subsequently rearrange at variable slow rates to the more stable in-cage complexes. The differences in kinetics noted above suggest that a phosphonate side-chain more efficiently assists in formation of an in-cage complex, even though the protonation constant of an amine bearing a phosphonate side-chain tends to be much higher than an amine bearing an acetate side-chain.

Slow complexation kinetics can become problematical whenever a macrocyclic ligand is covalently attached to a macromolecule prior to its loading with a metal ion. Li and Meares [67] have reported very low efficiency of loading Y^{3+} into DOTA-antibody conjugates, presumably due to slow reaction kinetics at the pH and other solution conditions where the preparation is most stable. In an early study, it was shown that the model DOTA–protein conjugate, DOTA–monopropylamide (PA), forms very slow complexes with Ln^{3+} cations at pH 4 (60 °C) [68], and the thermodynamic stability of GdDOTA–PA also dropped by about four orders of magnitude compared to $GdDOTA^-$. The latter effect can be traced to a less basic nitrogen in the amide derivative [68]. In another more recent study, Keire and Kobayaski [69] prepared another model DOTA–protein conjugate, i.e. DOTA–monobutylamide (BA), and compared the reaction kinetics of this ligand and DOTA with Y^{3+}. They confirmed that complex formation was slower with DOTA–BA than with DOTA under identical solution conditions but did not identify the reason(s) for the altered kinetic behavior of this system. This remains an interesting question since it has also been reported in a patent [70] that DOTA-based bifunctional ligands having two or more amido side-arms (instead of carboxylates) form complexes more readily with Ln^{3+} in aqueous solution. This suggests that the negative charge on the carboxylate side-arms is not a requirement for forming the initial weakly associated complex, but that, once formed, the lower pK_as of the macrocyclic nitrogen atoms in the amido derivatives allow more rapid rearrangement into the more stable in-cage species. This is entirely consistent with deprotonation of the monoprotonated species as the rate-determining step [22].

6 KINETIC INERTNESS OF ML COMPLEXES

As noted above, the fate of many Gd^{3+} complexes *in vivo* can not be predicted on the basis of competitive equilibria. Thus, it was realized that the kinetic stabilities of Gd^{3+}-complexes play a crucial role in determining the amount of Gd^{3+} released *in vivo*. The kinetic stability can be characterized by rate constants of exchange reactions that take place in plasma. The most important of these is probably the displacement of Gd^{3+} from GdL by the endogenous metals Cu^{2+} and Zn^{2+} in a metal exchange reaction:

$$\mathrm{GdL} + \mathrm{M}^{2+} \rightleftharpoons \mathrm{ML} + \mathrm{Gd}^{3+} \tag{6.15}$$

Metal-exchange reactions, like substitution reactions, can occur via associative or dissociative mechanisms. Exchange can take place by direct attack of the exchanging metal ion on the complex to form a dinuclear intermediate, GdLM, followed by transfer of ligand functional groups from the Gd^{3+} to M. Alternatively, any spontaneous or proton-assisted dissociation of the Gd^{3+} complex that occurs slowly via a dissociative pathway is followed by rapid reaction of the free ligand and endogenous metal ions.

In principle, ligand-exchange reactions can also occur with some ligands (Y) which form complexes of high stability with Gd^{3+}:

$$\mathrm{GdL} + \mathrm{Y} \rightleftharpoons \mathrm{GdY} + \mathrm{L} \tag{6.16}$$

The rates of the ligand-exchange reactions are rarely measured because the probability of such an exchange is very low in plasma. The stability constants of Gd^{3+} complexes formed with ligands naturally present in plasma are in general much lower than the stabilities of typical Gd^{3+}-based contrast agents. Furthermore, most ligand-exchange reactions occur via an associative pathway where formation of a ternary intermediate complex (where some functional groups of the attacking ligand are coordinated to the Gd^{3+}) is an important step. In realizing the importance of kinetic inertness in determining the toxicity of Gd^{3+} complexes, a qualitative method has been used to measure the extent of dissociation of a complex in the presence of excess of phosphate and carbonate ions over a fixed period of time [56]. The experimental data showed that the macrocyclic complexes, $GdDOTA^-$ and Gd(HP-DO3A), are kinetically inert (essentially no reaction) while about 20–25 % of $GdDTPA^{2-}$ dissociated and formed $GdPO_4$ in the presence of Zn^{2+} or Cu^{2+} [56]. Another approach has been to measure the rate of Gd^{3+}-complex dissociation in the presence of excess acid [1, 57, 71–74]. The rate of dissociation was found to be directly proportional to the concentration of a complex:

$$-\frac{\mathrm{R[GdL]}}{\mathrm{d}t} = k_{\mathrm{obs}}[\mathrm{GdL}] \tag{6.17}$$

The first-order rate constants, k_{obs} for the dissociation of a variety of complexes in 0.1 M HCl are summarized in Table 6.5, with the structures of the various ligands being shown in chart 6.3. These data show that complexes formed with macrocyclic ligands dissociate much more slowly than complexes of acyclic ligands, i.e. DTPA, DTPA-BMA and EDTA. Wedeking, *et al.* [1] found that long-term deposition of Gd^{3+} in the whole body was the lowest for $GdDOTA^-$ and Gd(HP-DO3A) and this correlated well with the dissociation rates of these macrocyclic complexes in acid (Table 6.5). Deposition of Gd^{3+} was the highest for $GdEDTA^-$, a complex that dissociates rather quickly in 0.1 M HCl. These findings suggested a correlation between the extent of Gd^{3+} deposition and the rates of proton-assisted dissociation of the complexes [1]. The rather high rate of release of Gd^{3+} from Gd(DTPA-BMA) in acid

Table 6.5 First-order rate constants for dissociation of Gd^{3+} complexes in 0.1 M HCl.

Complex	$k_{obs}{}^{a}$ (s^{-1})	$k_{obs}{}^{b}$ (s^{-1})	Reference
$GdEDTA^-$	1.4×10^{2}	—	1
$GdDTPA^{2-}$	1.2×10^{-3}	—	1
Gd(DTPA-BMA)	$> 2 \times 10^{-2}$	—	1
GdDO3A	2.3×10^{-3}	—	1
Gd(NP-DO3A)	$> 2 \times 10^{-2}$	—	1
Gd(HP-DO3A)	6.3×10^{-5}	—	1
$GdDOTA^-$	2.1×10^{-5}	3.2×10^{-6}	1, 72
$GdDOTMP^-$	—	1×10^{-5}	72
$GdDOTBrP^-$	—	2.39×10^{-5}	72
$GdDOTBuP^-$	—	3.69×10^{-5}	72
$GdDOTPP^-$	—	7.77×10^{-5}	72
Gd(DOTMP-MBBuA)	—	1.3×10^{-6}	72
Gd(DOTBuP-MMA)	—	4.1×10^{-6}	72
$GdDOTEP^-$	6.4×10^{-6}	—	74

[a] 1.0 M NaCl, 25 °C.
[b] $I = 0.1$, 37 °C.

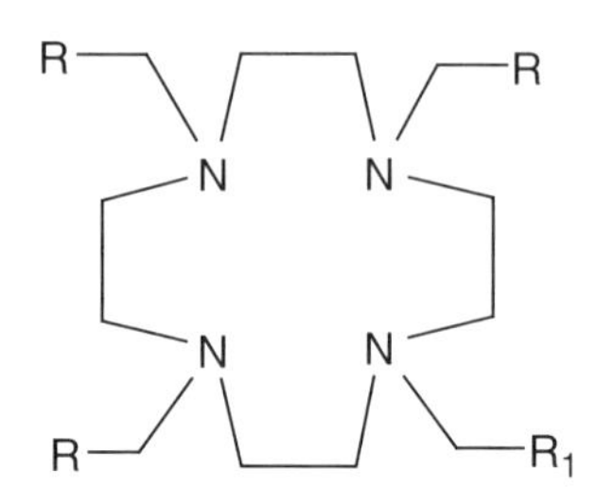

R	R_1	Ligand
R = — COOH	R_1 = — $CH(OH)CH_3$	HP-DO3A
R = — COOH	R_1 = — $CH_2(OH)$	HE-DO3A
R = — COOH	R_1 = *n*-butyl	DO3A-Bu
R = — COOH	R_1 = *n*-propyl	NP-DO3A
R = R_1 = — $PMeO_2H$		DOTMP
R = R_1 = — $PEtO_2H$		DOTEP
R = R_1 = — $PBzO_2H$		DOTBzP
R = R_1 = — $PBuO_2H$		DOTBuP
R = R_1 = — $PPhO_2H$		DOTPhP
R = — $PMeO_2H$	R_1 = — $CONBu_2$	DOTMP-MBBUA
R = — $PBuO_2H$	R_1 = — CONHMe	DOTBuP-MMA
R = R_1 = PO_3HBu		DOTPMB

[Chart 6.3]

suggested that this complex should be toxic *in vivo*, yet this agent is now widely used and is considered safe. In a more recent study, McMurry *et al.* [75] found no correlation between release of the ^{88}Y isotope from the ^{88}Y-complexes of several back-bone-substituted DTPA derivatives and the dissociation rates of the complexes measured in 0.1 M $HClO_4$. This lack of correlation between *in vivo* release of metal ions from complexes and the kinetic stability in acid argues that more detailed kinetic studies are needed, with involvement of the important endogenous metal ions, Cu^{2+} and Zn^{2+}.

Since the dissociation rates of Gd^{3+} complexes of macrocyclic versus acyclic ligands differ considerably and the rate laws governing those processes are also different, the kinetic properties of the two types of complexes will be discussed separately.

6.1 DECOMPLEXATION KINETICS OF GDDTPA^{2-} AND STRUCTURALLY SIMILAR SYSTEMS

Metal ion and ligand exchange reactions of the transition metal (II, III) and lanthanide (III) complexes formed with the acyclic polyamino-polycarboxylate ligands (EDTA, CDTA, DTPA, EGTA, etc.) occur relatively slowly in the pH range 4–8. The results of many early kinetic studies involving transition metal ion-exchange reactions have been reviewed by Margerum *et al.*[76]. As expected, it was found that exchange rates for reactions of this type are in general slower for complexes having higher thermodynamic stability constants. Although there was much less interest in the lanthanide ion complexes in the early literature, a few studies have been reported for LnEDTA$^-$ [76–78] and LnDTPA^{2-} complexes (where Ln=La, Ce, Nd, Ho, Lu and Y). [76, 79–81]. Only very recently have the kinetics of exchange [18, 82] between GdDTPA^{2-} and Eu^{3+} and between Gd(DTPA-BMA) and Eu^{3+}, Cu^{2+}, or Zn^{2+} been investigated in detail. The kinetics of metal ion exchange between GdDTPA^{2-} and Eu^{3+} has been described by the following pseudo-first-order rate expression [18]:

$$k_{\text{obs}} = \frac{k_1[\text{H}^+] + k_2[\text{H}^+]^2 + k_3^{\text{M}}[\text{M}] + k_4[\text{H}^+][\text{M}]}{1 + K_{\text{GdHL}}[\text{H}^+] + K_{\text{GdLM}}[\text{M}]} \tag{6.18}$$

where [M] is the concentration of the exchanging Eu^{3+} ion and K_{GdLM} is the stability constant of the dinuclear GdLM intermediate formed during the reaction. It was found that exchange can take place via proton-assisted dissociation of GdDTPA^{2-} (characterized by k_1 and k_2), followed by a fast reaction between the free ligand and Eu^{3+} (mono-and diprotonated complexes dissociate much faster than the non-protonated complexes) and by direct attack of Eu^{3+} on the non-protonated complex, GdL (k_3^{M}) and on the mono-protonated complex, GdHL (k_4). Once these dinuclear reaction intermediates are formed, the functional groups of the ligand are then gradually transferred from the Gd^{3+} to the attacking ion.

Exchange reactions between GdDTPA^{2-} and Cu^{2+}or Zn^{2+} occur much faster than exchange with Eu^{3+} [18]. At pH values higher than around 4.5, the reaction rates are proportional to the Cu^{2+} or Zn^{2+} concentration and are independent of pH. This indicates that exchange predominantly occurs via direct attack of a Cu^{2+} or Zn^{2+} on GdDTPA^{2-} and reactions taking place via dissociation of protonated complexes do not occur to any significant extent. Given this result, it is understandable why the amount of Gd^{3+} or Y^{3+} released *in vivo* does not necessarily correlate well with the rates of dissociation of their complexes in 0.1 M acid. If the acid-dissociation pathway does not contribute significantly to the observed exchange rates in the presence of Cu^{2+} or Zn^{2+}, then the expression for k_{obs} simplifies to the following:

$$k_{\text{obs}} = \frac{k_3^{\text{M}}[\text{M}]}{1 + K_{\text{GdLM}}[\text{M}]} \tag{6.19}$$

Table 6.6 Rate constants, characterizing the exchange reactions of $GdDTPA^{2-}$ [a] and Gd(DTPA-BMA) with Eu^{3+}, Cu^{2+} and Zn^{2+} ions (25 °C, 1.0 M KCl).

Ion	$k_1 (M^{-1}s^{-1})$		$k_3^M (M^{-1}s^{-1})$	
	$GdDTPA^{2-}$ [b]	Gd(DTPA-BMA) [c]	$GdDTPA^{2-}$ [b]	Gd(DTPA-BMA) [c]
Eu^{3+}	0.58	12.7 (8.65 [d])	4.9×10^{-4}	—
Cu^{2+}	—	—	0.93 (1.59 [e])	0.63
Zn^{2+}	—	—	5.6×10^{-2} (0.11 [e])	7.8×10^{-3}

[a] For $GdDTPA^{2-}$, $k_2 = 9.17 \times 10^4\ M^{-2}\ s^{-1}$ and $k_4 = 40\ M^{-2}\ s^{-1}$ [18].
[b] Reference [18].
[c] Reference [82].
[d] Gd(DTPA-BMEA) [83].
[e] Measured at 37 °C.

The rate constants (k_3^M) found for reactions of $GdDTPA^{2-}$ or Gd(DTPA-BMA) with Cu^{2+} or Zn^{2+} at 25 °C are summarized in Table 6.6. The binuclear complex stability constants, K_{GdLM}, were found to be around 10. If one considers the concentration of exchangeable Cu^{2+} (1×10^{-6}M) and Zn^{2+} (1×10^{-5}M) in plasma [52] and the exchange rates for the different possible pathways at pH = 7.4, one can quickly conclude that the proton-assisted dissociation of these complexes (the terms, $k_1[H^+] + k_2[H^+]^2$) is also negligible in plasma.

In a kinetic study of the exchange between Gd(DTPA-BMEA) and Eu^{3+}, Rothermel *et al.* [83] found that at pH < 6 the reaction predominantly occurs via dissociation of the monoprotonated complex with a rate constant, k_1, of 8.65 $M^{-1}\ s^{-1}$. Sarka *et al.* [82] found a similar value for k_1 for the exchange between Gd(DTPA-BMA) and Eu^{3+}, Cu^{2+} and Zn^{2+} (see Table 6.6). However, the protonation constant of Gd(DTPA-BMA) was found to be very low, $K_{GdHL} < 2$, while that for $GdDTPA^{2-}$ is $K_{GdHL} = 100$. This means that the concentration of protonated $Gd(HDTPA\text{-}BMA)^+$ is much lower than that of protonated $Gd(HDTPA)^-$, and consequently the rate of proton-assisted dissociation of Gd(DTPA-BMA) is not significantly higher than that of $GdDTPA^{2-}$. The dependence of the first-order rate constants on the concentration of Cu^{2+} or Zn^{2+} ions could be described by the rate expression shown in Equation (6.19). The rate constants characterizing the exchange reactions occurring with the direct attack of the Cu^{2+} and Zn^{2+} ions on the Gd(DTPA-BMA) are somewhat lower than the rate constants obtained for the similar reactions of $GdDTPA^{2-}$. When all possible exchange reactions are taken into account, the kinetic stabilities of the complexes $GdDTPA^{2-}$ and Gd(DTPA-BMA) are indeed very similar, a rather surprising result if one considers the greater stability of $GdDTPA^{2-}$ over Gd(DTPA-BMA). These results indicate that inclusion of amide side-chains in ligand structures (either acyclic or cyclic) is very useful. Although the binding strength of the amide group is somewhat weaker than that of a carboxylate, its affinity for a H^+, Cu^{2+} or Zn^{2+} is significantly lower. Thus, the extent of formation of protonated or dinuclear complexes is much less and this imparts greater kinetic stability to complexes formed with amide-containing side-chains.

The kinetics of exchange reactions between the Gd^{3+} complexes of the macrocyclic DTPA-bis(amide) ligands (DTPA-PAM, DTPA-BAM, etc.) and Cu^{2+} ions have been studied by Choi *et al.* [84]. The reactions occur both by proton-assisted dissociation and by direct attack of Cu^{2+} on the complex. However, at pH values > 6 the predominant pathway is via a direct reaction between the complex and a Cu^{2+} ion. The rate constants determined for this pathway for Gd(DTPA-PAM) and Gd(DTPA-BAM) were 1.27 and 1.24 M^{-1} s^{-1}, respectively. These constants are larger than the k_3^M values determined for Gd(DTPA-BMEA) and Gd(DTPA-BMA) by a factor of about 2 (see Table 6.6), thus indicating that the kinetic stability of the macrocyclic DTPA-bis(amide) derivatives is somewhat lower than that of the acyclic derivatives.

6.2 DECOMPLEXATION KINETICS OF Gd^{3+}-MACROCYCLIC LIGAND COMPLEXES

Complexes of Gd^{3+} formed with cyclen-based ligands functionalized with acetate, methylenephosphonate or methylene-phosphinate groups are all quite inert. The kinetic inertness of these complexes, mostly attributed to the rigidity of the 12-membered ring structure, [85], makes measurements of the dissociation rates near physiological pH conditions impossible. However, the complexes are thermodynamically unstable below pH ~2 so dissociation of the complexes can be examined in 0.01–1. M acid. Cation-exchange resins, [60], HPLC [71] and $GdPO_4$ precipitation [86] have all been used as methods to separate free and complexed Gd^{3+}. Dissociation of complexes can also be studied by measuring changes in water-proton-relaxation rates or changes in the Gd^{3+} fluorescence [26, 86]. Exchange reactions between Gd^{3+} complexes and some metal ions (e. g. Cu^{2+} and Eu^{3+}) can be followed by spectrophotometry in the pH range 3–5 by using high concentrations of complex (e. g. 0.1 M GdL) and low concentrations of exchanging metal ions (e.g. 0.01 M Eu^{3+} or Cu^{2+}) [62, 87, 88]. Under these conditions, the exchange rates are independent of the exchanging metal concentration and occur only through proton-assisted dissociation of the complexes. At low H^+ concentrations (<~ 0.2–0.3 M), the observed rates of dissociation can be expressed as follows [60, 62, 88]:

$$k_{obs} = k_o + k_1[H^+] \tag{6.20}$$

At higher H^+ concentrations, k_{obs} exhibits saturation behavior, probably due to accumulation of protonated complexes, GdH_xL. Since these protonation

Table 6.7 Protonation constants of the Gd^{3+} complexes (K_{GdHL}) and rate constants (k_1) characterizing the proton-assisted dissociation (at 25 °C).

Ligand[a]	log K_{GdHL}	$10^5 k_1 (M^{-1}s^{-1})$
DOTA	2.8[b], 2.3[c], 1.15[d], 1.35[c]	0.84[e], 3.64 (37 °C)[f], 2.0(37 °C)[g]
HP-DO3A	2.38[b], 1.1[h]	45[b], 26[h]
DO3A-Bu	1.1[h]	2.8[h]
HE-DO3A	2.04[i]	470[i]
HIP-DO3A	2.23[i]	58[i]
DO3A	2.06[b]	2.5×10^{3}[b], 1.17×10^{3}[j]
DOTP-MB	0.28[d]	21[d]

[a] Structures are shown in Chart 6.1.
[b] k_1 values were calculated from the reported k_{obs} values [27].
[c] Reference [11].
[d] Reference [51].
[e] Reference [60].
[f] Reference [85].
[g] Reference [62].
[h] Reference [34].
[i] Reference [86].
[j] Reference [87].

equilibrium constants are in the range ~ 10–100 (Table 6.7), monoprotonated complexes are present in significant quantities whenever $[H^+]$ is in the range 0.1–1.0 M. The rate laws derived from the experimental data suggested a proton-assisted dissociative mechanism, as found earlier for the LnNOTA complexes [88]:

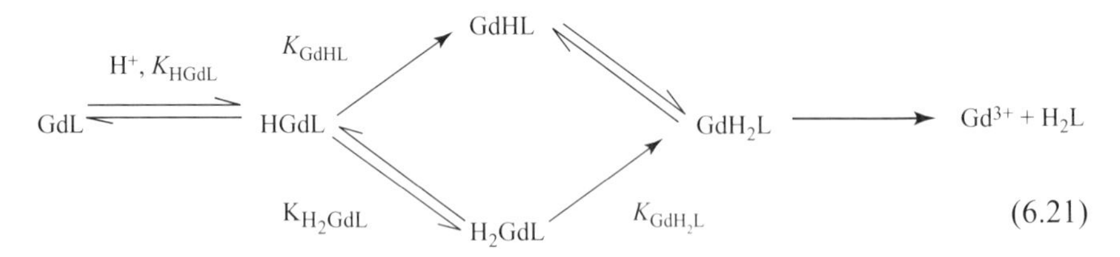

(6.21)

The scheme shown in Equation (6.21) has been used to interpret data for dissociation of the LnDOTA and various LnDO3A-derivative complexes. [26, 34, 62, 86]. Dissociation occurs via equilibrium formation of mono-and diprotonated complexes (HGdL and H_2GdL), presumably where the protons are attached to the oxygen atoms of the carboxylate groups (it has been shown that a non-coordinated carboxylate oxygen atom in $GdDOTA^-$ is protonated in 0.1–1.0 M $HClO_4$) [89]. Prior to dissociation of GdHL, the proton is transferred from a carboxyl to a nitrogen atom, Gd^{3+} steps out of the coordination cage, and a second nitrogen is immediately protonated to form GdH_2L. The structure of the diprotonated intermediate is similar to the structure of the intermediate detected in the formation reactions of the complexes [62, 63]. At high H^+ concentrations, this intermediate is unstable and rapidly dissociates.

The rate and equilibrium constants shown in Equation (6.21) have been calculated for several complexes [26, 62, 86], but only the pathway involving GdHL is considered important for dissociation near physiological pH values. For this pathway, it is quite easy to show that the observed rate constant, is given by $k_1 = k_{GdHl} \times K_{GdHL}$ [62]. Since the protonation constants of various DOTA complexes do not differ considerably, the rate constant k_1 is in most cases a direct measure of the kinetic stability of a complex.

A comparison of k_1 values (see Table 6.7) shows that the most inert complex is $GdDOTA^-$. Due to its high kinetic inertness, $GdDOTA^-$ is suitable for oral use as a gastrointestinal CA. Schwizer *et al.* [85] had demonstrated that, unlike $GdDTPA^{2-}$, $GdDOTA^-$ is stable both in HCl solution and gastric juice for prolonged periods of time. Among the DO3A derivatives, DO3A-butrol appears to form the most inert complex, being about tenfold more inert than Gd(HP-DO3A) [34]. This result was somewhat unexpected because the stability constant of Gd(HP-DO3A) is larger than that of Gd(DO3A-butrol) and, in general, the more stable polyaza-polycarboxylate complexes are also the most kinetically inert. The unusual kinetic stability of Gd(DO3A-butrol) may result from the bulkiness of the 'butrol' group. This bulky group containing three OH-groups may restrict rearrangement and dissociation of the protonated complex [34].

A number of phosphinate derivatives of cyclen have also been synthesized [71–74] and the Gd^{3+} complexes formed with these ligands have proven to have equally favorable (or perhaps even more favorable) kinetic stabilities as the DOTA derivatives (see Table 6.4). This likely reflects the reduced basicity of a phosphinate oxygen compared to a carboxylate oxygen. This is also reflected in the lower protonation constants of the free phosphinate ligands.

The rate constant, k_1, for dissociation of $Gd(DOTP\text{-}MB)^-$ [90] is similar to values measured for several Gd^{3+} complexes of DO3A derivatives. This finding is surprising, since the stability constant of $Gd(DOTP\text{-}MB)^-$ ($\log K_{GdL} = 12.3$) [51] is about ten orders of magnitude lower than the log K_{GdL} values of Gd^{3+} complexes of DO3A derivatives. The high kinetic inertness of the $Gd(DOTP\text{-}MB)^-$ species again reflects the low basicity of the phosphonate-ester oxygen atoms. Because of this low basicity, the protonation constant of the complex is also very low ($K_{GdHL} = 1.9$) [51] and thus the concentration of any protonated complex remains low at most pH values. If one compares these results with the dissociation rates of Gd^{3+} complexes of DTPA and its derivatives, one can conclude that the rates of dissociation of the complexes formed with acyclic and macrocyclic polyaza-polycaxboxylates, polyphosphinates and phosphonate esters strongly depend on the protonation constants of such complexes. This general rule should be kept in mind when designing new ligands for use *in vivo*.

7 IN VIVO TOXICITY OF Gd^{3+} COMPLEXES

Uncomplexed aqueous Gd^{3+}, its hydroxo species, and virtually all uncomplexed ligands are toxic to small animals, having LD_{50} values in the range 0.1–0.2 mmol/kg. [3, 4, 91]. Upon complexation, the LD_{50} increases dramatically from 6 mmol/kg for the meglumine salt of $GdDPTA^{2-}$ to 12 mmol/kg for the neutral complex, Gd(HP-DO3A). Even higher LD_{50} values have been reported for other systems, depending upon the formulation. Some agents are formulated with a small amount of Ca^{2+} to insure that the free Ca^{2+} in blood is not compromised upon injection of the CA. Since complexation of free Gd^{3+} reduces its LD_{50} by a factor of 60–120, it seemed reasonable to assume that toxicity should be directly proportional to the complex stability (the stability under blood pH conditions). However, the observation that Gd^{3+} complexes of amide derivatives of DTPA are less toxic than anticipated, as based upon conditional stability constants, led Cacheris *et al.* [8] to formulate a comprehensive model to correlate with the measured complex toxicity values. One key feature of this model is competition between other endogenous metal ions (Ca^{2+}, Zn^{2+}, and Cu^{2+}) for displacement of Gd^{3+} from L. For a series of linear amine ligand systems, the model successfully demonstrated a relationship between the complex LD_{50} values and the concentration of complex that would release about 20 μM Gd^{3+} *in vivo*. This value corresponds closely to the amount of Gd^{3+} that disrupts Ca^{2+} binding to the cardiac sarcoplasmic reticulum [92]. One practical consequence of this study was that it correctly predicted that the LD_{50} of GdDTPA-BMA should increase by the addition of ~ 5 % Na[CaDTPA-BMA] to act as a sacrificial complex to pre-empt displacement of Gd^{3+} from DTPA-BMA by adventurous Zn^{2+}. In practice, the measured LD_{50} values of GdDTPA-BMA increased significantly from 14.8 to 38.2 mmol/kg upon addition of the adjuvant Ca^{2+} complex. As a result, a slight excess of Na[CaDTPA-BMA] was included in the formulation of Omniscan™. As noted above in Section 2, Gd^{3+} complexes derived from macrocyclic ligands have a distinct kinetic advantage over ligands derived from linear amines. Kinetic inertness may not necessarily be an essential feature of low-molecular-weight CAs that are cleared from the body quickly, but this characteristic will likely become more important with the development of new macromolecular CAs [93–97] having prolonged clearance times.

Although the term toxicity is generally associated with LD_{50} values, there are other biological and physiological parameters that arguably are more sensitive indicators of CA toxicity. For example, Morris *et al.* [98] have demonstrated that above concentrations of 1 mM, both $GdDTPA^{2-}$ and $GdDOTA^{-}$ stimulated intermediary metabolism (increased CO_2 production from glucose) in rat hippocampus slices. Since these complexes are known to remain extracelluar *in vivo*, this observation suggests that the complexes can interact with cell membranes and initiate intracellular metabolic events. At concentrations of 10 mM, these same agents stimulated uptake of ${}^{32}PO_4^{3-}$ into phosphatidic acid (PA), phosphatydlinositol (PI), PI-phosphate, and PI-bis-phosphate, again pointing to interactions at the level of the cell membrane. However, it is not clear whether these metabolic changes have any clinical significance. The more highly charged $HTmDOTP^{4-}$ MR shift reagent causes a transient decrease in mean arterial blood pressure, an increase in liver resting transmembrane potential, and a decrease in free, blood plasma Ca^{2+} when infused into live rats [99]. These transient effects can all be traced to the relatively high affinity of this shift reagent for Ca^{2+}, which could likely be solved by proper formulation. It is interesting that highly charged agents such as $HTmDOTP^{4-}$ are generally dismissed as possible drugs for osmolality reasons, yet Ren *et al.* [100] have shown that a 0.5 M solution of $Na_4HTmDOTP$ has a measured osmolality (1700 mOsm/kg) which is somewhat less than the commercial product, Magnevist$^{TM)}$ (1960 mOsm/kg). This suggests that even highly charged agents could be formulated into safe, effective drug prepartions.

ACKNOWLEDGEMENTS

ADS wishes to thank the Robert A. Welch Foundation (AT-584) for financial support and Dr Carol Wiegers for expert technical assistance during the preparation of this chapter.

REFERENCES

1. Wedeking, P., Kumar, K. and Tweedle, M. F. *Magn. Reson. Imaging* 1992, **10**, 641–648.
2. Gries, H. and Miklauts, H. *Physiol. Chem. Phys.* Med. *Nucl. Magn. Reson.* 1984, **16**, 105–112.
3. Tweedle, M. F., 'Relaxation agents in NMR imaging', in Bünzli, J. -C. G. and Choppin, G. R. (Eds), *Lanthanide Probes in Life, Chemical and Earth Sciences: Theory and Practice*, Elsevier, Amsterdam, 1989, 127–79.
4. Weinmann, R. J., Laniado, M. and Muetzel, W. *Physiol. Chem. Phys.* Med. *Nucl. Magn. Reson.* 1984, **16**, 167–172.
5. Uggeri, F., Aime, S., Anelli, P. L., Botta, M., Brocchetta, M., de Haen, C., Ermondi, G., Grandi, M. and Paoli, P. *Inorg. Chem.* 1995, **34**, 633–642.
6. Schmitt-Willich, H., Brehm, M., Ewers, C. L. J., Michl, G., Müller-Fahrnow, A., Petrov, O., Platzek, J., Radüchel, B. and Sülzle, D. *Inorg. Chem.* 1999, **38**, 1134–1144.
7. Martell, A. E., Smith, R. M. and Motekaitis, R. J. *NIST Critical Stability Constants of Metal Complexes.* NIST Standard Reference Data, Gaithersburg, MD 1993.
8. Cacheris, W. P., Quay, S. C. and Rocklage, S. M. *Magn. Reson. Imaging* 1990, **8**, 467–481.
9. Ringbom, A. *Complexation in Analytical Chemistry*, Interscience Publishers (John Wiley & Sons), New York, 1963, pp. 35–74.
10. Geraldes, C. F. G. C., Sherry, A. D., Cacheris, W. P., Kuan, K. -T., Brown, R. D., Koenig, S. H. and Spiller, M. *Magn. Reson. Med.* 1988, **8**, 191–199.

11. Sherry, A. D., Cacheris, W. P. and Kuan, K. -T. *Magn. Reson. Med.* 1988, **8**, 180–190.
12. Tsukube, H., Adachi, H. and Morosawa, S. *J. Chem. Soc. Perkin Trans.* 1989, 1537–1538.
13. Geraldes, C. F. G. C., Urbano, A. M., Hoefnagel, M. A. and Peters, J. A. *Inorg. Chem.* 1993, **32**, 2426–2432.
14. Maumela, H., Hancock, R. D., Carlton, L., Reibenspies, J. H. and Wainwright, K. P. *J. Am. Chem. Soc.* 1995, **117**, 6698–6707.
15. Caravan, P., Ellison, J. J., McMurry, T. J. and Lauffer, R. B. *Chem. Rev.* 1999, **99**, 2293–2352.
16. Paul-Roth, C. and Raymond, K. N. *Inorg. Chem.* 1995, **34**, 1408–1412.
17. Martell, A. E. and Smith, K. M. (Eds), *Critical Stability Constants*, Vol. 1, Plenum, New York, 1974.
18. Sarka, L., Burai, L. and Brücher, E. *Chem. Eur. J.* 2000, **6**, 719–724
19. Carvalho, J. F., Kim, S. H. and Chang, C. A. *Inorg. Chem.* 1992, **31**, 4056–4068.
20. Inoue, M. B., Santacruz, H., Inoue, M. and Fernando, Q. *Inorg. Chem.* 1999, **38**, 1596–1602.
21. Desreux, J. F. *Inorg. Chem.* 1980, **19**, 1319–1324.
22. Burai, L., Fabian, I., Kiraly, R., Szilagyi, E. and Brücher, E. *J. Chem. Soc. Dalton Trans.* 1998, 243–248.
23. Martell, A. E. and Calvin, M. *Chemistry of Metal Chelate Compounds*, Prentice Hall, New York, 1952.
24. Irving, H. and Rossotti, H. *Acta Chem. Scand.* 1956, **10**, 72–93.
25. Choppin, G. R. *J. Less-Common Met.* 1985, **112**, 193–205.
26. Kumar, K., Chang, C. A. and Tweedle, M. F. *Inorg. Chem.* 1993, **32**, 587–593.
27. Kumar, K., Tweedle, M. F., Malley, M. F. and Gougoutas, J. Z. *Inorg. Chem.* 1995, **34**, 6472–6480.
28. Huskens, J., Torres, D. A., Kovacs, Z., Andre, J. P., Geraldes, C. F. G. C. and Sherry, A. D. *Inorg. Chem.* 1997, **36**, 1495–1503.
29. Cacheris, W. P., Nickle, S. K. and Sherry, A. D. *Inorg. Chem.* 1987, **26**, 958–960.
30. Chang, C. A., Chen, Y. -H., Chen, H. -Y. and Shieh, F. -K. *J. Chem. Soc. Dalton Trans.* 1998, 3243–3248.
31. Szilágyi, E., Tóth, E., Kovács, Z., Platzek, J., Radüchel, B. and Brücher, E. *Inorg. Chim. Acta* 2000, **298**, 226–234.
32. Burai, L., Ren, J., Kovacs, Z., Brücher, E. and Sherry, A. D. *Inorg. Chem.* 1998, **37**, 69–75.
33. Sherry, A. D., Ren, J., Huskens, J., Brücher, E., Tóth, E., Geraldes, C. F. C. G., Castro, M. M. C. A., and Cacheris, W. P. *Inorg. Chem.* 1996, **35**, 4604–4612.
34. Tóth, E., Király, R., Platzek, J., Radüchel, B. and Brücher, E. *Inorg. Chim. Acta* 1996, **249**, 191–199.
35. Delgado, R., Siegfried, L. C. and Kaden, T. A. *Helv. Chim. Acta* 1990, **73**, 140–148.
36. Dubost, J. -P., Leger, J. -M., Langlois, M. -H., Meyer, D., Schaefer, M. *C. R. Acad. Sci. Paris, Ser. II* 1991, **312**, 349–354.
37. Horrocks, W. DeW. and Sudnick, D. R. *J. Am. Chem. Soc.* 1979, **101**, 334–340.
38. Bryden, C. C. and Reilley, C. N. *Anal. Chem.* 1982, **31**, 610–615.
39. Zhang, X., Chang, C. A., Brittain, H. G., Garrison, J. M., Telser, J. and Tweedle, M. F. *Inorg. Chem.* 1992, **31**, 5597–5600.
40. Alpoim, M. C., Urbano, A. M., Geraldes, C. F. G. C. and Peters, J. A. *J. Chem. Soc. Dalton Trans.* 1992, 463–467.
41. Geiger, G. and Karlen, U. *Helv. Chim. Acta* 1971, **54**, 135–153.
42. Southwood-Jones, R. V. and Merbach, A. E. *Inorg. Chim. Acta* 1978, **30**, 135–143.
43. Király, R., Tóth, I. and Brücher, E. *J. Inorg. Nucl. Chem.* 1981, **43**, 345–349.

44. Spaulding, L. and Brittain, H. G. *Inorg. Chem.* 1985, **24**, 3692–3698.
45. Király, R., Tóth, I., Zékány, L. and Brücher, E. *Acta Chim. Hung.* 1998, **125**, 519–526.
46. Aime, S. and Nano, R. *Invest. Radiol.* 1988, **23**, 264–266.
47. Jackson, G. E., Wynchank, S. and Woudenberg, S. *Magn. Reson. Med.* 1990, **16**, 57–66.
48. Vander Elst, L., Van Haverbeke, Goudemant, J. F. and Muller, R. N. *Magn. Reson. Med.* 1994, **31**, 437–444.
49. Chang, C. A., Francesconi, L. C., Malley, M. F., Kumar, K., Gougoutas, J. Z. and Tweedle, M. F. *Inorg. Chem.* 1993, **32**, 3501–3508.
50. Burai, L., Hietapelto, V., Kiraly, R., Tóth, E., and Brücher, E. *Magn. Reson. Med.* 1997, **38**, 146–150.
51. Burai, L. *Ph. D. Dissertation*, Kossuth University, Debrecen, Hungary, 1997.
52. May, P. M., Linder, P. W. and Williams, D. R. *J. Chem. Soc. Dalton Trans.* 1977, 588–595.
53. Berthon, G., Hacht, B., Blais, M. and May, P. M. *Inorg. Chim. Acta* 1986, **125**, 219–227.
54. Harris, W. R. *Clin. Chem.* 1992, **38**, 1809–1818.
55. Puttagunta, N. R, Gibby, W. A. and Puttagunta, V. L. *Invest. Radiol.* 1996, **31**, 619–624.
56. Magerstadt, M., Gansow, O. A., Brechbiel, M. W., Colcher, D., Baltzer, L., Knop, R. H., Girton, M. E. and Naegele, M. *Magn. Reson. Med.* 1986, **3**, 808–812.
57. Tweedle, M. F., Hagan, J. J., Kumar, K., Mantha, S. and Chang, C. A. *Magn. Reson. Imaging* 1991, **9**, 409–415.
58. Kasprzyk, S. P. and Wilkins, R. G. *Inorg. Chem.* 1982, **21**, 3349–3352.
59. Brücher, E., Laurenczy, G. and Makra, Zs. *Inorg. Chim. Acta* 1987, **139**, 141–142.
60. Wang, X., Jin, T., Comblin, V., Lopez-Mut, A., Merciny, E. and Desreux, J. F. *Inorg. Chem.* 1992, **31**, 1095–1099.
61. Kumar, K., Tweedle, M. F. *Inorg. Chem.* 1993, **32**, 4193–4199.
62. Tóth, E., Brücher, E., Lazar, I. and Tóth, I. *Inorg. Chem.* 1994, **33**, 4070–4076.
63. Wu, S. L. and Horrocks, W. D. *Inorg. Chem.* 1995, **34**, 3724–3732.
64. Dale, J. *Isr. J. Chem.* 1980, **20**, 3–11.
65. Kumar, K. and Tweedle, M. F. *Pure Appl. Chem.* 1993, **65**, 515–520.
66. Albin, M., Horrocks, W. D. and Liotta, F. J. *Chem. Phys. Lett.* 1982, **85**, 61–64.
67. Li, M. and Meares, C. F. *Bioconjugate Chem.* 1993, **4**, 275–283.
68. Sherry, A. D, Brown, R. D., Geraldes, C. F. G. C., Koenig, S. H., Kuan, K. -T. and Spiller, M. *Inorg. Chem.* 1989, **28**, 620–622.
69. Keire, D. A. and Kobayashi, M. *Bioconjugate Chem.* 1999, **10**, 454–463.
70. Kruper, W. J. J., Fordyce, W. A. and Sherry, A. D. 'Carboxamide modified polyamine chelators and radioactive complexes thereof for conjugation to antibodies', *US Pat. 5 310 535*, 10 May 1994.
71. Broan, C. J., Cox, J. P. L., Craig, A. S., Kataky, R., Parker, D., Ferguson, G., Harrison, A. and Randall, A. M. *J. Chem. Soc. Perkin Trans. 2* 1991, 87–99.
72. Pulukkody, K. P., Norman, T. J., Parker, D., Royle, L. and Broan, C. J. *J. Chem. Soc. Perkin Trans. 2* 1993, 605–620.
73. Harrison, A., Royle, L., Walker, C., Pereira, C., Parker, D., Pulukkody, K. and Norman, T. J. *Magn. Reson. Imaging* 1993, **11**, 761–770.
74. Lazar, I., Sherry, A. D., Ramasamy, R., Brücher, E. and Kiraly, R. *Inorg. Chem.* 1991, **30**, 5016–5019.
75. McMurry, T. J., Pippin, C. G., Wu, C., Deal, K. A., Brechbiel, M. W., Mirzadek, S. and Gansow, O. A. *J. Med. Chem.* 1998, **41**, 3546–3549.

76. Margerum, D. W., Cayley, G. R., Weatherburn, D. C. and Pagenkopt, G. K. in *Coordination Chemistry*, Vol. 2, ACS Monograph 174, Martell, A. E. (Ed.), American Chemical Society, Washington, DC 1978.
77. Choppin, G. R. *J. Alloys Comp.* 1995, **225**, 242–245.
78. Brücher, E. and Laurenczy, G. *Inorg. Chem.* 1983, **22**, 338–341.
79. Glentworth, P., Wiseall, B., Wright, C. L. and Mahmood, A. J. *J. Inorg. Nucl. Chem.* 1968, **10**, 967–979.
80. Asano, T., Okada, S. and Taniguchi, S. *J. Inorg. Nucl. Chem.* 1970, **32**, 1287–1293.
81. Brücher, E. and Laurenczy, G. *J. Inorg. Nucl. Chem.* 1981, **43**, 2089–2096.
82. Sarka, L., Burai, L. and Brücher, E., (unpublished).
83. Rothermel, G. L., Rizkalla, E. N. and Choppin, G. R. *Inorg. Chim. Acta* 1997, **262**, 133–138.
84. Choi, K. Y., Kim, K. S. and Kim, J. C. *Polyhedron* 1994, **13**, 567–571.
85. Schwizer, W., Fraser, R., Maecke, H., Siebold, K., Funk, R. and Fried, M. *Magn. Reson. Med.* 1994, **31**, 388–393.
86. Kumar, K., Jin, T., Wang, X., Desreux, J. F. and Tweedle, M. F. *Inorg. Chem.* 1994, **33**, 587–593.
87. Cai, H. Z. and Kaden, T. A. *Helv. Chim. Acta* 1994, **77**, 383–398.
88. Brücher, E. and Sherry, A. D. *Inorg. Chem.* 1990, **29**, 1555–1559.
89. Szilágyi, E., Tóth, E., Brücher, E. and Merbach, A. E. *J. Chem. Soc. Dalton Trans.* 1999, 2481–2486.
90. Geraldes, C. F. G. C., Sherry, A. D., Lazar, I., Miseta, A., Bogner, P., Berenyi, E., Sumegi, B., Kiefer, G. E., McMillan, K., Maton, F. and Muller, R. N. *Magn. Reson. Med.* 1993, **30**, 696–703.
91. Chang, C. A. *Eur. J. Solid State Inorg. Chem.* 1991, **28**, 237–244.
92. Krasnow, N. *Biochem. Biophys. Acta* 1972, **282**, 187–194.
93. Brasch, R., Pham, C., Shames, D., Roberts, T., van Dijke, K., van Bruggen, N., Mann, J., Ostrowitzki, S. and Melnyk, O. *J. Magn. Reson. Imaging* 1997, **7**, 68–74.
94. Corot, C., Schaefer, M., Beauté, S., Bourrinet, P., Zehaf, S., Bénizé, V., Sabatou, M. and Meyer, D. *Acta Radiol.* 1997, **38** (Suppl. 412, 91–99.
95. Kroft, L. J. M., Doornbos, J., van der Geest, R. J. and de Roos, A. *J. Magn. Reson. Imaging* 1999, **10**, 170–177.
96. Kroft, L. J. M., Doornbos, J., Benderbous, S. and de Roos, A. *J. Magn. Reson. Imaging* 1999, **9**, 777–785.
97. Bryant, L. H. J., Brechbiel, M. W., Wu, C., Bulte, J. W. M., Herynek, V. and Frank, J. A. *J. Magn. Reson. Imaging* 1999, **9**, 348–352.
98. Morris, T. W., Ekholm, S. E., Marinetti, G. V., Prentice, L. I. and Leakey, P. *Invest. Radiol.* 1991, **26**, S209–S211.
99. Bansal, N., Germann, M. J., Seshan, V., Shires, G. T. III, Malloy, C. R. and Sherry, A. D. *Biochemistry* 1993, **32**, 5638–5643.
100. Ren, J., Springer, C. S. and Sherry, A. D. *Inorg. Chem.* 1997, **36**, 3493–3498.
101. White, D. A., de Learie, L. A., Moore, D. A., Wallace, R. A., Dunn, T. J., Cacheris, W. P., Imura, H. and Choppin, G. R. *Invest. Radiol.* 1991, **26**, 5226–5228.
102. Aime, S., Fasano, M., Paoletti, S. and Terreno, E. *Gazz. Chim. Ital.* 1995, **125**, 125–131.
103. Wang, Y. -M., Lin, S. -T., Wang, Y. -J. and Sheu, R. -S. *Polyhedron*, 1998, **17**, 2021–2028.
104. Wang, Y. -M., Lee, C. -H., Lin, S. -T., Wang, Y. -J. and Sheu, R. -S. *J. Chem. Soc. Dalton Trans.* 1998, 4113–4118.
105. Wiegers, C. B., Welch, M. J., Sharp, T. L., Brown, J. J., Perman, W. H, Sun, Y, Motekaitis, R. J. and Martell, A. E. *Magn. Reson. Imaging*, 1992, **10**, 903–911.

106. Jakab, S., Kovacs, Z., Burai, L. and Brücher, E. *Magy. Kem. Foly.* 1993, **99**, 391–396.
107. Sarka, L. and Brücher, E. *J. Chem. Soc.* Dalton Trans., 2000 (in press).
108. Clarke, E. T. and Martell, A. E. *Inorg. Chim. Acta*, 1991, **190**. 37–46.
109. Wu, S. L. and Horrocks, W. D. *Inorg. Chem.* 1996, **35**, 394–401.
110. Kodama, M., Koike, T., Mahatma, A. B. and Kimura, E. *Inorg. Chem.* 1991, **30**, 1270–1273.
111. Geze, C., Mouro, C., Hindré, F., Le Plouzennec, M., Moinet, C., Rolland, R., Alderighi, L., Vacca, A. and Simonneaux, G., *Bull. Soc. Chim. Fr.* 1996, **133**, 267–272.
112. Bianchi, A., Calabi, L., Ferrini, L., Losi, P., Uggeri, F. and Valtancoli, B. *Inorg. Chim. Acta* 1996, **249**, 13–15.

7 Computational Studies Related to Gd(III)-Based Contrast Agents

DETLEV SÜLZLE, JOHANNES PLATZEK, BERND RADÜCHEL and HERIBERT SCHMITT-WILLICH

Research Laboratories of Schering AG, Berlin, Germany

1 INTRODUCTION

In view of the importance of Gd(III)-based contrast agents for magnetic resonance imaging (MRI) [1–3], it is somewhat surprising that there are only a few reports in the literature about computational studies on gadolinium-containing compounds. Further more, the present knowledge about the interaction of the paramagnetic Gd(III) ion with water molecules, especially water protons, which is a process of fundamental importance for the application of Gd(III)-based compounds as contrast agents in MRI, does not allow a full theoretical description of the observed facts.

A rational design of improved MRI contrast agents requires a detailed understanding of the structure and dynamics of gadolinium-containing compounds, preferably in aqueous solution. Standard tools for the elucidation of the structure and dynamics of molecules in solution, like multi-nuclear magnetic resonance spectroscopy, are not applicable to MRI contrast agents due to the paramagnetic nature of these complexes. Alternatively, solid-state structures of Gd(III)-based compounds can be determined, at least in principle, by X-ray structural determination. However, only a limited number of gadolinium complexes has yet been characterized in the crystalline state by X-ray crystal and molecular structure determinations. These experimental data allow for the assessment of the quality of the theoretical predictions by comparison of the calculated structures with the experimental findings.

2 THEORETICAL METHODS

The whole panoply of theoretical methods available today was applied to create models of the structures of Gd(III)-based contrast agents, i.e. molecular mechanics, semi-empirical, density functional, and *ab initio* type calculations.

The Chemistry of Contrast Agents in Medical Magnetic Resonance Imaging
Edited by A. E. Merbach and É. Tóth.

The computational challenge at each level of theory is to expand the existing arsenal of methods, which have been developed for standard applications, such as the description of organic molecules, and to incorporate the gadolinium ion with its respective metal–ligand interactions. The aim at each level of theory is to calculate the energy of a particular molecular structure (spatial arrangement of atoms), and to perform optimizations, of the molecular energy with respect to the atomic positions. As a result of the geometry optimizations, stationary points are located on the corresponding potential energy hypersurface (PES) of the molecule. These points are characterized by a vanishing norm of the gradient of the energy of the molecule with respect to the Cartesian coordinates of the atoms. The nature of a stationary point on the PES is determined by evaluating the eigenvalues of the matrix of the second derivatives of the molecular energy with respect to the Cartesian coordinates of the atoms (Hesse matrix). Local minima or equilibrium structures on the PES are characterized by positive eigenvalues of the Hesse matrix. One or more negative eigenvalues are indicative of transition structures of the order of one or higher. Vibrational frequencies depend on the second derivatives of the molecular energy with respect to the nuclear coordinates. A knowledge about the Hesse matrix at a minimum point (equilibrium structure, molecular conformation) of the PES allows for the prediction of the vibrational spectrum of a molecule [4].

Throughout this chapter, energies and distances are given in the non-SI units of kcal/mol (1 cal = 4.184 J) and Å (1 Å= 10^{-10} m), respectively.

Ab initio methods compute solutions of the electronic Schrödinger equation solely on the basis of the laws of quantum mechanics and on the values of a small number of physical constants. Programs which enable the description of molecules by quantum mechanical *ab initio* methods are now routinely available (see, for example, Frisch and Frisch, 1998 [5]). These use a series of rigorous mathematical approximations, i.e. Hartree-Fock (HF) and *n*-th-order Møller–Plesset (MPn) perturbation theory. For a given configuration of the nuclei of a molecule (Born–Oppenheimer approximation), each electron moves in the mean-field of the remaining electrons. The electronic wavefunction is represented by a single wavefunction of determinantal form. Electron correlation, which is not inherent in the HF model, can be included by perturbation theory [6].

The electronic wavefunction of a molecule of arbitrary elemental composition can be calculated by the linear combination of atomic orbitals. The basic requirement for this type of calculation is the availability of basis sets for the elements of interest. Until recently, gadolinium-containing compounds were a great challenge, not only due to the size of the molecular systems which are interesting from the chemical and pharmacological point of view as MRI contrast agents, but also because of the number of electrons which has to be taken into account for gadolinium ($Z = 64$). However, the basis set can be restricted to the valence shell only by means of an effective core potential (ECP) which describes in an efficient way the incompletely filled 4f shell, as well as the large relativistic and correlation effects of the gadolinium atom. Several different ECPs were developed for gadolinium [7–9] and were tested on rather small gadolinium-containing compounds, e. g. oxides [7], halogenides [8, 10–13], and hydroxides [14]. In addition, complexes of Ln(III) ions and phosphoryl-containing ($O{=}PR_3$) ligands, where R = H, Me, Et and Ph, were described by this approach [15]. Even the interaction between adrenaline and Ln(III) ions was investigated by *ab initio* type calculations. It was shown that the metal ion interacts preferably with the phenolate oxygen donor atoms [16].

Density functional (DFT) methods implicitly include the effects of electron correlation by approximating the electronic energy as a functional of the electron density of the molecule [17]. The combination of DFT methods with a classical dynamics scheme enables the study of the dynamics of a molecular system at a finite temperature, typically over a time-scale of 10^{-12} s [18]. Structures and energetics of lanthanide trihalogenides in the gas phase were reasonably well described by application of a combination of DFT methods and ECP-type basis sets [19, 20].

Semi-empirical methods use parameters derived from experimental findings to simplify the solution of the Schrödinger equation, which is solved in an approximate form. Different semi-empirical methods, e. g. AM1, MNDO, PM3, MINDO/3, CNDO/2 and INDO/S are largely characterized by their differing parameter sets [21]. Electronic spectra of hydrated lanthanide ions were calculated for Pr(III), Nd(III) and Tm(III) within the INDO/S-CI approach [22].

For an excellent and exhaustive overview on electronic structure calculations for molecules containing lanthanide atoms, see Dolg and Stoll [7].

In molecular mechanics calculations, the laws of classical mechanics and electrostatics are used to calculate the energy of a molecule. The potential energy of a molecule is described as the energy of a set of atoms which are interacting via two-, three- and four-body potentials. The potential energy functions used in treating the two-, three- or four-body interactions are of empirical nature, and a great number of different molecular potential energy functions exists, e. g. Morse, Lennard–Jones and Buckingham potential energy functions [23, 24]. The parameters in the potential energy functions are fitted to the available experimental data, e. g. structures, vibrational wavenumbers and relative energies of conformers or diastereomers, in order to obtain an agreement between the calculated structures and properties and the corresponding experimental findings.

On the basis of the well-known molecular mechanics (MM) force field [25], Cundari *et al.* developed a tool for the description of gadolinium complexes [26]. In order to reproduce the experimentally determined structures of compounds of the type Gd(Schiff base)$(H_2O)_n$, Cundari *et al.* defined a new atom type and several new ligand types in an extension of the existing MM force field, namely a ninefold coordinated Gd, the oxygen-donating groups such as water, carboxylate, carbonyl and alkoxide, and the nitrogen-donating groups, imine and amine. Under the assumption that metal-independent parameters are transferable from the metal-free ligand to the metal–ligand complex, the newly developed force-field parameters were chosen. Electrostatic interactions were not treated explicitly in the model. According to Cundari *et al.*, the modified MM force field is capable of predicting bond lengths to within 0.1 Å, bond angles to within a few degrees and torsion angles to within five to seven degrees, in comparison to the experimentally determined X-ray structures [26].

In a more general approach, Beech *et al.* investigated different ways of applying force-field methods to model the structures of lanthanide complexes [27]. The structures chosen for comparison were experimentally determined X-ray structures of acyclic or cyclic poly(ethylene glycol ether) complexes of lanthanide nitrates or chlorides. In a first approach, the metal to ligand distances were examined, and the corresponding force constants were arbitrarily set to be 600 kcal/(mol Å). It was necessary in advance to decide on the number of donor atoms in the coordination sphere of the lanthanide. In a second approach, the lanthanide–ligand bonds were excluded, and the non-bonded interactions between pairs of atoms were described by Lennard–Jones potential energy functions. Finally, in a third approach, electrostatic interactions were incorporated by using arbitrarily adjusted charges on the metal ions and on the donor atoms of the ligand. To summarize, for all three methods the calculated structures are in reasonable agreement with the experimentally observed ones in terms of the root-mean-square (RMS) deviation of the Cartesian coordinates of the atoms C, N and O, and the lanthanide ions after superimposition of the force-field optimized with the experimentally determined structures. For the eleven compounds investigated, the RMS deviations were typically in the range from 0.06 to 0.28 Å, independent of the method used.

3 LN(III) WATER COMPLEXES

In view of the importance of the interactions between the Gd(III) ion and water molecules as a basis for the action of MRI contrast agents, there is great interest in the description of the structure and dynamics of Ln(III) water complexes.

3.1 $La(H_2O)_{1-3}{}^{3+}$

The potential energy hypersurfaces of $M(III)L_n$ complexes were investigated for the cations of the metals M = Sc, Y and La, with the neutral ligands, L = H_2O and NH_3, for $n = 1$, 2 and 3 [28]. For $n = 2$, the structures of lowest energies of the solvated La(III) ions are bent and possess C_{2v} symmetry, while for $n = 3$ they are pyramidal and belong to the point group C_3. For $La(H_2O)_2{}^{3+}$, there is a barrier to linearization of at least 3.7 kcal/mol, and for $La(H_2O)_3{}^{3+}$ the barrier to planarization is at least 0.4 kcal/mol. At the Hartree–Fock (HF) level of theory, the calculated La–O distances vary with the number of coordinated water molecules from 2.408 Å ($n = 1$), through 2.444 Å ($n = 2$), to 2.470 Å ($n = 3$). The variation of the La–N distances for L = NH_3 shows a similar behavior, i.e. 2.551 Å ($n = 1$), 2.594 Å ($n = 2$), and 2.632 Å ($n = 3$). A more detailed investigation of the ligand-binding energies reveals that the energy of interaction for ammonia is larger than the corresponding water-binding energies. According to the analysis of Kaupp and Schleyer [28], this is due to the fact that the ammonia dipole moment is located closer to the cation than the corresponding water dipole.

3.2 $Ln(H_2O)^{3+}$, $Ln(H_2O)_8^{3+}$ AND $Ln(H_2O)_9^{3+}$

Ab initio calculations of monoaqua complexes of the lanthanide ions, $La(H_2O)^{3+}$ – $Lu(H_2O)^{3+}$, and the corresponding octaaqua complexes, $La(H_2O)_8^{3+}$–$Lu(H_2O)_8^{3+}$, in the gas phase were performed by Hengrasmee and Probst [29]. An effective core potential and a contracted [5s4p3d] Gaussian-type orbital valence basis set were used for the lanthanide atoms, together with an ECP-type basis set for the oxygen atom, and a double-zeta basis set for the hydrogen atoms. At the restricted Hartree–Fock (RHF) level of theory only the Ln–O_{water} distances were optimized for the octaaqua complexes in the energetically unfavored cubic arrangement (see Figure 7.1) by using a fixed water geometry (bond distance, O–H 0.957 Å, bond angle H–O–H, 104.5°), whereas for the monoaqua complexes, full geometry optimizations were performed either at the RHF or the restricted Møller–Plesset correlation energy correction truncated at second order (RMP2) level of theory. The interaction of the metal ion with a water molecule in $Ln(H_2O)^{3+}$ results in a lengthening of the O–H bond. The mean elongation at each level of theory is 0.031 Å, in comparison to the calculated O–H bond distance of a free water molecule in the gas phase.

The calculated bond distances, Ln–O, for $La(H_2O)^{3+}$ and $Lu(H_2O)^{3+}$ are 2.4 and 2.15 Å, respectively, and a systematic shortening of the bond length by

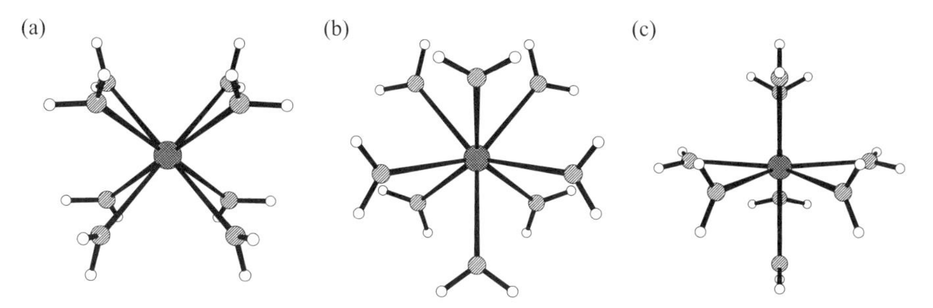

Figure 7.1 Different geometries of $Gd(H_2O)_8^{3+}$: (a) cube; (b) square-antiprism; (c) dodecadeltahedron.

going from La ($Z = 57$) to Lu ($Z = 71$) was observed. The calculated binding energy for $Ln(H_2O)^{3+}$ is also a linear function, which increases with the atomic number and varies at the RHF (RMP2) level of theory between −85 kcal/mol (−89 kcal/mol) for $La(H_2O)^{3+}$ and −108 kcal/mol (−115 kcal/mol) for $Lu(H_2O)^{3+}$.

The geometries of the octaaqua ions were optimized in the cubic arrangement with respect to the Ln–O distances. The calculated bond distances, Ln–O, for $La(H_2O)_8{}^{3+}$ and $Lu(H_2O)_8{}^{3+}$ are 2.66 and 2.47 Å, respectively, and a systematic shortening of the distances by going from La ($Z = 57$) to Lu ($Z = 71$) was observed. As in the case of the monoaqua ions, the binding energy per water molecule increases with increasing atomic number, and the binding energy per water molecule varies between −55 kcal/mol for $La(H_2O)_8{}^{3+}$ and −64 kcal/mol for $Lu(H_2O)_8{}^{3+}$, at the RHF level of theory.

Schafer and Daul investigated the structures of the octaaqua ions, $Ln(H_2O)_8{}^{3+}$, for Ln = Sm, Gd and Dy, and the nonaaqua ion, $Gd(H_2O)_9{}^{3+}$ [30]. At the local spin-density or generalized gradient approximation level of DFT theory, the atoms Sm, Gd, Dy, O and H are described by an uncontracted triple-zeta Slater-type orbitals (TZ STO) basis set. Polarization functions were added for O and H. Frozen-core approximation for oxygen (He core) and the lanthanides (Xe core) were applied. The calculations were performed by using spin-polarized orbitals with seven electrons in the 4f orbitals of the Gd(III) ion which corresponds to the ${}^8S_{7/2}$ ground state of the free Gd(III) ion.

Only a few different geometric arrangements of the octaaqua ion $Gd(H_2O)_8{}^{3+}$ were investigated in detail, i.e. square prism (cube), square antiprism and dodecadeltahedron (snub disphenoid) (see Figure 7.1). The geometry optimizations were performed in the point groups D_{4h}, D_{4d} and D_{2d}. Only one tri-capped trigonal prismatic (TTP) arrangement of D_{3h} symmetry was taken into account for $Gd(H_2O)_9{}^{3+}$.

In the course of the geometry optimizations, the Ln–O_{water} distance was (exclusively) varied by keeping the geometry of the water molecule frozen (bond distance, O–H 0.954 Å; bond angle, H–O–H 104.5°). In order to locate stationary points on the corresponding PESs, the total energies for selected geometric arrangements were calculated. The different geometric arrangements were chosen by a systematic variation of symmetrically independent internal coordinates. In the case of the square prismatic arrangement, only the Ln–O distance was varied, while in the case of the square antiprismatic arrangement both the Ln–O distance and the O–Ln–O bond angle were varied. The total energies were plotted as a function of the symmetrically independent coordinates in order to generate plots and contour maps which allow for the localization of stationary points by visual inspection. The structures thus located were not further characterized, neither by the corresponding norms of the gradients, nor by the eigenvalues of the corresponding Hesse matrices.

The relative energies of the different isomers of $Gd(H_2O)_8{}^{3+}$ give rise to the following ordering of the structures: square antiprism D_{4d} (± 0 kcal/mol), dodecadeltahedron D_{2d} (+ 3.4 kcal/mol) and square prism D_{4h} (+ 15.3 kcal/mol). The calculated total energy of the D_{3h} TTP arrangement for $Gd(H_2O)_9{}^{3+}$ is equivalent to the total energy of the D_{4d} square antiprism for $Gd(H_2O)_8{}^{3+}$ plus the total energy of a free water molecule in the gas phase. The calculated Gd–O distances are 2.52, 2.52 and 2.56 Å for the square antiprism, the dodecadeltahedron and the square prism, respectively. Values of 2.57 and 2.52 Å were obtained, respectively, for the equatorial and the axial Gd–O distances of the TTP [30].

Force-field structures and energetics of octa- and nonaaqua complexes of Ln(III) ions were reported for La(III) – Lu(III) by Hay and Wadt [31]. By extending the well-known molecular mechanics (MM) force field [25], these authors calculated the structures of $Ln(H_2O)_8{}^{3+}$ and $Ln(H_2O)_9{}^{3+}$. Ligand–metal–ligand interactions were replaced by non-bonded interactions between pairs of atoms of the ligands, and electrostatic interactions were ignored.

From the average geometry of the experimentally determined structures of water molecules in aqua Ln(III) ions, the metal-independent parameters were chosen to be consistent with the existing MM parameter set. An equilibrium value of 0.941 Å, together with a force constant of 662 kcal/(mol Å^2) were used for the O–H bond length, 109° for the H–O–H bond angle, and 125.5° for the H–O–Ln bond angle with a common bond angle deformation constant of 43 kcal/(mol Å rad^2). The non-bonded interactions between the atoms of the water molecules were described by the standard MM2 parameters for the atoms H and O; any non-bonded interactions between the metal ion and an atom of the ligand were omitted. Then, the solely remaining metal-dependent parameters were chosen in order to achieve an overall agreement between the calculated and the experimentally observed structures. The Ln–O bond distances and the corresponding force constants were adjusted through manual inspection. The chosen bond distances are a linearly decreasing function of the atomic number, i.e. 2.434 Å (La) to 2.092 Å (Lu), whereas the force constants show increasing values, i.e. 13 to 20 kcal/(mol Å^2).

Starting from the X-ray structures of $Ln(H_2O)_9(CH_3CH_2SO_3)_3$ and $Ln(H_2O)_9(CF_3SO_3)_3$, the force-field optimized structures for the nonaaqua ions, $Ln(H_2O)_9{}^{3+}$, exhibit a TTP arrangement of the ligand atoms around the central metal ion which belongs to the point group D_{3h}, (see Figure 7.2). The experimentally observed differences between the bond distances, Ln–O, of the water molecules in the equatorial and the axial planes of the TTP, Ln–O_{eq} and Ln–O_{ax}, could be reproduced, and in general, the bond distances Ln–O_{eq} and Ln–O_{ax}, were calculated to within ± 0.02 Å and the bond angles, O–Ln–O, to within ± 3°, over the entire lanthanide series.

However, starting from the X-ray structures of $Ln(H_2O)_9(BrO_3)_3$, the force-field optimized structures for the nonaaqua ions $Ln(H_2O)_9{}^{3+}$ exhibit a TTP arrangement of the water molecules around the central metal ion which belongs to the point group D_{3d}. This TTP arrangement differs from the D_{3h} arrangement by a rotation of the H–O–H plane around the Ln–O bond. According to

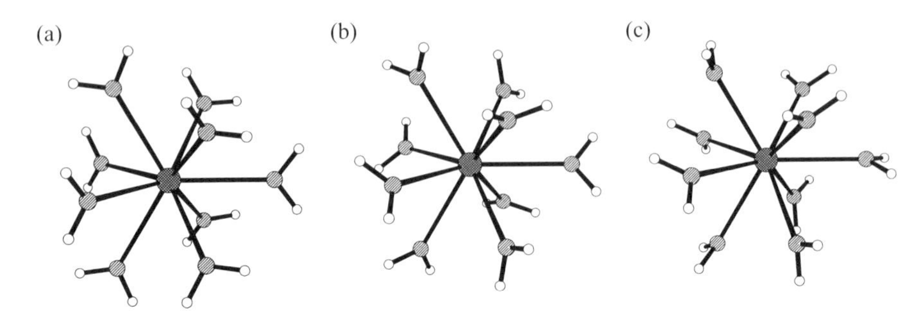

Figure 7.2 Tri-capped trigonal prismatic geometries of $Gd(H_2O)_9{}^{3+}$ ions with (a) D_{3h}, (b) D_3 and (c) C_{3h} symmetries.

the calculated force-field energies, the D_{3d} structures are 1–3 kcal/mol higher in energy than the D_{3h} structures. In this case the calculated bond distances, Ln–O_{eq}, are within ± 0.03 Å, but the calculated Ln–O_{ax} lengths are too long by 0.06 Å.

Experimentally observed structures of the octaaqua ions, $Ln(H_2O)_8{}^{3+}$, show either a square antiprismatic or a dodecadeltahedral arrangement of the LnO_8 polyhedron. Analogously, the force-field optimized structures of $Ln(H_2O)_8{}^{3+}$ belong either to the point group D_{4d} or to the point group D_{2d}. Calculated energy differences between these two arrangements are of the order of 0.05–0.2 kcal/mol in favor of the D_{4d} structure [31].

In a series of publications Kowall *et al.* (for the last one of the series see [32]) investigated the microscopic mechanisms of the water-exchange reaction between the hydration shells of Ln(III) ions and bulk water. A classical molecular dynamics simulation was performed for approximately 10^{-9} s on a system composed of one Ln(III) ion and several hundred water molecules. The electrostatic interactions between the Ln(III) ion and the water molecules, together with the interaction between two different water molecules, were explicitly taken into account. The coordination numbers thus obtained for the ions Nd(III) and Yb(III) are 9.0 and 8.0, respectively, whereas for Sm(III) an equilibrium value of 8.5 was obtained. The calculated coordination numbers are in agreement with the available experimental data for these ions.

Up until now these are the only theoretical studies about the structure and dynamics of second-sphere water molecules which are currently available in the literature.

3.3 $Gd(H_2O)_9{}^{3+}$

A detailed description of the interaction of the Gd(III) ion with nine water molecules was given by Cosentino *et al.* [33]. These authors investigated different tri-capped trigonal prismatic (TTP) arrangements of nine water molecules around the central metal ion in the gas phase, i.e. $Gd(H_2O)_9^{3+}$ ions of D_3, C_{3h}, C_3 and D_{3h} symmetry. These TTP arrangements differ in the orientation of the H–O–H plane with respect to the central GdO_9 polyhedron. The structure of lowest energy is of D_3 symmetry, either at the restricted Hartree–Fock (RHF) or at the restricted Møller–Plesset correlation energy correction truncated at second order (RMP2) levels of theory. Four different basis sets for the hydrogen and oxygen atoms of the coordinated water molecules (STO-3G, 3–21G, 6–31G(d), D95(d, p)) [34] and two different types of basis sets for the valence electrons of the gadolinium atom were used in the course of the geometry optimizations. The experimentally observed structures of the $Gd(H_2O)_9^{3+}$ ion in the salts $Gd(H_2O)_9(CF_3SO_3)_3$ [35] and $Gd(H_2O)_9(CH_3CH_2SO_3)_3$ [36] are both of approximate C_{3h} symmetry, and the positions of the hydrogen atoms of the water molecules are slightly distorted. However, at all investigated levels of theory the optimized structure of C_{3h} symmetry has a significantly higher total energy than the structure of D_3 symmetry. The difference between the total energies is calculated to be approximately 5 kcal/mol. Recent frequency calculations of the two $Gd(H_2O)_9^{3+}$ structures of lowest total energy, either at RHF or RMP2 levels of theory, predict that only the structure of D_3 symmetry is a true minimum; the stationary point of C_{3h} symmetry is a transition structure of order three [37].

Two different types of $Gd–O_{water}$ distances are the result of the geometry optimizations. The three $Gd–O_{eq}$ distances are significantly longer than the six $Gd–O_{ax}$ distances. The calculated distances for $Gd–O_{eq}$ and $Gd–O_{ax}$, of 2.531 and 2.499 Å are in good agreement with the mean values of the experimental findings, i.e. 2.53 and 2.40 Å, respectively. In contrast to the experimentally determined structures of the $Gd(H_2O)_9^{3+}$ ion, the positions of the hydrogen atoms of the water molecules are well defined, and the O–H distances and the H–O–H bond angles are of comparable size to the distance and the bond angle in free water molecules (see Figure 7.2). One possible explanation for the discrepancy between the character of the stationary point calculated in C_{3h} symmetry and the experimental finding could be that in the crystal the experimentally observed C_{3h} structure of the $Gd(H_2O)_9^{3+}$ ion is stabilized by the interaction with the corresponding anions. Two oxygen atoms of the respective anions are involved in hydrogen bonds between a hydrogen atom of one water molecule in the axial plane, as well as another water molecule in the equatorial plane [35, 36].

3.4 RELATED WORK OF INTEREST

There are only a few reports in the literature describing the interactions of metal ions other than Ln(III) ions with eight or more water molecules. These studies are described here in some detail, in particular to give some ideas about the computational difficulties involved, but also to illustrate the possibilities of the various theoretical methods.

3.4.1 $K(H_2O)_{1-10}^{+}$

A number of conformers of hydrated potassium ions, $K(H_2O)_n^+$, where $n = 1 - 10$, were investigated in the gas phase at the RHF and RMP2 levels of theory by using a triple-zeta plus two sets of polarization functions (TZ2P) basis set for the atoms of the water molecules and a (14s10p3d) contracted to [8s7p3d] basis set for potassium [38]. Isomeric structures of D_4 and S_8 symmetry were localized as stationary points on the PES of $K(H_2O)_8^+$. According to the Hesse matrices, these points correspond to minima on the PES with a negligible difference in total energies. For $n \geqslant 9$, the ninth and tenth water molecules are found to be located far from the metal ion, so the possible maximum coordination number is likely to be eight.

In an exhaustive *ab initio* study of the structure and energetics of hydrated alkali metal ions the $K(H_2O)_n^+$ ions were investigated by Feller *et al.* [39]. The calculated enthalpy of formation of -10.4 kcal/mol for the reaction $K(H_2O)_5^+ + H_2O \rightarrow K(H_2O)_6^+$ at the RMP2/ECP level of theory is in excellent agreement with the experimentally determined value of -10.0 kcal/mol. A comparison between the calculated and experimentally determined values was not given for $n > 6$. The optimized structure of $K(H_2O)_6^+$ belongs to the point group D_{3d}, whereas the optimized structures of $K(H_2O)_7^+$ and $K(H_2O)_8^+$ have C_1 and S_4 symmetry, respectively. The frequency analysis of these stationary points was performed at the RHF/6-31+G(d) level of theory, and both structures displayed seven very low frequency modes. The lowest frequency modes for each complex were imaginary ($-2i$ cm^{-1} and $-11i$ cm^{-1} for $K(H_2O)_7^+$ and $K(H_2O)_8^+$, respectively), and correspond to negative eigenvalues of the Hesse matrix. According to the authors, the numerical uncertainty associated with the finite-difference technique for constructing the Hesse matrix made it impossible to unambiguously characterize the stationary points either as minima or as transition structures [39].

In an *ab initio* molecular-dynamics simulation of the solvation of K(I) ions in water, the interaction of potassium ions with water molecules was described [40]. An infinitely diluted ionic solution is modeled by 59 water molecules and a single K(I) ion in a cubic supercell subject to periodic boundary conditions. The first solvation shell of the K(I) ion is very flexible and shows several exchanges of water molecules with the second shell, even on the short time-scale of 2×10^{-12} s.

4 LN(III) COMPLEXES OF POLYAMINOPOLYCARBOXYLIC ACIDS

4.1 STEREOCHEMICAL ASPECTS

Eight- and ninefold coordinated Gd(III) complexes are a challenging topic from the viewpoint of a dynamic stereochemical analysis. There is overwhelming experimental evidence that open-chain and macrocyclic complexes of lanthanide ions show a dynamic stereochemical behavior in solution, which is characterized by the presence of easily interconvertible diastereomers (see Chapter 8) [41, 42].

The interaction of the open-chain polyaminopolycarboxylic acid DTPA with the Gd(III) ion generates a new center of asymmetry, thus leading to enantiomeric complexes. The X-ray structure determination of $Gd(DTPA)(H_2O)^{2-}$ clearly reveals the presence of both enantiomers in the asymmetric unit [43]. The Gd(III) ion is eightfold coordinated by the three nitrogen atoms of the backbone and the five oxygen atoms of the pendant acetate groups of the DTPA ligand. The ninth coordination site is occupied by the oxygen atom of a water molecule (see Figure 7.3).

Eight coordination sites are also provided by the cyclic polyaminopolycarboxylic acids DOTA and DOTMA, and as expected, the oxygen and nitrogen atoms of the ligands DO3A and DO3MA occupy seven coordination sites. In the DOTA-type complexes, the ninth coordination site is occupied by a water molecule, while in the DO3A-type complexes two water molecules are located at the eighth and ninth coordination sites (see Figure 7.4).

In Figure 7.5, a superimposition of the experimentally determined structures of the complex anions $Gd(DTPA)(H_2O)^{2-}$ [43] and $Gd(DOTA)(H_2O)^{-}$ [44, 45] is presented which clearly demonstrates the geometrical similarity between the two chelates. The coordination polyhedra around the central metal ions are best described as a mono-capped square antiprism (SAP). In the case of the DTPA-type ligands, four coordinating oxygen atoms of the four carboxylate groups are found in a square plane, while the three nitrogen atoms and the coordinating oxygen atom of the fifth carboxylate group build the other plane. Alternatively, the coordination polyhedron of the Gd(III) ion in $Gd(DTPA)(H_2O)^{2-}$

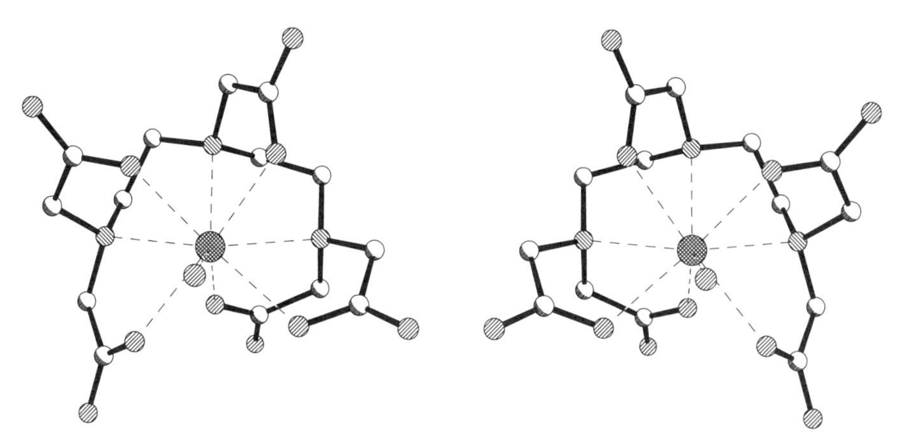

Figure 7.3 Enantiomers of $Gd(DTPA)(H_2O)^{2-}$ (top view); hydrogen atoms are omitted for clarity.

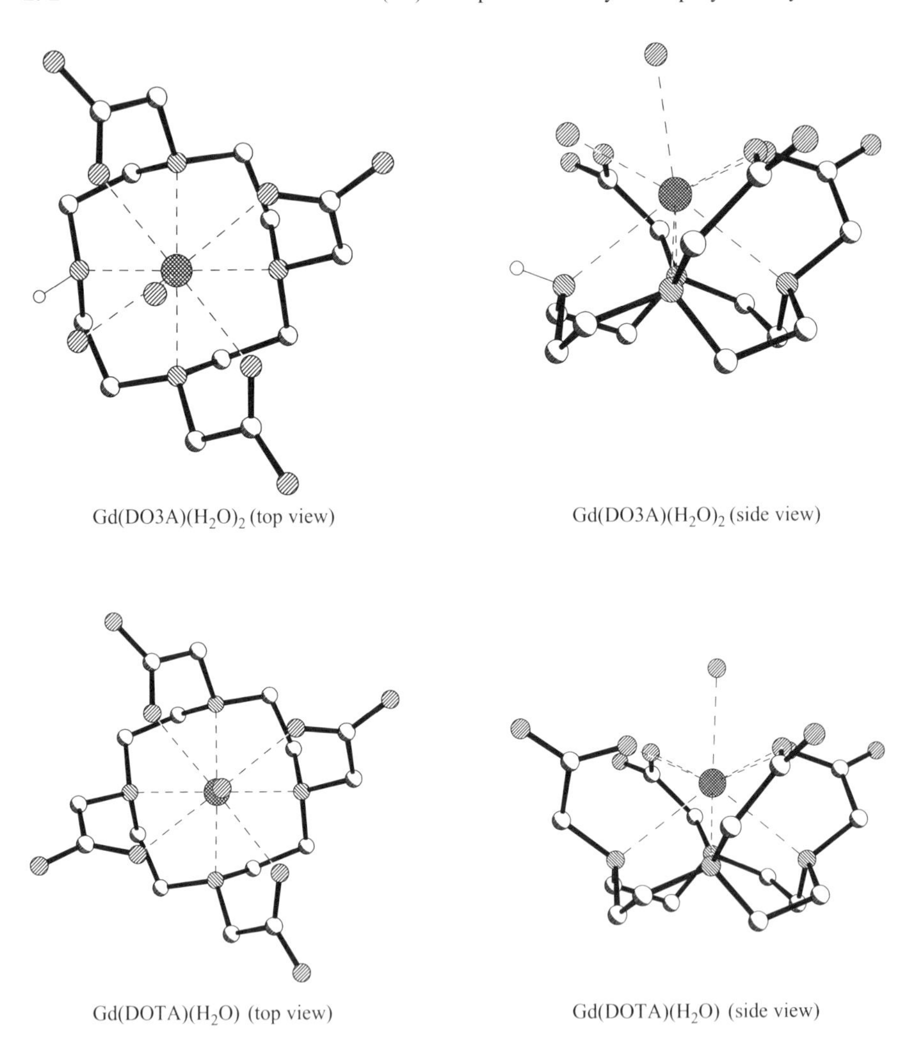

Figure 7.4 Top and side views of $Gd(DO3A)(H_2O)_2$ and $Gd(DOTA)(H_2O)^-$; hydrogen atoms are omitted for clarity, except for the N–H group.

was described as a tri-capped trigonal prism (TTP) [46]. This is a remarkable description because the three nitrogen atoms of DTPA do not occupy the symmetry-equivalent tri-capped positions of the TTP.

In the more symmetrical DOTA-type ligands, all of the coordinating nitrogen atoms are in one square plane and the oxygen atoms of the four acetate groups build the second plane of the SAP. Due to steric constraints, the tetraazamacrocycle in DOTA-type ligands adopts a square [3333] conformation [47]. In principle, for a given square [3333] conformation of the macrocyclic

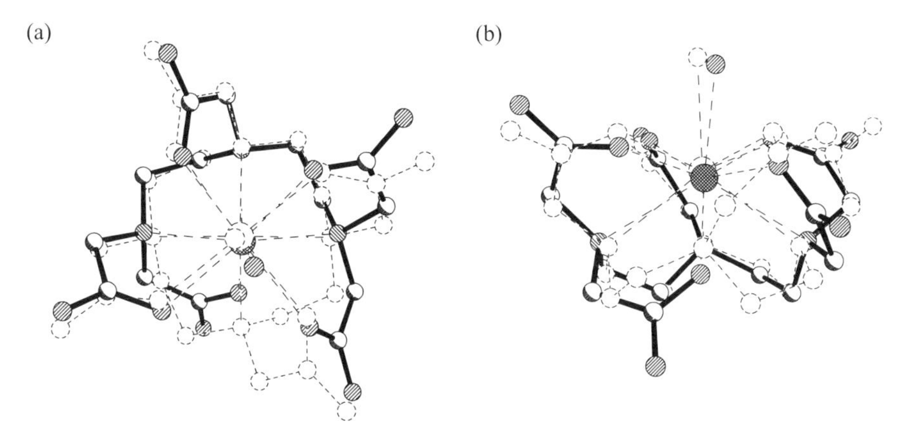

Figure 7.5 Superimposition of Gd(DTPA)$(H_2O)^{2-}$ and Gd(DOTA)$(H_2O)^-$, showing (a) top and (b) side views.

backbone, the SAP arrangement of the ligand in DOTA-type complexes gives rise to two different, i. e. clockwise and anticlockwise, orientations of the pendant acetate groups. The resulting isomers are diastereomers. However, the two resulting complexes differ not only in the orientation of the pendant acetate groups (see Figure 7.6(a)), but also in the height of the corresponding screw thread (see Figure 7.6(b)). In the X-ray structures of the complex anions $La(DOTA)^-$ [48] and $Gd(DOTA)^-$ [44, 45], the two different orientations of the screw were observed experimentally. Finally, the combination of the two possible orientations of the pendant acetate groups (Λ, Δ) [49] and those of the two square [3333] conformations ($\lambda\lambda\lambda\lambda$, $\delta\delta\delta\delta$) [50] of the tetraazamacrocycle gives rise to four stereoisomers. The enantiomers are the pairs [$\Lambda(\lambda\lambda\lambda\lambda)$, $\Delta(\delta\delta\delta\delta)$] and [$\Delta(\lambda\lambda\lambda\lambda)$, $\Lambda(\delta\delta\delta\delta)$], whereas the pairs [$\Delta(\lambda\lambda\lambda\lambda)$, $\Lambda(\lambda\lambda\lambda\lambda)$] and [$\Delta(\delta\delta\delta\delta)$, $\Lambda(\delta\delta\delta\delta)$] are diastereomers. Interconversion of the diastereomeric pairs [$\Lambda(\lambda\lambda\lambda\lambda)$, $\Delta(\lambda\lambda\lambda\lambda)$] and [$\Lambda(\delta\delta\delta\delta)$, $\Delta(\delta\delta\delta\delta)$] were observed experimentally, i. e. by NMR in aqueous solutions of Ln(DOTA) complexes (see Chapter 8) [41, 51].

In the crystalline state, the complex anion Gd(DOTA)$(H_2O)^-$ exists as the enantiomeric pair [$\Delta(\lambda\lambda\lambda\lambda)$, $\Lambda(\delta\delta\delta\delta)$] in the asymmetric unit [44, 45].

4.2 STRUCTURE AND ENERGETICS OF GD(III) COMPLEXES

4.2.1 Quantum Mechanical *ab initio* Methods

The same approach as in the case of $Gd(H_2O)_9^{3+}$ was used to describe Gd(III) complexes of polyaminopolycarboxylic acids in the gas phase [52]. The conformational properties of four anionic complexes, i.e. $Gd(DOTA)^-$,

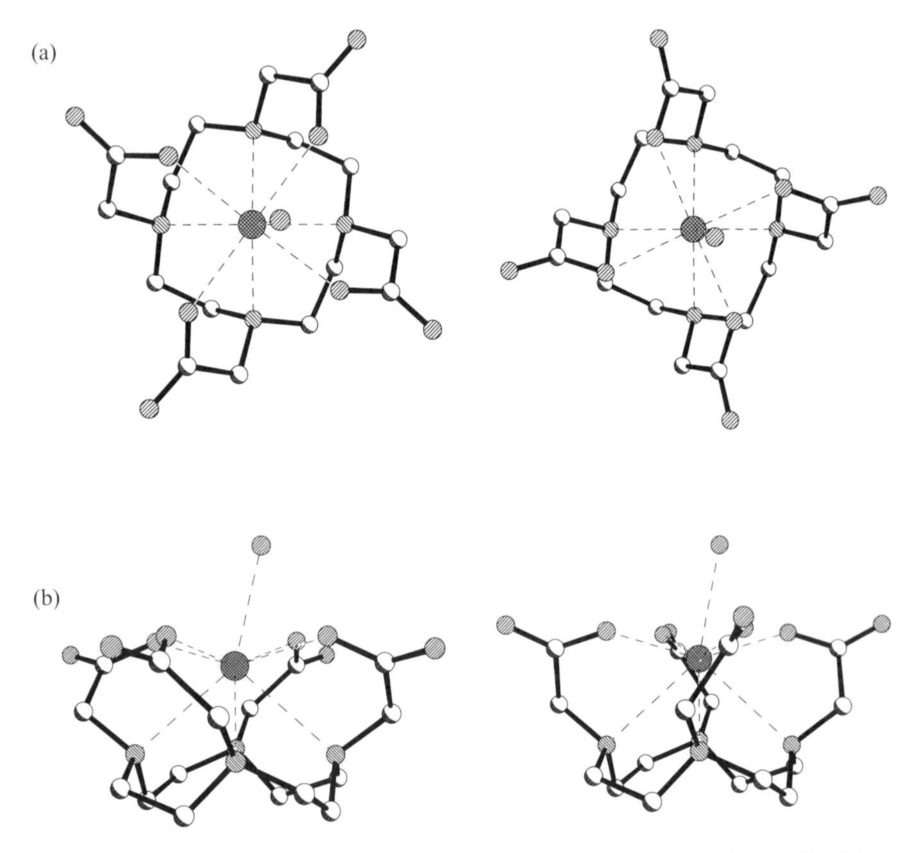

Figure 7.6 (a) Top and (b) side views of the diastereomeric pair [$\Delta(\delta\delta\delta\delta)$, $\Lambda(\delta\delta\delta\delta)$] of Gd(DOTA)($H_2O$)$^-$; hydrogen atoms are omitted for clarity.

Gd(DOTA)(H_2O)$^-$, Gd(DOTMA)(H_2O)$^-$ and Gd(DTPA)(H_2O)$^{2-}$, and of the two neutral complexes, Gd(DO3A)$(H_2O)_2$ and Gd(DO3MA)$(H_2O)_2$, were investigated at the RHF level of theory by using standard 3-21G and 6-31G(d) basis sets for the atoms of the ligand and an (1s-4d, $4f^7$) ECP with a contracted [5s4p3d] Gaussian-type orbital valence basis set for the gadolinium atom. The analysis of the conformational space by Cosentino *et al.* [52] was restricted to the calculation of stationary points on the corresponding PESs. The resulting stationary points were not further characterized by frequency calculations. In as much as experimental structures of the complexes were available, these data were used as starting points for the geometry optimizations. In the case of the water-containing complexes, no symmetry constraints were imposed on the geometry optimizations of the corresponding complexes, i.e. the calculations were performed in C_1 symmetry. However, the geometry optimizations of the water-free diastereomers, $\Lambda(\lambda\lambda\lambda\lambda)$ [Gd(DOTA)$^-$] and $\Delta(\lambda\lambda\lambda\lambda)$ [Gd(DOTA)$^-$] were performed in C_4 symmetry. The calculated differences of the total energies for $\Lambda(\lambda\lambda\lambda\lambda)$ and $\Delta(\lambda\lambda\lambda\lambda)$ $Gd(DOTA)^-$ at the RHF/3-21G, RHF/6-31G(d) and BLYP/6-31G(d) levels of theory are 3.72, 0.98 and 1.95 kcal/mol, respectively. The computationally demanding calculations of the two diastereomers at the BLYP/6-31G(d) density functional level of theory became only possible due to the C_4 symmetry of water-free $Gd(DOTA)^-$.

With respect to the calculated total energies for the diastereomers of $Gd(DOTA)(H_2O)^-$, it was shown that there is a clear preference for one diastereomer. In this case, $\Delta(\lambda\lambda\lambda\lambda)$ $Gd(DOTA)(H_2O)^-$ is energetically favored by at least 1.6 kcal/mol in comparison to $\Lambda(\lambda\lambda\lambda\lambda)$ $Gd(DOTA)(H_2O)^-$ at the RHF/6–31G(d) level of theory. Calculations with the smaller 3-21G basis set for the atoms of the ligand predict an energy difference of 4.7 kcal/mol.

To summarize, the calculated structures are in reasonable agreement with the experimentally observed ones in terms of the root-mean-square (RMS) deviation of the Cartesian coordinates of the atoms C, N, O and Gd, after superimposition of the optimized with the experimentally determined structure(s). For $Gd(DOTA)(H_2O)^-$, the superimposition with the available X-ray structures will be discussed in more detail. The structure optimization in C_1 symmetry at the RHF/3-21G and RHF/6-31G(d) levels of theory results in RMS deviations of 0.239 and 0.228 Å with respect to the X-ray data of Dudbost *et al.* [45]. Additionally, a geometry optimization of $Gd(DOTA)(H_2O)^-$ was performed in C_2 symmetry, which means that the plane of the water molecule at the ninth coordination site is perpendicular to the oxygen plane of the SAP. The optimized structure in C_2 symmetry shows RMS deviations of 0.198 and 0.219 Å at the RHF/6-31G(d) level of theory with respect to the available X-ray structures of Dudbost *et al.* [45] and Chang *et al.* [44], respectively. However, the RMS deviation between the two published X-ray structures of $Gd(DOTA)(H_2O)^-$ itself is 0.210 Å [53].

At the RHF/6-31G(d) level of theory, Gd–O distances of 2.366(30) and 2.573 Å are obtained for Gd–$O_{acetate}$ and Gd–O_{water}, respectively. The optimized Gd–N distance is 2.824(20) Å. These calculated interatomic distances at the RHF/6-31G(d) level of theory are in reasonable agreement with the experimentally observed distances for Gd–$O_{acetate}$ (2.362–2.463 Å), Gd–O_{water} (2.458–2.463 Å and Gd–N (2.645–2.689 Å [44, 45]. However, the Gd–N distances in the optimized structure of $Gd(DOTA)(H_2O)^-$ are significantly longer than in the experimentally determined structures, thus indicating deficiencies of the theoretical approach in the description of the Gd–N interaction in comparison to the Gd–O interaction. The origin of this discrepancy has not yet been elucidated. The introduction of a water molecule at the ninth coordination site of $Gd(DOTA)^-$ results in an elongation of the Gd–$O_{acetate}$ distances from 2.353 to 2.369(30) Å, and an even more pronounced lengthening of the Gd–N distances from 2.77 to 2.824(20) Å. Due to the interaction with the water molecule, the Gd(III) ion is moved from the nitrogen plane of the SAP in the direction of the oxygen plane and the ninth coordination site.

In addition, the optimized structures for Gd(DO3A)$(H_2O)_2$, Gd(DO3MA)$(H_2O)_2$ and Gd(DTPA)$(H_2O)^{2-}$ are in reasonable agreement with the experimental findings for these complexes. The optimized structures of the respective isomers thus obtained show RMS deviations of 0.292, 0.260 and 0.377 Å at the RHF/3-21G level of theory, whereas RMS values of 0.232, 0.229 and 0.291 Å are reported for the calculated structures at the RHF/6–31G(d) level of theory [52].

For the DOTA-type structures optimized in C_1 symmetry, a major geomerical deviation is observed concerning the orientation of the plane of the water molecule. Whereas the water plane in the available X-ray structures is typically perpendicular to the plane of the four oxygen atoms, and the hydrogen atoms of the coordinating water molecule are pointing away from the metal ion, the plane of the water molecule in the calculated structures is nearly parallel to the plane of the four oxygen atoms. The hydrogen atoms of the coordinating water molecule are pointing toward the oxygen atoms of two carboxylate groups, thereby establishing a hydrogen bond.

In conclusion, despite the greater computational effort when using the RHF/6-31G(d) model, the smaller RMS deviations in comparison to the results at the RHF/3–21G level of theory justify these computational expenses. Additionally, a comparison of the differences in total energies for diastereomeric pairs with the available experimental data gives a clear preference for the results of calculations with the larger 6–31G(d) basis sets for the atoms of the ligand.

4.2.2 Molecular Mechanics Calculations

In order to develop force-field parameters for the gadolinium atom in the Tripos atom force field as part of the SYBYL modeling package [54], the optimized structures of Gd(DOTA)$(H_2O)^-$ at the RHF/3-21G and the RHF/6-31G(d) levels of theory were taken by Cosentino *et al.* as a starting point for a systematic variation of the Gd–N and Gd–O bond distances and the Gd–N–C and Gd–O–C bond angles [52]. This was achieved by calculating the total energy as well as its first derivatives with respect to the Cartesian coordinates of the corresponding atoms for arbitrarily distorted Gd(DOTA)$(H_2O)^-$ structures. The Gd(III) ion was moved inside the frozen coordination cage of the complex. The randomly distorted geometries, together with the corresponding total energies and the first derivatives thereof, were used to determine the best set of force-field parameters for the gadolinium atom. This set is defined by the minimum of the sum of squares of the deviations between the *ab initio* and the Tripos force-field quantities, i. e. energy and gradient. This type of optimization problem belongs to the class of nonlinear least-squares problems and in general many different solutions exist. The parameter sets thus obtained show a strong dependency on the level of theory for both of the equilibrium bond distances Gd–$O_{acetate}$ and Gd–N and for the corresponding force constants. There is only a minor dependency on the molecular electrostatic potential which can be calculated from the electronic wavefunction.

In more detail, values of 2.628 and 2.310 Å were obtained for the equilibrium bond distances Gd–N and Gd–$O_{acetate}$, respectively, as well as 59.7 and 128.7 kcal/(mol Å^2) for the corresponding force constants from the *ab initio* data at the RHF/6-31G(d) level of theory, while taking into account the electrostatic contributions. For Gd–O_{water}, a bond distance of 2.495 Å and a surprisingly large value of 108.1 kcal/(mol Å^2) for the corresponding force constant results from the solution of the nonlinear least-squares problem. In order to achieve a better agreement between the experimentally determined and the calculated force-field derived structures, it was necessary to adjust some of the metal-independent parameters of the Tripos atom force field, especially the C–C bond distances and their corresponding force constants.

Finally, using the modified Tripos atom force field extended by the parameters described above, molecular mechanics calculations of the structures and energetics of the anionic complexes $Gd(DOTA)(H_2O)^-$, $Gd(DOTMA)(H_2O)^-$ and $Gd(DTPA)(H_2O)^{2-}$, and the two neutral complexes $Gd(DO3A)(H_2O)_2$ and $Gd(DO3MA)(H_2O)_2$, were performed. The force-field optimized structure of $Gd(DOTA)(H_2O)^-$ shows RMS deviations of 0.237 and 0.109 Å with respect to the experimentally determined structure and the structure optimized at the RHF/6-31G(d) level of theory. In general, the RMS deviations between force-field derived structures and optimized structures at the RHF/6-31G(d) level of theory are smaller than the RMS deviation between the force-field derived structures and the experimentally observed data. Calculated energy differences between diastereomers of DOTA-type complexes, either from force-field potential energies or from the corresponding *ab initio* total energies, show a RMS deviation of 0.78 kcal/mol with a maximum deviation of 1.52 kcal/mol. Neglecting the electrostatic contributions, a RMS deviation of 0.85 kcal/mol was obtained [52].

In order to develop force-field parameters for gadolinium and thulium atoms consistent with the CHARMM force field [55], the electrostatic potentials of $Gd(DOTA)^-$ and $Tm(DOTP)^{5-}$ at the RHF and B3LYP levels of density functional theory were determined [56]. The electrostatic potential was fitted to atom centered point charges. The atoms of the ligands were described by the 6-31G(d) basis set and the lanthanides by an (1s-4d, $4f^n$) ECP with a contracted [5s4p3d] Gaussian type orbital valence basis set. The calculations on the two diastereomers of the DOTA-type complexes were performed in C_4 symmetry. Alternatively for $Gd(DOTA)^-$, the 4f electrons were treated explicitly by using the ECP28MWB basis set. The derived atom centered point charges for $Gd(DOTA)^-$ show no significant dependency, neither on the level of theory, nor on the basis sets of the lanthanide atoms. The force constants of the metal–ligand parameters were empirically determined by trial-and-error in order to reproduce the available experimentally determined X-ray structures. Values of 2.665 Å and 95 kcal/(mol Å^2) were chosen for the Gd–O bond distance and force constant, while 2.365 Å and 105 kcal/(mol Å^2) were chosen for the Gd–N bond distance and force constant. In contrast to the force-field parametrizations discussed so far, ligand–metal–ligand interactions were explicitly taken into account, and values were assigned to the angle-bending terms N–Gd–N, N–Gd–O, and O–Gd–O. An equilibrium value of 100° and a force constant of 30 kcal/(mol rad^2) were chosen for all three angle-bending terms. The force field thus modified was capable of reproducing the experimentally determined structures of Gd(DOTA)(H_2O)$^-$ and Tm(DOTP)$^{5-}$. RMS deviations of 0.183 and 0.234 Å were obtained. However, the energy difference for $\Lambda(\lambda\lambda\lambda\lambda)$ and $\Delta(\lambda\lambda\lambda\lambda)$ Gd(DOTA)$^-$ was predicted to be 10.6 kcal/mol.

The structures and energetics of Gd(DOTAM) were calculated by Hancock and co-workers by applying the program SYBYL, using the Tripos atom force field. In a first attempt, arbitrarily chosen force constants of 200 kcal/(mol Å^2) were used for the Gd–N and Gd–O bonds [57]. The bending constants which describe the variation of the angles N–Gd–N, N–Gd–O and O–Gd–O were set to zero, and the geometry around the metal ion was mainly determined by van der Waals repulsion between the donor atoms and the forces within the ligand. A value of 5×10^{-3} kcal/(mol deg^2) was assigned to the Gd–O–C and Gd–N–C angle-bending constants; the corresponding bond angles were assumed to be tetrahedral, i.e. 109.5°. The selectivity of the ligand DOTAM for Gd(III) ions and other different metal ions, i.e. Zn(II), Cd(II) and Ca(II), was discussed in terms of coordinating properties of the amide oxygen donor and geometric requirements of the ligand.

4.2.3 Calculation of Stability Constants

In a second attempt, force-field parameters for gadolinium were developed from twelve experimentally determined structures of gadolinium-containing complexes: Gd(EDTA), Gd(DTPA-BEA), Gd(DOTA), Gd(DOTA-OH), Gd(DO3MA), Gd(HP-DO3A), Gd(HAM), Gd(18-CROWN-6), Gd(DO3A), Gd(BOPTA), Gd(DTPA-pn) and Gd(DTPA-en) (see Figure 7.7) [58]. It was demonstrated that the calculated structures are strongly influenced by the values of the Gd–O and Gd–N bond stretch parameters. Values of 2.965 and 2.420 Å were chosen for the Gd–N and Gd–$O_{acetate}$ distances, respectively, together with a force constant of 100 kcal/(mol Å^2) for both bond types. A value of 2.397 Å for the equilibrium distance and a force constant of 100 kcal/(mol Å^2) were assigned to the Gd–O_{water} bond. These parameters determine the bond length of the force-field optimized equilibrium structures as well as the ease with which the corresponding bond can be stretched. This parameter set for the gadolinium atom gives an average RMS deviation of 0.3624 Å of the Cartesian coordinates of the atoms C, N, O and Gd after superimposition of the calculated structures with the experimental data. In the force-field calculations, the effects of electrostatics were ignored, so that the structures of the complexes

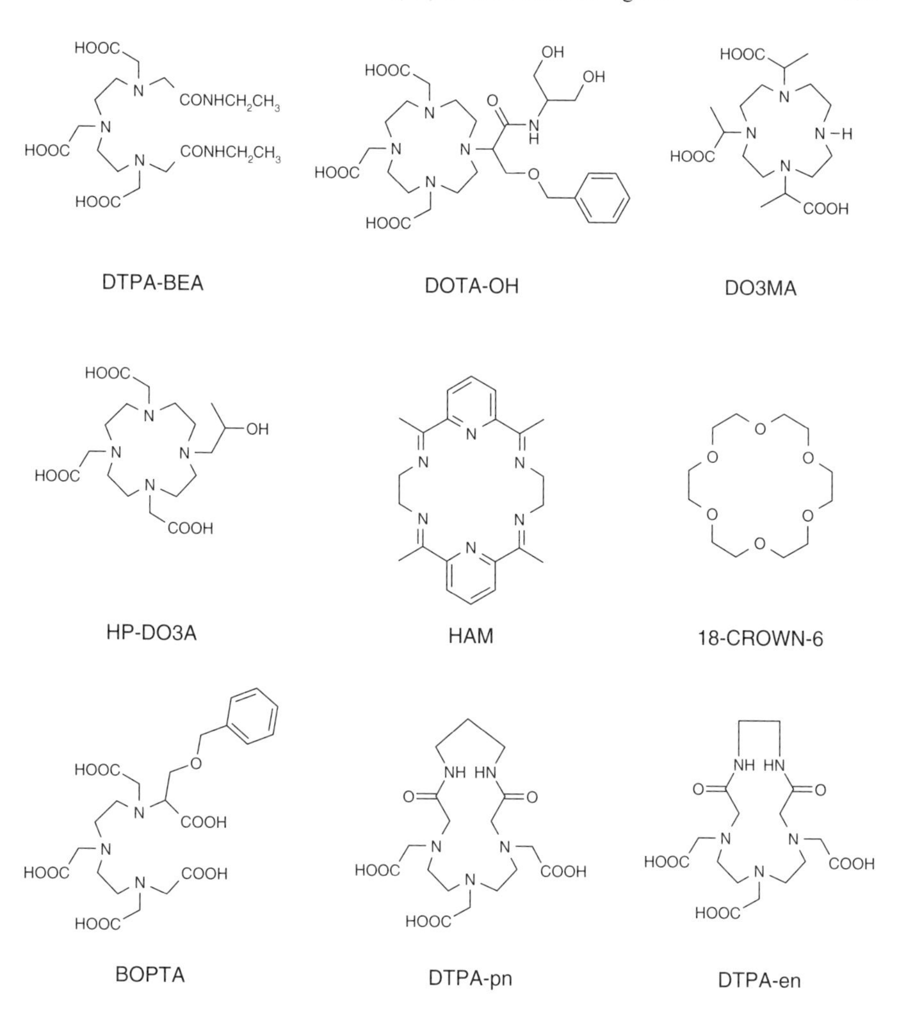

Figure 7.7 Chemical structures of various ligands used to complex bd (III ions).

are solely determined by steric effects. Consequently, the addition of a water molecule to a gadolinium-containing complex leads only to an increased steric interaction between the atoms of the complex and the atoms of the water molecule. Potential energies of the optimized structures obtained from the force-field calculations were used to predict complex stabilities. In parituclar, the potential energy differences between the ninefold coordinated complex containing an inner-sphere water molecule and the water-free eightfold coordinated complex, ΔE_{coord}, correlate with the experimentally observed stability constants, log K, of the corresponding complexes in water. A correlation between log K and ΔE_{coord} was noted for eightfold coordinating ligands, independent of the nature of the ligand, i.e. macrocyclic or linear, and independent of the nature of the ligating group, i.e. $-CO_2^-$, $-CO_2R$, $-CH_2OH$, $-CH_2OR$, $-C(O)NHR$ and $-C(O)NR^1R^2$. The excellent correlation between log K and ΔE_{coord}, which was reported with a correlation coefficient of $r^2 = 0.91$ by Reichert *et al.* is most probably only valid for the complexes carrying exclusively coordinating acetate groups [58].

In order to describe the molecular structures, energetics and dynamics of polyaminopolycarboxylate complexes of Gd(III) ions, Fossheim and co-workers performed molecular mechanics and molecular dynamics simulations of these complexes [59–62]. Electrostatic interactions were taken into account, and the corresponding molecular electrostatic potentials and the atomic point charges were taken from *ab initio* calculations at the HF/STO-3G level of theory. The complexes $Gd(DTPA)^{2-}$, $Gd(DOTA)^-$, Gd(DO3A), and its oxygen analogue Gd(OTTA), in which the NH group is replaced by an O atom, were investigated [59].

In a first step, geometry optimizations were performed in the gas phase, ignoring the electrostatic interaction terms, followed by a calculation of the atomic point charges. Taking these point charges into account, the geometry optimizations were continued until convergence was achieved. Molecular dynamic simulations were performed over a 10^{-11} s time-interval at a temperature of 300 K. The three structures of lowest potential energy were selected and used as starting points for geometry optimizations. Hydration effects were taken into account by submerging the complexes in a layer of water of approximately 8 Å thickness, followed by energy minimizations. One coordinated water molecule was obtained for the complexes $Gd(DTPA)^{2-}$ and $Gd(DOTA)^-$, whereas two coordinated water molecules were obtained for Gd(DO3A) and Gd(OTTA). The optimized structures of the complexes $Gd(DTPA)(H_2O)^{2-}$, $Gd(DOTA)(H_2O)^-$ and $Gd(DO3A)(H_2O)$ are in agreement with the experimental findings. The calculated Gd–$O_{acetate}$ distances for the three complexes are 2.408, 2.404 and 2.398 Å, whereby the corresponding experimental findings are in the range 2.363–2.437, 2.362–2.463, and 2.311–2.392 Å. The calculated Gd–N distances are 2.766, 2.683 and 2.662 Å. These are also in agreement with the experimentally determined distances, which are in the range of 2.629–2.728, 2.645–2.689, and 2.58–2.66 Å [46].

In addition to the calculation of geometry parameters, it is a great challenge for any theoretical method to predict complex stabilities from calculated reaction energies. Indeed, complex stabilities were derived from the calculated reaction energies in vacuum, $E_{R,VA}$, and in an aqueous environment [59]. First, the reaction energy in vacuum can be calculated according to the following:

$$E_{R,VA} = E_{GdL} - E_L \tag{7.1}$$

where E_{GdL} is the potential energy of the complex and E_L is the potential energy of the free ligand in the conformation of lowest energy. Alternatively, $E_{R,VA}$ can also be expressed as follows:

$$E_{R,VA} = E_I + E_{DL} \tag{7.2}$$

where E_I is the interaction energy between the Gd(III) ion and the ligand in its complex conformation, and E_{DL} is the ligand constraint energy which is defined as the difference in energy between the energies of the ligand in the conformation of the complex and the energy of the conformation of lowest energy.

Finally, the reaction energy in aqueous solution, $E_{R,AQ}$, can be calculated by taking two hydration effects into account, i.e. the hydration of the donor groups, E_{H1}, and the hydration of the Gd(III) ion by occupation of the vacant coordination sites in the complex by one or more water molecules, E_{H2}, as follows:

$$E_{R,AQ} = E_{R,VA} + E_{H1} + E_{H2}. \tag{7.3}$$

From these calculated reaction energies, the following order of complex stabilities was predicted: $Gd(DTPA)^{2-} > Gd(DOTA)^{-} \gg Gd(DO3A) >$ Gd(OTTA) [59].

However, the experimentally determined stability constants (log K) of these complexes in water (25°C and $\mu = 0.1$) are as follows: $Gd(DTPA)^{2-}$, 22.39; $Gd(DOTA)^{-}$, 24.0; Gd(DO3A), 21.0 [63].

More recently, a related approach was used by Fossheim *et al.* to examine the molecular structures and stabilities of Gd(III) complexes of nine cyclic and acyclic multidentate ligands in order to predict the thermodynamic stability constants of the corresponding complexes [61]. The DTPA-type ligands, DTPA-TRANS, DTPA-CIS, DTPA-BMA, DTPA-HMA, DTPA-BMPA and DTPA-HB, were investigated together with the cyclic ligands, NOTA, TETA and PHEA (see Figure 7.8). For the nine Gd(III) complexes investigated, a linear relationship was established between E_I and E_{DL} and a correlation coefficient of $r^2 = 0.71$ was reported. Analogously, a linear correlation between log K and $E_{R,AQ}$ was reported (correlation coefficient of $r^2 = 0.86$) and the following order of complex stabilities was predicted from the calculated $E_{R,AQ}$ values [61]: $Gd(DOTA)^{-} > Gd(DTPA)^{2-} >$ Gd(DO3A) $>$ Gd(DTPA-BMA) $>$ Gd(NOTA). In this case, the prediction is in line with the available experimentally determined stability constants (log K) of the complexes in water (25°C and $\mu = 0.1$): $Gd(DOTA)^{-}$, 24.0; $Gd(DTPA)^{2-}$, 22.4; Gd(DO3A), 21.0; Gd(DTPA-BMA), 16.85; Gd(NOTA), 13.7 [46, 63].

Even the molecular structures, energies and dynamics of the piperazine derivative 2, 3, 5, 6-tetrakis[*N*, *N*-bis(carboxymethyl)aminomethyl]piperazine-*N*, *N*′-diacetic acid (DIMER A), and the two isomeric cyclohexane derivatives 1, 5-bis[1, 1, 4-tris(carboxymethyl)-1, 4-diazabutyl]-2, 4-bis{[bis(carboxy-methyl)]amino}cyclohexane (DIMER B) and 1, 4-bis[1, 1, 4-tris(carboxy-methyl)-1, 4-diazabutyl]-2, 5-bis{[bis(carboxymethyl)]amino}cyclohexane (DIMER C) (see Figure 7.9) and their di-gadolinium complexes were investigated by molecular mechanics calculations and molecular dynamics simulations [60]. From the computational results, it was predicted that the piperazine

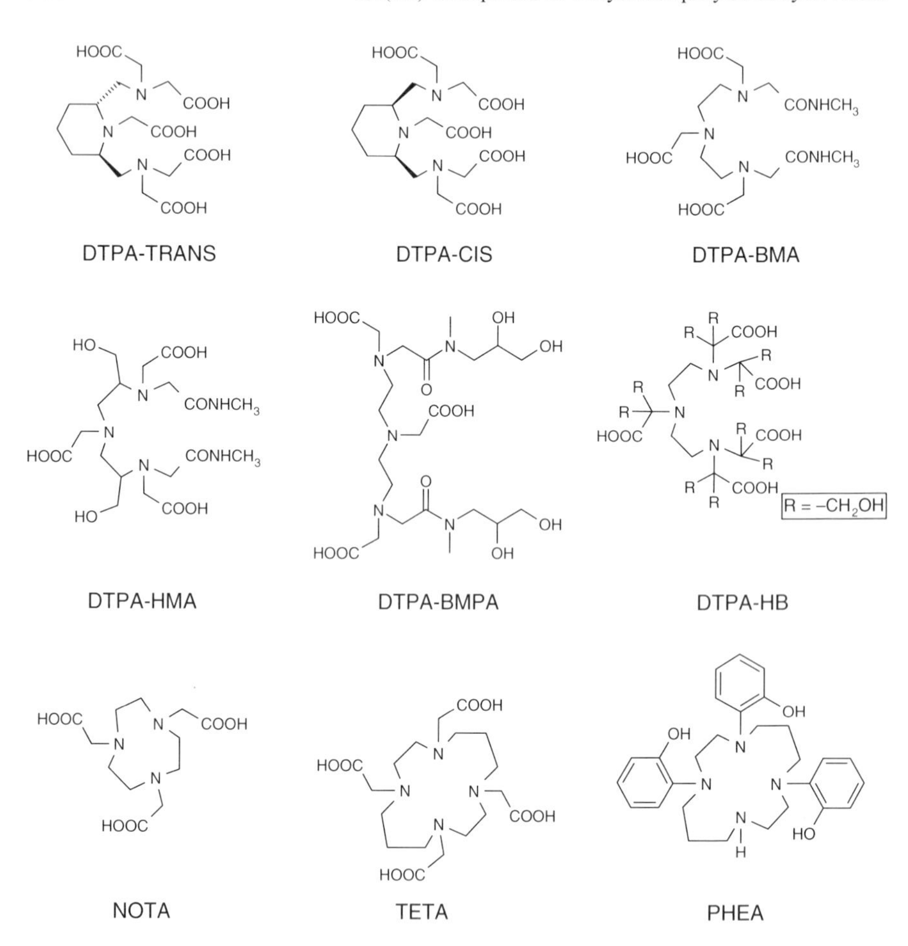

Figure 7.8 Chemical structures of various DTPA-type and cyclic ligands.

derivative would form the most stable binuclear complex among the three ligands. For the (*RSRS*)-, (*RRSS*)-, (*RRRS*)-, (*SRRS*)- and (*SSSS*)-isomers of DIMER A, similar reaction energies were calculated in aqueous solution, with $E_{R,AQ}$ ranging from −1111 to −1108 kcal/mol. Whereas for the di-gadolinium complexes of DIMER B and DIMER C, $E_{R,AQ}$ values in the range from −1099 to −1091 kcal/mol were obtained. The reaction energies of the (*SSRR*)-, (*RSSR*)- and (*SSSR*)-diastereomers of the two cyclohexane derivatives DIMER B and DIMER C were of similar magnitude. Initial model building by using the Tripos force field indicated that the (*SRSR*)- and the (*SSSS*)-diastereomers of the cyclohexane derivatives are less likely to form stable complexes, either due to steric crowding or to poor coordination properties.

In order to investigate the effects of the introduction of four hydroxymethyl, $-CH_2OH$, groups at the cyclen ring of DOTA on structure and stability, four

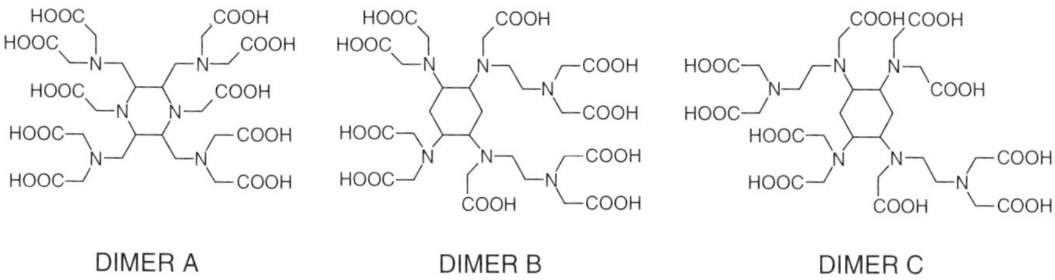

DIMER A DIMER B DIMER C

Figure 7.9 Chemical structures of the DIMERS A, B and C.

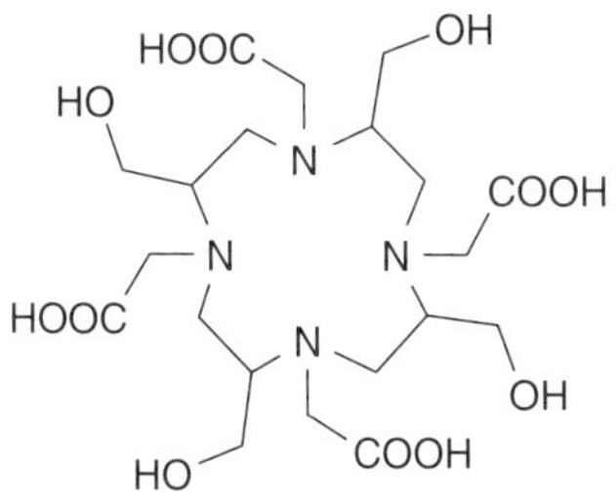

DOTA-THM

Figure 7.10 Chemical structure of DOTA-THM.

isomers of 1, 4, 7, 10-tetraaza-2, 5, 8, 11-tetra(hydroxymethyl)-cyclododecane-*N*, *N′*, *N″*, *N‴*-tetraacetic acid, (DOTA-THM), and their Gd(III) complexes were investigated (see Figure 7.10) [62]. The introduction of the hydroxymethyl substituent generates a chiral center and four diastereomers, i.e. (*RRRR*), (*RRRS*), (*RRSS*), and (*RSRS*), have to be taken into account for DOTA-THM. Combining them with the diastereomers $\Lambda(\lambda\lambda\lambda\lambda)$ and $\Delta(\lambda\lambda\lambda\lambda)$ of Gd(DOTA) generates a total of eight diastereomers for Gd(DOTA-THM). The calculated reaction energies in vacuum, $E_{R,VA}$, and the reaction energies in aqueous phase, $E_{R,AQ}$, suggest the following order for the complex stabilities: $Gd(DOTA)^- >$ (*RRRR*)-Gd(DOTA-THM)$^-\approx$(*RRRS*)-Gd(DOTA-THM)$^- >$ (*RRSS*)-Gd(DOTA-THM)$^-$ $\approx$(*RSRS*)-Gd(DOTA-THM)$^-$. Analyzing the different calculated quantities shows that the order of stabilities is mainly determined by the differences in the ligand constraint energy, E_{DL}. However, to the best of our knowledge, there are no experimentally determined data available yet to confirm the predicted order of complex stabilities.

For the ligands NOTA, DO3A and DOTA, Kumar and Tweedle were able to show a perfect linear correlation between the calculated ligand constraint energy E_{DL} and the activation free energy of reorganization of the transient species, $Gd(HNOTA)^+$, $Gd(HDO3A)^+$ and Gd(HDOTA), which are intermediates in the formation of the stable complexes Gd(NOTA), Gd(DO3A) and $Gd(DOTA)^-$ [64].

5 FORMATION OF Y(DOTA)$^-$

A theoretical investigation concerning the mechanism and the energetics of the complexation of Y(III) ions with DOTA was reported very recently in a detailed study by Jang *et al.* [65]. In view of the similarities of the coordination chemistry of the Y(III) and Gd(III) ions, this investigation gives some insight into the complexation reaction of Ln(III) ions with H_4DOTA. Geometry optimizations at the RHF level of theory were performed by using an ECP for yttrium [9], so that only the eleven valence electrons ($4s^24p^65s^24d^1$) and the 6-31G(d) basis set for the atoms of the ligand had to be treated explicitly. The stationary points located on the PES were further characterized by frequency calculations by using the approximate Hesse matrices which were obtained during the corresponding geometry optimizations. In an extension of the *ab initio* approaches reported so far, interactions with solvent molecules, especially water, were included in the calculation via the continuum-solvation approach [66] by solving the Poisson–Boltzmann equation numerically [67]. This approach can be summarized as follows. The solute is described as a low-dielectric cavity ($\epsilon = 1$) immersed in a high-dielectric solvent ($\epsilon = 80$ for water). The boundary between the solute and the solvent was defined as the surface that is generated by rolling a sphere of 1.4 Å (water radius) over the van der Waals envelope of the solute. The charge distribution of the solute is represented by a set of atom-centered point charges which were fitted to the electrostatic potential calculated from the electronic wavefunction. On the basis of the resulting charge distribution, the Poisson–Boltzmann equation is solved to obtain the reaction field of the solvent as a set of polarization charges located on the solute–solvent boundary surface. Then the reaction field is included in the RHF calculations and a new electronic wavefunction, as well as a new set of atom-centered point charges, is obtained. This procedure has to be repeated until self-consistent convergence is achieved to give the electrostatic contribution to the solvation energy. An additional contribution is accounted for by the introduction of a term proportional to the solvent-accessible surface area of the solute.

As a result, a van der Waals radius of 1.807 Å for the Y(III) ion leads to a solvation energy of −826.1 kcal/mol in water, in excellent agreement with the experimentally determined free energy of solvation of −826.2 kcal/mol [68]. For the DOTA ligand, van der Waals radii of 1.9 Å for C, 1.6 Å for both O and N, and 1.15 Å for H were used.

The geometry optimizations at the RHF/6–31G(d) level of theory for the diastereomers $\Lambda(\lambda\lambda\lambda\lambda)$ Y(DOTA)$^-$ and $\Delta(\lambda\lambda\lambda\lambda)$ Y(DOTA)$^-$ were performed in C_4 symmetry. In the gas phase ($\epsilon = 1$), the difference in total energies between the two optimized structures is 1.6 kcal/mol, whereas in aqueous solution ($\epsilon = 80$) the difference is predicted to be −0.9 kcal/mol. From the calculated differences of total energies it is predicted that the enantiomeric pair of lowest energy in the gas phase is [$\Delta(\lambda\lambda\lambda\lambda)$, $\Lambda(\delta\delta\delta\delta)$] Y(DOTA)$^-$ and that the enantiomeric pair of lowest energy in aqueous solution should be [$\Lambda(\lambda\lambda\lambda\lambda)$, $\Delta(\delta\delta\delta\delta)$] $Y(DOTA)^-$. However, the analysis of NMR experiments of aqueous solutions of $Y(DOTA)^-$ does not agree with the theoretical prediction that approximately 77% of the $Y(DOTA)^-$ exists as [$\Lambda(\lambda\lambda\lambda\lambda)$, $\Delta(\delta\delta\delta\delta)$], under the assumption that the difference in free enthalpies between these two species corresponds to the difference in total energies. In contrast, the analysis of the NMR data reveals that exclusively [$\Delta(\lambda\lambda\lambda\lambda)$, $\Lambda(\delta\delta\delta\delta)$] $Y(DOTA)(H_2O)^-$ exists in aqueous solution [51, 69]. Additionally, the X-ray structure determination of $Na[Y(DOTA)H_2O].4H_2O$ demonstrates the presence of the enantiomeric pair [$\Delta(\lambda\lambda\lambda\lambda)$, $\Lambda(\delta\delta\delta\delta)$] in the asymmetric unit [44, 70].

As in the case of the optimized structure in the corresponding $Gd(DOTA)(H_2O)^-$ complex, the interatomic metal oxygen distance, Y–$O_{acetate}$, of 2.272 Å is in reasonable agreement with the mean interatomic distances of 2.326(5) Å from the X-ray structure determination; again, the optimized metal–nitrogen distance of 2.725 Å is too long compared to the mean experimentally determined interatomic Y–N distance of 2.6545(18) Å [44, 70].

In order to describe the slow formation process of $Y(DOTA)^-$ in the reaction between Y(III) ions and H_4DOTA under physiological pH conditions [69, 71, 72], the structures of the possible intermediates $Y(H_2DOTA)^+$ and Y(HDOTA) were investigated for the two possible diastereomeric orientations of the pendant acetate groups of $Y(DOTA)^-$. Furthermore, different structural isomers of the protonated complexes were taken into account during the geometry optimizations. The structure of the isomer of lowest energy of the double-protonated complex $Y(H_2DOTA)^+$ reveals that the Y(III) ion is only fourfold coordinated by the four acetate oxygens of the ligand. The two protons are located at two nitrogen atoms of the tetraazamacrocycle in a *trans*-arrangement, pointing inwards to the direction of the Y(III) ion. The resulting Y–N distances in the two diastereomers, Δ and $\wedge$, are greater than 3.8 Å, while the two Y–H distances are approximately 3.04 Å. Structural isomers that have an even smaller coordination number, i.e. a two- or threefold coordination of the Y(III) ion by acetate oxygens of the DOTA ligand, are 39.9 and 109.8 kcal/mol higher in energy for the optimized structures in the gas phase ($\epsilon = 1$), and are still 17.8 and 37.9 kcal/mol higher for the optimized structures in aqueous solution ($\epsilon = 80$) than the fourfold-coordinated $Y(H_2DOTA)^+$, respectively. However, no protonated structures other than these *trans*-arrangements mentioned above were taken into account by Jang *et al.* for the description of $Y(H_2DOTA)^+$ [65].

Upon removal of one proton from the ring nitrogen of $Y(H_2DOTA)^+$ to form Y(HDOTA), the optimized structure indicates that the Y(III) ion is moved toward the macrocycle and interacts with the nitrogen atom which is *trans* to the protonated ring nitrogen, thus establishing a fivefold coordination of the metal ion with one Y–N distance of 2.74 Å and four Y–O distances in the range 2.175–2.278 Å. An extended isomer of Y(HDOTA), in which the Y(III) ion is fourfold coordinated by one ring nitrogen and three oxygen atoms of three acetate groups, lies 27.4 kcal/mol higher in energy (in the gas phase) and 25.4 kcal/mol higher in energy (in aqueous solution) than the fivefold-coordinated isomer. In order to explain the deprotonation step of Y(HDOTA) towards $Y(DOTA)^-$, which is known to be very slow in aqueous solution according to kinetic studies [72], an intramolecular proton transfer from the ring nitrogen atom to the oxygen atom of an acetate group has to be proposed. The 'buried' proton in Y(HDOTA) is thereby exposed to the surface of the complex. The transition structure for the intramolecular proton transfer was located, starting from the fivefold-coordinated structure of Y(HDOTA) in which one ring nitrogen atom is protonated, while the hydrogen atom is pointing inwards to the Y(III) ion. In the optimized structure, the metal hydrogen distance Y–H is 2.76 Å. The position of the proton is relatively close to the position of the one $O_{acetate}$ atom which belongs to the pendant acetate arm bound to the protonated ring nitrogen. The N–$O_{acetate}$ distance in the optimized structure is 2.5 Å and the corresponding H–$O_{acetate}$ distance is 1.8 Å. The N–H–$O_{acetate}$ bond angle is 124°, which is a favorable arrangement for an intramolecular hydrogen transfer between the N and O atoms. The resulting eightfold-coordinated isomer of Y(HDOTA) containing one protonated acetate group is 29.8 kcal/mol in the gas phase and 5.6 kcal/mol more stable in aqueous solution than the isomer of lowest energy for the fivefold-coordinated Y(HDOTA), in which the proton is located at one nitrogen atom of the macrocyclic ring. In the gas phase, the energy of the transition structure is 9.2 kcal/mol higher, and in aqueous solution 20.7 kcal/mol above the fivefold-coordinated isomer of Y(HDOTA). Upon inclusion of electron correlation by single-point calculations at the B3LYP density functional level of theory at the geometries of the RHF/6–31G(d) optimized structures, the relative energies of the transition structure are lowered to 2.0 kcal/mol in the gas phase and to 12.2 kcal/mol in the aqueous phase, respectively. At the B3LYP level of theory, the differences in the total energies between the optimized structures for the eight- and fivefold-coordinated Y(HDOTA) isomers were calculated to be −25.2 kcal/mol in the gas phase and −8.8 kcal/mol in aqueous solution, respectively.

The calculated energy barrier was corrected by the change in zero-point energy, and the thermodynamic functions $(-TS)_{298\,K}$ and $\Delta H_{0\rightarrow 298\,K}$ were obtained from the vibrational analysis by using the approximate Hesse matrix as described above. This leads to an activation free energy of 8.4 kcal/mol at 25 °C in aqueous solution at the B3LYP/6–31G(d)//RHF/6–31G(d) level of theory. This is in good agreement with the available experimental activation free energies as estimated from the rate constants for DOTA complexes, i.e. Eu(III) 8.1 kcal/mol, Gd(III) 8.2 kcal/mol and Ce(III) 9.3 kcal/mol [64, 73, 74].

To summarize, Jang *et al.* proposed the following reaction pathway for the formation of $Y(DOTA)^-$ from Y(III) ions and H_2DOTA^{2-} under physiological pH conditions. First, the formation of a fivefold-coordinated isomer of Y(HDOTA), which is characterized by an intramolecular hydrogen bond,

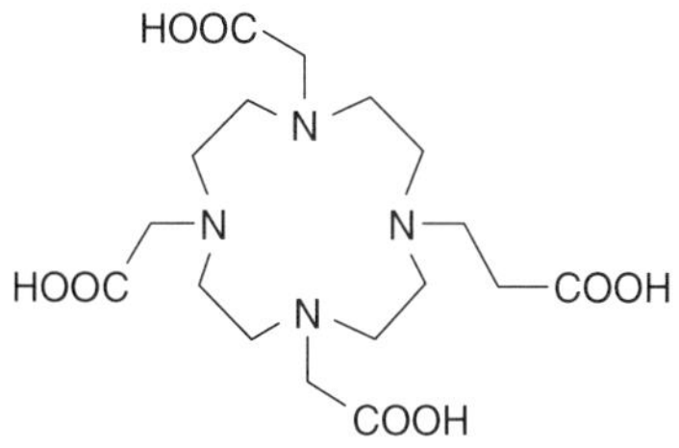

Figure 7.11 Chemical structure of the ligand DO3A1Pr.

followed by an intramolecular proton transfer from the ring nitrogen to the adjacent oxygen atom of the acetate group, leading to an eightfold-coordinated isomer of Y(HDOTA) in which the proton is exposed to the solvent and easily deprotonated [65].

In order to lower the barrier of the intramolecular hydrogen transfer, and thereby enhancing the formation rate of the corresponding yttrium complex, Jang *et al.* suggested replacing one of the acetate groups with a propionate group, to give the ligand DO3A1Pr (see Figure 7.11). In aqueous solution, the calculated difference of total energies of the optimized structures, with respect to the fivefold-coordinated isomer of Y(HDO3AlPr) for the transition structure of the intramolecular proton transfer, is 4.5 kcal/mol and -7.8 kcal/mol for the eightfold-coordinated isomer of Y(HDO3A1Pr). The estimated free energy of activation is 4.5 kcal/mol, which is significantly lower than the 8.4 kcal/mol for Y(HDOTA). A possible explanation for this difference in energies is the greater flexibility of the propionate pendant arm which enables a closer contact between the hydrogen atom localized at the ring nitrogen and the oxygen atom of the propionate, thereby lowering the barrier of the intramolecular hydrogen transfer [65]. The introduction of a propionate arm may change the forward rates of association but it would also compromise the stability of the complex. A six-ring chelate will lower the stability of the complex significantly. However, to our knowledge there are no experimental data available yet to confirm this hypothesis.

6 STABILITY OF CA(II) COMPLEXES OF POLYAMINOPOLYCARBOXYLIC ACIDS

In order to predict the free energy of complexation in the reaction between Ca(II) ions and chelating ligands, Bakken and Schöffel [75] performed RHF and RMP2 calculations with the all-electron basis sets (STO-3G and 3-21G) for the ligand atoms and the calcium atom [34]. In view of the similarities of the coordination chemistry of the Ca(II) and Gd(III) ions, this investigation gains some insight into the free energy of complexation of the Gd(III) ion in its reaction with chelating ligands. All 20 electrons of the Ca atom ($Z = 20$) were treated explicitly in the calculations. The total energies of the free Ca(II) ion, the anionic complexes $Ca(NTA)^-$, $Ca(NTA)_2^{4-}$, $Ca(EDTA)^{2-}$, $Ca(HEDTA)^-$, $Ca(DTPA)^{3-}$, Ca(1, 2-PDTA)$^{2-}$, Ca(1, 3–PDTA)$^{2-}$, and $Ca(EGTA)^{2-}$, the neutral Ca(II) complex of 1, 4, 10, 13–tetraoxa-7, 16-diazacyclooctadecane-*N*, *N′*-diacetic acid (CROWN), and the cationic Ca(II) complex of 1, 10–bis(2–pyridylmethyl)-1, 4, 7, 10-tetraazadecane (BPTETA), were all calculated in the gas phase. The conformational space for each ligand and every complex had been searched at the semi-empirical AM1 level of theory. Then, the conformers of lowest energy were investigated. The total energies of the corresponding ligand anions, i.e. NTA^{3-}, $EDTA^{4-}$, $HEDTA^{3-}$, $DTPA^{5-}$, 1, 2–PDTA^{4-}, 1, 3–PDTA^{4-}, $EGTA^{4-}$ and $CROWN^{2-}$, and the neutral ligand BPTETA (see Figure 7.12), together with the total energies of the corresponding complexes, were calculated at the RHF/3–21G//RHF/STO-3G level of theory. The binding energy of the ligand, ΔE, was then obtained from these data by the following relationship between the total energies:

$$\Delta E = E_{\mathrm{tot}}(\mathrm{Ca\ L}) - E_{\mathrm{tot}}(\mathrm{Ca(II)}) - E_{\mathrm{tot}}(\mathrm{L}). \tag{7.4}$$

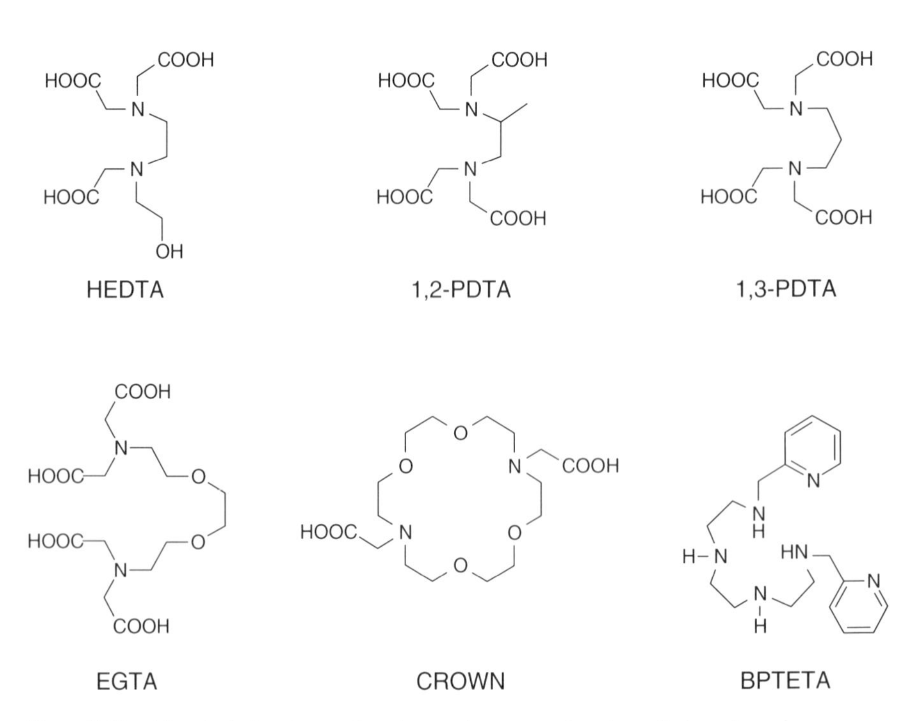

Figure 7.12 Chemical structures of various polyaminopolycarboxylic-type ligands.

From the calculated binding energies for the different ligands, the following order of stability for the corresponding complexes was predicted: $Ca(DTPA)^{3-}$ < $Ca(EDTA)^{2-}$ < Ca(1, 2-PDTA)$^{2-}$ < Ca(1, 3–PDTA)$^{2-}$ < $Ca(EGTA)^{2-}$ < $Ca(NTA)^{-}$ < Ca(NTA)2^{4-} < $Ca(HEDTA)^{-}$ < Ca(CROWN) < $Ca(BPTETA)^{2+}$. However, the available experimentally determined stability constants (log K) of these complexes in water (25 °C and $\mu = 0.1$) are as follows: $Ca(DTPA)^{3-}$, 10.75; $Ca(EDTA)^{2-}$, 10.65; Ca(1, 2–PDTA)$^{2-}$, 11.54j $Ca(EGTA)^{2-}$, 10.86; $Ca(NTA)^{-}$ 6.44; $Ca(HEDTA)^{-}$, 8.2; Ca(CROWN), 8.57 [63]. The ordering of the calculated binding energies for the different ligands is not in total agreement with the experimental data. The calculated binding energies do not allow a correct prediction of complex stabilities in all cases for the ligands discussed here. The same applies to the comparison of ΔE with the experimentally determined enthalpies of formation ΔH (kcal/mol) of the complexes: $Ca(DTPA)^{3-}$, −6.0; $Ca(EDTA)^{2-}$, −6.1; Ca(1, 2–PDTA)$^{2-}$, −3.0; $Ca(EGTA)^{2-}$, −7.9; $Ca(NTA)^{-}$, −1.36; $Ca(HEDTA)^{-}$, −6.5; Ca(CROWN), −8.3 [63].

These large differences, especially in the case of the CROWN ligand, is probably due to the fact that the global minimum on the corresponding PES for the ligand and the complex were not explicitly taken into consideration.

In the optimized structures of the complexes, there is a clear preference for the interaction of charged carboxylate groups with the metal ion. The comparison of the binding energies for 1, 2–PDTA and 1, 3–PDTA discloses, from an energetic point of view, that the five-membered ring incorporating the Ca(II) ion is preferred to the six-membered ring by a difference of 47 kcal/mol. This difference in binding energies had been described previously by Hancock for Ca(II) ions and rationalized in terms of the size of the metal ion which is complexed [76].

In the gas phase, Ca(NTA)2^{4-} is not stable, with respect to $Ca(NTA)^{-}$ and NTA^{3-}. This can be easily explained by the charge repulsion between the two chelating NTA^{3-} groups. The destabilization energy is approximately 17 kcal/mol.

Binding energy calculations were performed on the Ca(II) complexes of NTA^{3-} and those derivatives in which the three acetate groups, $-CH_2CO_2^{-}$, had been replaced by $-CH_2CH_2OH$, $-CH_2CH_2OCH_3$, $-CH_2CH_2NH_2$, $-CH_2CH_2F$ and $-CH_2CH_2S^{-}$ as substituents of the central nitrogen. The following order of binding energies for the chelating groups was established: $-CO_2^{-}$ > $-OCH_3$ > $-NH_2$ > $-F$ > $-S^{-}$ [75].

7 SUMMARY AND OUTLOOK

In order to summarize the computational results discussed so far, it is obvious that the whole panoply of theoretical methods is capable of reproducing the experimentally observed structures of gadolinium (or lanthanide) containing compounds fairly well. The methods applied differ in the degree of arbitrariness. Only a limited number of methods are capable of reproducing the experimentally observed energy differences between the isomers. Whereas force-field methods are easy to use and should be applied on a routine base, the force-field parameter sets which have been deduced and applied so far seem to be not very well-defined and show a large variation. On the other hand, the well-defined quantum mechanical *ab initio* methods suffer from the drawback that they are computationally very expensive, and moreover their extension to real-world applications, by describing lanthanide- containing complexes in the aqueous phase, has not yet been too successful. In the future, it will become probably more and more important to include solvation effects into the calculations, either due to an improved parametrization of force-field methods or by incorporating solvents or solvent effects into *ab initio* type calculations. This will probably also give a better access to the calculation of stability constants of lanthanide complexes in the aqueous phase.

The final goal of any theoretical calculation in this field should be the prediction of the relaxivities of gadolinium-containing complexes, a priori. A first step in this direction might be the work of Tan *et al.* in which a combination of molecular dynamics and molecular mechanics methods was used to describe the structures and properties of Gd(III)-based complexes, and in which for the first time the relaxivity of an MRI contrast agent was calculated by using the results of a classical molecular dynamic simulation of the complex in water over 5×10^{-9}s [77].

ACKNOWLEDGEMENTS

The authors gratefully acknowledge the excellent support by the members of the Information Services and Library, Schering AG, especially the help of Mrs Dr Ortrud Lammer and Mr Volkert Herzog in the preparation of this overview.

REFERENCES

1. Watson, A. D., Rocklage, S. R. and Carvlin, M. J. in *Contrast Agents in Magnetic Resonance Imaging*, Stark, D. D., Bradley Jr, W. G. and Mosby, C. V. (Eds), Mosby Yearbooks, Inc, St. Louis, MO, 1991, Chapter 14, pp. 372–381.
2. Brady, T. J. and Reimer, P. in *Encyclopedia of Nuclear Magnetic Resonance*, Grant, D. M. and Harris, R. K. (Eds), John Wiley & Sons, Chichester, 1996, Vol. 3, pp. 1432–1438.
3. Weinmann, H. J., Mühler, A. and Radüchel, B. in *Encyclopedia of Nuclear Magnetic Resonance*, Grant, D. M. and Harris, R. K. (Eds), John Wiley & Sons, Chichester, 1996, Vol. 4, pp. 2166–2173.

4. Mezey, P. G. *Potential Energy Hypersurfaces*, Elsevier, Amsterdam, 1987.
5. Frisch, Æ. and Frisch, M. J. *Gaussian 98 User's Reference*, Gaussian, Inc., Pittsburgh, PA, 1998.
6. Szabo, A. and Ostlund N. S. *Modern Quantum Chemistry*, McGraw-Hill, New York, 1982.
7. Dolg, M. and Stoll, H. *Handbook on the Physics and Chemistry of Rare Earths* 1995, Vol. *22*, 1.
8. Cundari, T. R., Sommerer, S. O., Strohecker, L. A. and Tippett, L. *J. Chem. Phys.* 1995, **103**, 7058.
9. Hay, P. J. and Wadt, W. R. *J. Chem. Phys.* 1985, **82**, 270.
10. Tsuchiya, T., Taketsugu, T., Nakano, H. and Hirao, K. *Theochem* 1999, **461–462**, 203.
11. Joubert, L; Picard, G and Legendre, J. J. *Inorg. Chem.* 1998, **37**, 1984.
12. Kovács, A., Konings, R. J. M. and Booji, A. S. *Chem. Phys. Lett.* 1997, **268**, 207.
13. Gutowski, M., Boldyrev, A. I., Simons, J., Rak, J. and Blazejowski, J. *J. Am. Chem. Soc.* 1996, **118**, 1173.
14. Kitao, O., Shiozaki, R. and Kera, Y. *Kidorui* 1996, 332.
15. Troxler, L., Dedieu, A., Hutschka, F. and Wipff, G. *Theochem* 1998, **431**, 151.
16. Wu, Z. J., Meng, Q. B., Gao, F., Niu, C. J. and Zhang, S. Y. *Chin. Chem. Lett.* 1997, **8**, 1073.
17. Parr, R. G. and Yang, W. *Density-Functional Theory of Atoms and Molecules*, Oxford University Press, New York, 1989.
18. Car, R. and Parrinello, M. *Phys. Rev. Lett.* 1985, **55**, 2471.
19. Adamo, C. and Maldivi, P. *Chem. Phys. Lett.* 1997, **268**, 61.
20. Joubert, L., Picard, G. Legendre, J. J. *J. Alloys Compd.* 1998, **275–277**, 934.
21. Zerner, M. C. *Rev. Comput. Chem.* 1991, **2**, 313.
22. Kotzian, M., Fox, T. and Rösch, N. *J. Phys. Chem.* 1995, **99**, 600.
23. Maitland, G. C., Rigby, M., Smith, E. B. and Wakeham, W. A. *Intermolecular Forces*, Clarendon Press, Oxford, 1981.
24. Murrell, J. N., Carter, S., Farantos, S. C., Huxley, P. and Varandas, A. J. C. *Molecular Potential Energy Functions*, John Wiley & Sons, Chichester, 1984.
25. Burkert, U. and Allinger, N. L. *Molecular Mechanics*, ACS Monograph 177, American Chemical Society, Washington, DC, 1982.
26. Cundari, T. R., Moody, E. W. and Sommerer, S. O. *Inorg. Chem.* 1995, **34**, 5989.
27. Beech, J., Drew, M. G. B. and Leeson, P. B. *Struct. Chem.* 1996, **7**, 153.
28. Kaupp, M. and Schleyer, P. v. R. *J. Phys. Chem.* 1992, **96**, 7316.
29. Hengrasme, S., Probst, M. M. *Z. Naturforsch A.* 1991, **46a**, 117.
30. Schafer, O. and Daul, C. *Int. J. Quantum Chem.* 1997, **61**, 541.
31. Hay, P. J. and Wadt, W. R. *J. Chem. Phys.* 1985, **82**, 270.
32. Kowall, T., Foglia, F., Helm, L. and Merbach, A. E. *Chem. Eur. J.* 1996, **2**, 285.
33. Cosentino, U., Moro, G., Pitea, D., Calabi, L. and Maiocchi, A. *Theochem* 1997, **392**, 75.
34. Hehre, W. J., Radom, L., Schleyer, P. v. R. and Pople, J. A. *Ab Initio Molecular Orbital Theory*, John Wiley & Sons, New York, 1986.
35. Chatterjee, A., Maslen, E. N. and Watson, K. J. *Acta Crystallogr. Sect. B.* 1988, **44**, 381.
36. Gerkin, R. E. and Reppart W. J. *Acta Crystallogr. Sect. C* 1984, **40**, 781.
37. Sülzle, D., unpublished results, 1999.
38. Lee, H. M., Kim, J., Lee, S., Mhin, B. J. and Kim, K. W. *J. Chem. Phys.* 1999, **111**, 3995.
39. Feller, D., Glendening, E. D., Woon, D. E. and Feyereisen, M. W. *Chem. Phys.* 1995, **103**, 3526.

40. Ramaniah, L. M., Bernasconi, M. and Parinello M. *J. Chem. Phys.* 1999, **111**, 1587.
41. Meyer, M., Dahaoui-Gindrey, V., Lecomte, C. and Guilard, R. *Coord. Chem. Rev.* 1998, **178–180**, 1313.
42. Schmitt-Willich, H., Brehm, M., Ewers, C. L. J., Michl, G., Müller-Fahrnow, A., Petrov, O., Platzek, J., Radüchel, B. and Sülzle, D. *Inorg. Chem.* 1999, **38**, 1134.
43. Gries, H. and Miklautz, H. *Physiol. Chem. Phys. Med. NMR* 1984, **16**, 105.
44. Chang, C. A., Francesconi, L. C., Malley, M. F., Kumar, K., Gougoutas, J. Z., Tweedle, M. F., Lee, D. W. and Wilson, L. J. *Inorg. Chem.* 1993, **32**, 3501.
45. Dudbost, J. P., Leger, J. M., Langlois, M. H. and Schaefer, M. *Compt. Rend. Acad. Sci. Paris Ser. 2* 1991, **312**, 349.
46. Caravan, P., Ellison, J. J., McMurry T. and Lauffer R. B. *Chem. Rev.* 1999, **99**, 2293.
47. Dale, J. *Top. Stereochem.* 1976, **9**, 199.
48. Aime, S., Barge, A., Benetollo, F., Bombieri, G., Botta, M. and Uggeri, F. *Inorg. Chem.* 1997, **36**, 4287.
49. Von Zelewsky A. *Stereochemistry of Coordination Compounds*, John Wiley & Sons, Chichester, 1986.
50. Corey, E. J. and Baillar Jr, J. C. *J. Am. Chem. Soc.* 1959, **81**, 2620.
51. Aime, S., Botta, M., Fasano, M., Marques, M. P. M., Geraldes, C. F. G. C., Pubanz D. and Merbach A. E. *Inorg Chem* 1997, **36**, 2059.
52. Cosentino, U., Moro, G., Pitea, D., Villa, A., Fantucci, P. C., Maiocchi, A. and Uggeri, F. *J. Phys. Chem. A* 1998, **102**, 4606.
53. Sülzle, D., unpublished results, 1995.
54. SYBYL®, Tripos, Inc, St Louis, MO.
55. MacKerell Jr, A. D., Bashford, D., Bellott, M., Dunbrack Jr, R. L., Evanseck, J. D., Field, M. J., Fischer, S., Gao, J., Guo, H. Ha, S., Joseph-McCarthy, D., Kuchnir, L., Kuczera, K., Lau, F. T. K., Mattos, C., Michnik, S., Ngo, T., Nguyen, D. T., Prodhom, B., Reiher III, W. E., Roux, B., Schlenkrich, M., Smith, J. C., Stote, R., Straub, J., Watanabe, M., Wiórkiewicz-Kuczera, J., Yin, and D., Karplus, M. *J. Phys. Chem. B* 1998, **102**, 3586.
56. Henriques, E. S., Bastos, M., Geraldes, C. F. G. C. and Ramos, M. J. *Int. J. Quantum Chem.* 1999, **73**, 237.
57. Maumela, H., Hancock, R. D., Carlton, L., Reibenspies, J. H. and Wainwright, K. P. *J. Am. Chem. Soc.* 1995, **117**, 6698.
58. Reichert, D. E., Hancock, R. D. and Welch, M. J. *Inorg. Chem.* 1996, **35**, 7013.
59. Fossheim, R. and Dahl, S. G. *Acta Chem. Scand.* 1990, **44**, 698.
60. Fossheim, R., Dahl, S. G. and Dugstad, H. *Eur. J. Med. Chem.* 1991, **26**, 299.
61. Fossheim, R., Dugstad, H. and Dahl, S. G. *J. Med. Chem.* 1991, **34**, 819.
62. Fossheim, R., Dugstad, H. and Dahl, S. G. *Eur. J. Med. Chem.* 1995, **30**, 539.
63. Smith, R. M., Martell, A. E. and Motekaitis, R. J. *NIST Critically Selected Stability Constants of Metal Complexes Database*, National Institute of Standards and Technology, Gaithersburg, MD, 1997, Version 3.0.
64. Kumar, K. and Tweedle, M. F. *Inorg. Chem.* 1993, **32**, 4193.
65. Jang, Y. H., Blanco, M., Dasgupta, S., Keire, D. A., Shively, J. E. and Goddard III, W. A. *J. Am. Chem. Soc.* 1999, **121**, 6142.
66. Tannor, D. J., Marten, B., Murphy, R., Friesner, R. A., Sitkoff, D., Nicholls, A., Ringnalda, M., Goddard III, W. A. and Honig, B. *J. Am. Chem. Soc.* 1994, **116**, 11875.
67. Nicholls, A. and Honig, B. *J. Comput. Chem.* 1991, **12**, 435.
68. Marcus, Y. *Ion Solvation*, John Wiley & Sons, New York, 1985.
69. Broan, C., Cox, J. P., Craig, A. S., Kataky, R., Parker, D., Harrison, A., Randall, A. M. and Ferguson, G. *J. Chem. Soc. Perkin Trans. 2* 1991, 87.

70. Parker, D., Pulukkody, K., Smith, F. C., Batsanov, A. and Howard, J. A. K. *J. Chem. Soc. Dalton Trans.* 1994, 689..
71. Kasprzyk, S. P. and Wilkins, R. G. *Inorg. Chem.* 1982, **21**, 3349.
72. Keire, D. A. and Kobayashi, M. *Bioconjugate Chem.* 1999, **10**, 454.
73. Wang, X., Jin, T., Comblin, V., Lopez-Mut, A., Merciny, E. and Desreux, J. F. *Inorg. Chem.* 1992, **31**, 1095.
74. Wu, S. L. and Horrocks Jr, W. D. *Inorg. Chem.* 1995, **34**, 3724.
75. Bakken, V. and Schöffel, K. *Rev. Inst. Fr. Pet.* 1996, **51**, 151.
76. Hancock, R. D. *J. Chem. Educ.* 1992, **69**, 615.
77. Tan, Y. T., Judson, R. S., Melius, C. F., Toner, J. and Wu, G. *J. Mol. Model.* 1996, **2**, 160.

8 Structure and Dynamics of Gadolinium-Based Contrast Agents

JOOP A. PETERS, EMRIN ZITHA-BOVENS, DANIELE M. CORSI
Delft University of Technology, Delft, The Netherlands

and

CARLOS F. G. C. GERALDES
University of Coimbra, Coimbra, Portugal

1 INTRODUCTION

The successful introduction of Gd^{3+}-based contrast agents for magnetic resonance imaging (MRI) and the fast evolution of the MRI technique have given rise to an increasing demand for more effective and specific contrast agents. NMR relaxation theory predicts the possibility of having contrast agents with relaxivities up to 50 times higher than those presently in use. Therefore, much research effort has been directed towards gaining insight into the parameters that govern the relaxivities of these agents. It appears that each of the parameters concerned is related to the solution structure and/or dynamics of the complexes involved. For example, the rotational correlation time, τ_R, is determined by the molecular tumbling and segmental motions, whereas the bound water lifetime, τ_M, is dependent on the overall charge of the complex and on the steric strain around the bound water molecule. The electronic relaxation rate of the Gd^{3+} ion, τ_{S0}, has been shown to be related to the flexibility of the complex (see Chapter 2). Evidently, extensive knowledge of the solution structure and dynamics is indispensable for further developments in this field.

In this chapter, we will focus on the elucidation of the solution structures of the Ln^{3+} complexes of relevance to MRI. Major tools in this field are NMR spectroscopy and luminescence, both of which give direct information on species in solution, and X-ray crystallography, which provides strong support by means of the corresponding solid-state structures. A review of the NMR techniques used and the supporting theory will be included here. Luminescence will be dealt with in detail in Chapter 11.

The Ln^{3+} ions form a unique series in the periodic system. The first and the last members (La^{3+} and Lu^{3+}, respectively) are diamagnetic, whereas all others

The Chemistry of Contrast Agents in Medical Magnetic Resonance Imaging
Edited by A. E. Merbach and É. Tóth.

have one to seven unpaired $4f$ electrons and are thus paramagnetic. The $4f$ electrons are shielded by the $5s$ and $5p$ electrons and are, therefore, not readily available for covalent interactions with ligands. Interactions are largely electrostatic and consequently the geometry of Ln^{3+} complexes is usually determined by steric rather than by electronic factors. Another consequence of this shielding is a large similarity in chemical behavior among the 15 Ln^{3+} ions. Differences in chemical behavior, if any, may be ascribed to the decrease in ionic radius from La^{3+} to Lu^{3+} (1.36–1.17 Å) [1]. Commonly, the various Ln^{3+} complexes of a particular ligand are nearly isostructural. Each of the Ln^{3+} ions, however, has its own characteristic effects on the NMR parameters of the nuclei in its proximity. The pertubations of the NMR parameters of the ligand nuclei of the same ligand by various Ln^{3+} ions can be exploited for the elucidation of the molecular structures in solution of the complexes concerned.

Lanthanide complexes are among the most exchange-labile metal complexes. For example, first-order water-exchange rate constants for the aquo ions between 4.7×10^7 and $8.3 \times 10^8\ s^{-1}$ have been reported [2]. This contributes to the success of Gd^{3+} complexes as contrast agents for MRI (see Chapter 2). An important requirement that a Ln^{3+} complex must meet for *in vivo* applicability is high thermodynamic and kinetic stabilities. Practically, this usually means that the organic ligand used to sequester the Ln^{3+} ion should be hepta- or octadentate. The high denticity and stability of these complexes give rise to specific features and hence problems with structural analysis. For example, several conformations and/or configurations of a given complex, which interconvert slowly on the NMR time-scale, may be present. Studies relating to these issues were few prior to the emergence of MRI contrast agents.

2 LANTHANIDE-INDUCED SHIFTS

The lanthanide-induced shift (LIS) for a nucleus of a ligand upon coordination to a Ln^{3+} cation (Δ) can be expressed as the sum of three terms, i.e. the diamagnetic (Δ_d) the contact (Δ_c), and the pseudocontact shift (Δ_p). Each of these, but particularly the contact and pseudocontact shifts, contain useful information regarding the structure of the concerning Ln^{3+} complex [3, 4].

$$\Delta = \Delta_d + \Delta_c + \Delta_p \qquad (8.1)$$

2.1 DIAMAGNETIC SHIFTS

The diamagnetic or complexation shift is usually small and is often neglected. It originates from effects such as conformational changes, inductive effects, and direct field effects. In saturated ligands, the diamagnetic shifts are usually insignificant, with the exception of those nuclei which are directly coordinated to the Ln^{3+} cation [5].

2.2 CONTACT SHIFTS

Contact shifts arise from through-bond transmission of unpaired electron-spin density from the lanthanide *f*-orbital to the nucleus under study. It is given by Equation (8.2), where $< S_z >$ is the reduced value of the average spin polarization, β the Bohr magneton, k the Boltzmann constant, γ_I the gyromagnetic ratio of the nucleus in question, and $A/\hbar$ the hyperfine coupling constant (in rad/s), with Δ_c being expressed in ppm. Golding and Halton have presented calculated values for $< S_z >$ at 300 K [6]. Pinkerton *et al.* have performed refined calculations and given values for all reasonable temperatures [7].

$$\Delta_c = < S_z > F = < S_z > \frac{\beta}{3kT\gamma_I}\frac{A}{\hbar}10^6 \tag{8.2}$$

The contact shift, in general, rapidly decreases its magnitude upon an increase of the number of bonds between Ln^{3+} and the nucleus under study[5], with the largest contact shifts being generally observed for the directly Ln^{3+}-bonded donor sites. Therefore, donor sites in a ligand can be easily identified by means of their relatively large contact shifts.

The tabulated $< S_z >$ values show a linear relationship with, for example, the ^{17}O LIS values for water [8, 9] in both the Ln^{3+}-aquo complexes and the first coordination sphere of $Ln(EDTA)^-$ [10], as well as in the ^{14}N LIS values for pyridine [11]. This suggests that the LIS for a donor site is predominantly of contact origin and that the hyperfine coupling constant does not vary significantly along the lanthanide series [6, 10, 12].

For coordinated oxygen atoms, the magnitude of F is always in a narrow range ($F = -70 \pm 11$ ppm, at 73 °C, corresponding to $A/\hbar = -3.9 \times 10^6$ rad s^{-1}), irrespective of the ligand in question and of the other ligands coordinated to the Ln^{3+} ion. It may be deduced that if the experimental paramagnetic shifts are extrapolated to a molar ratio of Ln^{3+}/ligand $\rho = 1$, the separation of contact-and pseudocontact shifts (see below) affords nF, where n is the number of coordinated ligands. Division of nF by -70 then gives n. This has been applied to a variety of oxygen-containing functional groups, i.e. carboxyl, carbonyl, hydroxyl, ether and phosphate. In the field of MRI contrast agents, this may be exploited for the determination of the number of Ln^{3+}-bound water molecules in the Ln^{3+} complexes (see also below) [13]. The hyperfine coupling constant, $A/\hbar$, has been evaluated for bound-water ^{17}O nuclei in many Gd^{3+} complexes of (potential) MRI contrast agents after the fitting of data from variable-temperature ^{17}O transversal relaxation studies, often in a simultaneous fitting procedure on longitudinal relaxation rates and nuclear magnetic resonance dispersion (NMRD) data. Usually it is assumed that the complex concerned has a single hydration water. The obtained hyperfine coupling constants are all in the range $(-3.9 \pm 0.3) \times 10^6$ rad s^{-1} [14], which is consistent with the generally obtained values mentioned above.

It should be noted that the large LIS of the donor site nucleus is accompanied by extensive line-broadening. As a result, the signal for the bound state is often not observable if exchange between the bound and free states is slow on the NMR time-scale, which is usually the case for the organic ligands in MRI contrast agents. Under these circumstances, information may be obtained from the decrease in intensity of the signal of the donor sites in the free ligand upon titration with a paramagnetic Ln^{3+} ion.

The unpaired spin-density at a nucleus, I, is reflected in the hyperfine coupling constant, according to the following equation:

$$A_I = \frac{8\pi}{3} \gamma_I h g \beta |\Psi(0)|_I^2 \rho_I \tag{8.3}$$

Here, $|\Psi(0)|_I^2$ is the probability of finding the unpaired electron at the nucleus I, ρ_I is the unpaired spin density at I, h is the Planck constant and g the electron Lande factor. The relative values of $|\Psi(0)|_I^2$ for ^{1}H, ^{13}C, ^{14}N, ^{17}O, ^{19}F, and ^{31}P are 1.000, 8.693, 14.985, 23.995, 37.592, and 17.671, respectively [15]. The unpaired spin-density may be transmitted by two mechanisms, i.e. (i) direct spin delocalization and (ii) spin polarization. The latter mechanism gives rise to alternating signs of the corresponding contact shifts. Reilley *et al.* have critically evaluated the literature data on contact shifts of various Ln^{3+} complexes [12]. These data suggest that spin delocalization dominates in straight carbon chains, whereas spin polarization dominates in cyclic compounds. This has been supported by *ab initio* calculations of long-range hyperfine interactions in the propyl radical, which show that the spin delocalization is much larger in carbon chains with a W planar rearrangement than in those with an anti-W planar rearrangement [16].

Since the prediction of contact shifts for nuclei not directly coordinated to the Ln^{3+} ion is rather cumbersome, these shifts are commonly separated off and ignored, or are treated only in a qualitative manner.

2.3 PSEUDOCONTACT SHIFTS

The pseudocontact or dipolar shift results from the local magnetic field induced in the nucleus under study by the magnetic moment of the lanthanide ion. If no assumptions are made regarding the location of the principal magnetic axes or the symmetry of the complex, the pseudocontact shift can be expressed as shown in the following [17]:

$$\Delta_p = \frac{1}{2N\hbar\gamma_I}(\chi_{zz} - \bar{\chi})\left\langle\frac{3\cos^2\theta - 1}{r^3}\right\rangle + \frac{1}{2N\hbar\gamma_I}(\chi_{xx} - \chi_{yy})\left\langle\frac{\sin^2\theta\cos 2\varphi}{r^3}\right\rangle + \frac{1}{N\hbar\gamma_I}\chi_{xy}\left\langle\frac{\sin^2\theta\sin 2\varphi}{r^3}\right\rangle + \frac{1}{N\hbar\gamma_I}\chi_{xz}\left\langle\frac{\sin 2\theta\cos\varphi}{r^3}\right\rangle + \frac{1}{N\hbar\gamma_I}\chi_{yz}\left\langle\frac{\sin 2\theta\sin\varphi}{r^3}\right\rangle \tag{8.4}$$

Here, N is the Avogadro number, $\bar{\chi} = (1/3)Tr_\chi$ (trace of the tensor χ), where $\chi_{xx}, \chi_{yy}, \chi_{zz}, \chi_{xy}, \chi_{xz}$ and χ_{yz} are the components of the tensor χ in the molecule-fixed, ligand system, and r, θ and ϕ are the spherical coordinates of the observed nucleus in the coordinate system fixed in the molecule with respect to Ln^{3+} at the origin. The last three terms of Equation (8.4) vanish when the principal magnetic axis is taken as the coordinate system. Furthermore, only the first term remains for the special case of axial symmetry ($\chi_{xx} = \chi_{yy}$).

Bleaney has calculated pseudocontact shifts with the assumption that the ligand field splittings for the lowest J state in the lanthanide complexes are small compared to kT [18]. If the principal magnetic axes system is used, Equation (8.4) can be written as either of the following:

$$\Delta_p = \frac{C_j\beta^2}{60k^2T^2}\left[\frac{\langle r^2\rangle A_2^0(3\cos^2\theta - 1)}{r^3} + \frac{\langle r^2\rangle A_2^2\sin^2\theta\cos 2\varphi}{r^3}\right] \tag{8.5}$$

$$\Delta_p = D_1\frac{3\cos^2\theta - 1}{r^3} + D_2\frac{\sin^2\theta\cos 2\varphi}{r^3} \tag{8.6}$$

Here C_j is Bleaney's constant, characteristic of the Ln^{3+} ion, while $< r^2 > A_2{}^0$ and $< r^2 > A_2{}^2$ are ligand field coefficients of the second degree. Usually, it is assumed that the ligand field coefficients do not vary along the lanthanides for a series of isostructural complexes.

Horrocks *et al.* [19] and Golding and Pyykkö [20] have presented alternative approaches by using a more general field theory with all of the J multiplets of the ground state serving as a basis. This leads to deviations of the T^2 dependence of Δ_p (except for Eu^{3+} and Sm^{3+}) [21]. However, McGarvey has shown that Equation (8.5) is adequate for estimating shifts to an accuracy of 10–20 % [22].

The constants D_1 and D_2 for several complexes have been determined independently from magnetic anisotropy measurements on single crystals by assuming that the solid-state structures persist in solution [23] and from luminescence measurements [24–26]. Estimates of the relative values of these constants have been made with the use of a model in which the complexes are regarded as a set of point charges [24–27].

For systems with an n-fold axis of symmetry ($n \geq 3$) the second term in Equation (8.6) is zero. Frequently in structural analyses, only the first term of Equation (8.6) is needed to calculate the LIS properly, even if the complex has no axial symmetry. Briggs *et al.* have shown that any n-fold rotational axis ($n \geq 3$) in the system may serve as an effective axial magnetic axis for the part of the system concerned, provided that for a particular nucleus, r and θ are equal for each of the rotamers and that the populations of the rotamers are equal [28]. Then, Equation (8.6) reduces to:

$$\Delta_p = C\frac{3\cos^2\theta - 1}{r^3} \tag{8.7}$$

$$C = \frac{1}{2}D_1(3\cos^2\alpha - 1) + \frac{1}{2}D_2\sin^2\alpha\cos 2\beta \tag{8.8}$$

The new proportionality constant C is related to D_1 and D_2 via Equation (8.8), where α and β are the Euler angles that define the position of the effective axis with respect to the (old) principal magnetic susceptibility axis. It should be noted that Equations (8.7) and (8.8) are also valid when the rotation takes place via dissociation–association as long as the fast-exchange conditions with respect to the NMR time-scale are met.

Horrocks assumed that the Ln^{3+} complex is present as an ensemble of a large number of conformations [29]. Random variations of the axis of the susceptibility tensor then lead to the vanishing of the second term in Equation (8.6). The variation of the position of the ligand with respect to the extremum of the susceptibility axis was restricted, whereas that with respect to the other axes was varied randomly. As such, a situation was calculated that resembles an n-fold rotation around the Ln–donor bond. Completely unbiased variations would lead to cancellation of the contributions to the induced shift and result in $\Delta_p = 0$.

2.4 EVALUATION OF BOUND SHIFTS

Some caution is required when the LIS values of paramagnetic compounds are measured with an external standard without frequency-locking of the NMR spectrometer. In these cases, the measured shift has a considerable contribution of the difference in bulk magnetic susceptibility between the reference sample (δ_{BMS}) which needs to be subtracted. To a good approximation, δ_{BMS} is given by Equation (8.9) [30], where c is the concentration of the paramagnetic Ln^{3+} complex in mol/l, s is dependent on the shape and the position in the magnetic field, and μ_{eff} is the effective magnetic moment for the particular Ln^{3+} ion (values for μ_{eff} [31] are compiled in Table 8.1 below). The factor s has values of 0, 1/3 and $-1/6$ for a sphere, a cylinder parallel to the main magnetic field, and a cylinder perpendicular to the magnetic field, respectively.

$$\delta_{BMS} = \frac{1558.02cs\mu_{eff}^2}{T} \tag{8.9}$$

In a single compartmental sample, the BMS effect is effectively canceled by locking the spectrometer on, for instance, the ^2H resonance, since the BMS effect is the same for all nuclei in the sample. Since s in Equation (8.9) is zero for spherical sample tubes, it may be advantageous to use such tubes when locking and/or the use of an internal reference is not possible [32]. Alternatively, BMS effects can be removed by placing the sample tube under the magic angle (54° 44′) with respect to the magnetic field [33]. Since the value of μ_{eff} is practically independent of the ligation of the Ln^{3+} ion [34], measurement of δ_{BMS} may be a convenient method to determine the total Ln^{3+} concentration in a sample [35].

Since the exchange of nuclei of the organic ligand between the bound and free state is commonly slow on the NMR time-scale for MRI contrast agents, the evaluation of the bound LIS values is straightforward. The observation of signals, particularly for complexes with Ln = Dy → Yb is sometimes hampered by extreme line-broadening mainly due to exchange broadening and to the effect of the Curie relaxation mechanism (see below). Measuring at low magnetic field strengths and at high temperatures may alleviate the latter problem. The assignment of the NMR signals may be a difficult task, particularly because they are usually shifted substantially from their diamagnetic positions as a result of the paramagnetic effects and due to the possibility of multiple isomers. Longitudinal relaxation rates (see below) and two-dimensional techniques, particularly COSY and 2D-exchange spectroscopy [36], are useful to obtain the assignments. Furthermore, plotting of the chemical shifts for various Ln^{3+} ions according to equations similar to Equations (8.13) and (8.14) may be helpful to establish the correlation of the signals among complexes for the various Ln^{3+} ions (see below). Forsberg *et al.* have developed a computer program to analyze LIS data, including an assignment routine that permutes the LIS values over a number of selected nuclei until the best fit between calculated and observed LIS values is achieved [37].

The exchange of water ^1H nuclei between Ln^{3+} complexes and the bulk is usually rapid on the NMR time-scale. Since the measurements are always performed in a large excess of water, the bound shift (Δ) is then given by Equation (8.10), where δ and δ_0 are the observed chemical shifts in the presence and the absence of the Ln^{3+} complex (after correction for any BMS effects), respectively, q is the hydration number of the Ln^{3+} complex, and ρ is the Ln^{3+} complex/water molar ratio.

$$\Delta = \frac{q}{\rho}(\delta - \delta_0) \tag{8.10}$$

For water ^{17}O nuclei in Gd^{3+}-based contrast agents, the exchange is often in a regime intermediate between rapid and slow. The bound shift can then only be evaluated by means of a variable temperature study (see Chapter 2).

2.5 SEPARATION OF SHIFT CONTRIBUTIONS

Diamagnetic shifts are usually small, and can be determined directly or by interpolation from the shifts induced by the diamagnetic members of the series (La^{3+} and Lu^{3+}). Subtraction of the diamagnetic contributions gives the paramagnetic shift Δ'. If the principal magnetic axis system is used as the coordinate system, combination of Equations (8.2) and (8.6) gives Equations (8.11) and (8.12):

$$\Delta' = \Delta_c + \Delta_p = < S_z > F + C_j D \tag{8.11}$$

$$D = A_2^0 < r^2 > G_1 + A_2^2 < r^2 > G_2 \tag{8.12}$$

It should be noted that $< S_z >$ and C_j are terms that are characteristic of the Ln^{3+} ion but independent of the ligand, whereas F and D are characteristic of the nucleus under study, but independent of the Ln^{3+} ion.

Very often, the various Ln^{3+} complexes of a particular ligand are isostructural and the crystal field coefficients are invariant along the Ln^{3+} series. Then, F and D can be determined by linear regression if the Δ' values of a ligand nucleus are known for two or more Ln^{3+} cations [12]. To test the assumption that the complex structures and the crystal field coefficients do not vary along the Ln^{3+} series, it is convenient to rewrite Equation (8.11) in two linear forms, as follows [38]:

$$\frac{\Delta'}{< S_z >} = F + \frac{C_j}{< S_z >} D \tag{8.13}$$

$$\frac{\Delta'}{C_j} = \frac{< S_z >}{C_j} F + D \tag{8.14}$$

When these assumptions hold, the F and D values for the various ligand nuclei are independent of the Ln^{3+} cation and, consequently, plots according to Equations (8.13) and (8.14) are straight lines. Since the ionic radii of the Ln^{3+} ions decrease across the series from 1.36 to 1.17 Å, slight changes in the orientation of the ligands around the Ln^{3+} often occur, which is reflected in small changes of D only. The values of C_j for the first part of the series (Ce → Eu) are much smaller (0.7–11) than those of the second part (Tb → Yb) (22–100). Small changes of D are thus magnified by using Equation (8.13) and breaks are frequently observed in plots according to this equation [39]. However, the F values are not affected by these slight structural changes, and plots according to Equation (8.14) show no breaks in these cases. Non-linearity of *both* the plots of the LIS values according to Equations (8.13) and (8.14) indicates that either a significant structural change occurs across the Ln^{3+} series (i.e., a change of coordination mode of a ligand) or that the crystal field coefficients vary across the series. A typical example of such a behavior is shown in Figure 8.1 (see also Table 8.1) [40].

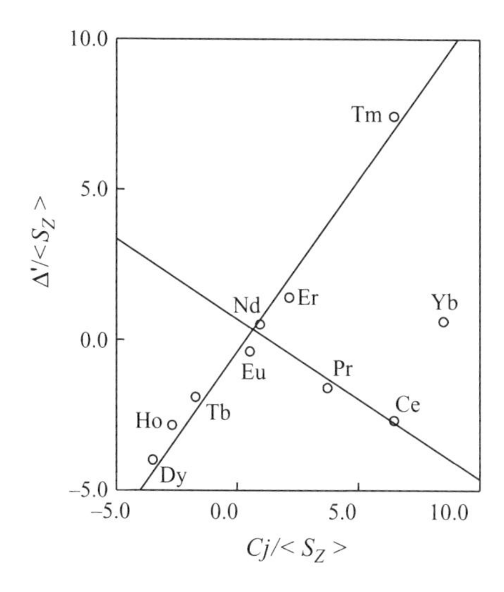

Figure 8.1 Plot of LIS data of the protons of the acetate pendant arms for Ln-EGTA at 300 K according to (8.13); data from [40], are plotted by using the C_j and $< S_z >$ values shown in Table 8.1.

Table 8.1 Values of $< S_z >$, relative values of C_j at room temperature, longitudinal relaxation times (T_{le}) in Ln(III) aquo complexes at 312 K and 2.1 T, and theoretical effective magnetic moments (μ_{eff}) of Ln(III) ions.

Ln	$< S_z >^a$	C_j^b	T_{le}^c (10^{-13}s)	$\mu_{eff}{}^d$ (BM)
Ce	−0.974	−6.3	0.9	2.56
Pr	−2.956	−11.0	0.57	3.62
Nd	−4.452	−4.2	1.15	3.68
Sm	0.224	−0.7	0.45	1.55–1.65
Eu	7.569	4.0	0.09	3.40–3.51
Gd	31.500	0	10^4–10^{5e}	7.94
Tb	31.853	−86	2.03	9.7
Dy	28.565	−100	2.99	10.6
Ho	22.642	−39	1.94	10.6
Er	15.382	33	2.38	9.6
Tm	8.210	53	3.69	7.6
Yb	2.589	22	1.37	4.5

[a] Reference [7].
[b] References [18] and [20], with values scaled to −100 for Dy.
[c] References [51] and [52]; when zero-field spliting (ZFS) effects are included in the processing of the data, the values are three times as large.
[d] Reference [31].
[e] Reference [50].

Discrimination between these two possibilities is possible for complexes with (effective) axially symmetric magnetic susceptibilities ($G_2 = 0$). Then it is possible to factor out the remaining crystal field coefficient. Several procedures to achieve this have been outlined in the literature [41–43]. For example, Equation (8.15) can be derived [43], where, $\Delta_{i,j}$ and $\Delta_{k,j}$ denote the paramagnetic-induced shifts (after subtraction of the diamagnetic contributions) of two given ligand nuclei, j and k, by the Ln^{3+} ion, i, where $R_{ik} = G_i/G_k$:

$$\frac{\Delta_{i,j}}{< S_z >} = (F_i - R_{ik}F_k) + \frac{R_{ik}\Delta_{k,j}}{< S_z >} \tag{8.15}$$

Accordingly, plots of $\Delta_{i,j}/ < S_z >_j$ versus $\Delta_{k,j}/ < S_z >_j$ should give straight lines for isostructural complexes. The slopes give the relative G values (R_{ik}) and the intercepts give the values of $(F_i - R_{ik}F_k)$.

Alternatively, if it is known that the geometry of a Ln^{3+} complex is invariant along the Ln^{3+} series, relative G values can be obtained by using Equation (8.11) and optimization of F, G, and C_j [5]. First, the values of C_j are kept constant, while F and G are fitted. Then, in a second step, the optimized values of F and G are held constant while C_j for each Ln^{3+}, scaled for convenience to an arbitrary value of -100 for Dy^{3+}, is optimized by linear regression. This two-step procedure is repeated until no further improvement of the fit between calculated and observed Δ' values is found. In this way, relative G values are obtained, while any variations in the crystal field coefficients are compensated for by inclusion of C_j in the fitting procedure.

Kemple *et al.* have included contact shifts as variable parameters in a fitting procedure of LIS values using Equation (8.4) [17]. Forsberg *et al.* have further expanded this procedure and demonstrated its usefulness in the field of MRI contrast agents by a study on the solution structure of a DOTA derivative [37].

3 LANTHANIDE-INDUCED RELAXATION RATE ENHANCEMENTS

3.1 EVALUATION OF BOUND RELAXATION RATES

Upon coordination of a ligand to a paramagnetic lanthanide ion, both the longitudinal ($1/T_1$) and the transverse relaxation rates ($1/T_2$) are enhanced. The inversion-recovery method is usually applied to measure longitudinal relaxation times [44], and the transverse relaxation times are usually measured with the Carr–Purcell–Meiboom–Gill pulse sequence [45] or from the line-widths of the resonances with the use of the relationship, $1/T_2 = \pi\Delta\nu_{1/2}$. The relaxation-enhancing effects consist of inner-sphere and outer-sphere contributions. The latter are usually relatively small and therefore are often neglected. The effects of Gd^{3+} on the relaxation rate of water nuclei are extensively discussed in Chapter 2. For structural investigations on the bound organic ligand, use of this lanthanide is not very practical, because of the huge line broadening in the NMR spectra of ligand nuclei. Therefore, investigations which involve the organic ligand nuclei are usually carried out with the other Ln^{3+} ions.

The first step in the analysis of the experimental data is an evaluation of the relaxation rates in the bound ligand ($1/T_{i,M}, i = 1, 2$). In studies of contrast agents, the relaxation rates are usually measured on samples with the pure Ln^{3+} complex, for which the dissociation of the organic ligand is negligibly small. Then the bound relaxation rates are related to the measured ones ($1/T_{i,obs}$) via Equation (8.15) [46–48], where τ_M is the residence time of the ligand in the complex:

$$\frac{1}{T_{i,obs}} = \frac{1}{T_{i,M}} + \frac{1}{\tau_M} \tag{8.15}$$

Therefore, $T_{i,M}$ can easily be determined using Equation (8.15) when $\tau_M >> T_{i,M}$. Otherwise, τ_M should be included in the procedure for the fitting of the relaxation rates to the structural model (see below). Other equations are required for the evaluation of bound relaxation rates if the exchange between bound and free ligand is rapid on the NMR time-scale [13, 46–48].

3.2 INNER-SPHERE RELAXATION

After evaluation of $1/T_{i,M}$, these values have to be corrected for any diamagnetic contributions, for example, by subtraction of the relaxation enhancements observed for La^{3+}, Lu^{3+} or Y^{3+}. The resulting relaxation may be attributed to the interaction between the fluctuating magnetic field of the unpaired electrons of the lanthanide ion and the nuclear spins [49, 50]. This electron–nuclear interaction is randomly modulated by electron spin relaxation, molecular tumbling, and chemical exchange. The relaxation rates are composed of contributions from the contact or hyperfine ($1/T_{i,c}$), the dipolar ($1/T_{i,p}$), and the Curie ($1/T_{i,\chi}$) mechanisms:

$$\frac{1}{T_{i,M}} = \frac{1}{T_{i,c}} + \frac{1}{T_{i,p}} + \frac{1}{T_{i,\chi}} \tag{8.16}$$

The contact (also known as the scalar or hyperfine) contribution is the result of a through-bond effect and is negligible for Ln^{3+} ions other than Gd^{3+} [51].

The dipolar contribution ($1/T_{i,p}$) results from through-space interactions due to the random fluctuations of the electronic field. If zero-field splittings are neglected [52], it can be described by the Solomon–Bloembergen Equations (8.17) and (8.18), when it is assumed that the complexes undergo isotropic reorientation [53].

$$\frac{1}{T_{1,p}} = \frac{2}{15}\left(\frac{\mu_0}{4\pi}\right)^2 \frac{\gamma_I^2 \mu_{eff}^2 \beta^2}{r^6}\left(\frac{3\tau_c}{1 + \omega_I^2\tau_c^2} + \frac{7\tau_c}{1 + \omega_S^2\tau_c^2}\right) \tag{8.17}$$

$$\frac{1}{T_{2,p}} = \frac{1}{15}\left(\frac{\mu_0}{4\pi}\right)^2 \frac{\gamma_I^2 \mu_{eff}^2 \beta^2}{r^6}\left(4\tau_c + \frac{3\tau_c}{1+\omega_I^2\tau_c^2} + \frac{13\tau_c}{1+\omega_s^2\tau_c^2}\right) \tag{8.18}$$

Here, $\mu_0/4$ is the magnetic permeability of a vacuum, γ_I is the magnetogyric ratio, μ_{eff} is the effective magnetic moment of the Ln^{3+} ion, β is the Bohr magneton, ω_s is the Larmor frequency of the nucleus under study and τ_c is the correlation time. The latter is given by Equation (8.19), where τ_R is the rotational tumbling time of the complex and T_{1e} is the longitudinal electronic relaxation times.

$$\frac{1}{\tau_c} = \frac{1}{T_{1e}} + \frac{1}{\tau_M} + \frac{1}{\tau_R} \tag{8.19}$$

Typically, $\tau_R = 10^{-10} - 10^{-11}s$ for low-molecular-weight complexes in water, and $\tau_M > 10^{-9}$s. For lanthanide ions other than Gd^{3+}, T_{1e} is much smaller ($\sim 10^{-13}$ s, see Table 8.1) and then Equation (8.17) and (8.18) simplify to the following:

$$\frac{1}{T_{1,p}} = \frac{1}{T_{2,p}} = \frac{4}{3}\left(\frac{\mu_0}{4\pi}\right)^2 \frac{\gamma_I^2 \mu_{eff}^2 \beta^2}{r^6} T_{1e} \tag{8.20}$$

The Curie or susceptibility mechanism is another dipolar effect, which arises from the interaction of the nuclear spin with the thermal average of the electronic spin. This process is modulated by the rotational motion of the complex and can be expressed as shown in Equations (8.21) and (8.22) [54, 55]:

$$\frac{1}{T_{1,\chi}} = \frac{6}{5}\left(\frac{\mu_0}{4\pi}\right)^2 \frac{\gamma_I^2 H_0^2 \mu_{eff}^4 \beta^4}{(3kT)^2 r^6}\left(\frac{\tau_R}{1+\omega_I^2\tau_R^2}\right) \tag{8.21}$$

$$\frac{1}{T_{2,\chi}} = \frac{1}{5}\left(\frac{\mu_0}{4\pi}\right)^2 \frac{\gamma_I^2 H_0^2 \mu_{eff}^4 \beta^4}{(3kT)^2 r^6}\left(4\tau_R + \frac{3\tau_R}{1+\omega_I^2\tau_R^2}\right) \tag{8.22}$$

Here, H_0 is the magnetic field strength. The contribution of this mechanism to the total longitudinal relaxation can be substantial for slowly tumbling molecules (macromolecules). It has been shown that at higher magnetic fields, the Curie contribution to the ^1H relaxation of even small molecules, such as the Ln^{3+} aquo ions, becomes significant [52].

The dipolar contribution to the transverse relaxation rates ($1/T_{2,p}$) of ligand nuclei in the proximity of a Ln^{3+} ion ($Ln \neq Gd$) equals $1/T_{1,p}$ (Equation (8.20)). The Curie spin contribution, however, is given by Equation (8.22), and, usually, can not be neglected. As pointed out by Aime *et al.*, this can be exploited by using the difference of the transverse and the longitudinal relaxation rates [56]:

$$\frac{1}{T_{2,M}} - \frac{1}{T_{1,M}} = \frac{1}{5}\left(\frac{\mu_0}{4\pi}\right)^2 \frac{\gamma_I^2 H_0^2 \mu_{eff}^4 \beta^4}{(3kT)^2 r^6}\left(4\tau_R - \frac{3\tau_R}{1+\omega_I^2\tau_R^2}\right) \tag{8.23}$$

When τ_R is known, the r values can be determined directly from the relaxation rates by using this equation. The advantage is that no knowledge of the magnitude of T_{1e} is required. It should be noted, however, that at high magnetic fields, T_{1e} of the Ln^{3+} ions is generally not very sensitive to the ligation of these ions [51, 57, 58]. A good estimate can be obtained from, for example, the electronic relaxation rates of the Ln^{3+}-aquo ions [52, 57]. Because of the $1/r^6$ relationship, the accuracy of the estimated T_{1e} does not need to be very high in order to obtain accurate r values; an error of 30 % in $1/T_{1e}$ corresponds to an error of only 5 % in r.

Alternatively, Equations (8.20) and (8.21) may be combined to give the following:

$$\frac{1}{T_{1,M}} = \frac{k}{r^6} \tag{8.24}$$

where k is a constant, provided that τ_R is constant for all nuclei of the ligand. Application of this equation allows the determination of relative distances between the Ln^{3+} ion and the ligand nuclei, without the need for good estimates of $1/T_{1e}$ or τ_R.

The relaxation-enhancing effect diminishes steeply upon increase of the distance to the lanthanide ion, due to the $1/r^6$ dependence. Consequently, relaxation data are particularly powerful in obtaining information on geometrical parameters of nuclei in the proximity of the Ln^{3+} ion.

It should be noted that the Solomon–Bloembergen and the Curie equations discussed above, which are commonly applied for all lanthanides, are derived with the assumption of magnetic isotropy. This holds strictly for Gd^{3+} only, which has $S = 0$. More complex expressions have been developed for systems where magnetic anisotropy is present [52, 59–61].

3.3 OUTER-SPHERE RELAXATION

Outer-sphere relaxation is a complex problem that is often neglected in relaxation studies on organic ligands. There are, however, several situations in which contributions by this mechanism may be important. For instance, in the more remote nuclei of an organic ligand, the outer-sphere contribution may be relatively important [62]. Outer-sphere relaxation contributions can be estimated by using a compound that has a similar shape and size as the complex under study, but which does not coordinate to the species concerned. For example, in a study of the solution structure of inclusion compounds of $Tm(DOTA)^-$ in γ-cyclodextrin, β-cyclodextrin was applied in order to estimate the outer-sphere contribution to the relaxation-rate enhancements [63].

4 GEOMETRY CALCULATIONS

A large amount of literature exists on fitting of lanthanide-induced shift (LIS) and lanthanide-induced relaxation (LIR) values to expressions, which relate these parameters to molecular structures [13]. Usually, only the pseudocontact contribution to the total LIS is considered. These are obtained by dissecting the LIS into the various contributions (see above). Often, only Yb^{3+}-induced ^1H shifts are considered, which are assumed to be almost completely of pseudo-contact origin. In some cases, contact contributions have been included as a variable into the fitting procedure [17, 37]. Lisokowski *et al.*, however, have pointed out that omitting positions with large contact contributions produces similar results in shift calculations [64]. It is important to realize that shift and relaxation data provide different types of information. Relaxation rates are proportional to r^{-6}, whereas shifts are proportional to r^{-3} and contain, in addition, angular information.

Generally, the agreement between the observed and the calculated values is evaluated by using Hamilton's crystallographic agreement factor AF (see Equation (8.25)) [65]. Here f_{oi} and f_{ci} are the observed and calculated values, respectively, and w_i are weighting factors. Structural determinations are commonly carried out by minimization of AF or of the summed (weighted) squared errors.

$$AF = \sqrt{\frac{\sum_i (f_{oi} - f_{ci})^2 w_i}{\sum_i f_{oi}^2 w_i}} \tag{8.25}$$

The number of variables in structure elucidations is rather high, with three parameters being required to define the position of a rigid ligand with respect to the Ln^{3+} ion and the magnetic axis if there is a unique position and if the (effective) axial model applies. When the latter is not true, another two parameters are needed to specify the position relative to the axes. If several isomers occur, even more variables are required to define the various species and their populations. Usually, the factors D_1 and D_2 in Equation (8.6) or C in Equation (8.7) for the induced shifts are unknown, and have to be treated as variables. Similarly, relaxation rates are often fitted by using Equation (8.24), since the correlation times associated with the relaxation processes are not determined. Shifts and relaxations can provide a maximum of three observations per ligand nucleus (Δ, T_1, and T_2). Consequently, the number of variables is often large with respect to the number of observations. It is therefore useful to introduce as many constraints as possible.

The analysis of LIS data is quite complicated in cases where the effective axially symmetric model is not valid. Usually, the data are fitted by using Equation (8.7). The directions of the principal axis are varied in the molecule-fixed frame. After each variation on the location of the axes, the angular functions must be recalculated, which makes the procedure non-linear. Kemple *et al.* have proposed another procedure, in which the ligand is placed in an arbitrary axis system with the Ln^{3+} ion at the origin [17]. In this case, Equation (8.4) applies and no assumptions are made regarding the directions of the principal magnetic axes. The fitting is performed *linearly* with $(\chi - \chi_{zz})$, $(\chi_{xx} - \chi_{yy})$, χ_{xy}, χ_{xz}, and χ_{yz} as fitting parameters. These five parameters allow determination of the principal magnetic axis of the anisotropic part of the magnetic susceptibility tensor. Forsberg *et al.* have further elaborated this procedure and developed an integrated approach for the analysis of LIS data, combining molecular mechanics with LIS calculations, assignment via the permutation method (see above) and a separation of contact and pseudocontact shifts (^{1}H) [37, 66]. The procedure was exemplified with a study on Ln^{3+} complexes of 1,4,7,10-tetrakis(*N*,*N*-diethylacetamido)-1,4,7,10-tetraazacyclo-dodecane. An accurate description was obtained of the equilibrium of two isomeric Ln^{3+} complexes which occur in solution.

5 TWO-DIMENSIONAL NMR

Two-dimensional NMR spectroscopy of paramagnetic complexes is becoming an increasingly important technique [67], particularly in the study of exchange processes. The difficulties that may be envisaged with two-dimensional NMR have been discussed by Jenkins and Lauffer in a study on the structure and dynamics of $Ln(DTPA)^{2-}$ complexes [68]. The problems that may arise are related to the large induced shifts and to the relaxation-rate enhancements induced by the paramagnetic ion. The large chemical shift ranges require large spectral windows, which make it difficult to produce uniform pulse-widths across the entire NMR spectrum [69]. Furthermore, in COSY spectra an adequate digitization is required in order to avoid cancellation of the anti-phase absorptions in cross-peaks. This cancellation problem is reinforced by the relatively large linewidths of the signals. It is most likely that a better approach to study *J*-correlations is with TOCSY experiments since these produce in-phase multiplets in the cross-peaks, and therefore cancellation does not occur when the multiplets are unresolved. Another problem in COSY spectra is that a short T_2 will lead to a rapid decay of phase coherence and thus to low intensities of cross-peaks [68]. As soon as the evolution time approaches the magnitude of T_2, little cross-peak intensity will accumulate. It is, therefore, important to minimize contributions resulting from Curie relaxation by working at high temperatures and at as low a field as possible. Similarly, the transverse magnetization in magnetization-transfer experiments, such as EXSY, will be destroyed by spin-lattice relaxation before appreciable cross-peak intensity can build up if $1/T_1$ is much greater than the exchange rate.

Several applications of COSY [37, 68–70] and EXSY NMR [68–72] are described in the literature. The last technique in particular has proven to be very valuable in the study of the dynamics of complex conformational equilibria of the multidentate ligands in Ln^{3+} complexes in the field of MRI contrast agents. For example, Jacques and Desreux have completely solved the dynamic matrix and determined the activation parameters of the inter- and intramolecular dynamic processes that occur in the $Yb(DOTA)^-$ system with the use of variable-temperature EXSY spectroscopy [69].

Lisokowski *et al.* have employed COSY, ROESY, HMQC, and HMBC spectra for the assignment of the ^1H and ^{13}C resonances of Ln^{3+} complexes of a texaphyrin derivative [64, 73].

6 ^{139}La AND ^{89}YNMR

Lanthanum has two stable isotopes, ^{138}La and ^{139}La , both of which are NMR-sensitive and have been observed [74]. The latter has by far the more favorable NMR properties, and therefore, it is the only one that has found practical applications. The ^{139}La nucleus has a natural abundance of 99.91 %, spin $I = 7/2$, quadrupole moment $Q = 0.21 \times 10^{-28} m^2$, and a high receptivity (336 relative to ^{13}C). Consequently, observation is easy, when the ion is in a highly symmetric environment. Pronounced relaxation effects occur upon coordination, which can be utilized for the study of both the structure and dynamics of La^{3+} compounds. The chemical shift range is about 1700 ppm, indicating also that this parameter is highly sensitive to the coordination of Ln^{3+}. An empirical relationship for the prediction of ^{139}La chemical shifts of polyhydroxycarboxylate Ln^{3+} complexes [75, 76] has been extended for polyaminocarboxylates [77] (Equation (8.26)):

$$\delta = 30n_{COO} + 50n_N \tag{8.26}$$

Here, n_{COO} is the number of carboxylate donor sites and n_N is the number of bound ethylenediamine nitrogen atoms. Although these rules are very useful for the interpretation of shift data, some caution is needed, since the experimental chemical shift of $La(DTPA)^{2-}$ does not agree with the coordination structure as obtained more recently by other methods [78, 79]. More experimental data of structurally well-defined compounds would be needed to develop more refined relationships which are generally applicable. The chemical shifts of various complexes in the field of MRI contrast agents are compiled in Table 8.2.

The large difference in chemical shift between free and bound La^{3+} has been exploited in a study on the transmetallation of La^{3+} complexes of DTPA and bisamide derivatives with Zn(II). The liberated La^{3+} gives rise to a broad signal ($\Delta\nu_{1/2} = 2$–5 kHz) at 20–80 ppm, which leads to the conclusion that interaction occurs with the negatively charged La^{3+} and Zn(II) complexes of the DTPA derivatives [80].

The chemical properties and the ionic radius of Y^{3+} (0.89 Å) are comparable to that of Ln^{3+} ions (1.06–0.85 Å), and thus, Y^{3+} complexes have a

Table 8.2 ^{139}La chemical shifts of various La^{3+} complexes in the field of MRI contrast agents.

Compound	T (°C)	δ (ppm)	$\delta\nu_{1/2}$ (Hz)	Reference
$La(D_2O)_9^{3+}$	73	0	80	75
$La(EDTA)^-$	67	193	7 600	77
$La(DTPA)^{2-}$	67	258	6 300	77
$La(BOPTA)^{2-}$	70	246	9 300	120
La(DTPA-BPA)	80	210	10 000	80
La(DTPA-BGLUCA)	80	210	9 850	80
La_2(30-DTPA-en-DTPA-en)	80	240	17 000	82
$La(TTHA)^{3-}$	67	224	5 570	77
La(NOTA)	67	223	10 000	77
$La(DOTA)^-$	67	323	3 120	77

Table 8.3 ^{89}Y chemical shifts of various Y^{3+} complexes in the field of MRI contrast agents.

Compound	Concentration (M)	δ (ppm)	Reference
$Y(Cl)_{3aq}^{3+}$ (1 M)	1.0	0	81
$Y(EGTA)^-$	0.35	68.5	81
$Y(DTPA)^{2-}$	0.30	82.2	81
Y(DTPA-BEA)	0.1	82.5	82
Y(DTPA-BGLUCA)	0.1	82	82
Y_2(30-DTPA-en-DTPA-en)	0.1	81	82
YTm(30-DTPA-en-DTPA-en)	0.1	81	82
$Y(TTHA)^{3-}$	0.30	101.5	81
Y_2(TTHA)	0.30	109.4	81

coordination geometry that is closely related to the geometry of the Ln^{3+} complexes. ^{89}Y is a spin $I = 1/2$ nucleus, and consequently, the NMR signals are usually very narrow (< 5 Hz). The NMR receptivity is very low (0.007 with respect to ^{13}C), but this is compensated by a natural abundance of 100%. Several chemical shifts of Y^{3+} polyaminocarboxylates have been reported (see Table 8.3). No clear relationship between the chemical shifts and the structures has been deduced up until the present day [81]. The small linewidths of the ^{89}Y resonances allow the resolution of the various isomers present in solutions of Y^{3+} complexes of DTPA-bisamides [81, 82]. The observation of geminal coupling between ^{89}Y and ^{15}N in Y^{3+} complexes of EDTA, DTPA, and TTHA affords direct proof of binding of the metal ion to the nitrogen atoms in these ligands [83].

7 WATER HYDRATION NUMBERS

For MRI contrast agents, the number of water molecules present in the first coordination sphere of Gd^{3+} is an important factor which contributes to the relaxivity of the complex. Several methods have been used to determine the hydration numbers of MRI contrast agents. Ln^{3+} induced ^{17}O contact shifts of bound water molecules are essentially independent of structure (see above); as a result, these shifts are proportional to the hydration number. Thus, comparison of the contact shift of a Ln^{3+} complex with that of the corresponding Ln^{3+}-aquo ion affords the hydration number [13]. In principle, the contact shifts may be directly determined with the use of Gd^{3+} ($C_j = 0$); however in practice, LIS measurements with this cation may be difficult due to extreme line-broadening and exchange phenomena. Alternatively, fitting of variable-temperature LIS and LIR data may afford the required shifts (see Chapter 2). A favorable ratio of LIS and line-broadening may be obtained with Dy^{3+}. For Dy^{3+}, the LIS is dominated by the contact contribution ($> 85\%$) and a laborious procedure to dissect the observed LIS into the various contributions is not needed. In this way, the number of inner-sphere water molecules of Dy^{3+} analogs of MRI contrast agents can be determined very rapidly [84]; usually, only 1 min of spectrometer time is required for such a measurement and ^{17}O enrichment is not necessary. In order to ensure that rapid exchange takes place between the bound and free state, the measurements are preferably performed at high temperatures and at low field. Concentrations of 20–50 mM are required for sufficient accuracy. Since the ^{17}O water signal is very strong, measurements are possible with sample volumes as small as 50 μL [85]. If frequency-locking of the spectrometer is not possible, a correction for the BMS shift (e.g. calculated with Equation (8.9)) or use of an internal standard (e.g. the ^1H signal of *tert*-butanol [85]) is required.

Alternatively, hydration numbers can be determined by using the luminescence method of Horrocks and Sudnick which is based on the difference in decay rate of the Eu^{3+} or Tb^{3+} luminescence between solutions of the complex concerned in H_2O and D_2O [86]. Recently, Parker and co-workers have introduced a refinement of this method by considering the contributions from water molecules diffusing close to the Ln^{3+} complex and of other proton-exchangeable oscillators [87]. In general, the results obtained with the ^{17}O technique are in good agreement with those obtained from luminescence measurements (see Chapter 2, Table 2.1).

A third way to determine hydration numbers involves NMRD. The hydration number may be included as a variable in the fitting procedure. Chang *et al.* have reported that, for a series of Ln^{3+} complexes, the relaxivity at a Larmor frequency of 20 MHz shows a linear relationship with the hydration number [88]. Recent insights suggest that erroneous results may be obtained in this way, due to the limitation of the observed relaxivity by τ_M or τ_R (see Chapter 2). It is common practice now to determine the hydration number independently and then enter the value into the fitting procedure of NMRD curves to generate the other parameters that govern the relaxivity.

Hydration numbers measured using any of the techniques are frequently not integral numbers but rather fractions. This may be the result of either the accuracy (± 0.1 – ±0.5) or the existence of an equilibrium of two or more complexes with different hydration numbers. The latter may be checked by UV–Vis spectrophotometric measurements of the $^7F_0 \rightarrow ^5D_0$ transition for the Eu^{3+} complex. Species with different hydration numbers usually give rise to multiple absorption bands [89].

Since the Ln^{3+} coordination number of an aqueous Ln^{3+} complex is usually 8–9, the readily accessible water solvation number may give a good indication of the coordination of the other donor groups in the coordinated ligand.

8 CHIRALITY IN LANTHANIDE COMPLEXES OF POLYAMINOCARBOXYLATES

Many Ln^{3+} complexes of relevance for MRI contrast agents contain 1,2-ethylenediamine moieties. Upon binding to a Ln^{3+} ion in a bidentate fashion, the resulting five-membered chelate may adopt several conformations, of which those with gauche ethylene groups are by far the most stable [90, 91]. Thus, two forms may occur, δ and λ, defined by the helicity of the C–C bond relative to the Ln^{3+}–N–N plane (see Figure 8.2). An inspection of the crystal structures of Ln^{3+} complexes of DTPA and DOTA derivatives shows that Ln^{3+}-bound diethylenetriamine moieties in these complexes always occur either in the $\delta\delta$ or in the $\lambda\lambda$ conformation. In these conformations, the steric interactions are

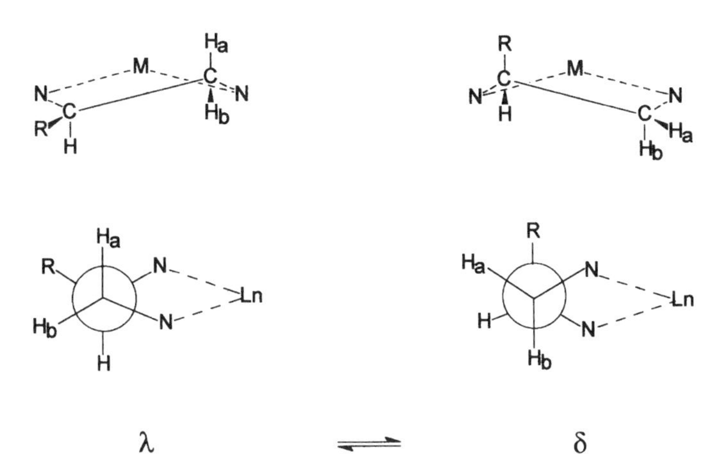

Figure 8.2 Equilibrium between the two gauche conformations of ethylenediamine moieties in polyaminocarboxylates.

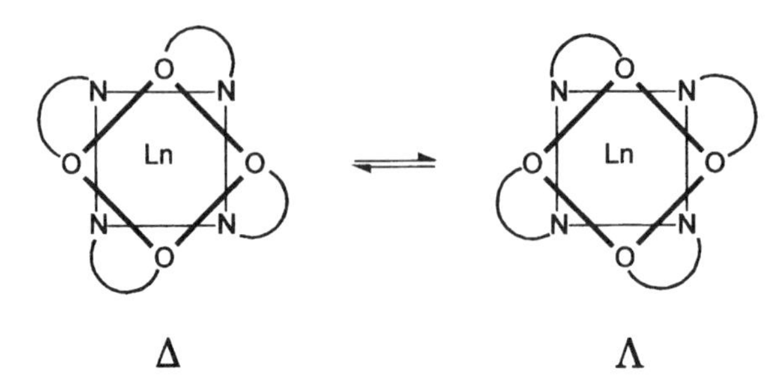

Figure 8.3 Chirality in macrocyclic polyaminocarboxylates.

minimized. As a result of the helicity, the central nitrogen atom is chiral when it is coordinated by a metal ion, as, for example, in $Ln(DTPA)^{2-}$.

Upon coordination of a nitrogen atom, its inversion is precluded. Consequently, it will become chiral when it has three different substituents.

A common feature of cyclen-based macrocyclic Ln^{3+} complexes is the formation of various isomers which may display dynamic behavior on the NMR time-scale (see below). The inherent structural features of these complexes give rise to two independent motions which provide sources of chirality ($\lambda\lambda\lambda\lambda$ versus $\delta\delta\delta\delta$ conformation of the ethylene rings and Δ versus Λ orientation of the pendant arms (see Figure 8.3) and may in principle generate four stereoisomers (see below).

9 LANTHANIDE COMPLEXES OF LINEAR POLYAMINOCARBOXYLATES

9.1 DTPA

The ligation of Ln^{3+} ions by DTPA has been deduced from Nd^{3+}-induced ^{13}C relaxation-rate enhancements [78] and from ^{1}H NMR spectra of various $Ln(DTPA)^{2-}$ complexes [68, 79]. From the T_1 values for the ^{13}C nuclei in the diamagnetic La^{3+} complex, the rotational correlation time τ_R of the $Ln(DTPA)^{2-}$ complexes has been estimated to be 4×10^{-11} s [78]. By using Equations (8.17) and (8.21), the ratio of the Curie and the dipolar relaxation rates may be estimated as ≤ 0.05 for ^{13}C nuclei in the Nd^{3+} complexes, under the experimental conditions (where $H_0 = 4.7$ T and $T = 346$ K). Therefore, the Curie contributions were neglected in further calculations. After correction for diamagnetic contributions by subtraction of the relaxation rates in the corresponding La^{3+} complex, the Nd^{3+}–C distances in the complexes were calculated by the use of Equation (8.20). The distances obtained (see Table 8.4) show that the ligand is coordinated in an octadentate fashion via all carboxylate groups and the three nitrogen atoms of the diethylenetriamine backbone. Ln^{3+}-induced

Table 8.4 Nd^{3+}–C distances in the isomers of $Nd(DTPA)^{2-}$ and Nd(DTPA-BPA), and comparison with the solid-state structures of Gd(DTPA-BEA).

	r(Å)		
Nucleus	$Nd(DTPA)^{2-}$ (soln) [78]	Nd(DTPA-BPA) (soln) [104]	Gd(DTPA-BEA) (X-ray) [109]
*C*O	3.15–3.20	3.08–3.19	3.25–3.31
CH_2COO	—[a]	3.25–3.37	3.40–3.56
$N^1\mathit{CH_2}CH_2N^3$ [b]	3.21	3.42–3.43	3.46–3.54
$N^1CH_2\mathit{CH_2}N^3$ [b]	3.48	3.36–3.36	3.46–3.51
α-*C* (amide chain)	—	5.33–5.47	5.13–5.23
β-*C* (amide chain)	—	6.46–6.89	5.26–6.45
γ-*C* (amide chain)	—	5.91–6.44	—

[a] Not measured because of overlapping resonances.
[b] N^1 and N^3 are the terminal and the central nitrogen atoms, respectively.

^{17}O NMR shift measurements show that the coordination sphere is completed by one water, which is in agreement with the results of Eu^{3+} luminescence studies [92, 93]. The water-coordination number appears to be independent of the Ln^{3+} ion [78]. Therefore, the total coordination number in all of these complexes is nine.

At low temperatures (0–25 °C), the ^1H spectra for Ln = Pr, Eu, and Yb displayed 18 signals, which coalesced to 9 signals upon increase of the temperature [68, 79]. From the presence of separate resonances for all acetate groups, it was concluded that the central acetate group is probably coordinated. The signals were assigned and the exchange was studied with 2D COSY and EXSY spectroscopy (see Table 8.5).

As outlined above, Ln^{3+} binding of the three nitrogens of the diethylenetriamine backbone results in chirality of the central nitrogen atom. Obviously, the two enantiomers can not be discriminated by NMR directly. However, from the exchange phenomena observed, it may be concluded that the $\delta\delta$ and the $\lambda\lambda$ enantiomers interconvert, which results at high temperatures in an average situation with an effective mirror plane through the central glycine unit [78, 79].

Calculations on separated pseudocontact ^{13}C LIS values [78] and Yb^{3+} induced ^1H shifts (assumed to be of fully pseudocontact origin) [68], are consistent with calculated shifts for the proposed structure by using a non-axial model (Equation (8.6)). From LIS and LIR measurements on ^6Li counter-ions, it has been concluded that these ions reside in the second coordination sphere near the Ln^{3+}-bound carboxylate groups [78].

Further support for the proposed solution structure of $Ln(DTPA)^{2-}$ complexes has been obtained from T_1 values for methylene and carboxylate ^{13}C nuclei in the diamagnetic $La(DTPA)^{2-}$ complex, which were very similar for all

Table 8.5 A selection of thermodynamic data for exchanges in Ln^{3+} complexes.

Complex	Dynamic process	$\Delta G^{\ddagger}$ (kJ/mol)	$\Delta H^{\ddagger}$ (kJ/mol)	$\Delta S^{\ddagger}$ (J/mol K)	Reference
$Pr(DTPA)^{2-}$	Racemization central N	56.5 (298 K)	35.2	−71.4	79
$Eu(DTPA)^{2-}$	Racemization central N	55.4 (298 K)	38.5	−56.8	79
$Yb(DTPA)^{2-}$	Racemization central N	49.4 (298 K)	37.0	−41.7	79
Nd(DTPA-BPA)	Racemization central N	53 (283 K)	—	—	104
La(DTPA-BPA)	Racemization terminal Ns	71 (283 K)	47	−84	104
Lu(DTPA-BPA)	Racemization terminal Ns	67 (283 K)	42	−88	104
La(DTPA-BGLUCA)	Racemization terminal Ns	66 (283 K)	34	−116	105
La(DTPA-BENGALAA)	Racemization terminal Ns	65 (283 K)	37	−100	105
$Gd(EOB\text{-}DTPA)^{2-}$	Racemization between enantiomers	75.3[a]	—	—	121
$Eu(DTPA\text{-}dienH^{+})$	Racemization central N	57.5 (299 K)	—	—	70
$Yb(DOTA)^{-}$	$M_1 \rightleftharpoons M_2$	65.9 (298 K)	82	52	69
	$M_1 \rightarrow m_1$; $M_2 \rightarrow m_2$	65.1 (298 K)	79	46	69
	$m_1 \rightarrow M_1$; $m_2 \rightarrow M_2$	61.2 (298 K)	64	9	69
	$M_1 \rightarrow m_2$; $M_2 \rightarrow m_1$	65.7 (298 K)	80	49	69
	$m_1 \rightarrow M_2$; $m_2 \rightarrow M_1$	61.6 (298 K)	66	14	69
$Lu(TETA)^{-}$	Exchange between two dodecahedra	63.7 (298 K)	71.7	27	145
$La(DOTP)^{5-}$	Ring inversion	101	—	—	163

[a] Activation energy

five acetate arms [94]. This suggests that each of the arms is coordinated, since an uncoordinated acetate would result in shorter rotational correlation times, and thus shorter relaxation rates for the atoms in the arm concerned. Furthermore, proton T_1 values for the Eu^{3+} complex agree with the octadentate coordination of the DTPA ligand.

Similar structures have been observed for solid complexes of Ln-DTPA derivatives from X-ray crystallography [95–97]. In all of these structures, the DTPA ligand is bound to the Ln^{3+} ion via the diethylenetriamine nitrogens, and the five carboxylate groups and the coordination sphere is completed with one water (see Figure 8.4). The coordination polyhedra can be described as (distorted) tri-capped trigonal prims (TPP) [98]. The distances between the DTPA carbon nuclei and the Ln^{3+} ion are about the same as those found in solution and both $\delta\delta$ and $\lambda\lambda$ conformations of the diethylenetriamine backbone are observed. Furthermore, some dimeric structures have been reported in which a carboxylate is bridging two Ln^{3+} ions [99–101]. These complexes have no water in the first coordination sphere of Ln^{3+}.

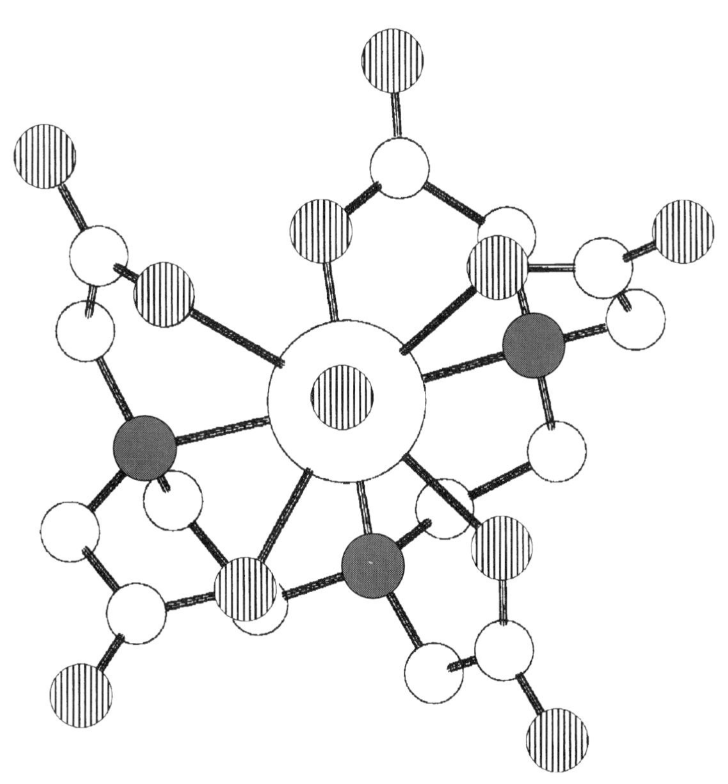

Figure 8.4 Crystal structure of $Nd(DTPA)^{2-}$ viewed looking down from the water–Nd bond.

A detailed X-ray absorption spectroscopic study of $Gd(DTPA)^{2-}$ showed that the local structure around the Gd^{3+} ion in aqueous solution is similar to that found in the crystals [102].

9.2 DTPA-BIS(AMIDES)

The ^{13}C NMR spectra of Ln^{3+} complexes of DTPA-BPA [103, 104], DTPA-BGLUCA and DTPA-BENGALAA [105] (see Chart 8.1) revealed a large number of resonances, indicating the presence of various isomers. Nd–C distances were calculated from Nd^{3+}-induced longitudinal relaxation rates by using Equation (8.20) (see Table 8.4). From these distances, it may be concluded that the Ln^{3+} coordination of these DTPA-bis(amides) is similar to that of DTPA. The organic ligand is bound via the three nitrogen atoms of the diethylenetriamine backbone, the three carboxylate groups, and the two amide oxygens. Ln^{3+}-induced ^{17}O NMR measurements show that the coordination sphere is completed by one water for all Ln^{3+} ions. The bound terminal nitrogen atoms of the diethylenetriamine unit are also chiral, thus resulting in a total of three chiral nitrogen atoms. Consequently, four diastereomeric pairs of enantiomers should be expected, which is in agreement with the number of signals observed at low temperatures (see Figure 8.5). The coordination

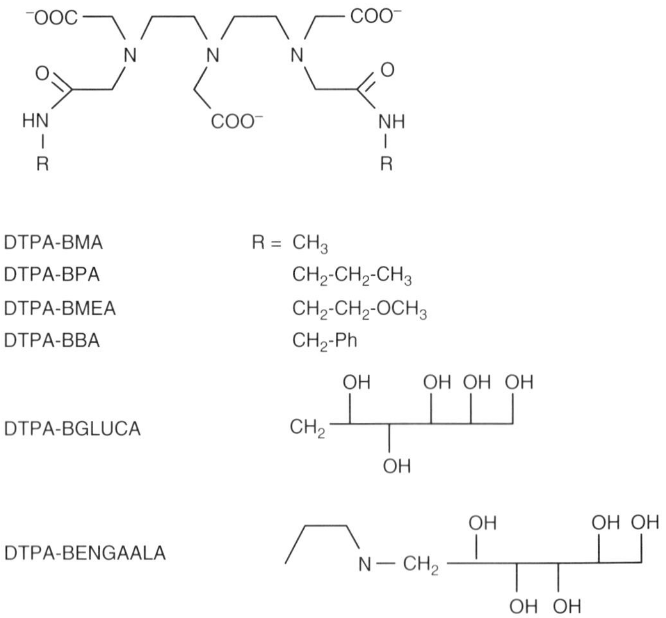

DTPA-BMA R = CH_3

DTPA-BPA CH_2-CH_2-CH_3

DTPA-BMEA CH_2-CH_2-OCH_3

DTPA-BBA CH_2-Ph

DTPA-BGLUCA

DTPA-BENGAALA

[Chart 8.1]

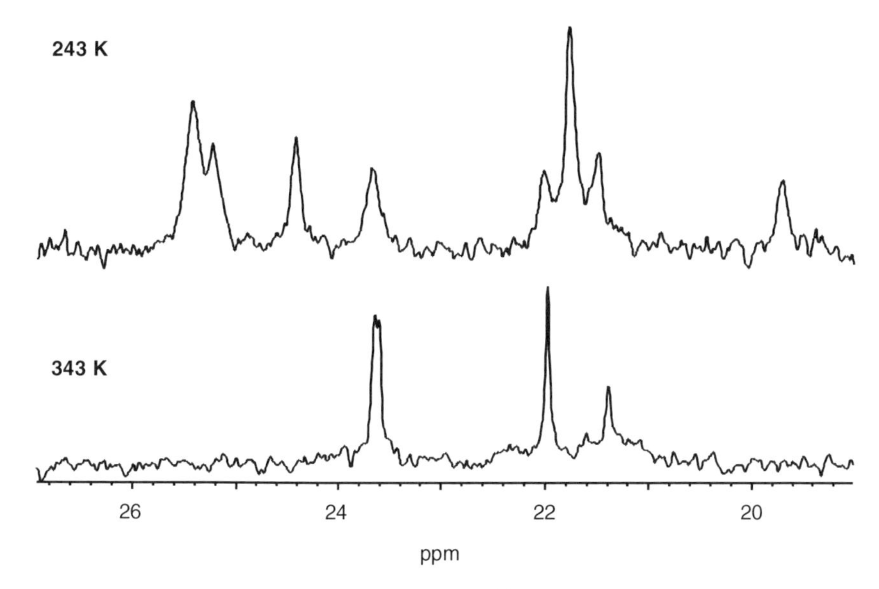

Figure 8.5 Influence of the temperature on the ^{13}C NMR resonances of the *β*-methylene carbons (–N–CH_2–*C*H_2–CH_3) of Nd(DTPA-BPA) (100.6 MHz) in CD_3OD–D_2O (1:1, v/v).

polyhedron of acyclic nine-coordinate Ln^{3+} complexes may be described by a distorted TPP. An inspection of molecular models shows that, for such a coordination polyhedron, a DTPA-bis(amide) can only be arranged around the Ln^{3+} ion when the middle nitrogen atom (N_3, see Figure 8.6) is on a prismatic position, whereas the two other nitrogens (N_1, N_2) must occupy the capping positions. The eight possible coordination configurations, assuming a TPP, are given in Figure 8.6.

The dynamics of the interchange of the various isomers in these complexes have been studied by line-shape analysis (see Table 8.5). Two isomerization processes could be discerned, including a slow process that was ascribed to racemization of the terminal nitrogen atoms, requiring decoordination of the concerned N atom and its neighboring acetate groups, and a relatively rapid racemization of the central N atom (see Figure 8.6). The latter process is associated with the interconversion of the ethylene groups of the diethylenetriamine unit ($\delta\delta \rightleftharpoons \lambda\lambda$). During this rearrangement, the nitrogen atoms of the backbone and the water oxygen remain at the same locations, whereas the other binding sites shuffle along the coordination polyhedron (for example in isomer $1 \rightarrow 1'$: $A_1 \rightarrow B_1 \rightarrow A_3 \rightarrow B_2 \rightarrow A_2 \rightarrow A_1$, (see Figure 8.7).

The occurrence of various interconverting isomers has also been observed in variable-temperature ^{13}C NMR studies of the diamagnetic Lu^{3+} complexes of DTPA-BMEA [106] and DTPA-BBA [107] (see Chart 8.1), which, in both cases, showed the presence of at least three isomers under slow-exchange

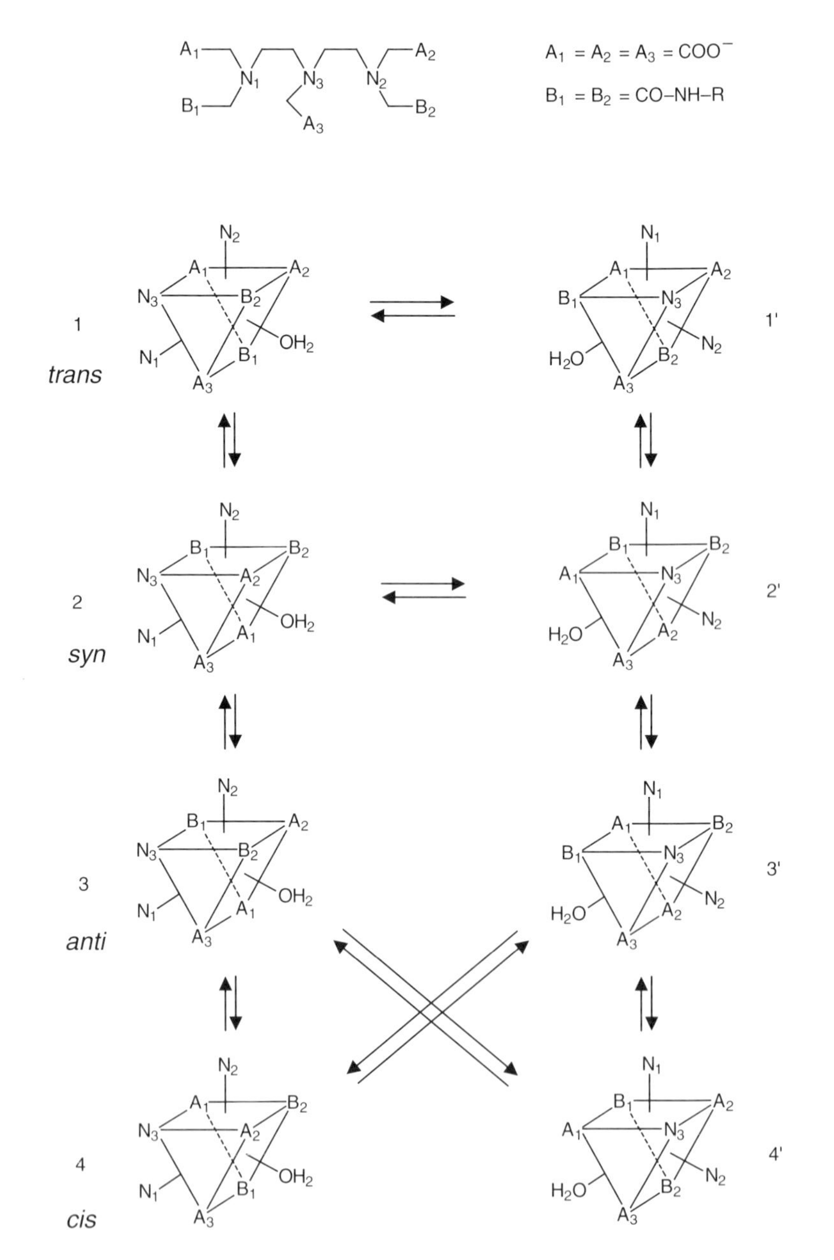

Figure 8.6 Coordination polyhedra of the eight enantiomers of Ln[DTPA-bis(amide)] complexes, assuming that the geometry is a tri-capped trigonal prism. Interconversions between the two columns correspond to the 'wagging' process (racemization at N_3), while the interconversions between rows in a column result in racemization at N_1 and N_3.

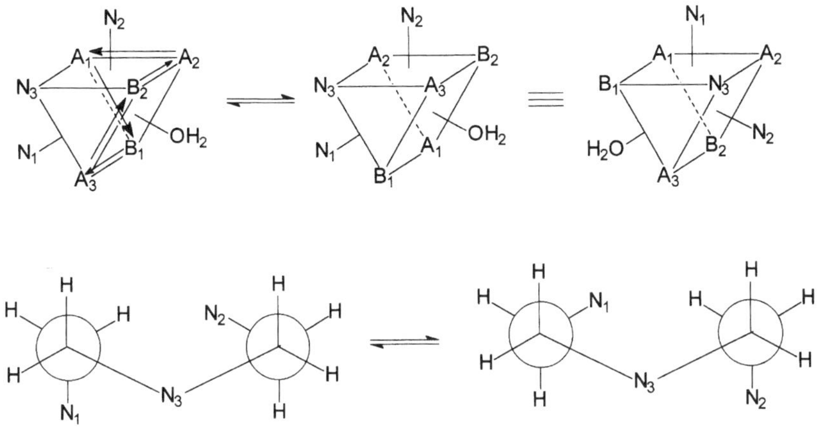

Figure 8.7 The shuffle of the coordination sites during the racemization of the central N atom of Ln[DTPA-bis(amide)] complexes (isomers 1 and 1').

conditions. A ^1H NMR study on La^{3+} and Lu^{3+} complexes of DTPA-BMA showed the presence of at least two isomers [108].

X-ray structures show that in crystals the structures are similar to those in solution [107, 109–112]: the DTPA-bisamide ligand is bound through the diethylenetriamine nitrogens, the amide oxygens, and the three carboxylates, and all complexes have one coordinated water. The Ln^{3+} complexes of DTPA-BBA have the *cis*-configuration [107, 111], whereas all other structures adopt the *trans*-configuration [109, 110, 112].

9.3 DTPA-BIS(AMIDES) INCORPORATED IN MACROCYCLES

Franklin and Raymond have studied Ln(18-DTPA-dien) complexes, in which the DTPA-bisamide unit is part of an 18-membered macrocycle [70] (see Chart 8.2). In the crystal structures of the La^{3+} and Eu^{3+} complexes, oligomerization was observed. The coordination of the Ln^{3+} ion by each DTPA unit, however, was similar to that in the linear bisamides. Vapor pressure osmometry indicated that oligomerization in solution does not occur under the conditions applied (0.2–0.5 M). Because of the constraints imposed by linking the two amide functions in the macrocycle, two of the isomers depicted in Figure 8.6 (**1**, **1**′ and **3**, **3**′) are sterically very unfavorable. In line with this, resonances for only two pairs of enantiomers were observed in the ^1H spectra of the Ln(18-DTPA-dien) complexes. The dynamic behavior showed that one pair was rather static, whereas the other was rapidly interconverting. The $\Delta G^{\ddagger}$ for this interconversion, determined with the use of a 2D-EXSY experiment, was about the same as those observed for the $\delta\delta \rightleftharpoons \lambda\lambda$ isomerization in the other DTPA derivatives (see Table 8.5). In the Ln(18-DTPA-dien) complex, the isomerization via this process is only

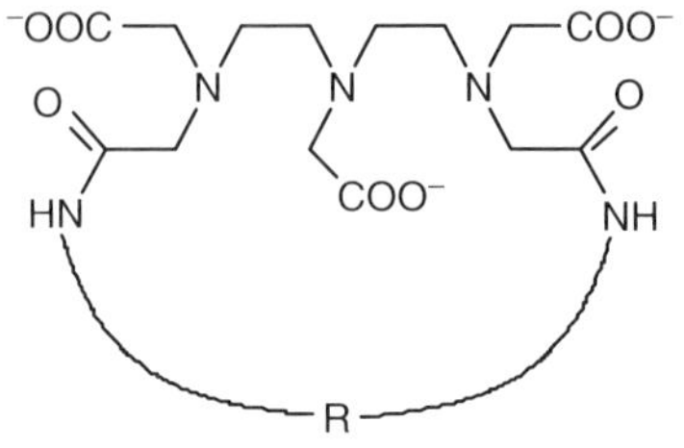

^{-}OOC N N N COO^{-}
O O
HN COO^{-} NH
R

15-DTPA-EAM	R = CH_2CH_2
16-DTPA-PAM	$CH_2CH_2CH_2$
16-DTPA-HPAM	$CH_2CHOHCH_2$
17-DTPA-BAM	$CH_2CH_2CH_2CH_2$
18-DTPA-dien	$CH_2CH_2NHCH_2CH_2$

30-DTPA-en-DTPA-en (n=2)

30-DTPA-bn-DTPA-en (n=4)

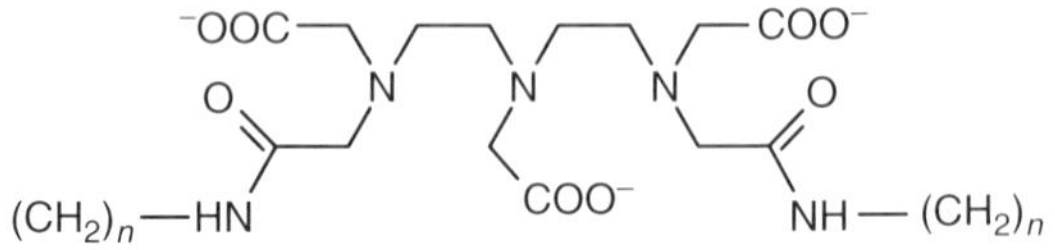

^{-}OOC N N N COO^{-}
O O
$(CH_2)_n$—HN COO^{-} NH—$(CH_2)_n$

[Chart 8.2]

possible between the mirror images **2** and **2′**. All other isomerizations would require decoordination and, therefore, are much slower. The latter process is acid-catalyzed. At 95 °C and pH 2, a slow isomerization takes place via partial decoordination. The EXSY spectrum showed cross-peaks corresponding to interconversion between **4** and **4′**, via the 'dynamic isomers' **2** and **2′** as intermediates.

Similar species occur in solutions of Ln^{3+} complexes of a 30-membered macrocycle Ln_2(30-DTPA-en-DTPA-en) constructed of two DTPA units linked by two ethylenediamine bridges [82]. For the 15-membered macrocycle 15-DTPA-EAM, however, only the equilibrium **2** $\rightleftharpoons$ **2′** was observed for the heavier Ln^{3+} ions [82]. Dy^{3+}-induced ^{17}O NMR shifts showed that the hydration number of this complex is 0.8 ± 0.2. A Eu^{3+} luminescence study indicated that $q = 2.3 \pm 0.5$ for the corresponding Eu^{3+} complex at low concentrations (5 μM) [113]. The results of the ^{17}O NMR method and luminescence are usually in good agreement with each other, and therefore it is assumed that the hydration number of the Ln(15-DTPA-EAM) complexes changes between Eu^{3+} and Dy^{3+} [82]. Unfortunately, the complexes of the lighter Ln^{3+} ions were insoluble under the conditions required for NMR analysis, probably due to the formation of dimeric complexes, similar to those observed in the solid state for the Gd^{3+} [114], La^{3+} [115] and Y^{3+} [115] complexes of this ligand. It should be noted that Ln^{3+} complexes of similar 16- [116] and 17-membered [117] macrocycles (16-

DTPA-PAM, 16-DTPA-HPAM, and 17-DTPA-BAM) crystallized as mononuclear complexes with a *syn*-configuration of the amides. The Gd^{3+} complex of a 34-membered macrocycle containing two DTPA units (DTPA-bn-DTPA-bn), however, was monomeric and showed coordination polyhedrons with the amide functions in a *cis*-configuration [118]. The structures of the various ligands referred to above are shown in Chart 8.2.

The stoichiometries, hydration numbers, stabilities, conformations, and Eu^{3+}–Eu^{3+} distances of a series of Eu^{3+} complexes of amide-based macrocycles, including 30-DTPA-en-DTPA-en and 15-DTPA-EAM, have been studied in solution by Eu^{3+} luminescence and molecular mechanics [113]. A single peak was observed in the ${}^7F_0 \rightarrow {}^5D_0$ excitation spectrum, which leads to the conclusion that a single isomeric form is present in solution for all of the complexes. This is, however, in conflict with the results of NMR studies [70, 82]. Apparently, it is not possible to discriminate such closely related isomers by using luminescence spectroscopy.

The formation of the complex between Eu^{3+} and the 18-DTPA-dien ligand has been studied by laser-excited luminescence [119]. Immediately upon mixing the components, both the final complex and a long-lived intermediate were observed. It has been proposed that the formation of this intermediate is a 'blind alley' in the reaction pathways. Based on the wavelength in the ${}^7F_0 \rightarrow {}^5D_0$ excitation spectrum and its excited lifetime (in H_2O and D_2O), it has been suggested that in the intermediate the ligand is coordinated to Eu^{3+} with three carboxylates (in an 'up–down–up' configuration with respect to the macrocycle), three amino nitrogen atoms, and a water molecule. Transformation of this intermediate into the final product requires some decomplexation, which explains the relatively low rate of this reaction.

9.4 DTPA DERIVATIVES WITH SUBSTITUENTS ON THE ETHYLENE BRIDGES OR ON THE ACETATE GROUPS

For DTPA derivatives with a single substituent on one of the ethylene or acetate groups, the interconversion between the $\delta\delta \rightleftharpoons \lambda\lambda$ conformers does not lead to an effective plane of symmetry through the central glycine unit. In this case, the isomers **1′**, **2′**, **3′**, and **4′** in Figure 8.6 are no longer the mirror images of **1**, **2**, **3**, and **4**. In addition to the eight isomers depicted, another eight isomers are possible which are their mirror images. Discrimination between these pairs of mirror images is not possible by using NMR spectroscopy and interconversion between them will be an extremely slow process since it requires decoordination of at least seven donor sites of the organic ligand. Due to the lower symmetry and the larger amount of isomers, the number of resonances for these complexes is relatively large. This is, for example, observed in the ^{13}C NMR spectra of the La^{3+} and Lu^{3+} complexes of BOPTA which display about twenty resonances in the aliphatic region at 60 °C [120]. Upon lowering

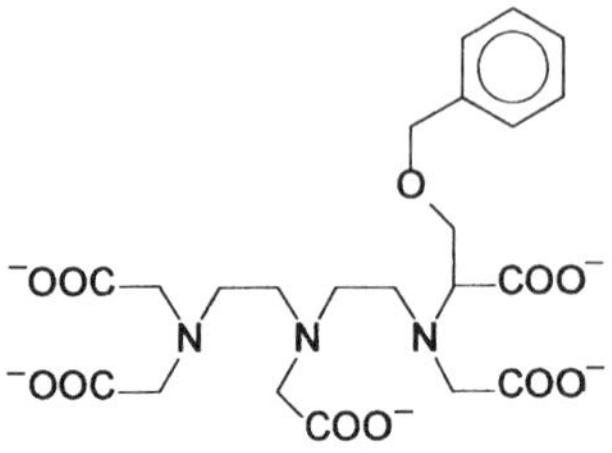

BOPTA

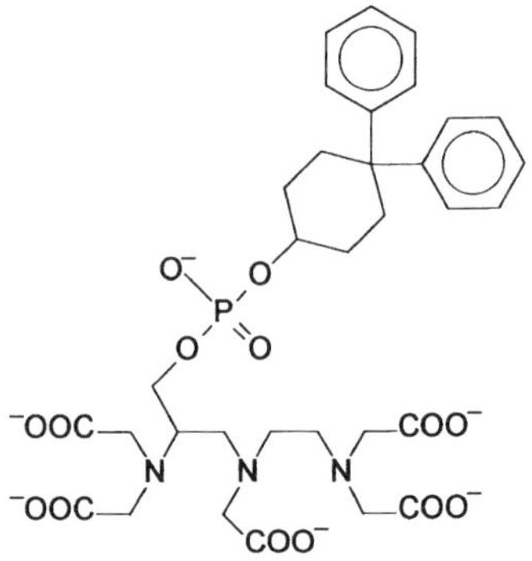

MS-325

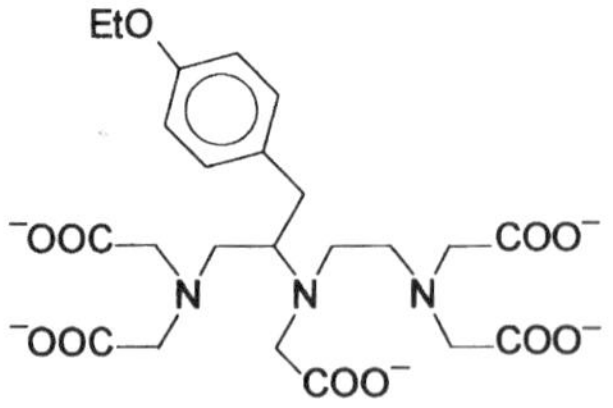

EOB-DTPA

[Chart 8.3]

of the temperature, the resonances broaden, thus showing that a dynamic process is slowing down. It can be concluded that this compound occurs as at least two pairs of interconverting isomers [120]. If it is assumed that the ligand is bound in a similar way as DTPA, two chiral centers are present in the DTPA backbone of BOPTA (one of the terminal nitrogen atoms and the central one). Therefore, two diastereomeric pairs of isomers should be expected, which is in agreement with the observations. In Figure 8.6, $A_1 = B_1$ for this compound and, consequently, the *trans* (**1**, **1**′)-and *cis* (**4**, **4**′)-isomeric pairs are identical with the *anti* (**3**, **3**′)-and *syn* (**2**, **2**′)-ones, respectively. As stated above, mirror images of these pairs are also possible but these can not be discriminated by using NMR. In principle, the chirality in the benzyloxymethyl chain can give rise to a doubling of the number of isomers; however, they are probably not observable in the NMR spectra of the diamagnetic La^{3+} and Lu^{3+} complexes. No kinetic data were reported, but the dynamic behavior is similar to that observed for the racemization of the terminal nitrogen atoms in the La(DTPA-BPA) complexes [103]. Apparently, the bulky side-chain has no influence on this process. In the solid-state structure of $Na_2[Gd(BOPTA)]$ [120] the BOPTA ligand is bound through the three amine nitrogen atoms and the five carboxylates. The first coordination sphere is completed by one water molecule. The coordination polyhedron can be described as a distorted tri-capped trigonal prism. Two enantiomeric forms (*trans*-, see Figure 8.6) occur in each unit cell.

Two diastereomeric Gd(*S*-EOB-DTPA) complexes have been isolated by preparative HPLC [121]. The interconversion between these complexes is very slow; the half-lifetime of one of the isomers of the Gd^{3+} complex is 13 100 h at pH 9 and 25 °C. At the thermodynamic equilibrium, the ratio of the isomers is 65:35. The first-order rate constant for the interconversion between the isomers is pH-dependent, thus suggesting that the isomerization is an acid-catalyzed process similar to the dissociation of the complex. The related activation energy has been estimated to be 75.3 kJ mol^{-1}. NMR studies on the two corresponding diamagnetic La^{3+} complexes gave no evidence for the presence of enantiomers that differ in the orientation of the acetate groups [121]. In principle, four isomers are possible for these systems [121]; however, two of these are mirror images of the other two and, therefore, discrimination by NMR is impossible. Dy^{3+} induced water ^{17}O shifts indicated that $q = 1.2$ for $Dy(EOB-DTPA)^{2-}$, which is similar to $Dy(DTPA)^{2-}$ [122, 123].

Relaxivity studies on the Gd^{3+} complexes of BOPTA [120], EOB-DTPA [122, 123] and MS-325 [124] indicated that the relaxivity of these compounds is higher than would be expected on the basis of the relevant correlation times. This has been attributed to a reduced distance between Gd^{3+} and the inner-sphere water protons (about 0.29 nm), as compared to the value usually reported for $Gd(DTPA)^{2-}$ (0.31 nm) [123]. This shortening only concerns the Gd–H distance because similar Gd–O distances are found by crystallography for $Gd(DTPA)^{2-}$ and $Gd(BOPTA)^{2-}$ [120]. The structures of the various ligands are shown in Chart 8.3.

9.5 TTHA

The crystal structure of $La(HTTHA)^{2-}$ was reported by Ruloff *et al.* [125] and Wang *et al.* [126]. The La^{3+} ion was coordinated in a decadentate fashion by the monoprotonated TTHA ligand (see Chart 8.4) via the four nitrogen atoms and six carboxylate oxygen atoms. The coordination polyhedron was described as a distorted bi-capped square antiprism. The coordinating carboxylate oxygen atoms exhibit bond distances of about 2.5 Å from the metal ion, whereas one of the terminal carboxylate groups (carrying the proton) is bound more weakly, thus resulting in a bond length of about 2.8 Å. The $Dy(HTTHA)^{2-}$ [125], $Gd(HTTHA)^{2-}$ [127] and $Yb(TTHA)^{3-}$ [128] complexes show ninefold co-ordination with the TTHA ligand in a distorted mono-capped square

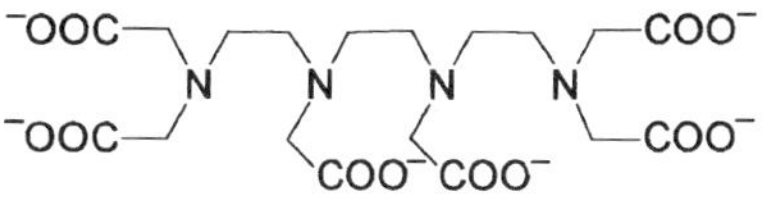

^{-}OOC N N N N COO^{-}
^{-}OOC COO^{-} COO^{-} COO^{-}

TTHA

[Chart 8.4]

antiprismatic (SAP) geometry. In these complexes, a terminal carboxylate group was not coordinating. It is this group that is protonated in the Dy^{3+} and the Gd^{3+} complexes. An X-ray study on $Nd(TTHA)^{3-}$ showed a coordination geometry with a decadentate binding of the ligand arranged in a distorted bi-capped square antiprism [129]. Here, the Nd–O distances are all comparable (about 2.5 Å). The crystal structure of a dimeric $Nd_2(TTHA)_2^{6-}$ complex [130] revealed a nonadentate coordination of both of the Nd^{3+} ions in a TTP geometry. Each ligand molecule was bound via three nitrogen atoms and six carboxylate oxygen atoms, from which two were coordinating the second metal ion. The fourth nitrogen atom remains uncoordinated. In all reported Ln-TTHA crystal structures, no water molecules were found in the first coordination sphere of the metal ion.

The solution structure was elucidated by various NMR and luminescence techniques. A ^1H and ^{13}C NMR study on the $La(TTHA)^{3-}$ complex indicated that all of the carboxylate and amine groups in the ligand molecule were directly interacting with the metal ion [131], as was observed in the crystal structure. An ^{15}N NMR spectrum of this system [132] clearly shows two signals, thus indicating a twofold symmetry in the complex. Luminescence studies of the Eu-TTHA complex in solution [88, 131], indicated the presence of 1:1 and 2:1 metal:ligand species. Both of these species exist in two isomeric forms. The mononuclear compound was found at pH values above 6 and had no water molecules in the first coordination sphere, while the binuclear species is formed mainly in acidic medium ($pH < 3$) and has three or four inner-sphere water molecules. For this species, a structure was proposed with Eu^{3+} coordinating at opposite ends of the TTHA ligand, binding to two nitrogen and three carboxylate oxygen atoms. In the pH range 3–6, oligomeric species of the protonated complexes were identified that dissociated to form mononuclear complexes at $pH > 6$. The pH dependence of the structure of the complex and the hydration numbers (from 3–4 at low pH to 0 at $pH > 6$) was confirmed by variable-pH relaxivity studies described for $Gd(TTHA)^{3-}$ [88]. The ^1H and ^{13}C NMR spectra of $Lu(TTHA)^{3-}$ were very complicated [131] due to fluxional processes such as wrapping/unwrapping of the ligand or exchange between 1:1 and 2:1 metal:ligand species.

9.6 EGTA

The solid-state structures of the Nd^{3+} and Er^{3+} complexes of EGTA (see Chart 8.5) show different coordinations of the EGTA ligand. The $[Nd(EGTA)(H_2O)]^-$ complex displays a ten fold coordination [133], in which the coordination polyhedron can be described as a distorted bi-capped square antiprism where the two capping positions are the two nitrogen atoms of EGTA. The square faces are linked by the diether bridge and each is formed by one of the ether oxygens, two oxygen atoms of adjacent carboxylate groups,

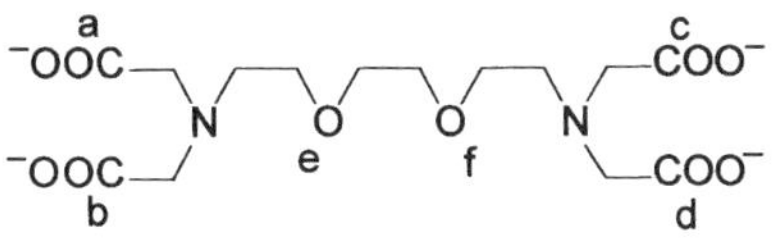

EGTA

[Chart 8.5]

respectively, and the water molecule or a carboxylate oxygen of an adjacent molecule. It may be expected that the latter is replaced by water in solution or that the coordination number is reduced to nine. The solid-state structure of the Er^{3+} complex displays a nine-coordinate Er^{3+} ion with EGTA bound in an octadentate fashion through the two nitrogen atoms, the two ether oxygens, and the four carboxylate groups [133]. The polyhedron can be described as a TTP.

A plot of ^{1}H LIS data for one of the acetate resonances according to Equation (8.13) shows a clear break near Eu^{3+}, whereas Yb^{3+} lies outside each of the two lines (see Figure 8.1) [40]. This suggests that the structure of the Ln(EGTA) complexes changes at Eu^{3+} or Sm^{3+}, whereas possibly another change of structure occurs at Yb^{3+}. This is supported by variable-temperature and variable-pressure UV–vis measurements on the Eu^{3+} complex showing a single band for the $^{7}D_0 \rightarrow ^{5}D_0$ transition at 580 nm, which was assigned to the nine-coordinated complex. By contrast, the UV–vis spectra of the Ce^{3+} complex from 220 to 350 nm confirmed the presence of two species which were assigned to an equilibrium of nine- and ten-coordinate species [40]. A structural change along the lanthanide series is also reflected in differences in dynamic behavior between the lighter and the heavier lanthanides [40]. The combined experimental data are explained by equilibria of two ten- and two nine-coordinated complexes for the light and heavy Ln^{3+} ions, respectively (see Figure 8.8). The cross-over between the ten- and nine-coordination takes place between Sm^{3+} and Eu^{3+}.

9.7 TREN-ME-3,2-HOPO

Raymond and co-workers have reported that the Gd^{3+} complex of TREN-Me-3,2-HOPO (see Chart 8.6) has promising potential for MRI contrast applications [134]. The X-ray structure of this complex shows that it is eight-coordinate. The organic ligand is coordinated in a hexadentate fashion through the hydroxypyridone oxygen atoms, with two water molecules completing the coordination sphere. Although the organic ligand is only hexadentate, the thermodynamic and kinetic stabilities of the complex are good (e.g. log $\beta_{110} = 20.3$), which may be ascribed to some degree of preorganization and stability due to internal hydrogen bonds.

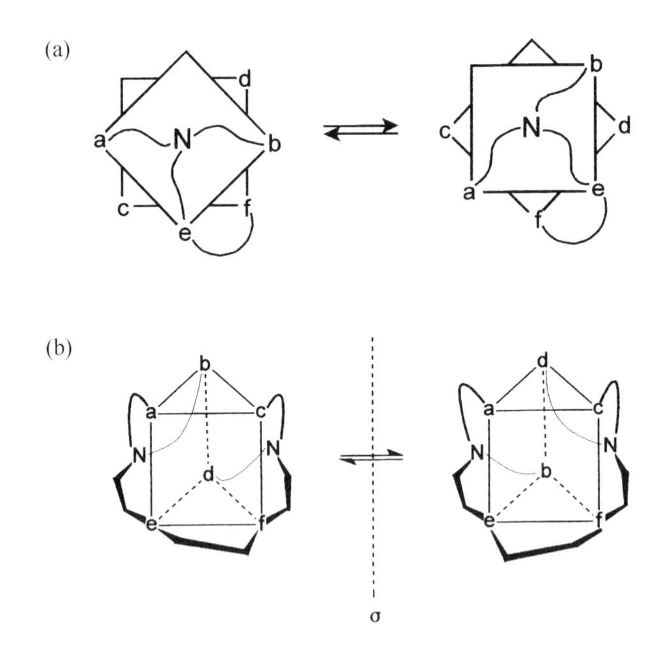

Figure 8.8 Fluxional processes in Ln-EGTA complexes leading to effective C_2 symmetry: (a) ten-coordinate species; (b) nine-coordinate species. (the oxygen atoms are identified in Chart 8.5).

10 LANTHANIDE COMPLEXES OF MACROCYCLIC LIGANDS

10.1 NOTA

A study of Ln^{3+}-induced water ^{17}O shifts of solutions for complexes formed with the triazamacrocyclic ligand NOTA (see Chart 8.7), Ln(NOTA) [10], has

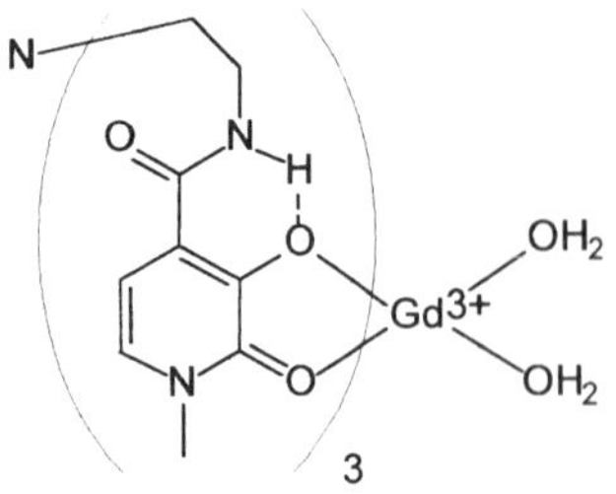

TREN-Me-3,2-HOPO

[Chart 8.6]

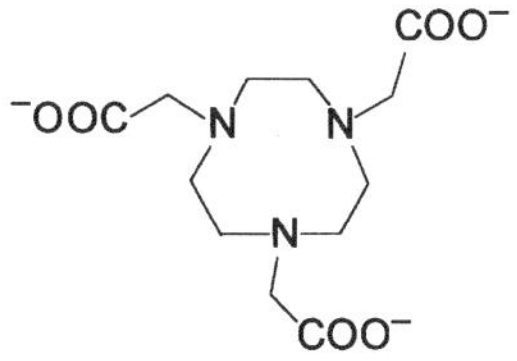

[Chart 8.7]

shown that the hydration number changes across the lanthanide series: it is three in the second half (Dy–Yb) and between three and four in the first half (Ce–Eu). The ^{1}H and ^{13}C NMR spectra of the whole series of Ln(NOTA) complexes were also analyzed in terms of their structure and dynamics in solution [135, 136]. It was shown that the Ln^{3+} ion does not fit inside the internal cavity but is located outside of it, being bound to the three nitrogen atoms and the carboxylate oxygens. All of the ethylene groups in the nine-membered macrocyclic ring adopt an identical gauche conformation, either δ or λ, leading to one of two possible square [333] conformations. The fast interconversions between the two gauche conformations of the ethylene bridges, corresponding to a high conformational flexibility of the macrocycle, explain their AA'XX' proton spectrum at all temperatures. The ^{1}H and ^{13}C LIS values indicated that the later Ln(NOTA) chelates (Dy–Yb) have a structure where the ligand is hexacoordinate with three inner-sphere water molecules, while the earlier chelates (Ce–Eu) adopt a structural mixture of hexa-and pentacoordinate (with one free acetate arm) ligands with three and four inner-sphere water molecules, respectively.

10.2 DOTA AND DERIVATIVES

A study of Ln^{3+} induced water ^{17}O shifts of $Ln(DOTA)^-$ (see Chart 8.8) solutions has shown that the hydration number is one across the lanthanide series [10]. The substantial pseudocontact contribution to its LIS indicated that this water ligand has a preferred location in the complex. Two sets of peaks have been observed in ^{1}H and ^{13}C NMR spectra of $Ln(DOTA)^-$ complexes at room temperature, thus showing the presence of two slowly interconverting structural isomers [69, 71, 72, 137]. These structural features have been confirmed by luminescence studies [92, 138]. In the spectra of the paramagnetic complexes, one isomer has larger induced shifts than the other. From the temperature dependence of the ^{1}H and ^{13}C NMR spectral features of both the dia- and paramagnetic Ln^{3+} complexes, Desreux concluded that the twelve-membered macrocyclic ring of the isomer which is present in the higher percentage ('major') is very rigid [137]. The Ln^{3+} ion does not fit inside the internal cavity and is, therefore, located outside of it, being bound to the four nitrogen atoms and to the four carboxylate groups. All ethylene groups adopt an

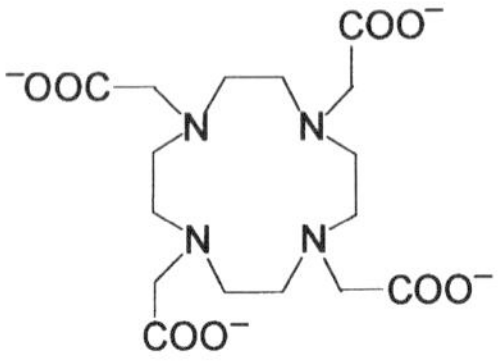

DOTA 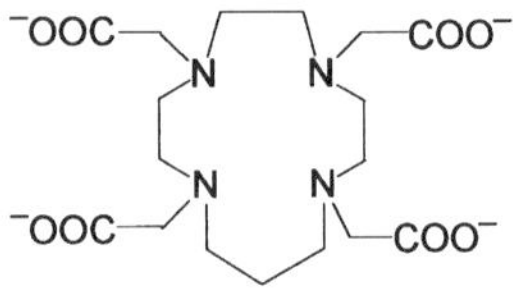 TRITA 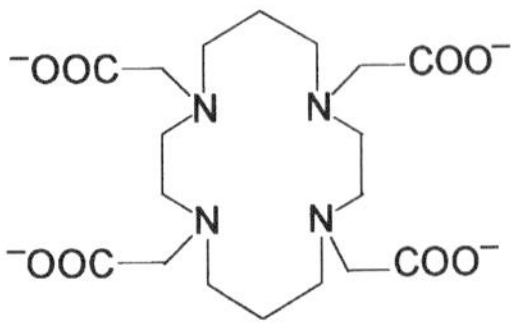 TETA

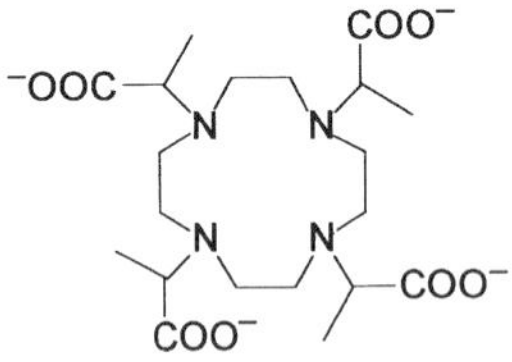

DOTMA 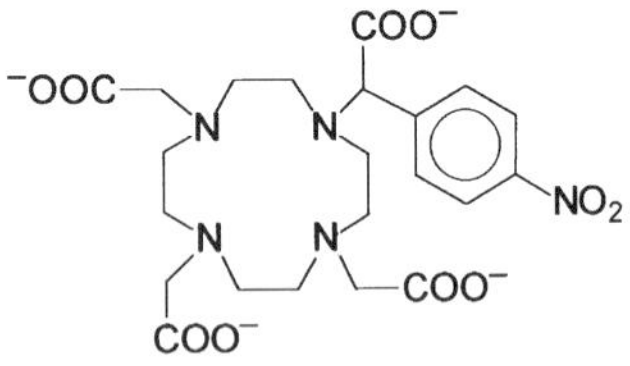 DOTA-pNB

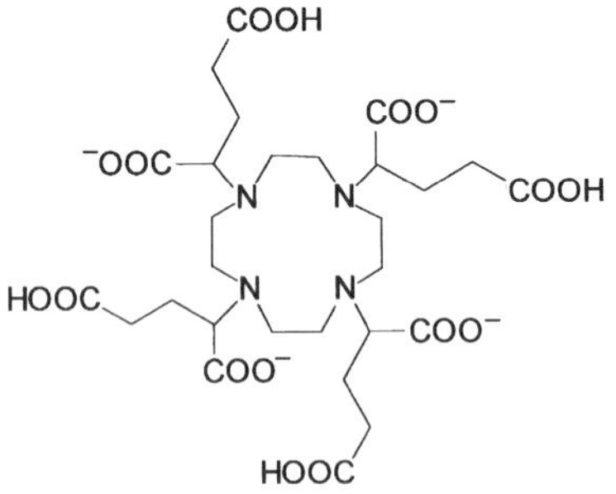

TCE-DOTA

[Chart 8.8]

identical gauche conformation, either δ or λ, leading to one of two possible square [3333] conformations of the macrocyclic ring. The interconversions between the two gauche conformations of the ethylene bridges explain the dynamic phenomena observed in the NMR spectra at higher temperatures. This solution structure is similar to the solid-state structure of the $Eu(DOTA)^{-}$ complex which shows the octadentate binding [139]. The coordination polyhedron is a capped square antiprism (CSAP) with two opposite parallel faces occupied by the ligand nitrogen (N_4 plane) and oxygen (O_4 plane) donors and a water molecule at the capping position. The twist angle between the N_4 and O_4 planes is $\theta = 38.9°$. The similarity between the solid-state and the solution structure was further supported by the excellent agreement between the Yb^{3+}-induced ^{1}H shifts (of the 'major' isomer), which are assumed to be almost purely of pseudocontact origin, and shifts calculated from the X-ray structure of the Eu^{3+} complex [137]. Since this structure has a C_4 symmetry axis, the axial model could be applied (Equation (8.7)). Originally, the other set of resonances was tentatively assigned to an isomer in which one of the pendant acetate arms was not coordinated to the Ln^{3+} ion. More recently, by using higher magnetic fields, in combination with COSY and EXSY, Aime *et al.* [72] obtained a full spectral description of both isomers, which showed a great similarity apart from a difference in the magnitude of the LIS values. Based on the similarity of the vicinal couplings in the ethylene bridges, it was concluded that the structures of the macrocyclic rings are the same and that the difference between the two isomers is in the layout of the acetate arms. The distances between the Ln^{3+} ion and the ligand protons in the Yb^{3+} complexes were evaluated by exploiting the Curie relaxation (see Equation (8.23)) [56, 72]. The chemical shifts of the main isomer, which were assumed to be completely of pseudocontact origin, agree with the crystal structure described above. By using the structure of the main isomer, the torsion angle Ln–N–C–COO, was altered stepwise until an optimal fit was obtained between the calculated values (Equation (8.7)) and the observed LIS (^{1}H) for the minor isomer. This difference in the arrangement of the acetates led to a twisted capped square antiprismatic structure (twisted or inverted CSAP) which contained a layout of the acetate arms that is inverted with respect to that in the main isomer, corresponding to a negative and smaller twist angle between the N_4 and O_4 planes. With the use of variable-temperature ^{13}C NMR on the Nd^{3+} complex and ^{1}H EXSY on the Yb^{3+} complex, it was shown that exchange processes occur between the isomers. Hoeft and Roth came to the same conclusions based on a similar study with the Eu^{3+} and Yb^{3+} DOTA complexes [71].

The structure and dynamics of this system are summarized in Figure 8.9. The difference in the arrangements of the acetates, which leads to CSAP and twisted CSAP geometries for the two isomers, can be described by a twist angle of ca 40° for the CSAP geometry and of ca −30° for the twisted CSAP geometry. Thus, there are four stereoisomers, i.e. two pairs of enantiomers, which can interconvert in solution by either ring inversion ($(\delta\delta\delta\delta) \rightleftharpoons (\lambda\lambda\lambda\lambda)$) or acetate-arm rotation ($\Delta \rightleftharpoons \Lambda$). Either process alone results in exchange between the CSAP and twisted CSAP geometries, while both processes combined, either in succession or concerted, result in an exchange between the enantiomeric pairs (see Figure 8.9).

Jacques and Desreux have performed a thorough quantitative analysis of variable-temperature EXSY spectra of $Yb(DOTA)^{-}$ by completely solving the dynamic matrix [69]. It was shown that each species in the dynamic equilibrium is exchanging with all other species. The activation parameters ($\Delta G^{\ddagger}_{298}$) for enantiomerization and arm rotation are 65.9 and 65.1 kJ/mol, respectively

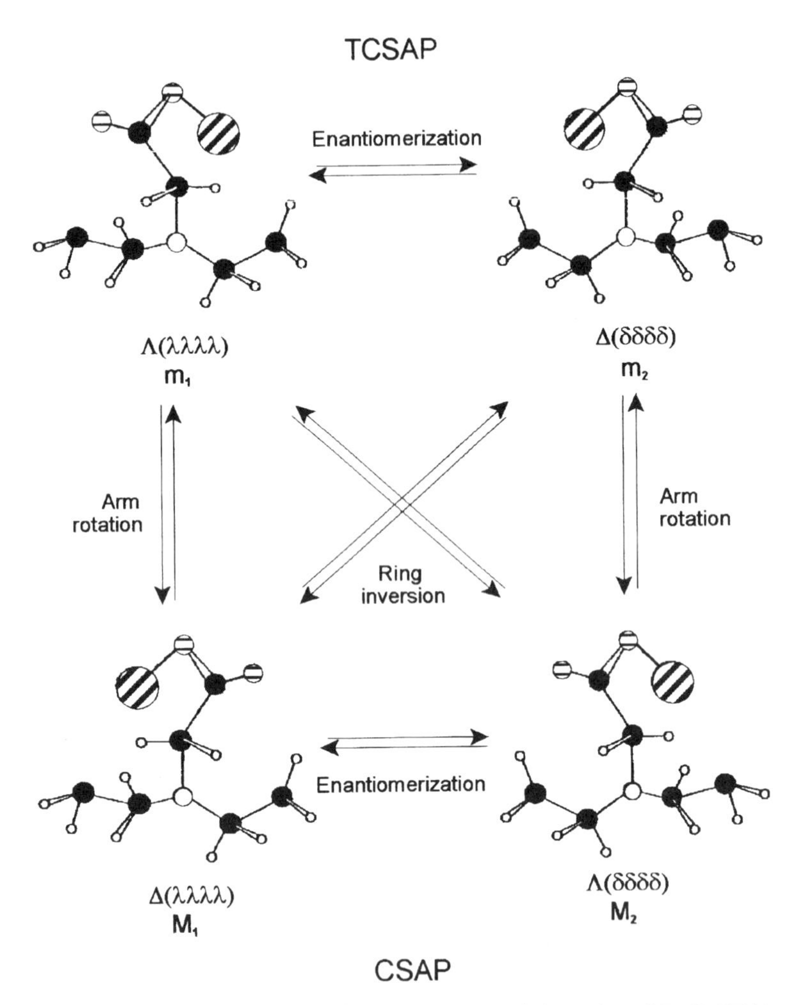

Figure 8.9 Schematic representation of the structures and dynamics of $Ln(DOTA)^-$ complexes.

[69], while the corresponding value for ring inversion in $Yb(DOTA)^-$ is 60.7 kJ/mol. These values show that, for this system, arm rotation is faster than ring inversion which was confirmed by other ^{1}H EXSY [71, 72] and variable-temperature ^{13}C NMR studies [72, 140]. They also reflect the high rigidity of the Ln^{3+} complexes of DOTA as compared to those of DTPA derivatives.

The relative concentrations of the two species depend on the size of the Ln^{3+} ion, temperature, pressure and on the concentration of added inorganic salts [71, 72, 141]. While the twisted CSAP geometry is the 'major' isomer for the complexes of the larger cations, La^{3+}–Nd^{3+}, the CSAP geometry becomes the most stable for the smaller cations, i.e. Sm^{3+}–Er^{3+}. In all of these cases, the isomerization process is purely conformational, as shown by the near–zero reaction volumes obtained by high-pressure NMR [141]. However, for the complexes of the smallest cations (Tm^{3+}–Lu^{3+}), the large positive isomerization volumes obtained show that the 'minor' isomer results from a fast water dissociation process superimposed on the conformational rearrangement, thus leading to an eight-coordinate square antiprismatic (SAP) geometry [141]. High concentrations of non-coordinating salts stabilize the twisted CSAP geometry relative to the CSAP form due to preferential weak ion binding and water solvent stabilization of the former ligand geometry. Fluoride ions preferentially replace the coordinated water in the first geometry [141].

The solution structure of the CSAP isomer is consistent with the X-ray structures of the Eu^{3+} [139], Gd^{3+} [142, 143], Y^{3+} [143] and Lu^{3+} [140] complexes of DOTA where the twist angle has a value of *ca* 39° (see Figure 8.10). The structure of the twisted CSAP isomer is consistent with the X-ray structure of the La^{3+} complex of DOTA where the twist angle has a value of ca −22° [144]. An EXAFS study confirmed that the local environments of the Gd^{3+} ions are similar in solution and in crystals of $Gd(DOTA)^-$ [102].

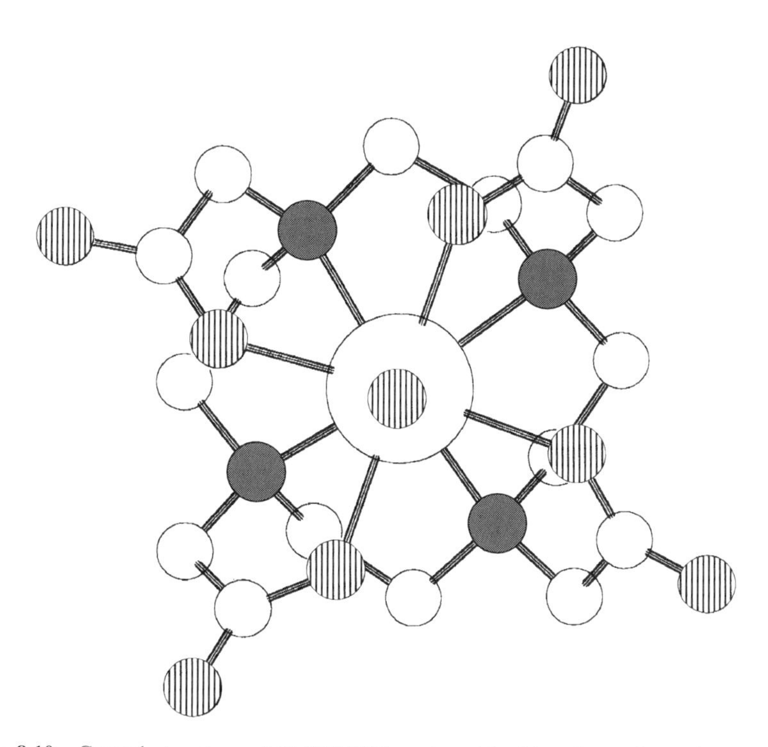

Figure 8.10 Crystal structure of $Gd(DOTA)^-$, viewed looking down from the water–Gd bond.

The solution structure of the homologous TETA (see Chart 8.8) complexes has been elucidated by Desreux and Loncin in a similar way as described above for the DOTA complexes [145]. The $Ln(TETA)^-$ complexes contain no inner-sphere water and the complexes are also very rigid. Here, an exchange between two equivalent eight-coordinate dodecahedral geometries occurs ($\Delta G^{\ddagger}_{298} = 63.7$ kJ/mol). Ascenso *et al.* have studied the corresponding thirteen-membered system (TRITA, see Chart 8.8) by using ^{1}HNMR techniques [146]. The macrocyclic backbone prefers a [12433] conformation. From the coalescence temperature, the free energy of activation for the interconversion between the two equivalent conformations was determined to be $\Delta G^{\#}_{357} = 55$ kJ/mol.

In the case of α-substituted DOTA derivatives, the introduction of chiral centers of equal configuration to all four acetate arms of DOTA (*RRRR* or *SSSS*) results in four possible diastereomers upon chelation. However, the occurrence of only two isomers appears to be a general phenomenon of Ln^{3+} complexes of such DOTA derivatives. This is observed in $Ln(DOTMA)^-$ complexes (Chart 8.8) where the configuration at each chiral carbon is *R* [147]. The ^{1}H NMR spectrum of $Yb(DOTMA)^-$ and the high-resolution luminescence spectrum of the Eu^{3+} complex show only two species in solution. The 'major' isomer of the Yb^{3+} complex has a structure corresponding to the 'minor' isomer of $Yb(DOTA)^-$. The crystal structure of the Eu^{3+} complex of the ligand TCE-DOTA (Chart 8.8), synthesized as a mixture of stereoisomers defined by the absolute configuration of the chiral carbon, consists of two co-crystallized enantiomers, the (*RRRR*) and (*SSSS*) complexes, with opposite helicities which have the same CSAP geometry [148]. However, two solution isomers were also observed for the (*RRRR*) form of the Eu(TCE-DOTA) complex [148].

The introduction of only one chiral center in DOTA by derivatizing one acetate α-carbon with a *p*-nitrophenyl group (DOTA-pNB) (Chart 8.8) [149] also results in four possible diastereomers upon chelation. The Ho^{3+} and Yb^{3+} complexes of DOTA-pNB again give ^{1}H NMR spectra showing the presence of only two isomers [149]. In these cases, as well as for Eu(TCE-DOTA) [148] ^{1}H EXSY has shown that the two isomers exchange through inversion of the macrocyclic ring, whereas no rearrangement of the acetate groups is observed. Thus, such derivatization drastically slows down the acetate-arm rotation process so that ring inversion now becomes the faster isomerization process. The ring inversion results in an exchange between CSAP and twisted CSAP geometries of the same helicity while the arm rotation would lead to opposite helicities. The absence of the second process suggested that the configuration of the stereogenic center at the carbon determines the least sterically hindered helical form of the complex. In fact, the crystal structures of the (*RRRR*) and (*SSSS*) TCE-DOTA Eu^{3+} complexes show that in both enantiomers the substituent is equatorially positioned, pointing away from the coordination cage [148]. The same observation was made from the structural analysis of the LIS data for the $Yb(DOTA\text{–}pNB)^-$ complex [149]. It is this preference for a particular helicity that reduces the number of diastereomers from four to two, in agreement with the NMR data for the DOTMA, TCE-DOTA and DOTA-pNB complexes [147–149].

10.3 DO3A AND DERIVATIVES

The success of $Gd(DOTA)^-$ as a contrast agent for MRI has initiated an intensive search for derivatives with improved properties [150]. The non-ionic reagents, Gd(HP-DO3A) and Gd(DO3A-butrol) (see Chart 8.9), in which one of the acetate groups of DOTA is replaced by a hydroxyalkyl group, have recently been introduced.

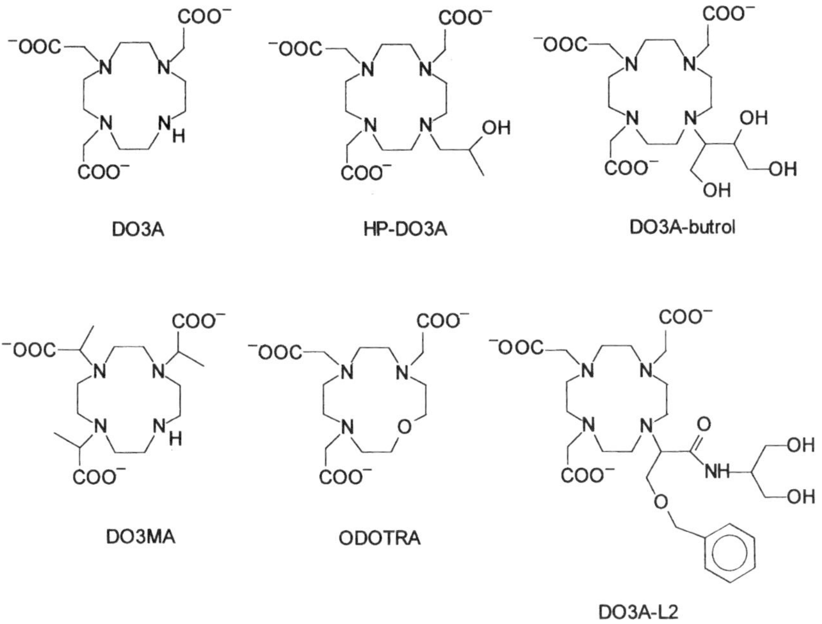

DO3A HP-DO3A DO3A-butrol

DO3MA ODOTRA DO3A-L2

[Chart 8.9]

The crystal structures of the parent Gd(DO3A) [143] and of the Gd(DO3MA) [151] derivatives (Chart 8.9) have been reported. The nine-coordinate Gd^{3+} is bound by the ligands in a heptadentate fashion. The coordination geometry is CSAP ($\theta = 39°$) for the DO3A ligand and both CSAP ($\theta = 38°$) and twisted CSAP ($\theta = -30°$) for the DO3MA ligand. The crystal structure of Gd(DO3A-L2) (in which one of the acetate arms of DOTA is replaced by an amide hydroxyalkyl group, and its α-carbon derivatized with a benzyloxy function, see Chart 8.9) [152] shows the nine-coordinate Gd^{3+} bound by the ligand in an octadentate fashion, including the amide oxygen, and by one water molecule, in a CSAP geometry ($\theta = 39°$). In the crystal structures of the Gd(HP-DO3A) and Y(HP-DO3A) complexes, the Ln^{3+} ion is again nine-coordinate, and is bound by the ligand in an octadentate fashion, including the hydroxyalkyl oxygen, and by one water molecule, with both the CSAP ($\theta = 38°$) and twisted CSAP ($\theta = -26°$) geometries being present [153]. However, the similar Gd(DO3A-butrol) complex has the nine-coordinate Gd^{3+} exclusively in the twisted CSAP ($\theta = -28°$) geometry [154].

Among the various Ln(DO3A) derivatives, only Ln(HP-DO3A) has so far been studied in solution [155]. Replacement of an acetate arm by a hydroxypropyl group, which contains a chiral carbon center, leads to eight possible stereoisomers (four pairs of enantiomers). While the ^{1}H NMR spectrum of Y(HP-DO3A) had a resolution which was too poor to confirm the presence of multiple diastereoisomers by counting peaks, observation of exchange processes by ROESY spectra confirmed the presence of multiple species. Quantitative analysis of the cross-peaks in terms of the exchange dynamics gave the exchange rates of the various processes present. These have shown that the exchange of the methylene groups during ring inversion is faster than the exchange of pendant arms, and that the exchange of the hydroxypropyl pendant arm is faster than for the acetate arms. The two diastereoisomers found in the crystal structures of the Gd^{3+} and Y^{3+} complexes [153] are also found in solution, and interconvert by ring inversion. Observation of an exchange process involving rearrangement of the acetate arms indicated that diastereomers of opposite helicities are present.

Spirlet *et al.* reported the crystal structure of the Gd^{3+} complex of ODOTRA (the analog of DO3A with the macrocyclic NH replaced by an ether oxygen, see Chart 8.9) [156]. The Gd^{3+} ion is nine-coordinate, and is bound to three amine nitrogens, one ether oxygen, four carboxylate oxygens (one of which is bridging from a neighboring molecule in an infinite chain) and one water molecule, with a twisted CSAP geometry ($\theta = -31°$).

10.4 OTHER POLYAZAPOLYCARBOXYLATES

The solution structures of Ln^{3+} complexes of some polyazamacrocyclic acetate ligands containing pyridine have been studied [157–159]. In particular, a temperature-dependent ^{1}H and ^{13}C NMR study of the La^{3+}, Lu^{3+}, Eu^{3+} and Yb^{3+} chelates of a series of pyridine-containing triaza triacetate ligands (PCTA), which act as heptacoordinating chelators, with 12-, 13- and 14-membered macrocyclic rings (PCTA-[12], PCTA-[13] and PCTA-[14], respectively, see Chart 8.10) was undertaken [158, 159]. The various complexes show a large variability of solution structures and non-rigidities dictated by the matching of size between the Ln^{3+} ion and the macrocyclic cavity. To some extent, the changes observed in going from the 12- to the 14-membered ring complexes parallels the behaviors shown by the octacoordinated Ln^{3+} complexes of

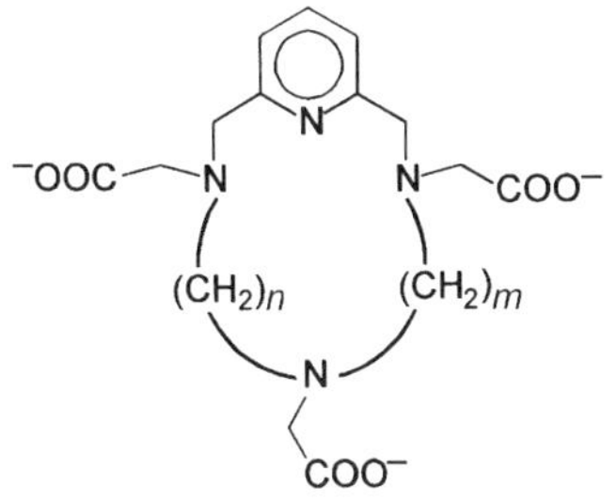

PCTA-[12] n=2, m=2
PCTA-[13] n=3, m=2
PCTA-[14] n=3, m=3

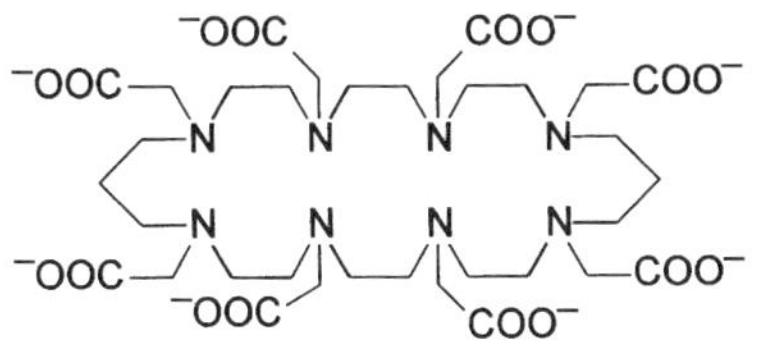

OHEC

[**Chart 8.10**]

DOTA and TETA. The PCTA-[12] and PCTA-[13] chelates seem to have nine-coordinate DOTA-like structures with two coordinated water molecules forming CSAP isomeric species, while the less stable PCTA-[14] chelates presumably have an eight-coordinate TETA-like structure with a reduced inner-sphere hydration forming antiprismatic (AP) or dodecahedral (DOD) structures [158, 159].

Some structural studies of Ln^{3+} complexes of larger polyazamacrocyclic acetate ligands containing five or more nitrogen atoms in the macrocyclic ring have been undertaken. The octaaza-octa-acetic ligand H_8OHEC (see chart 8.10) forms homo-dimer chelates $[Ln_2(OHEC)(H_2O)_2]^{2-}$ (Ln = La, Eu, Gd and Y) which have been characterized by X-ray diffraction in the solid state [160]. Each Ln^{3+} ion is nine-coordinated by eight donor atoms of the ligand and one water molecule, forming an irregular polyhedron built up by a trigonal plane and a mono-capped pentagonal plane. ^1H NMR studies of the diamagnetic complex of Y^{3+} show the existence of two isomers, i.e. a centrosymmetric major isomer with a structure very similar to the one present in the crystal form and an asymmetric minor isomer. The two isomers are in dynamic exchange as shown by ROESY cross-peaks. The loss of centrosymmetry of the major isomer involves independent dynamic processes occurring in its two halves. These processes have a ΔG_{300} of 65.0 kJ/mol and probably involve the rotation of acetate groups.

10.5 DOTP

Over the past decade, lanthanide DOTP complexes (see Chart 8.11) have been thoroughly investigated by using an array of multinuclear NMR techniques

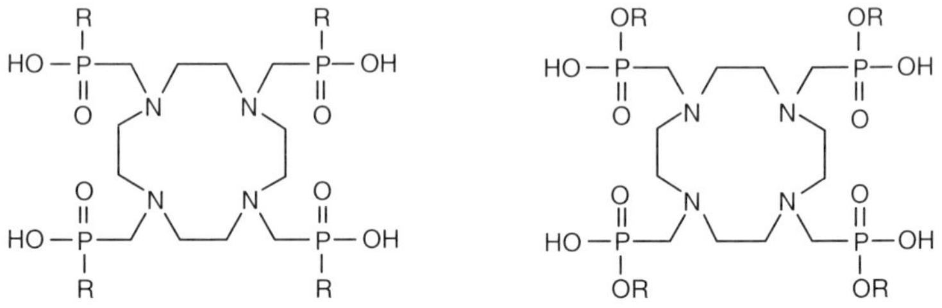

R =	
Me	DOTMP
Et	DOTEP
Bu	DOTBuP
Ph	DOTPP
CH_2Ph	DOTPBzP

R =	
OH	DOTP
Et	DOTPME
Bu	DOTPMBu
CH_2CF_3	F-DOTPME

[Chart 8.11]

[42, 161–167]. In these anionic complexes, the octadentate DOTP ligand coordinates to Ln^{3+} ions in a square antiprismatic (SAP) arrangement via the four nitrogen atoms of the macrocyclic ring and four phosphonate oxygen atoms of the pendant arms. Structurally very similar to $Ln(DOTA)^{-}$ complexes, the methylene phosphonate arms of the DOTP ligand are arranged in a propeller-like fashion above the basal plane made up of the four N donor atoms which encompass the Ln^{3+} ion, thereby generating a C_4 symmetry axis in these complexes. Four of the residual negative charges are localized on the phosphonate oxygen atoms which are directed away from the lanthanide coordination site, with the fifth averaged over the bound oxygens in the coordination cage. There has been some controversy over the hydration number in $Ln(DOTP)^{5-}$ complexes. Phosphorescence measurements with $Eu(DOTP)^{5-}$ indicated a smaller hydration number ($q = 0.7 \pm 0.2$) compared to DOTA complexes ($q = 1.2 \pm 0.2$) [163]. Furthermore, Ren and Sherry reported ^{17}O NMR measurements of $Dy(DOTP)^{5-}$ which revealed that this complex lacks an inner-sphere water molecule [42]. In the early 1990s, Aime *et al.* suggested that the NMRD profile of $Gd(DOTP)^{5-}$ is consistent with the presence of one exchangeable water molecule [165]. However, these authors offered an alternative explanation of the data in which two species are present in solution, one with $q = 1$ and another with $q = 0$, thus yielding an average hydration number of 0.7. During the course of luminescence and NMRD investigations of tetra-aza macrocyclic complexes incorporating one carboxamide and three phosphonate arms, Aime *et al.* identified a behavior that implicated the presence of a second hydration sphere which contributes to the total relaxivity [168]. These findings provoked a reinvestigation of the $Gd(DOTP)^{5-}$ system. The results indicated that the relaxivity profile fittings are consistent with a model in which the outer-sphere component, as well as two water molecules in the second hydration sphere, contribute to the overall relaxivity [169]. Since Ln^{3+} aquo ions and other polyaminocarboxylate complexes are known to have varying hydration numbers throughout the series, a variant *q*-value may also be expected in $Ln(DOTP)^{5-}$ complexes, depending upon the Ln^{3+} cation.

The DOTP ligand coordinates the trivalent Ln^{3+} ions, producing anionic complexes with various net charges. Potentiometric studies have indicated four protonation steps between pH 2 and 10, with $H[Ln(DOTP)]^{4-}$ existing as the predominant species at pH 7.4 [166].

^{1}H and ^{13}C NMR studies of the diamagnetic $La(DOTP)^{5-}$ and $Lu(DOTP)^{5-}$ complexes revealed a high degree of stereochemical rigidity in these compounds [164]. The activation energy for interconversion of the ethylenediamine rings was determined to be 101 ± 11 kJ/mol for DOTP complexes, thus making them considerably less flexible than the analogous DOTA derivatives ($E_a = 60.7 \pm 1.2$ kJ/mol) [137]. In addition, $Ln(DOTP)^{5-}$ complexes possess a remarkable thermodynamic stability with long-lived coordinate bonds which renders them extremely inert chelates. Geraldes *et al.* have examined solution structures of eleven paramagnetic lanthanide complexes of DOTP by ^{1}H, ^{13}C, ^{31}P and ^{23}Na NMR spectroscopy as well as molecular mechanics calculations [164]. Separation of contact and pseudocontact contributions to the LIS was accomplished by the Reilley method using Equations (8.13) and (8.14). Inspection of this data suggests only minor structural changes throughout the Ln series. After dissection of the LIS, the dipolar contributions were analyzed for an axial symmetry model (Equation (8.7)). The geometric factors for each nucleus were determined from the minimized structure by using the MMX force field (average Ln–N and Ln–O bond lengths of 2.70 and 2.35 Å, respectively). Using this as an initial geometry, an optimum fit of the LIS data was determined. Comparison of the experimental LIS values and those calculated for the MMX minimized structure exhibited good agreement.

Results obtained from ^{23}Na NMR studies with the $Tm(DOTP)^{5-}$ complex revealed two possible binding sites for the Na^{+} countercations [166, 167]. One site (A), observed at low $Na^{+}/Tm(DOTP)^{5-}$ ratios and under basic conditions, gave rise to an extremely large ^{23}Na LIS ($\approx$ 420 ppm), indicating a binding position near the four-fold symmetry axis. MMX calculations generated a model where the Na^{+} ion interacts with one inner Ln-bound oxygen atom and an axial oxygen atom of an adjacent phosphonate group ($\theta = 26°$). At high $Na^{+}/Tm(DOTP)^{5-}$ ratios or under acidic conditions, there is evidence for at least three sites of a second type (B) which impart significantly smaller bound shifts ($\approx$ 160 ppm). An MMX model suggested that the Na^{+} ions are associated with two unbound axial oxygens of adjacent phosphonate groups with an average dipolar angle of 34°. The sign of the LIS for ^{23}Na in both cases is the same as with the $Ln(DOTA)^{-}$ complexes, indicating that the two binding locations are located within the positive shift cone.

Although the direction of the lanthanide-induced ^{23}Na shifts in $Ln(DOTP)^{5-}$ compounds are identical to those in DOTA complexes, the magnitudes were found to be substantially greater at identical concentrations, most likely due to the formation of strong ion-pairs resulting from the increased negative charge on the DOTP complexes [161]. This behavior was exemplified in $Tm(DOTP)^{5-}$ where extraordinarily large ^{23}Na shifts defied theoretical expectations. Therefore, $Ln(DOTP)^{5-}$ complexes have been successfully applied as shift reagents for the separation of NMR signal degeneracy which is normally observed for intra- and extracellular compartments since they are impermeable to cell membranes. $Tm(DOTP)^{5-}$ complexes have proven to be extremely effective shift reagents in perfused organs and *in vivo* [170–174].

Similar to other cyclen-based macrocycles (e.g. DOTA), two enantiomeric forms defined by a clockwise or counter-clockwise spiraling of the methylenephosphonate arms may be envisaged for lanthanide DOTP complexes. However, unlike DOTA complexes, $Ln(DOTP)^{5-}$ complexes are believed to exist in solution as one enantiomeric pair. An isomeric form of $Tm(DOTP)^{5-}$ corresponding to the minor isomer observed for $Yb(DOTA)^{-}$ (eight-coordinate SAP geometry) was inferred by comparison of the paramagnetic ^{1}H shifts of the two species [166]. At temperatures below 50°C, the coordination cage appears to be locked into a single conformation; however, variable-temperature ^{1}H and ^{13}C NMR spectra suggest a dynamic behavior which is related to the interconversion of the ethylenediamine chelate rings, coupled with the concerted flipping motion of the arms [164]. As a result, $Ln(DOTP)^{5-}$ complexes exist as racemic mixtures in solution with the two enantiomers furnishing indistinguishable NMR signals at room temperature when using conventional NMR techniques. Chiral NMR resolution, using the formation of diastereomeric adducts between the two enantiomers of $Ln(DOTP)^{5-}$ and a chiral substrate, has provided indirect, albeit conclusive, evidence for the existence of these two enantiomers [175, 176]. The ion-pair interactions between lanthanide DOTP complexes and the chiral organic base, *N*-methyl-D-(−)-glucamine (Meg), were investigated by Aime *et al.* through the use of ^{1}H, ^{13}C and ^{31}P NMR spectroscopies [176]. Addition of Meg to a solution of $Eu(DOTP)^{5-}$ lifted the signal degeneracy of the NMR spectra, thus resulting in doubling of the corresponding signals. Similar spectral resolution in the ^{1}H NMR spectrum of $Eu(DOTP)^{5-}$ was achieved by the addition of the chiral transition metal complex, $(+) - Co(en)_3{}^{3+}$, as reported by Sherry and co-workers [175]. The formation of strong electrostatic interactions between lanthanide complexes and organic substrates, such as valeric acid, butanol, 1-adamantylamine, cyclen, and *N*-methyl-D-glucamine are well documented [10, 176, 177].

The ability of $Ln(DOTP)^{5-}$ complexes to interact in this manner has been exploited to probe charged areas in proteins [178, 179]. For example, $Gd(DOTP)^{5-}$ was used in assigning the structure of the fd gene V DNA binding protein.

An X-ray structure of the $Tm(DOTP)^{5-}$ complex has been reported by Paulus et al [180].

10.6 PHOSPHINATES AND PHOSPHONATE ESTERS

In principle, lanthanide complexes of alkyl- (phosphinates) or alkoxy- (phosphonate esters) DOTP derivatives may give rise to 32 stereoisomers, existing as 16 enantiomeric pairs, which are indistinguishable by NMR spectroscopy. The isomers originate from chiral elements inherent in these complexes, including the (*R*)- or (*S*)- configurations at each phosphorus and the helicity defined by the pendant arm orientations (Δ/Λ). Various Ln^{3+} complexes of phosphinate and phosphonate ester ligands derived from 1,4,7,10-tetraazacyclododecane (cyclen) have been described in the literature [181–184] (see Chart 8.12).

In the crystal structure of the $La(DOTPP)^-$ complex, the phosphinate groups are orientated corresponding to an (*RSRS*) diastereoisomer [185], while for $La(DOTBzP)^-$ [186] and $Y(DOTMP)^-$ [183] only the (*RRRR*) and (*SSSS*) isomers were found in the solid state. All Ln^{3+} complexes of DOTBzP and DOTMP have the organic ligand bound in an octadentate fashion (four nitrogen atoms of the cyclen ring and four phosphinate oxygen atoms). The larger cations (e.g. La^{3+}) have an additional water molecule bound in the first coordination sphere, while for smaller metal ions, such as Eu^{3+} and Gd^{3+}, no inner-sphere water molecules were detected. The coordination geometry is similar for all phosphinate complexes studied and can be described by a (mono-capped) square antiprism with a twist angle of about 25° (for DOTPP) or 29° (for DOTBzP and DOTMP). In aqueous solution, various diastereoisomers can be formed, as described in the previous paragraph. Based on the integrals of the signals observed in the ^{31}P NMR spectrum, it was concluded that the (*RRRR*) isomer is the dominant species, which was a general feature for all ligands studied [187]. The abundance of the other isomers decreases with the number of *S*-oriented phosphinates. Aime and co-workers performed a ^{1}H NMR study on the $Yb(DOTBzP)^-$ complex and only one set of signals was found in the spectrum which corresponds to the minor isomer of the related DOTA complex. The sterically more demanding phosphinate group (in comparison to the acetate group) favors the square antiprismatic coordination geometry. The spectra of the $Yb(DOTBzP)^-$ complex, and the Y^{3+} and Eu^{3+} analogues, remained unchanged on temperature variations (from 278 to 353 K), indicating a highly rigid coordination cage. Luminescence and relaxometry studies indicated that no water molecules were bound in the first coordination sphere of the metal ion [183, 185–187]. An LIS ^{31}P NMR study on the DOTBzP complexes, performed to test the isostructurality of the different Ln^{3+} complexes, indicated a slightly different structure for the lower lanthanide ions (such as Ce^{3+}, Pr^{3+} and Nd^{3+}), compared to the ions in the middle and second half of the series. This phenomenon was explained by the presence of a bound water molecule, exclusively in the inner sphere of the complexes with the first elements of the series [187]. This water molecule is not present in complexes with $Ln^{3+} = Eu^{3+} \rightarrow Lu^{3+}$. The $Ln(DOTPP)^-$ complexes did not show this irregularity [185] and can be considered as isostructural with an overall hydration number of $q = 0$.

NMRD studies of Gd^{3+} complexes of DOTPME and DOTPMB indicate $q < 1$, suggesting that the inner coordination sphere of these complexes is obstructed due to the steric encumbrance of the alkoxy substituents [181]. Recently, Sherry and co-workers reported multinuclear NMR studies of Ln^{3+} complexes (Ln = La, Gd, Dy, Tm and Yb) with a fluorinated ethyl ester analog of DOTP (F-DOTPME) [182]. The ^{19}F NMR spectra reveal up to sixteen resonances, which demonstrate that these complexes exist in aqueous solution as a mixture of stereoisomers. $Gd(F{-}DOTPME)^{-}$ afforded a water proton relaxivity typical of non-hydrated complexes. ^{17}O NMR of the Dy^{3+} complex confirmed the lack of a bound water molecule.

Parker, Aime and co-workers have described structural studies for a large number of lanthanide (Eu, Gd, Tb and Yb) and Y complexes with macrocyclic ligands containing three phosphinate and one carboxamide arms (see Chart 8.12) [168, 188–192]. Among the 32 possible isomers which may exist, one stereoisomer predominates in aqueous solution, as verified by ^{1}H, ^{13}C and ^{31}P NMR spectroscopy [168, 189, 190]. Luminescence studies with the Eu^{3+} and Tb^{3+} complexes are in agreement with the presence of one major isomer in aqueous solution. By comparison with crystal structures of the related tetra-(benzylphosphinate) complexes [186] and considering the steric requirements, it has been suggested that these complexes adopt the twisted SAP geometry. Spectral resolution of the ^{31}P NMR resonances was achieved by the addition of a chiral solvating agent (β-cyclodextrin) which demonstrated that the isomer is present as a 1:1 mixture of enantiomers [190]. Evaluation of the Curie contribution to the longitudinal relaxation rates of the ^{31}P resonances allowed for the determination of Yb–P distances in the Yb(DOTMP-MBBzA), Yb(DOTMP-MBMeA) and Yb(DOTMP-MHBzA) complexes [190]. The results suggest a certain distortion in the coordination polyhedron compared to the related tetra(benzylphosphinate) Y^{3+} complex [186]. The degree of distortion induced by the presence of the carboxamide substituent is dependent on the size of this group. Analysis of the variable-temperature NMRD profiles indicates no coordinated water molecules in the Gd^{3+} complexes. Non-integral q values ($q < 1$) determined by luminescence measurements have been explained by the existence of a well-defined 'second hydration sphere', promoted by the proximity of the amide carbonyl functionality which is capable of forming hydrogen bonds with local water molecules [168].

Introducing a chiral center in the amide functionality renders all 32 potential isomers diastereomeric and thus discernable (in principle) by NMR spectroscopy. In practice, the lanthanide (Eu, Gd and Tb) complexes formed with macrocyclic monoamide tris(phosphinate) ligands bearing a chiral center on the amide group exist as only two non-interconverting diastereomers in a ratio of 2:1 and 4:1 for the α-phenylethyl and α-1-napthylethyl derivatives, respectively (DOTMP-MPMeA and DOTMP-MNaphMeA) [191]. The configuration at the

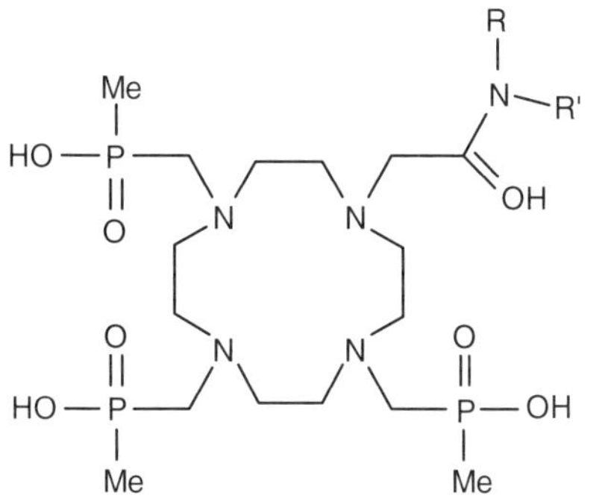

R = H R' = Me DOTMP-MHMeA

R = H R' = Ph DOTMP-MHBzA

R = R' = Me DOTMP-MBMeA

R = R' = N-Bu DOTMP-MBBuA

R = R' = CH_2Ph DOTMP=MBBzA

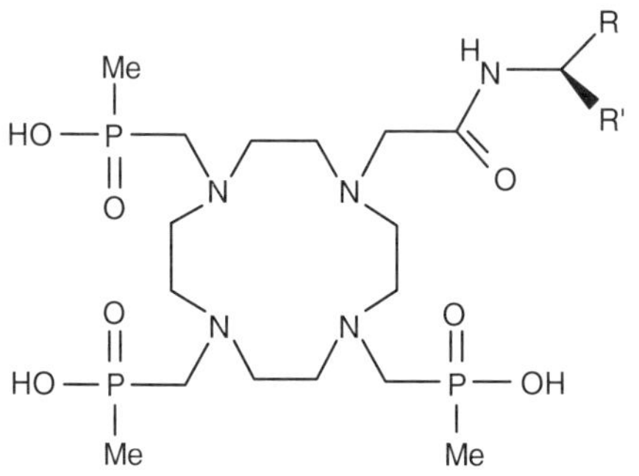

R	R'	
R = 1-$C_{10}H_7$	R' = Me	(R)-DOTMP-MnaphMeA
Me	1-$C_{10}H_7$	(S)-DOTMP-MnaphMeA
Me	Ph	(S)-DOTMP-MPMeA

[Chart 8.12]

chiral carbon center (*R* or *S*) determines the helicity of the pendant arms and the conformation of the macrocyclic ring, as demonstrated by ^{1}H and ^{31}P NMR spectroscopy and circularly polarized luminescence studies. Aime *et al.* suggest that the major isomers observed in these complexes adopt a twisted SAP geometry with the P–Me groups projected away from the cyclen ring and an *RRR* or *SSS* configuration at the phosphorus [191].

10.7 CATIONIC MACROCYCLIC LANTHANIDE COMPLEXES

Neutral *N*-derivatized octadentate ligands (see Chart 8.13) based on cyclen (1,4,7,10-tetraazacyclododecane) form tripositive cationic complexes with the trivalent lanthanides. These complexes show potential as artificial nucleases and NMR shift reagents for anions in aqueous solution, while various chiral Eu^{3+} and Tb^{3+} complexes are of interest as emissive chiral probes in biological media [193–201]. Furthermore, the *N*-substituted tetraamide derivatives have proven useful in understanding the relationship between the solution structure of the Ln^{3+} complex and the water exchange rate, which is a critical issue in attaining optimal relaxation efficiency of contrast agents for MRI [87, 202–208].

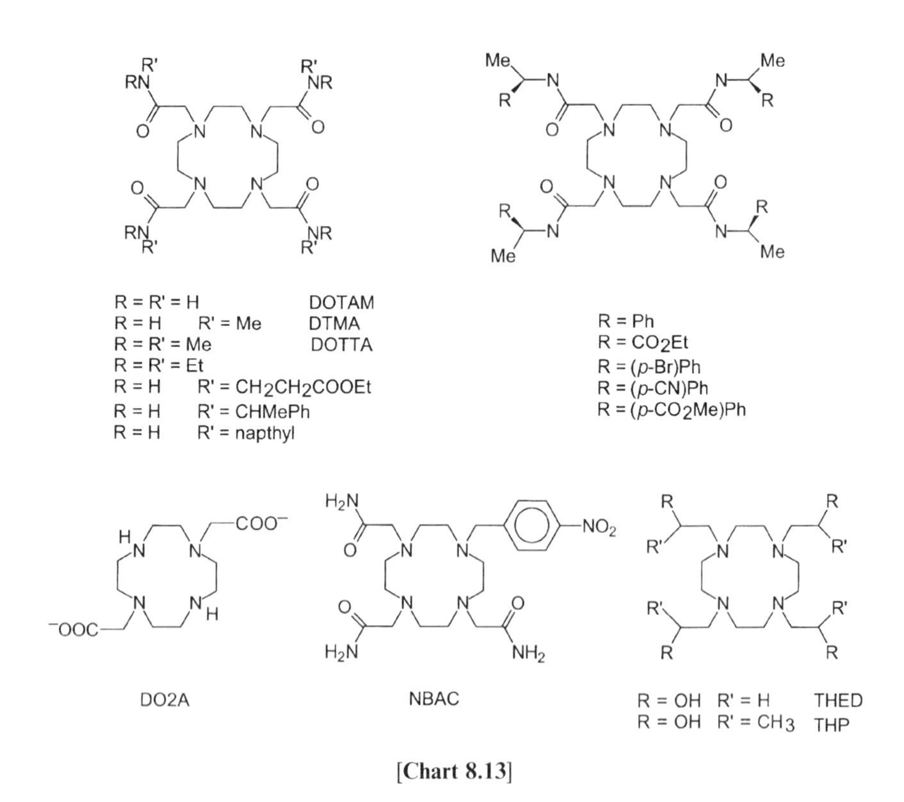

[Chart 8.13]

The solution structure and dynamics of the positively charged Ln^{3+} chelates containing the tetraaza dicarboxylate ligand DO2A, $Ln(DO2A)^+$ [157], have been studied by ^{1}H and ^{17}O NMR, relaxometry, UV–Vis spectroscopy and luminescence [209–211]. The ^{17}O water shifts show that an increased number of inner-sphere water molecules replaces the two missing acetate arms of DOTA in the $[Ln(DO2A)(H_2O)_q]^+$ chelates, but this changes, however, from three to two along the Ln^{3+} series [210, 211]. One of the inner-sphere water molecules of the Eu^{3+} chelate hydrolyses at slightly basic pH conditions, with a pK_h of 8.1 ± 0.3 (as shown by luminescence) [210]. The smaller extent of encapsulation of the Ln^{3+} ions by the DO2A ligand leads to a less rigid macrocyclic [3333] structure than in the DOTA chelates, as indicated by the ^{1}H NMR resonances of its CH_2 protons [209].

Detailed solution structure and dynamics studies have been reported for the tetraamide Ln^{3+} complexes of DOTAM, DTMA and DOTTA [193, 194, 197, 205–208]. These achiral tetraamide ligands form Ln^{3+} complexes which may exist as two diastereomeric species in solution, similar to the Ln(DOTA) complexes. The ^{1}H NMR spectra of the corresponding Eu^{3+} complexes reveals the existence of major (M) and minor (m) isomers at 273 K, where exchange is slow on the NMR time-scale. For $Eu(DOTAM)^{3+}$ and $Eu(DTMA)^{3+}$, the major isomer (M) is defined by a square antiprismatic geometry, similar to $Eu(DOTA)^{3+}$, with m/M isomer ratios of 0.25, 0.31 and 0.19, respectively [208], whereas the $Eu(DOTTA)^{3+}$ has a m/M ratio of 2 with the predominant isomer having the twisted square antiprismatic geometry. These results indicate that a twisted SAP structure is preferred with increasing steric demand at the metal center. In addition, the ratio of the two isomers for $Ln(DOTAM)^{3+}$ complexes was shown to be sensitive to temperature and to the nature of the Ln^{3+} ion. Analysis of the variable-temperature ^{13}C NMR spectra of $La(DOTAM)^{3+}$ revealed a high degree of structural rigidity with an activation energy for the ethylenediamine ring inversion of 56.1 ± 0.5 kJ/mol ($T_c = 22°$ C) [194]. Luminescence lifetime measurements of the DOTAM, DTMA and DOTTA Eu^{3+} complexes revealed the existence of one bound water molecule [197, 205, 208]. Furthermore, ^{1}H NMR spectroscopy allowed for the direct observation of the bound water molecules for both isomers at low temperatures in dry CD_3CN [207].

Crystal structures have been reported for $[Eu(DOTAM)(H_2O)](CF_3SO_3)_3$ [194], $[Gd(DTMA)(H_2O)](ClO_4)_3$ [206] and $[Dy(DTMA)(H_2O)](PF_6)_3$ [208]. Four stereoisomers (two enantiomeric pairs) appear in the crystal of $[Eu(DOTAM)(H_2O)]^{3+}$ with the two essentially identical diastereomers at each site having very similar torsion angles (30.2° and 30.4°). The coordination geometry at the Eu^{3+} site of either diastereomer is best described as a mono-capped twisted SAP. In $[Gd(DTMA)(H_2O)]^{3+}$, the Gd^{3+} ion is nine-coordinate with four nitrogen atoms of the cyclen ring and four amide carbonyl oxygen atoms located at the vertices of a distorted SAP and capped by the coordinated water molecule. The $[Dy(DTMA)(H_2O)]^{3+}$ complex is also nine-coordinate; however, the Dy^{3+} ion adopts a mono-capped SAP geometry (twist angle = 39°). Two enantiomeric complexes which possess the $\Delta(\lambda\lambda\lambda\lambda)$ and $\Lambda(\delta\delta\delta\delta)$ configurations are present in the unit cell. In addition, an X-ray structure was reported for $[La(DOTAM)(EtOH)(CF_3SO_3)](CF_3SO_3)_2$, [194], where the La^{3+} ion was found to be ten-coordinate, producing an unusual 1,5,4 geometry. The six oxygen atoms (four from the amidic arms, one from ethanol and one from the coordinated triflate anion) form a pentagonal pyramid with ethanol at the capping position. This pyramid is stacked base upon apex above a square pyramid made up of four nitrogen atoms from the macrocycle at the base and the La^{3+} ion at the vertex.

Variable-temperature ^{1}H and ^{13}C studies of La(DOTAM) in CD_3CN revealed a high degree of structural rigidity, with the activation energy for ethylenediamine ring inversion being determined as 58.9 ± 0.3 kJ/mol. The ^{1}H NMR spectrum in D_2O closely resembles that in CD_3CN; however, the complex is kinetically unstable and readily dissociates in water, presumably due to the formation of less favorable six-membered chelate rings [193]. An X-ray crystallographic study of $La(DOTAM)^{3+}$ showed one cation in the asymmetric unit with an eight-coordinate La^{3+} ion with four La–N and four La–O bonds originating from the cyclen ring and amide groups, respectively [193]. The chiral complex possesses C_4 symmetry with the amide moieties arranged in a clockwise fashion, producing a twisted SAP structure (twist angle = 26.5°).

A detailed LIS study for ten paramagnetic Ln^{3+} complexes of 1,4,7,10-tetrakis(*N*,*N*-diethylacetamido)-1,4,7,10-tetraazacyclododecane in CD_3CN was carried out in conjunction with variable-temperature ^{1}H and ^{13}C NMR investigations of the diamagnetic La^{3+} complex [37]. NMR data established the existence of a single pair of enantiomers in solution ($\Delta G^{\neq} = 58.8$ kJ/mol for ethylenediamine ring inversion). Molecular mechanics calculations combined with analysis of the LIS data determined a solution structure where the Ln^{3+} ion is eight-coordinate with the four nitrogen atoms of the ring and four carbonyl oxygen atoms at the vertices of a distorted square antiprism. The complexes exhibit C_4 symmetry where a λ configuration of the ethylenediamine rings and a Δ rotation of the pendant arms, or its enantiomeric form, is favored in solution.

Lanthanide complexes of several chiral tetraamide ligands have been reported [197, 198, 200, 201]. Introducing a chiral center α to the amide N (δ to the ring N) imparts ample conformational rigidity to hinder pendant arm rotation. Solution NMR studies of these Eu^{3+} complexes in CD_3OD and D_2O show no evidence of exchange broadening (Δ/Λ interconversion) from 200 to 320 K, consistent with the presence of only one isomer having average C_4 symmetry [197, 200, 201]. The observed resonances in the ^{1}H NMR spectrum of a Yb^{3+} complex ($CH_2C(O)NH$–$CH(Me)(Ph)$ amide substituents) correspond closely to those obtained for the related chiral tetraphosphinate complexes which exist exclusively as one diastereomer in solution having a twisted SAP geometry [201]. Four crystal structures have been reported for Eu^{3+}, Dy^{3+} and Yb^{3+} complexes of the (*R*)-and (*S*)-derivatives of the tetraamide – $[CH_2C(O)NH$–$CH(Me)(Ph)]_4$-substituted ligand [197, 200, 201]. In each of these structures, the metal center is bound by four nitrogen atoms of the macrocycle and four amidic oxygen atoms, with a single water molecule in the capping position. The geometry of the coordination polyhedron varies from slightly distorted to regular SAP with the chirality of the amide stereocenter ultimately determining the helicity of the complexes and the configuration of the macrocycle ring (with the (*R*)- and (*S*)- isomers producing the $\Lambda(\delta\delta\delta\delta)$ and $\Delta(\lambda\lambda\lambda\lambda)$ configurations, respectively). The (*S*)- Eu and (*S*)- Dy complexes are isostructural. A crystal structure of the Eu^{3+} complex with the tetraamide (*S*)-$[CH_2C(O)NH\text{-}CH(Me)(CO_2Et)]_4$-derived ligand shows a mono-capped regular SAP geometry with the $\Delta(\lambda\lambda\lambda\lambda)$ absolute configuration [198].

Additional cationic *N*-substituted amide complexes having mixed pendant groups have been reported [195, 196, 212].

Morrow and co-workers investigated the ^1H and ^{13}C NMR spectra of Ln^{3+} complexes formed with a single stereoisomer of THP having an *S*-configuration at all α-carbons (*S,S,S,S*-THP). In these (*S,S,S,S*)-THP complexes [213], the additional methyl group on the α-carbons adds four chiral centers, thus producing two diastereomers rather than an enantiomeric pair. ^1H and ^{13}C NMR spectra of La^{3+} and Lu^{3+} complexes of *S,S,S,S*-(−)-THP do not display the fluxional behavior as seen in $Ln(DOTA)^-$ complexes over the temperature range from 18 to 100 °C, indicating the presence of only one diastereomer in solution. Molecular models suggest that a clockwise orientation of the pendant arms may be sterically favored. In the crystal structure of the racemic $Eu(THP)^{3+}$ complex, the unit cell was shown to consist of two discrete $[Eu(THP)(H2O)]^{3+}$ cations which were diastereomers of each other, differing in the configuration at the chiral carbon (i.e. *R,R,R,S* versus *S,S,S,R*) and consequently the handedness of the helix. The THP ligand coordinated the Eu^{3+} ion through four nitrogen atoms of the macrocycle and four hydroxyl oxygen atoms of the pendant arms, with the ninth coordination site filled by a bound water molecule. The coordination geometry is best described as an inverted mono-capped SAP with an average twist angle of ca 20°.

The ^1H NMR spectra of the related $La(THED)^{3+}$ as a function of temperature reveal a dynamic process at room temperature similar to that observed for $Ln(DOTA)^-$ complexes [214]. At ambient temperature, the ^{13}C NMR spectra (methanol-d_4) consist of two sharp resonances assigned to the pendant arms and one broad resonance attributed to the ethylene ring carbons, which sharpens as the fast-exchange limit is approached (ca 50 °C). Likewise, at −20 °C the broad resonance resolves into two peaks. An activation energy of 52 ± 0.7 kJ/mol for the ethylenediamine ring exchange displays the decreased rigidity of $La(THED)^{3+}$ as compared to DOTA complexes and suggests that the pendant groups contribute to the structural rigidity of the macrocyclic ring.

10.8 LANTHANIDE COMPLEXES OF TEXAPHYRINS

A texaphyrin (HTx) is an expanded porphyrin ligand with five nitrogen donor atoms and a cavity that encompasses the Gd^{3+} ion almost ideally resulting in a high stability (see Chart 8.14) [215]. The remaining axial positions are probably occupied by water molecules, which is in agreement with a higher relaxivity compared to $Gd(DTPA)^{2-}$ and $Gd(DOTA)^-$ (3–4 times greater). In the

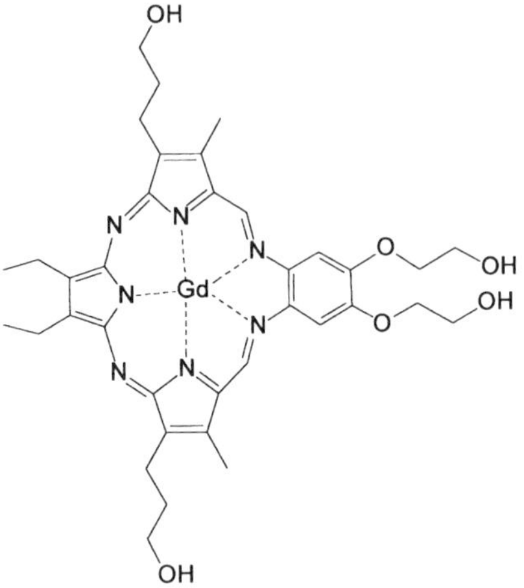

[Chart 8.14]

presence of phosphate, the relaxivity is considerably reduced, thus revealing that phosphate competes with water for binding to the Gd^{3+} ion [216]. It has been suggested that these reagents are tissue-specific and may be applied to tumor diagnosis with MRI, while an analogous diamagnetic Lu^{3+} complex may be used as a photosensitizer to destroy tumors photodynamically [217].

Lisokowski *et al.* reported on an extensive ^{1}H NMR study of paramagnetic Ln^{3+} texaphyrins in $CDCl_3/CD_3OD$ [64]. The signals in the spectra of $Ln(Tx)(NO_3)_2$ were assigned by using NOE-difference, COSY and ROESY techniques in combination with T_1 measurements and comparison of the pseudocontact shifts for the various Ln^{3+} complexes. The LIS values were fitted with Equation (8.4) using solid state geometries [216] and assuming that the x- and y-axes are in the symmetry plane of the molecule. When the imino protons were excluded from the fitting procedure, excellent agreement was obtained for all Ln^{3+} complexes. All other LIS values were considered to be mainly of pseudocontact origin. The pseudocontact shifts of the imino protons were evaluated by using the parameters obtained from the fitting procedure. Subtraction of the pseudocontact shifts from the total LIS values afforded the contact shifts. These show a linear relationship with the theoretical $< S_z >$ values which is consistent with a constant hyperfine coupling along the Ln^{3+} series. The factors D_1 and D_2, however, show large deviations from linearity which has been ascribed to changes in the metal-centered axial ligation along the lanthanide series. The dramatic effect of the axial ligation on the magnetic susceptibility tensor was demonstrated by substitution of the nitrate ligands with phosphates. From the D_1 and D_2 values obtained and from the solid state structures, pseudocontact LIS values for the ^{13}C nuclei were calculated for the Ce^{3+}, Pr^{3+}, Nd^{3+} and Eu^{3+} complexes of Tx [73]. The contact shifts, calculated by subtraction of the pseudocontact shifts from the experimental ones, appeared to be dominating and to result from through-bond spin delocalizations.

11 TARGETING CONTRAST AGENTS

Targeting contrast agents are able to recognize specific target molecules. In a recent review by Lauffer and co-workers, an overview of disease-specific MRI agents is given [98]. Aime *et al.* [218] have described the targeting properties of a DTPA derivative that, because of its boronic acid functionality, is able to map the glycation level of proteins in the blood. The boronic function in the DTPA-bis(*m*-boroxyphenyl)amide (DTPA-BMPA) ligand (see Chart 8.15) binds sugars with a *syn*-diol moiety. High levels of glucose in blood serum can induce a non-enzymatic glycation of albumin. Binding of Gd(DTPA-BMPA) to the glycated albumin induces an enhancement of the water proton relaxation rate as a consequence of the increased rotational correlation time τ_R.

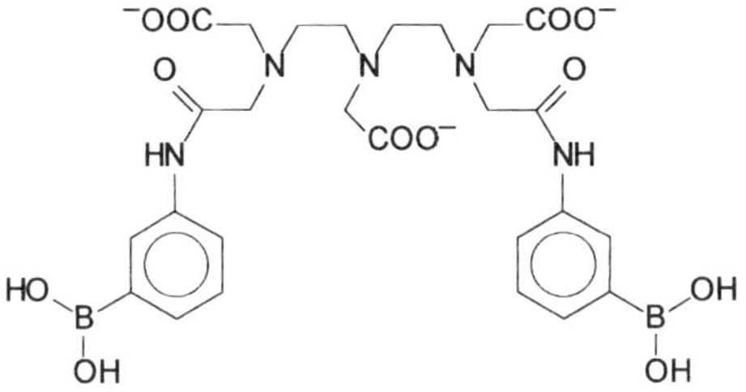

DTPA-BMPA

[Chart 8.15]

Recently, Aime *et al.* reported that the Gd(DTPA-BMPA) complex also has a strong interaction with a non-glycated protein, namely oxygenated human hemoglobin [219]. The interaction involved the formation of coordinative N $\rightarrow$ B bonds at two histidine residues of different β-chains of the peptide. The water-proton relaxation-rate enhancement was more than three times higher in case of the hemoglobin adduct compared to the albumin system described in the previous paragraph. It was shown that the interaction of the Gd^{3+} complex with hemoglobin was site-specific, and resulted in a conformational switch (from the low-affinity T to the high-affinity R state) in the tetrameric protein.

12 RESPONSIVE OR 'SMART' CONTRAST AGENTS

Over the last few years, much interest has been directed towards the development of paramagnetic complexes whose relaxivity is dependent on a certain biochemical variable. For this new class of MRI contrast agents

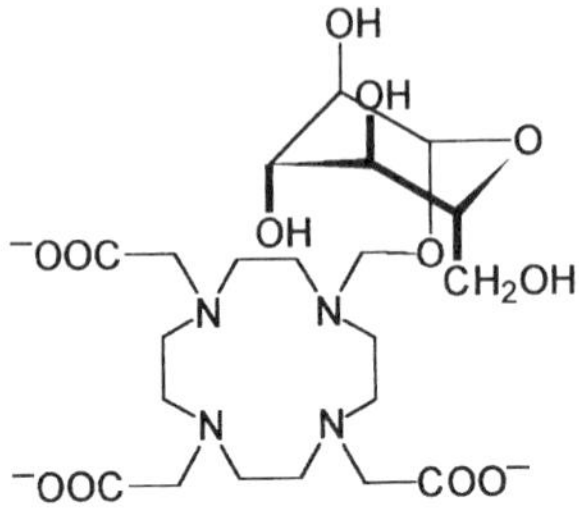

DO3A-gal

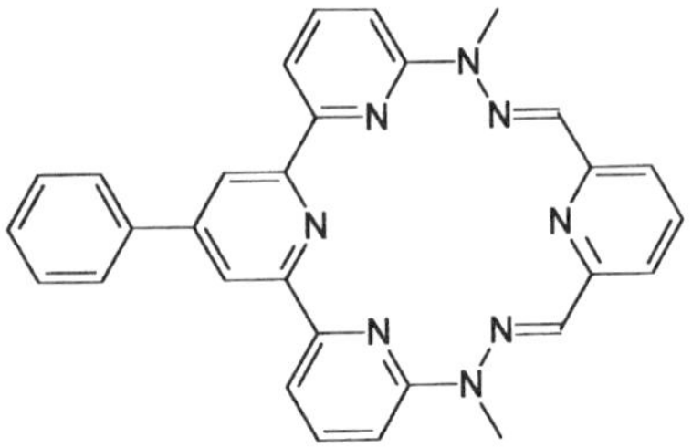

cy(TPyDAPy)

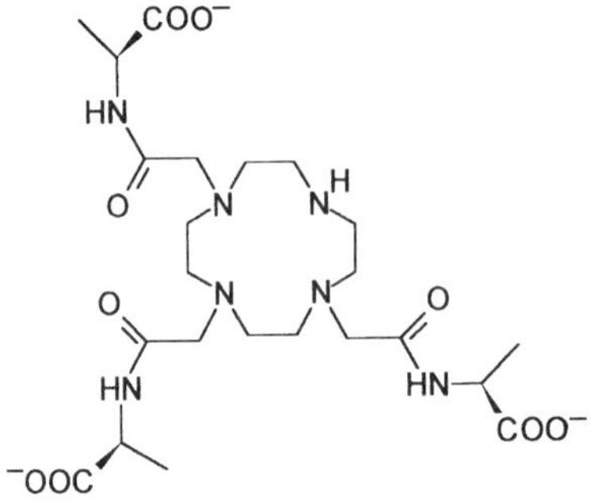

DO3MA-Ala

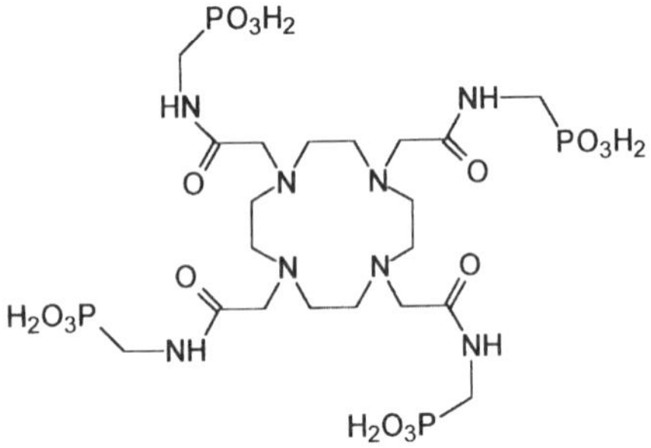

DOTAM-MP

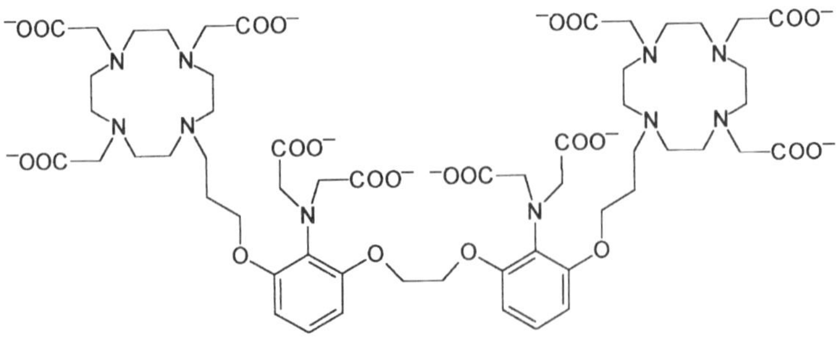

DOPTA

[Chart 8.16]

(see Chart 8.16), the solution structure is affected by factors such as pH or concentration of metal ions (e.g. Ca^{2+}) resulting in a change of the hydration number and thus the relaxivity of the specific complexes.

The first 'smart' contrast agent reported in the literature was Gd(DO3A-gal), a DOTA type ligand in which one of the acid groups is replaced by a galactose moiety [220]. This sugar residue blocks the metal ion for water binding. The galactose unit can be cleaved enzymatically, thus making the inner-sphere coordination site accessible again for coordination of a water molecule. It was demonstrated that the change in hydration number from 0 to 1, resulting in a 20 % increase in the measured water T_1 values, was indeed due to the enzymatic cleavage of the galactose residue.

Aime, Parker and co-workers reported a Gd^{3+} complex of a hexaazamacrocyclic ligand, cy(TPyDAPy) [221], displaying a pH-dependent relaxivity in the range $6 < pH < 10$ which may be applied to study *in vivo* pH gradients. The relaxivity values measured varied from about 13 $mM^{-1}s^{-1}$ (pH < 6) to 2 $mM^{-1}s^{-1}$ (pH = 11). In previous studies, these values were measured for compounds with hydration numbers of $q = 2$–3 and for purely outer-sphere complexes, respectively. The authors demonstrated the formation of an OH-bridged dimeric complex at higher pH values that indeed has no water molecules (or hydroxyl groups) bound in the first coordination sphere.

The interaction of hydrogen carbonate with the metal ion in a macrocyclic tetraazatriamide complex, Gd(DO3A-Ala), under ambient conditions [199] prevents any water molecule from directly binding to the paramagnetic center. In more acidic medium (pH < 6), the hydrogen carbonate ion will be protonated and displaced by a water molecule, thus resulting in an increase of the relaxivity. Complexes that show a pH-dependent relaxivity around physiological pH values are of particular interest since they may afford a way to distinguish tumor tissue (pH = 6.9) from healthy tissue (pH = 7.4). Protonation of a donor atom (O or N) can lead to competitive binding of a water molecule and thus affect the relaxivity. The development of a ligand exhibiting a pH-dependent coordination number has not yet been successful [222], but is definitely a promising approach. Aime *et al.* reported a macromolecular system based on a poly(amino acid) chain substituted with DO3A chelates [223] that showed a pH-dependent relaxivity. In this case, the pH dependence was not originating from the complex solution structure or ionic interactions at the paramagnetic metal ion; instead, it was the macromolecular chain that reflected conformational changes as the pH was varied from 4 to 8. These conformational changes affected the overall and local rotational motions in the macromolecule, which influence the water-proton relaxation-rate-enhancement.

Recently, Sherry and co-workers have described a novel pH-sensitive MRI contrast agent based on $Gd(DOTAM)^{3+}$ [224]. Extension of the amide pendant arms with methylenephosphonate chains resulted in a complex, $Gd(DOTAM\text{-}MP)^{3+}$, that shows a very unusual pH-dependent relaxivity. Between pH 4 and 6, the relaxivity increases from about 5 to 10 $mM^{-1}s^{-1}$. Upon further increase of the pH, the relaxivity gradually decreases to 3.8 $mM^{-1}s^{-1}$ at pH 8. The ^{31}P NMR spectra of all of the $Ln(DOTAM)^{3+}$ complexes showed single resonances with chemical shifts close to that of the free ligand, thus indicating that the phosphonate groups are not coordinated to the Ln^{3+} ion. The ^{1}H and ^{13}C NMR spectra indicated one main molecular species having a high stereochemical rigidity. The LIS values of the Yb^{3+} complex were similar to those of the DOTP and DOTA complexes. This indicates that the Yb^{3+} ion is bound by the four amide oxygen atoms and the four nitrogen atoms of the cyclen backbone. Dy^{3+}-induced water ^{17}O shifts at 75°C showed that the complex contains one inner-sphere water molecule. The unique pH dependence of the relaxivity of the Gd^{3+} complex is ascribed to the protonation of the (uncoordinated) phosphonate groups. It is suggested that the hydrogen-bonding network created by protonation of the phosphonates provides a catalytic pathway for exchange of the water protons between the complex and the bulk.

Meade and co-workers reported a calcium-sensitive MRI contrast agent, Gd_2(DOPTA) [225]. The ligand consists of two DO3A units that are linked via an EGTA-type spacer carrying two aromatic iminoacetate groups. In the absence of Ca^{2+}, these iminoacetate functions are coordinating the Gd^{3+} ions. However, in the presence of Ca^{2+}, these carboxylate groups will rearrange and bind the divalent metal ion. Consequently, two coordination sites on each of the Gd^{3+} ions are vacant for binding of water molecules, thus resulting in an increase of the relaxivity from 3.26 to 5.76 $mM^{-1}s^{-1}$. The interference of Ca^{2+} binding with Mg^{2+} or H^{+} was determined to be minimal within physiological pH ranges. This compound offers a large and reliable change in relaxivity upon exposure to Ca^{2+}.

13 ZEOLITE-BASED MRI CONTRAST AGENTS

A suspension of Gd^{3+}-exchanged zeolite NaY (gadolite) has been shown to be a good oral contrast agent for MRI on the gastrointestinal tract [226–228]. Zeolite NaY is a microporous aluminosilicate with a well-defined channel system consisting of spherical 11.8 Å cavities (super-or α-cages) connected by twelve-membered rings with a diameter of 7.4 Å (see Figure 8.11). Each supercage is surrounded by ten sodalite units containing the smaller β-cages (6.6 Å internal diameter). The negative charge of the framework is balanced by Na^{+} counterions. In gadolite, part of the Na^{+} ions are exchanged carefully with Gd^{3+} [229]. X-ray diffraction (XRD) powder patterns and Fourier-transform infrared spectroscopy (FT-IR) of the exchanged material show that the crystallinity of the material was maintained during the exchange and that no dealumination took place. Gd^{3+} loadings of up to about 8.2 wt % were obtained. The resulting material is stable at pH 2.5–5 over a period of at least 8 h; however, at lower pH values some leaching of Gd^{3+} occurs. It should be noted, however, that the acute toxicity of orally administered $GdCl_3$ is low (LD_{50} mice > 2000 mg/kg).

At room temperature, the hydrated Gd^{3+} ions are restricted to the supercages because of their size (radius of hydrated Gd^{3+}, 3.18 Å). The most

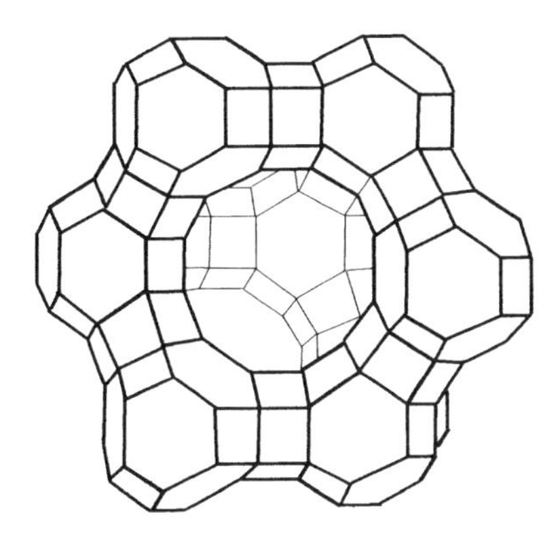

Figure 8.11 The zeolite framework of gadolite.

probable location is at the center of the twelve-membered ring. At higher temperatures, Ln^{3+} ions start moving irreversibly to the smaller β-cages [230]. It has been shown that water molecules enter the β-cages of Y zeolites but these water molecules do not exchange with water molecules in the supercages during the T_1 time interval and, therefore, do not contribute to the NMR signal produced by the bulk water [231–234].

The relaxivity of GdNaY increases with decreasing Gd^{3+} content in the zeolite [229]. This has been attributed to magnetic dilution of the intrazeolite gadolinium at low loadings, as well as an increased mobility of the water molecules. At higher loadings, Gd^{3+} ions may be located in the center of the twelve-ring windows and in this way hinder the exchange of water between the supercages and the bulk. Movement of the Gd^{3+} ions to the small cages upon dehydration should enhance the intrazeolite exchange of water between supercages, which is consistent with the higher relaxivities observed for a material that was heated at 373 K.

Suspending a Gd^{3+}-exchanged NaY zeolite in a solution of DTPA resulted in a zeolite containing Gd(DTPA) in the supercages [235]. Surprisingly, the NMRD curve of the GdNaY did not change significantly upon complexation with DTPA. A possible reason for this phenomenon could be the decrease of the amount of water in the supercages, as a result of the volume occupied by the complex or to limitation of the relaxivity by the exchange rate of water into the cage.

A hexagonal Y-type zeolite has been synthesized using Gd^{3+} complexes of 18-crown-6 as a template [236]. The material has been characterized by XRD, FT-IR spectroscopy and scanning electron microscopy. Its relaxivity at 40 MHz is similar to that of GdNaY.

Clays modified with paramagnetic ions have also been shown to be effective oral MRI contrast agents [237, 238]. Clays are crystalline aluminosilicates with a layered structure. The charges are balanced by cations between the layers. Various suspending agents appeared to have an enhancing effect on the relaxivities of Gd^{3+}-exchanged hectorite clays [239]. Poly(ethylene oxide) was shown to intercalate into the clay, which resulted in an increase of the relaxivity. The increased relaxivities of Gd^{3+}-hectorite in the presence of other suspending agents has been attributed to an increased viscosity.

REFERENCES

1. Shannon, R. D. *Acta Crystallogr. Sect. A* 1976, **32**, 751.
2. Cossy, C., Helm, L. and Merbach, A. E. *Inorg. Chem.* 1989, **28**, 2699.
3. Horrocks, W. DeW. Jr In *NMR of Paramagnetic Molecules*, La Mar, G. N., Horrocks, W. DeW. Jr and Holm, R, (Eds), Academic Press New York 1973, pp. 475–519.
4. Bertini, I., Turanom, P. and Vila, A. J. *Chem. Rev.* 1993, **93**, 2833, and references therein.
5. Peters, J. A., Nieuwenhuizen, M. S. and Raber, D. J. *J. Magn. Reson.* 1985, **65,** 417.
6. Golding, R. M. and Halton, M. P. *Aust. J. Chem.* 1972, **25**, 2577.
7. Pinkerton, A. A., Rossier, M. and Spiliadis, S. *J. Magn. Reson.* 1985, **64**, 420.
8. Lewis, W. B., Jackson, J. A., Lemons, J. F. and Taube, H. *J. Chem. Phys.* 1962, **36**, 694.
9. Reuben, J. and Fiat, D. *J. Chem. Soc. Chem. Commun.* 1967, 729.
10. Bryden, C. C., Reilley, C. C. and Desreux, J. F. *Anal. Chem.* 1981, **53**, 1418.
11. Witanowski, M., Stefaniak, L. and Januszewski, H. *J. Chem. Soc. Chem. Commun.* 1971, 1573.
12. Reilley, C. N., Good, B. W. and Allendoerfer, R. D. *Anal. Chem.* 1976, **48**, 1446.
13. Peters, J. A., Huskens, J. and Raber, D. J. *Progr. Nucl. Magn. Reson. Spectrosc.* 1996, **28**, 283.
14. Powell, D. H., Ni Dhubhghaill, O. M., Pubanz, D., Helm, L., Lebedev, Y. S., Schlaepfer, W. and Merbach, A. E. *J. Am. Chem. Soc.* 1996, **118**, 9333.
15. McGarvey, B. M., Kurland, R. J. in *NMR of Paramagnetic Molecules*, La Mar, G. N. and Horrocks, W.deW. Jr and Holm, R., (Eds), Academic Press, New York, 1973, pp. 555–593.
16. Russell, G. A. in *Radical Ions*, Kaiser, E. T. and Kevan, L. (Eds), Interscience Publishers (John Wiley & Sons), New York, 1968, p. 87.
17. Kemple, M. D., Ray, B. D., Lipkowitz, K. B., Prendergast, F. G. and Rao, B. D. N. *J. Am. Chem. Soc.* 1988, **110**, 8275.
18. Bleaney, B. *J. Magn. Reson.* 1972, **8**, 91.
19. Horrocks, W. DeW. Jr, Sipe, J. P. III and Sudnick, D. R. in *Nuclear Magnetic Resonance Shift Reagents*, Sievers, R. E. (Ed.), Academic Press, New York, 1973, pp. 53–86.
20. Golding, R. M. and Pyykkö, P. *Mol. Phys.* 1973, **26**, 1389.
21. Horrocks, W. DeW. Jr *J. Magn. Reson.* 1977, **26**, 333.
22. McGarvey, B. R. *J. Magn. Reson.* 1979, **33**, 445.
23. Horrocks, W. DeW. Jr and Sipe, J. P. III *Science* 1972, **177**, 994.
24. Zolin, V. F. and Koreneva, L. G. *Zh. Strukt. Khim. (Engl. Transl.)* 1980, **21**, 51.
25. Zolin, V. F., Koreneva, L. G., Obukhov, A. E., Zvolinskii, V. P. and Kordova, I. R. *Theor. Exp. Chem. (Engl. Transl)* 1982, **18**, 166.
26. Babushkina, T. A., Zolin, V. F. and Koreneva, L. G. *J. Magn. Reson.* 1983, **52**, 169.
27. Görller-Walrand, C. and Binnemans, K. in *Handbook on the Physics and Chemistry of Rare Earths,* Vol. 23, Gschneider, K. A. Jr and Eyring L., (Eds), Elsevier, Amsterdam, 1996, Ch. 155, pp. 121–283.

28. Briggs, J. M., Moss, G. P., Randall, E. W. and Sales, K. D. *J. Chem. Soc. Chem. Commun.* 1972, 1180.
29. Horrocks, W. DeW. Jr *J. Am. Chem. Soc.* 1974, **96**, 3022.
30. Chu, S. C.-K., Xu, J., Balschi, J. A. and Springer, C. S. Jr *Magn. Reson. Med.* 1990, **13**, 239.
31. Gysling, H, and Tsutsui, M. *Adv. Organomet. Chem.* 1970, **9**, 361.
32. Micskei, K., Helm, L., Brücher, E. and Merbach, A. E. *Inorg. Chem.* 1993, **32**, 3844.
33. Mao, X., Shen, L., Gao, Y. and Ni, J. *Huaxue Xuebao* 1985, **43**, 566; *Chem. Abstr.* 1985, **103**, 114834a.
34. Fossheim, S., Sæbø, K. B., Fahlvik, A. K., Rongved, P. and Klaveness, J. *J. Magn. Reson. Imaging* 1997, **7**, 251.
35. Corsi, D. M., van Bekkum, H. and Peters, J. A., to be published.
36. Bertini, I. and Luchinat, C. *Coord. Chem. Rev.* 1996, **150**, 221.
37. Forsberg, J. H., Delaney, R. M., Zhao, Q., Harakas, G. and Chandran, R. *Inorg. Chem.* 1995, **34**, 3705.
38. Reuben, J. and Elgavish, G. A. *J. Magn. Reson.* 1980, **39**, 421.
39. Peters, J. A. *J. Magn. Reson.* 1986, **68**, 240.
40. Aime, S., Barge, A., Borel, A., Botta, M., Chemerisov, S., Merbach, A. E., Müller, U. and Pubanz, D. *Inorg. Chem.* 1997, **36**, 5104.
41. Reuben, J. *J. Magn. Reson.* 1982, **50**, 233.
42. Ren, J. and Sherry, A. D. *J. Magn. Reson.* 1996, **B 111**, 178.
43. Platas, C., Avecilla, F., de Blas, A, Geraldes, C. F. G. C., Rodríguez-Blas, T., Adams, H. and Mahía, J. *Inorg. Chem.* 1999, **38**, 3190.
44. Vold, R. L., Waugh, J. S., Klein, M. P. and Phelps, D. E. *J. Chem. Phys.* 1968, **43**, 3831.
45. Meiboom, S. and Gill, D. *Rev. Sci. Instr.* 1958, **29**, 688.
46. Swift, T. J. and Connick, R. E. *J. Chem. Phys.* 1962, **37**, 307.
47. Leigh, J. S. Jr *J. Magn. Reson.* 1971, **4**, 308.
48. McLaughlin, A. C. and Leigh, J. S. Jr *J. Magn. Reson.* 1973, **9**, 296.
49. Kowalewski, J., Nordenskiöld, L., Benetis, N. and Westlund, P.-O. *Prog. Nucl. Magn. Reson. Spectrosc.* 1985, **17**, 141.
50. Bertini, I. and Luchinat, C. *Coord. Chem. Rev.* 1996, **150**, 77.
51. Alsaadi, B. M., Rossotti, F. J. C. and Williams, R. J. P. *J. Chem. Soc. Dalton Trans.* 1980, 2147.
52. Bertini, I, Capozzi, F., Luchinat, C., Nicastro, G. and Xia, Z. *J. Phys. Chem.* 1993, **97**, 6351.
53. Solomon, I. *Phys. Rev.* 1955, **99**, 559.
54. Gueron, M. *J. Magn. Reson.* 1975, **19**, 58.
55. Vega, A. J. and Fiat, D. *Mol. Phys.* 1976, **31**, 347.
56. Aime, S., Barbero, L., Botta, M. and Ermondi, G. *J. Chem. Soc. Dalton Trans.* 1992, 225.
57. Alsaadi, B. M., Rossotti, F. J. C. and Williams, R. J. P. *J. Chem. Soc. Dalton Trans.* 1980, 813.
58. Burns, P. D. and La Mar, G. N. *J. Magn. Reson.* 1982, **46**, 61.
59. Sternlicht, H. *J. Chem. Phys.* 1965, **42**, 2250.
60. Vasavada, K. V. and Rao, B. D. N. *J. Magn. Reson.* 1989, **81**, 275.
61. Bertini, I., Luchinat, C. and Vasavada, K. *J. Magn. Reson.* 1990, **89**, 243.
62. Peters, J. A., van Bekkum, H. and Bovée, W. M. M. J. *Tetrahedron* 1982, **38**, 331.
63. Zitha-Bovens, E., van Bekkum, H., Peters, J. A. and Geraldes, C. F. G. C. *Eur. J. Inorg. Chem.* 1999, 287.
64. Lisokowski, J., Sessler, J.L, Lynch, V. and Mody, T. D. *J. Am. Chem. Soc.* 1995, **117**, 2273.

65. Willcott, M. R. III, Lenkinski, R. E. and Davis, R. E. *J. Am. Chem. Soc.* 1972, **94**, 1742.
66. Forsberg, J. H. in *Handbook on the Physics and Chemistry of Rare Earths,* Vol. 23, Gschneider, K. A. Jr, and Eyring, L., (Eds), Elsevier, Amsterdam, 1996, Ch. 153, pp. 1–68.
67. Bertini, I. and Luchinat, C. *Coord. Chem. Rev.* 1996, **150**, 185.
68. Jenkins, B. G. and Lauffer, R. B. *J. Magn. Reson.* 1988, **80**, 328.
69. Jacques, V. and Desreux, J. F. *Inorg. Chem.* 1994, **33**, 4048.
70. Franklin, S. J. and Raymond, K. N. *Inorg. Chem.* 1994, **33**, 5794.
71. Hoeft, S. and Roth, K. *Chem. Ber.* 1993, **126**, 869.
72. Aime, S., Botta, M. and Ermondi, G. *Inorg. Chem.* 1992, **31**, 4291.
73. Lisokowski, J., Sessler, J. L. and Mody, T. D. *Inorg. Chem.* 1995, **34**, 4336.
74. Lutz, O. and Oehler, H. *J. Magn. Reson.* 1980, **37**, 261.
75. Vijverberg, C. A. M., Peters, J. A., Kieboom, A. P. G. and van Bekkum, H. *Recl. Trav. Chim. Pays-Bas* 1980, **99**, 287.
76. Peters, J. A. and Kieboom, A. P. G. *Recl. Trav. Chim. Pays-Bas* 1983, **102**, 381.
77. Geraldes, C. F. G. C. and Sherry, A. D. *J. Magn. Reson.* 1986, **66**, 274.
78. Peters, J. A. *Inorg. Chem.* 1988, **27**, 4686.
79. Jenkins, B. G. and Lauffer, R. B. *Inorg. Chem.* 1988, **27**, 4730.
80. Lammers, H., van der Heijden, A. M., van Bekkum, H., Geraldes, C. F. G. C. and Peters, J. A. *Inorg. Chim. Acta* 1998, **268**, 249.
81. Holz, R. C. and Horrocks, W. deW. Jr *J. Magn. Reson.* 1990, **89**, 627.
82. Bovens, E., Hoefnagel, M. A., Boers, E., Lammers, H., van Bekkum, H. and Peters, J. A. *Inorg. Chem.* 1996, **35**, 7679.
83. Lee, S.-G. *Bull. Korean Chem. Soc.* 1996, **17**, 589.
84. Alpoim, M. C., Urbano, A. M., Geraldes, C. F. G. C. and Peters, J. A. *J. Chem. Soc. Dalton Trans* 1992, 463.
85. Peters, J. A., unpublished results.
86. Horrocks, W. DeW. Jr and Sudnick, D. R. *J. Am. Chem. Soc.* 1979, **101**, 334.
87. Beeby, A., Clarkson, I. M., Dickins, R. S., Faulkner, S., Parker, D., Royle, L., de Sousa, A. S., Williams, J. A. G. and Woods, M. *J. Chem. Soc., Perkin Trans 2* 1999, 493.
88. Chang, C. A., Brittain, H. G., Telser, J. and Tweedle, M. F. *Inorg. Chem.* 1990, **29**, 4468.
89. Graeppi, N., Powell, H., Laurenczy, G., Zékány, L. and Merbach, A. E. *Inorg. Chim. Acta* 1995, **235**, 311, and references therein.
90. Corey, E. J. and Bailar, J. C. Jr *J. Am. Chem. Soc.* 1959, **81**, 2620.
91. Beattie, J. M. *Acc. Chem. Res.* 1971, **4**, 253.
92. Bryden, C. C. and Reilley, C. N. *Anal. Chem.* 1982, **54**, 610.
93. Geraldes, C. F. G. C., Sherry, A. D., Cacheris, W. P., Kuan; Brown, R. D. III, Koenig, S. H. and Spiller, M. *Magn. Reson. Med.* 1988, **8**, 191.
94. Aime, S. and Botta, M. *Inorg. Chim. Acta* 1990, **177**, 101.
95. Stekowski, J. J. and Hoard, J. L. *Isr. J. Chem.* 1984, **24**, 323.
96. Gries, H. and Miklautz, H. *Physiol. Chem. Phys. Med. NMR* 1984, **16**, 105.
97. Jin, T.-Z., Zhao, S.-F., Xu, G.-X., Ha, Y.-Z., Shi, N.-C. and Ma, Z.-S. *Huaxue Xuebao* 1991, **49**, 569; *Chem. Abstr.* 1991, **115**, 221774a.
98. Caravan, P., Ellison, J. J., McMurry, T. J. and Lauffer, R. B. *Chem. Rev.* 1999, **9**, 2293.
99. Sakagami, N., Homma, J-I., Konno, T. and Okomoto, K.-I. *Acta Crystallogr. Sect. C 1997,* **53**, *1378.*
100. Ruloff, R., Gelbrich, T., Hoyer, E., Sielwer, J. and Beyer, L. *Z. Naturforsch. B 1998,* **53**, *955.*

101. Inoue, M. B., Inoue, M. and Fernandeo, Q. *Inorg. Chim. Acta* 1995, **232**, 203.
102. Bénazeth, S., Purans, J., Chalbot, M.-C., Kim Nguyen-van-Duong, M., Nicolas, L., Keller, F. and Gaudemer, A. *Inorg. Chem.* 1998, **37**, 3667.
103. Geraldes, C. F. G. C., Urbano, A. M., Alpoim, M. C., Hoefnagel, M. A. and Peters, J. A. *J. Chem. Soc. Chem. Commun.* 1991, 656.
104. Geraldes, C. F. G. C., Urbano, A. M., Hoefnagel, M. A. and Peters, J. A. *Inorg. Chem.* 1993, **32**, 2426.
105. Lammers, H., Maton, F., Pubanz, D., van Laren, M. W., van Bekkum, H., Merbach, A. E., Muller, R. N. and Peters, J. A. *Inorg. Chem.* 1997, **36**, 2527.
106. White, D. H., deLearie, L. A., Dunn, T. J., Rizkalla, E. N., Imura, H. and Choppin, G. R. *Invest. Radiol.* 1991, **26**, S229.
107. Aime, S., Benetollo, F., Bombieri, G., Colla, S., Fasano, M. and Paoletti, S. *Inorg. Chim. Acta* 1997, **254**, 63.
108. Rizkalla, E. N., Choppin, G. R. and Cacheris, W., *Inorg. Chem.* 1993, **32**, 582.
109. Konings, M. S., Dow, W. C., Love, D. B., Raymond, K. N., Quay, S. C. and Rocklage, S. M. *Inorg. Chem.* 1990, **29**, 1488.
110. Ehnebom, L. and Pedersen, B. F. *Acta Chem. Scand.* 1992, **46**, 126.
111. Bligh, S. W. A., Chowdhury, A. H. M. S., McPartlin, M., Scowen, I. J. and Bulman, R. A. *Polyhedron* 1995, **14**, 567.
112. Wang, Y.-M., Cheng, T.-H., Sheu, R.-S., Chen, I.-T. and Chiang, M. Y. *J. Chin. Chem. Soc. (Tapei)* 1997, **44**, 123.
113. Frey, S. T., Chang, C. A., Carvalho, J. F., Varadarajan, A., Schultze, L. M., Pounds, K. L. and Horrocks, W.DeW. Jr *Inorg. Chem.* 1994, **33**, 2882.
114. Inoue, M. B., Inoue, M., Muñoz, I. C., Bruck, M. A. and Fernando, Q. *Inorg. Chim. Acta* 1993, **209**, 29.
115. Inoue, M. B., Inoue, M. and Fernando, Q. *Acta Crystallogr., Sect. C* 1994, **50**, 1037.
116. Inoue, M. B., Navarro, R. E., Inoue, M. and Fernando, Q. *Inorg. Chem.* 1995, **34**, 6074.
117. Inoue, M. B., Oram, P., Inoue, M., Fernando, Q., Alexander, A. L. and Unger, E. C. *Magn. Reson. Imaging* 1994, **12**, 429.
118. Inoue, M. B., Santacruz, H., Inoue, M. and Fernando, Q. *Inorg. Chem.* 1999, **38**, 1596.
119. Wu, S. L., Franklin, S. J., Raymond, K. N. and Horrocks, W. DeW. Jr *Inorg. Chem.* 1996, **35**, 162.
120. Uggeri, F., Aime, S., Anelli, P. L., Botta, M., Brochetta, M., DeHaën, C., Ermondi, G., Grandi, M. and Paoli, P. *Inorg. Chem.* 1995, **34**, 633.
121. Schmitt-Willich, H., Brehm, M., Ewers, Ch. L. J., Michl, G., Müller-Fahrnow, A., Petrov, O., Platzek, J., Radüchel, B. and Sülzle, D. *Inorg. Chem.* 1999, **38**, 1134.
122. Vander Elst, L., Maton, F., Laurent, S. and Muller, R. N. *Acta Radiol.* 1997, **38 (S412)**, 135.
123. Vander Elst, L., Maton, F., Laurent, S., Seghi, F., Chapelle, F. and Muller, R. N. *Magn. Reson. Med.* 1997, **38**, 604.
124. Muller, R. N., Radüchel, B., Laurent, S., Platzek, J., Piérart, C., Mareski, P. and Vander Elst, L. *Eur. J. Inorg. Chem.* 1999, 1949.
125. Ruloff, R., Prokop, P., Sieler, J., Hoyer, E. and Beyer, L. *Z. Naturforsch. B* 1996, **51**, 963.
126. Wang, R.-Y., Li, J.-R., Jin, T.-Z., Xu, G.-X., Zhou, Z. Y. and Zhou, X. G. *Polyhedron* 1997, **16,** 1361.
127. Ruloff, R., Gelbrich, T., Sieler, J., Hoyer, E. and Beyer, L. *Z. Naturforsch. B* 1997, **52**, 805.

128. Wang, R.-Y., Li, J.-R., Jin, T.-Z., Xu, G.-X., Zhou, Z. Y. and Zhou, X. G. *Polyhedron* 1997, **16,** 2037.
129. Mondry, A. and Starynowicz, P. *Inorg Chem.* 1997, **36**, 1176.
130. Mondry, A. and Starynowicz, P. *J. Chem. Soc. Dalton Trans.* 1998, 859.
131. Holz, R. C. and Horrocks, W. DeW. Jr *Inorg, Chim. Acta* 1990, **171,** 193.
132. Lee, S. G. *Bull. Korean Chem. Soc.* 1997, ***18***, 1231.
133. Schauer, C. K. and Anderson, O. P. *J. Chem. Soc. Dalton Trans.* 1989, 185.
134. Xu, J., Franklin, S. J., Whisenhunt, W. D. Jr and Raymond, K. N. *J. Am. Chem. Soc.* 1995, **117**, 7245.
135. Geraldes, C. F. G. C., Alpoim, M. C., Marques, M. P. M., Sherry, A. D. and Singh, M. *Inorg. Chem.* 1985, **24**, 3876.
136. Sherry, A. D., Singh, M. and Geraldes, C. F. G. C. *J. Magn. Reson.* 1986, **66**, 511.
137. Desreux, J. F. *Inorg. Chem.* 1980 and **19**, 1319.
138. Albin, M., Horrocks, W. deW. Jr and Liotta, F. J. *Chem. Phys. Lett.* 1982, **85**, 61.
139. Spirlet, M.-R., Rebizant, J., Desreux, J. F. and Loncin, M. F. *Inorg. Chem.* 1984, **23**, 359.
140. Aime, S., Barge, A., Botta, M., Fasano, M., Ayala, J. D. and Bombieri, G. *Inorg. Chim.Acta* 1996, **246**, 423.
141. Aime, S., Botta, M., Fasano, M., Marques, M. P. M., Geraldes, C. F. G. C., Pubanz, D. and Merbach, A. E. *Inorg. Chem.* 1997, **36**, 2059.
142. Dubost J. P., Leger, J. M., Langlois, M. H., Meyer, D. and Schaefer, M. *C. R. Acad. Sci., Ser. II Univers.* 1991, **312**, 329.
143. Chang, C. A., Francesconi, L. C., Malley, M. F., Kumar, K., Gougoutas, J. Z., Tweedle, M. F., Lee, D. W. and Wilson, L. J. *Inorg. Chem.* 1993, **32**, 3501.
144. Aime, S., Barge, A., Benetollo, F., Bombieri, G., Botta, M. and Uggeri, F. *Inorg. Chem.* 1997, **36**, 4287.
145. Desreux, J. F. and Loncin, M. F. *Inorg. Chem.* 1986, **25**, 69.
146. Ascenso, J. R., Delgado, R. and Fraústo da Silva, J. J. R. *J. Chem. Soc. Dalton Trans.* 1986. 2395.
147. Brittain, H. G. and Desreux, J. F. *Inorg. Chem.* 1984, **23**, 4459.
148. Howard, J. A. K. and Kenwright, A. M., Moloney, J. M., Parker, D., Woods, M., Port, M., Navet, M. and Rousseau, O. *J. Chem. Soc. Chem. Commun.* 1998, 1381.
149. Aime, S., Botta, M., Ermondi, G., Terreno, E., Anneli, P. L., Fedeli, F. and Uggeri, F. *Inorg. Chem.* 1996, **35**, 2726.
150. Alexander, V. *Chem. Rev.* 1995, **95**, 273.
151. Kang, S. I., Ranganathan, R. S., Emswiler, J. E., Kumar, K., Gougoutas, J. Z., Malley, M. F. and Tweedle, M. F. *Inorg. Chem.* 1993, **32**, 2912.
152. Aime, S., Anneli, P. L., Botta, M., Fedeli, F., Grandi, M., Paoli, P. and Uggeri, F. *Inorg. Chem.* 1992, **31**, 2422.
153. Kumar, K., Chang, C. A., Francesconi, L. C., Dischino, D. D., Malley, M. F., Gougoutas, J. Z. and Tweedle, M. F. *Inorg. Chem.* 1994, **33**, 3567.
154. Platzek, J., Blaszkiewicz, P., Gries, H., Luger, P., Michl, G., Mueller-Fahrnow, A., Raduechel, B. and Suelzle, D. *Inorg. Chem.* 1997, **36**, 6086.
155. Shukla, R. B. *J. Magn. Reson. Ser. A* 1995, **113**, 196.
156. Spirlet, M.-R., Rebizant, J., Wang, X., Jin, T., Gilsoul, D., Comblin, V., Maton, F., Muller, R. N. and Desreux, *J. Chem. Soc. Dalton Trans.* 1997, 497.
157. Kim, W. D., Kiefer, G. E., Maton, F., McMillan, K., Muller, R. N. and Sherry, A. D. *Inorg. Chem.* 1995, **34**, 2233.
158. Aime, S., Botta, M., Crich, S. G., Giovenzana, G. B., Jommi, G., Pagliarin, R. and Sisti, M. *J. Chem. Soc. Chem. Commun.* 1995, 1885.
159. Aime, S., Botta, M., Crich, S. G., Giovenzana, G. B., Jommi, G., Pagliarin, R. and Sisti, M. *Inorg. Chem.* 1997, **36**, 2992.

160. Schumann, H., Böttger, U. A., Weisshoff, H., Ziemer, B. and Zschunke, A. *Eur. J. Inorg. Chem.* 1999, 1735.
161. Sherry, A. D. and Geraldes, C. F. G. C., Cacheris, W. P. *Inorg. Chim. Acta* 1987, **139**, 137.
162. Sherry, A. D., Malloy, C. R., Jeffrey, F. M. H., Cacheris, W. P. and Geraldes, C. F. G. C. *J. Magn. Reson.* 1988, **76**, 528.
163. Geraldes, C. F. G. C., Brown, R. D., III Cacheris, W. P., Koenig, S. H., Sherry, A. D. and Spiller, M. *Magn. Reson. Med.* 1989, **9**, 94.
164. Geraldes, C. F. G. C., Sherry, A. D. and Kiefer, G. E. *J. Magn. Reson.* 1992, **97**, 290.
165. Aime, S., Botta, M., Terreno, E., Anelli, P. L. and Uggeri, F. *Magn. Reson. Med.* 1993, **30**, 583.
166. Sherry, A. D., Ren, J., Huskens, J., Brücher, E., Tóth, E., Geraldes, C. F. C. G., Castro, M. M. C. A. and Cacheris, W. P. *Inorg. Chem.* 1996, **35**, 4604.
167. Ren, J. and Sherry, A. D. *Inorg. Chim. Acta* 1996, **246**, 331.
168. Aime, S., Botta, M., Parker, D. and Williams, J. A. G. *J. Chem. Soc. Dalton Trans.* 1996, 17.
169. Aime, S., Botta, M., Fasano, M., Terreno, E., Anelli, P. L., Calabi, L. and Uggeri, F. in *Proceedings of the 1st Scientific Meeting of the International Society for Magnetic Resonance in Medicine.* International Society for Magnetic Resonance in Medical, New York, 1996, Vol. 3, p. 1688.
170. Buster, D. C., Castro, M. M. C. A., Geraldes, C. F. G. C., Malloy, C. R., Sherry, A. D. and Siemers, T. C. *Magn. Reson. Med.* 1990, **15**, 25.
171. Malloy, C. R., Buster, D. C., Castro, M. M. C. A., Geraldes, C. F. G. C., Jeffrey, F. M. H. and Sherry, A. D. *Magn. Reson. Med.* 1990, **15**, 33.
172. Bansal, N., Germann, M. J., Seshan, V., Shires G. T., III Malloy, C. R. and Sherry, A. D. *Biochemistry* 1993, **32**, 5638.
173. Bansal, N., Germann, M. J., Lázár, I., Malloy, C. R. and Sherry, A. D. *J. Magn. Reson. Imag.* 1992, **2**, 385.
174. Seshan, V., Germann, M. J., Preisig, P., Malloy, C. R., Sherry, A. D. and Bansal, N. *Magn. Reson. Med.* 1995, **34**, 25.
175. Ren, J., Springer, C. S. and Sherry, A. D. *Inorg. Chem.* 1997, **36**, 3493.
176. Aime, S., Botta, M., Crich, S. G., Terreno, E., Anelli, P. L. and Uggeri, F. *Chem. Eur. J.* 1999, **5**, 1261.
177. Carvalho, R. A., Peters, J. A. and Geraldes, C. F. G. C. *Inorg. Chim. Acta* 1997, **262**, 167.
178. Dick, L. R., Geraldes, C. F. G. C., Sherry, A. D., Gray, C. W. and Gray, D. M. *Biochemistry* 1989, **28**, 7896.
179. van Duynhoven, J. P. M., Nooren, I. M. A., Swinkels, D. W., Folkers, P. J. M., Harmsen, B. J. M., Konings, R. N. H., Tesser, G. I. and Hilbers, C. W. *Eur. J. Biochem.* 1993, **216**, 507.
180. Paulus, E. F., Juretschke, P. and Lang, J., *Proceeding of the 3rd Annual Meeting of the German Society for Crystallography*, Germany, Darmstadt, March 6–8, 1995, book of abstracts.
181. Geraldes, C. F. G. C., Sherry, A. D., Lázár, I., Miseta, A., Bogner, P., Berenyi, E., Sumegi, B., Kiefer, G. E., McMillan, K., Maton, F. and Muller, R. N. *Magn. Reson. Med.* 1993, **30**, 696.
182. Kim, W. D., Kiefer, G. E., Huskens, J. and Sherry, A. D. *Inorg. Chem.* 1997, **36**, 4128.
183. Broan, C. J., Jankowski, K. J., Kataky, R., Parker, D., Randall, A. M. and Harrison, A. *J. Chem. Soc. Chem. Commun.* 1990, 1739.
184. Broan, C. J., Jankowski, K. J., Kataky, R., Parker, D., Randall, A. M. and Harrison, A. J. *Chem. Soc. Chem. Commun.* 1991, 204.

185. Rohovec, J., Vojtíšek, P., Hermann, P., Mosinger, J., Zák, Z. and Lukeš, I. *J. Chem.Soc. Dalton Trans.* 1999, 3585.
186. Aime, S., Batsanov, A. S., Botta, M., Howard, J. A. K., Parker, D., Senanayake, K. and Williams, G. *Inorg. Chem.* 1994, **33**, 4696.
187. Aime, S., Batsanov, A. S., Botta, M., Dickins, R., Faulkner, S., Foster, C. E., Harrison, A., Howard, J. A. K., Moloney, J. M., Norman, T. J., Parker, D., Royle, L. and Williams, J. A. G. *J. Chem. Soc., Dalton Trans.* 1997, 3623.
188. Parker, D. J., Pulukkody, K., Norman, T. J., Harrison, A., Royle, L. and Walker, C. *J. Chem. Soc. Chem. Commun.* 1992, 1441.
189. Pulukkody, K. P., Norman, T. J., Parker, D., Royle, L. and Broan, C. J. *J. Chem. Soc. Perkin Trans. 2* 1993, 605.
190. Aime, S., Botta, M., Parker, D. and Williams, J. A. G. *J. Chem. Soc. Dalton Trans.* 1995, 2259.
191. Aime, S., Botta, M., Dickins, R. S., Maupin, C. L., Parker, D., Riehl, J. P. and Williams, J. A. G. *J. Chem. Soc. Dalton Trans.* 1998, 881.
192. Harrison, A., Norman, T. J., Parker, D., Royle, L., Pereira, K. A., Pulukkody, K. P. and Walker, C. A. *Magn. Reson. Imaging* 1993, **11**, 761.
193. Morrow, J. R., Amin, S., Lake, C. H. and Churchill, M. R. *Inorg. Chem.* 1993, **32**, 4566.
194. Amin, S., Morrow, J. R., Lake, C. H. and Churchill, M. R. *Angew. Chem. Int. Ed. Engl.* 1994, **33**, 773.
195. Amin, S., Voss, D. A. Jr. Horrocks, W. DeW. Jr. Lake, C. H., Churchill, M. R. and Morrow, J. R. *Inorg. Chem.* 1995, **34**, 3294.
196. Amin, S., Voss, D. A. Jr, Horrocks, W. DeW. Jr and Morrow, J. R. *Inorg. Chem.* 1996, **35**, 7466.
197. Chappell, L. L., Voss, D. A. Jr, Horrocks, W. DeW. Jr and Morrow, J. R. *Inorg. Chem.* 1998, **37**, 3989.
198. Dickens, R. S., Howard, J. A. K., Lehmann, C. W., Moloney, J., Parker, D. and Peacock, R. D. *Angew. Chem. Int. Ed. Engl.* 1997, **36**, 521.
199. Aime, S., Barge, A., Botta, M., Howard, J. A. K., Kataky, R., Lowe, M. P., Moloney, J. M., Parker, D. and de Sousa, A. S. *J. Chem. Soc. Chem. Commun.* 1999, 1047.
200. Dickens, R. S.,Howard, J. A. K., Maupin, C. L., Moloney, J. M., Parker, D., Riehl, J. P., Siligardi, G. and Williams, J. A. G. *Chem. Eur. J.* 1999, **5**, 1095.
201. Batsanov, A. S. Beeby, A., Bruce, J. I., Howard, J. A. K., Kenwright, A. M. and Parker, D. *J. Chem. Soc. Chem. Commun.* 1999, 1011.
202. Kataky, R., Matthes, K. E., Nicholson, P. E. and Parker, D. *J. Chem. Soc. Perkin Trans. 2* 1990, 1425.
203. Carlton, L., Hancock, R. D., Maumela, H. and Wainwright, K. P. *J. Chem. Soc. Chem. Commun.* 1994, 1007.
204. Dickins, R. S., Parker, D., de Sousa, A. S. and Williams, J. A. G. *J. Chem. Soc. Chem. Commun.* 1996, 697.
205. Aime, S., Barge, A., Botta, M., Parker, D. and De Sousa, A. S. *J. Am. Chem. Soc.* 1997, **119**, 4767.
206. Alderighi, L., Bianchi, A., Calabi, L., Dapporto, P., Giorgi, C., Losi, P., Paleari, L., Paoli, P., Rossi, P., Valtancoli, B. and Virtuani, M. *Eur. J. Inorg. Chem.* 1998, 1581.
207. Aime, S., Barge, A., Botta, M., De Sousa, A. S. and Parker, D. *Angew. Chem. Int. Ed. Engl.* 1998, **37**, 2673.
208. Aime, S., Barge, A., Bruce, J. I., Botta, M., Howard, J. A. K., Moloney, J. M., Parker, D., de Sousa, A. S. and Woods, M. *J. Am. Chem. Soc.* 1999, **121**, 5762.

209. Huskens, J., Torres, D., Kovacs, Z., André, J. P., Geraldes. C. F. G. C. and Sherry, A. D. *Inorg. Chem.* 1997, **36**, 1495.
210. Chang, C. A., Chen, Y.-H., Chen, H.-Y. and Shieh, F.-K. *J. Chem. Soc., Dalton Trans.* 1998, 3243.
211. Yerly, F., Dunand, F., Tóth, E., Figueirinha, A., Kovacs, Z., Sherry, A. D., Geraldes, C. F. G. C. and Merbach, A. E. *Eur. J. Inorg. Chem.*, 2000, 1001.
212. Parker, D. and Williams, J. A. G. *J. Chem. Soc. Perkin Trans. 2* 1995, 1305.
213. Chin, K. O. A., Morrow, J. R., Lake, C. H. and Churchill, M. R. *Inorg. Chem.* 1994, **33**, 656.
214. Morrow, J. R. and Chin, K. O. A. *Inorg. Chem.* 1993, **32**, 3357.
215. Sessler, J. L., Mody, T. D., Hemmi, G. W., Lynch, V., Young, S. W. and Miller, R. A. *J. Am. Chem. Soc.* 1993, **115**, 10368.
216. Sessler, J. L., Mody, T. D., Hemmi, G. W. and Lynch, V. *Inorg. Chem.* 1993, **32**, 3175.
217. Sessler, J. L., Hemmi, G., Mody, T. D., Murai, T., Burell, A. and Young, S. W. *Acc. Chem. Res.* 1994, **27**, 43.
218. Aime, S., Botta, M., Dastrú, W., Fasano, M., Panero, M. and Arnelli, A. *Inorg. Chem.* 1993, **32**, 2068.
219. Aime, S., Digilio, G., Fasano, M., Paoletti, S., Arnelli, A. and Ascenzi, P. *Biophys. J.* 1999, **76**, 2735.
220. Moats, R. A., Fraser, S. E. and Meade, T. J. *Angew. Chem. Int. Ed. Engl.* 1997, **36,** 726.
221. Hall, J., Haner, R., Aime, S., Botta, M., Faulkner, S., Parker, D. and de Sousa, A. S. *New J. Chem.* 1998, **22** 627.
222. Aime, S., Batsanov, A. S., Botta, M., Howard, J. A.K; Lowe, M. P. and Parker, D. *New J. Chem.* 1999, **23**, 669.
223. Aime, S., Botta, M., Geninatti-Crich, S., Giovenzana, G., Palmisano, G. and Sisti, M. *J. Chem. Soc. Chem. Commun.* 1999, 1577.
224. Zhang, S., Wu, K. and Sherry, A. D. *Angew. Chem. Int. Ed.* 1999, **38**, 3192.
225. Li, W.-H., Fraser, S. E. and Meade, T. J. *J. Am. Chem. Soc.* 1999, **121**, 1413.
226. Balkus, K. J. Jr, Sherry, A. D. and Young, S. W. *US Pat. No. 5 122 363*, 1992; *Chem. Abstr.* 1992, **117**, 56079j.
227. Balkus, K. J. Jr and Bresinska, I. *J. Alloys Compd.* 1994, **207/208**, 25.
228. Young, S. W., Qing, F., Rubin, D., Balkus, K. J. *d* Jr, Engel, J. S., Lang, J., Dow, W. C., Mutch, J. D. and Miller, R. A. *J. Magn. Reson. Imaging* 1995, **5**, 499.
229. Bresinska, I. and Balkus, K. J. Jr *J. Phys. Chem.* 1994, **98**, 12989.
230. Lee, E. F. T. and Rees, L. V. C. *Zeolites* 1987, **7**, 446.
231. Pfeifer, H. *Surf. Sci.* 1975, **52**, 434.
232. Basler, W. D. *ACS Symp. Ser.* 1976, **34**, 291.
233. Winkler, H., Steinberg, K. H. and Kapphahn, G. J. *J. Colloid Interface Sci.* 1984, **98**, 144.
234. Winkler, H. and Steinberg, K. H. *Zeolites* 1989, **9**, 445.
235. Sur, S. K., Heinsbergen, J. F. and Bryant, R. G. *J. Magn. Reson.* 1993, **A103**, 27.
236. Balkus, K. J. Jr and Shi, J. *Microporous Mater.* 1997, **11**, 325.
237. Balkus, K. J. Jr *US Pat. 5 277 896*, 1994; *Chem. Abstr.* 1994, **120**, 186777h.
238. Balkus, K. J. Jr and Shi, J. *J. Phys. Chem.* 1996, **100**, 16429.
239. Balkus, K. J. Jr and Shi, J. *Langmuir* 1996, **12**, 6277.

9 Multi-Frequency and High-Frequency EPR Methods in Contrast Agent Research: Examples from Gd^{3+} Chelates†

R. B. CLARKSON, A. I. SMIRNOV, T. I. SMIRNOVA and R. L. BELFORD
University of Illinois, Urbana, IL, USA

1 INTRODUCTION

The technique of multi-frequency electron paramagnetic resonance (EPR), in which the same sample is studied at several different spectrometer frequencies, can provide information that is not available at any single experimental frequency. The approach is critical for testing models of structure and dynamics that are used to simulate EPR spectra, and it can greatly improve the precision of computer-derived spectral parameters by fitting experimental results from several frequencies with a single parameter set [1, 2]. Multi-frequency EPR also can imply the experimental option to select the most appropriate frequency for the particular system and question under investigation, as a review by Belford and Clarkson surveys [3]. The multi-frequency approach has been developing rapidly in the last decade, in part because of the increased availability of EPR spectrometers operating at frequencies other than 9.5 and 35 GHz; both higher and lower frequencies can be valuable, and instruments exist which range from less than 100 MHz to more than 500 GHz [4].

It is customary to speak of *multi-frequency* methods, because it is typical to use several instruments with different fixed frequencies in a continuous-wave (CW) EPR experiment. The frequency of any single instrument is held constant while the external magnetic field $\boldsymbol{B}$ is swept to register the spectrum ($\boldsymbol{B} = \boldsymbol{B}_0 \pm$ sweep); the approach could equally well be described as a *multi-field* method, where $\boldsymbol{B}_0$ changes with spectrometer frequency. The multi-frequency approach

† Part of this chapter is adapted from the article 'Multi-frequency EPR determination of zero-field splitting of high-spin species in liquids: Gd(III) chelates in water', R. B. Clarkson, A. I. Smirnov, T. I. Smirnova, H. Kang, R. L. Belford, K. Earle and J. H. Freed, *Mol. Phys.*, 1998, **95**, 1325–1332, and is used with permission of the publisher, Taylor & Francis, Ltd (http://www.tandf.co.utc/journals).

The Chemistry of Contrast Agents in Medical Magnetic Resonance Imaging
Edited by A. E. Merbach and É. Tóth.

makes use of the fact that the spin Hamiltonian for electrons may contain terms which do not depend on the strength of the applied magnetic field, as well as some that do. A typical spin Hamiltonian for a system where $S > 1/2$ might look like the following expression:

$$\mathcal{H}_{spin} = \beta_e \boldsymbol{BgS} + \beta_n \boldsymbol{Bg_nI} + \boldsymbol{SAI} + \boldsymbol{SDS} \tag{9.1}$$

Here, the first two terms represent the field-dependent electron and nuclear Zeeman contributions, respectively, while the third and fourth terms, without any explicit $\boldsymbol{B}$-dependence, respectively represent the electron–nuclear hyperfine interaction and the zero field splitting (ZFS). The EPR spectrum from this system depends on contributions from all four terms, so performing the spectroscopy at different frequencies/fields alters the relative importance of field-dependent and field-independent effects, thus significantly changing the spectral line shape. By analyzing spectra taken at different frequencies, the contributions from various terms in the Hamiltonian can be deconvoluted, and information not available at any one frequency can be obtained.

The study of Gd(III) chelates is facilitated by using the multi-frequency EPR approach. Since this ion has $S > 1 (S = 7/2)$, its spin Hamiltonian can include a significant ZFS contribution. Although gadolinium has two isotopes with nuclear spin (^{155}Gd, 14.73 %; ^{157}Gd, 15.68 %; both $\boldsymbol{I} = 3/2$), the hyperfine interaction is small, and is usually neglected. Thus, the spin Hamiltonian contains only electronic Zeeman and ZFS terms. By studying Gd(III) chelates at different spectrometer frequencies, the relative magnitude of these terms can be altered, and much can be learned about the ZFS and its relationship to chelate structure and environment.

Electronic relaxation is also affected by the strength of the applied magnetic field, which is controlled by the choice of experimental frequency, and multi-frequency/multi-field methods can be used to study $T_{1e,2e}$ and their relationship to the ZFS, chelate structure, and environment. Early applications of the multi-frequency approach, when the two commercially available frequencies, i.e. the X-band (9.5 GHz) and Q-band (35 GHz), were the only choices available, helped to develop the field [5]. Contemporary multi-frequency studies, making use of millimeter-wave instruments, began with the work of Merbach and co-workers, in particular Ya. S. Lebedev. Lebedev pioneered the development of very high frequency EPR instrumentation [6], and Merbach's group, in collaboration with Lebedev's, pioneered its application to Gd^{3+} chelates [7]. Today, studies making use of three, four or five different frequencies are beginning to appear. Caravan *et al.* have recently used this approach to study relaxation in $Eu^{2+}_{(aq)}$ [8], and Borel *et al.* have examined the Gd^{3+} chelates of DOTA and DTPA-BMA, as well as the Gd^{3+} aquo ion [9]. Clarkson and co-workers have used the multi-frequency method to study T_{1e} and T_{2e} in several Gd^{3+} chelates [10–12]. They have also used different EPR frequencies to explore chelate interactions with membranes and proteins. In this present chapter, some of these applications will be explored in order to illustrate the utility of this approach in the study of Gd(III) chelates and paramagnetic contrast agents useful in magnetic resonance imaging (MRI).

2 THE Gd(III) SYSTEM

As pointed out by Abragam and Bleaney [13], the ground state of the Gd(III) ion ($4f^7$) is nearly pure $^8S_{7/2}$, with a slight admixture of $^6P_{7/2}$ through intermediate coupling. Perturbations from ligands cause a splitting of the electronic states, thus producing a fine structure in the resonance spectrum that can be represented by a simplified spin Hamiltonian of the following form:

$$\mathcal{H}_{spin} = \beta \boldsymbol{BgS} + \sum_{m,n} B_n^m O_n^m \tag{9.2}$$

where the second term describes the fine structure in terms of spin operators characterizing the effects of the ligand fields. While we may anticipate a number of higher-order terms for the $4f^7$ (ground state $S = 7/2$) ion, the very small orbital angular momentum in this system causes terms of higher degree to become rapidly smaller [13], and for this discussion they will be neglected. In single crystals, the expansion in (m, n) can often be evaluated for specific ligand symmetries; in solutions, this term is less well defined, and is generally time dependent. Odelius *et. al.*, when discussing the analogous problem for Ni(II) ($3d^8, {}^3F_1$), rewrite the Hamiltonian as follows:

$$\mathcal{H}_{spin} = \mathcal{H}_{Zeeman} + \mathcal{H}_{ZFS} = \beta \boldsymbol{BgS} + \boldsymbol{SD}_{\boldsymbol{ZFS}}(t)\boldsymbol{S} \tag{9.3}$$

where $\boldsymbol{D}_{ZFS}(t)$ is the time-dependent zero field splitting (ZFS) tensor, which describes the second-order coupling of the electron spins with the orbital angular momentum, and which removes some of the degeneracy in the spin states even in the absence of a magnetic field [14]. Depending on the symmetry of the system, $\mathbf{D}_{ZFS}$ also may contain a static term, in which case, $\mathbf{D}_{ZFS}(t) = D_0 + D(t)$.

For many years, the time-dependent fluctuations of the ZFS term have been thought to dominate electron spin relaxation in Gd(III) systems [5, 7, 15–18]. Earlier workers had adopted a form of the Bloch–Wangsness–Redfield (BWR) theory [19] to characterize the time-dependent fluctuations of $\boldsymbol{D}_{ZFS}$ by means of an exponentially decaying correlation function, which defines a correlation time τ_v for the modulation of the ZFS. The theory characterizes the *x*-, *y*-, and *z*-components of the fluctuating magnetic field $H_q(t)$ produced by D_{ZFS} in terms of spectral density functions $J_{\alpha\beta\alpha'\beta'}(\omega)$, where:

$$\frac{1}{2\hbar^2} J_{\alpha\beta\alpha'\beta'(\omega)} = \gamma_n^2 \sum_q (\alpha|\mathrm{S}_q|\beta)(\beta'|S_q|\alpha')k_{qq}(\omega) \tag{9.4}$$

In the above equation, $q = x, y$, and z; α, α', β and β' represent the eigenstates of the operator S_q, and:

$$k_{qq}(\omega) = 1/2 \int_{-\infty}^{+\infty} \overline{H_q(t)H_q(t+\tau)} \exp(-i\omega\tau)\mathrm{d}\tau \tag{9.5}$$

For the case where spins are subjected to a random fluctuating field, different at each spin, Slichter evaluates Equation (9.5) as follows [19]:

$$k_{qq}(\omega) = \overline{H_q^2} \frac{\tau_v}{1 + \omega^2 \tau_v^2} \tag{9.6}$$

This approach is valid when $\tau_v \ll T_{2e}$, a condition which Slichter characterizes as 'not asking for information over time intervals comparable with τ_v' [19]. The inequality is not strongly satisfied for Gd(III) chelates at 9.5 GHz, where τ_v/T_{2e} can approach 0.1, and it may be violated at lower frequencies. This point will be considered later in the discussion. It has also been pointed out by several authors that when the ZFS term is large compared with the electronic Zeeman energy, a contribution to electron relaxation (and nuclear relaxation enhancement) can be expected from a *static* ZFS component [20, 21], as well as from a *transient* ZFS (described in Equation (9.3)). Bertini *et al.* [22] have considered the case of $S = 1$ systems, and write designations for static ZFS (Δ_s) and transient ZFS (Δ_t). For many Gd(III) complexes, including the ones that we have studied, the magnitude of the static ZFS is of the order of 0.1 cm^{-1}. At values of the magnetic field corresponding to X-band EPR (0.34 T), the static ZFS term is therefore less than the electronic Zeeman interaction, and its relative contribution to the Hamiltonian becomes proportionally smaller at higher fields. Since $\mathcal{H}_{Zeeman} > \mathcal{H}_{ZFS}$ at the magnetic field values used in our work (the Zeeman limit), we interpret D_{ZFS} only in terms of a single ZFS interaction parameter for considerations of electron spin relaxation and dynamic frequency shifts. As Bertini points out, this approximation is strictly valid only for cases when the rotation of the Gd(III) complex is slow compared with electron relaxation ($\tau_R \gg \tau_S$); for rapid rotation, the spin-dynamics approach of Abernathy and Sharp can be employed [23]. In the examples we will discuss, this rotation/relaxation inequality is only weakly satisfied for small Gd(III) chelates in low-viscosity solvents. Nevertheless, for all of the Gd(III) chelates that we have studied, whether observed by CW EPR methods in frozen solutions at liquid helium temperatures or by multi-frequency EPR in aqueous solutions, the parameter identified as the squared ZFS magnitude, Δ^2 or $(2/3D^2 + 2E^2)$, is always about the same for each complex, regardless of whether the system is in a frozen or liquid state. This suggests that the effects producing powder EPR spectral lineshapes, dynamic frequency shifts, and frequency-dependent spin relaxation have a common origin.

The EPR spectrum of a Gd(III) chelate in water is typically observed as a single resonance line whose *effective* g-factor depends upon the experimental value of B_0. Dynamic frequency shifts in EPR spectra have been considered previously and treated by using the Redfield relaxation matrix approach[24, 25]. Theoretical results were compared with experimental X- and Q-band EPR data. In another study, Smirnova *et al.*[10] calculated the frequency shift by averaging the anisotropic shift derived from a static spin Hamiltonian over all orientations. This treatment is based on the assumptions that molecular motion neither changes the spin precession rate nor perturbs the states and, thus, that the center of gravity of the spectrum is invariant even in the presence of some motional averaging. For the allowed $|1/2\rangle \leftrightarrow \langle -\frac{1}{2}|$ transition under perturbation theory, with expressions valid up to the third order, this shift is given by the following:

$$\Delta B = \frac{1}{20 g \beta \omega_0}(4S(S+1)-3)(2/3D^2 + 2E^2) \tag{9.7}$$

so the shift is directly proportional to the ZFS term $(2/3D^2 + 2E^2)$, which is usually denoted by the term Δ^2. A more rigorous approach to analyzing multi-frequency EPR data of this type would be to follow the treatment of Poupko *et al.* [25], which involves the diagonalization of a *complex* relaxation matrix and then simultaneously using the real part of the eigenvalues for fitting the EPR line shapes and the imaginary part for analyzing the apparent *g*-shifts. This analysis of the *g*-shift is especially important at X-band frequencies and lower. The result of Equation (9.7) is identical to the shift expression for a spin of 5/2, which Baram *et al.* [26] derived by using the semi-classical stochastic Liouville general formalism [27] in the limit when motional effects are negligible. For the Gd(III) ion (*S*=7/2), the field shift (Equation (9.7)) can be recast for an effective *g*-factor, g_{eff}, in terms of the squared ZFS matrix $\Delta^2 = D_{xx}^2 + D_{yy}^2 + D_{zz}^2$, as follows:

$$g_{eff} = g\left(1 - \frac{3\Delta^2}{\omega_0^2}\right) = g\left(1 - \frac{3\Delta_f^2}{\nu^2}\right) \tag{9.8}$$

where ν is the resonance frequency; Δ is in units of rad/s, while Δ_f is in units of hertz (s^{-1}). Thus, the variation of g_{eff} with experimental frequency, observed by multi-frequency EPR, can be analyzed to obtain Δ^2. Borel *et al.* [9] have also considered this shift of g_{eff} with frequency, by employing an approach similar to Poupko *et al.* [25], and have checked Equation (9.8) for its region of validity.

Not only will the effective *g*-factor shift with changes in the experimental frequency, but the electron spin relaxation in Gd(III) chelates will also be strongly affected. Powell *et al.* have given an excellent review of several approaches to interpret the frequency dependence of T_{1e} and T_{2e} in these systems [7]. We have adopted the approach of Hudson and Lewis [18], who showed that the eigenvalues ξ_i of the relaxation matrix $\mathcal{R}$ are functions of τ_v and the experimental frequency ω, and are related to the relaxation time T_{2ei} of the *i*th allowed electron spin transition by the following expression:

$$\frac{1}{T_{2ei}} = -\Delta^2 \tau_v \xi_i \tag{9.9}$$

In applications of the BWR theory, Hudson and Lewis assume that the dominant line-broadening mechanism is provided by the modulation of a second-rank tensor interaction (i.e. ZFS); higher-rank tensor contributions are assumed to be negligible. In addition, $\mathcal{R}$ is a 7×7 matrix for the $S = 7/2$ system, with matrix elements written in terms of the spectral densities $J(\omega, \tau_v)$ (see [18] and [19] for details). The intensity of the ith transition can also be calculated from the eigenvectors of $\mathcal{R}$. In general, there are four transitions with non-zero intensity at any frequency, thus raising the prospect of a multi-exponential decay of the transverse magnetization. There is not a one-to-one correspondence between the relaxation rates and the degenerate $|m_s\rangle \leftrightarrow |m_s \pm 1\rangle$ transitions, and generally one transition is predicted to have a T_{2ei} that is much longer than the other three. Figure 9.1(a) shows a plot of the eigenvalues (ξ_i) of the four non-zero transitions as a function of frequency. Particularly at frequencies greater than 35 GHz, the differences between the longest T_{2ei} and the other three values are predicted to be greater than a factor of 10, which suggests that in continuous-wave EPR experiments, the spectrum of Gd(III) will consist of a narrow line superimposed on very much broader, weaker lines. The narrow lineshape is likely to dominate any fitting of the spectrum by

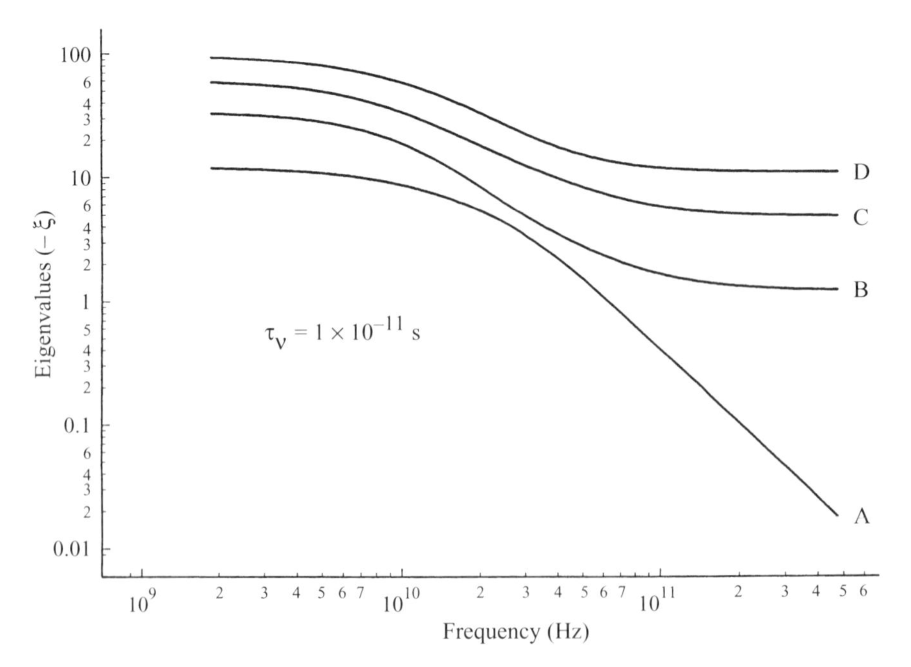

Figure 9.1 (a) Frequency variation of the eigenvalues (ξ_i) of the four spin transitions with non-zero intensity (A – D) for an $S = 7/2$ system, as calculated by the method of Hudson and Lewis [18], assuming $\tau_v = 1 \times 10^{-11}$ s. The ξ_i are directly related to the predicted values of T_{2ei} by Equation (9.9).

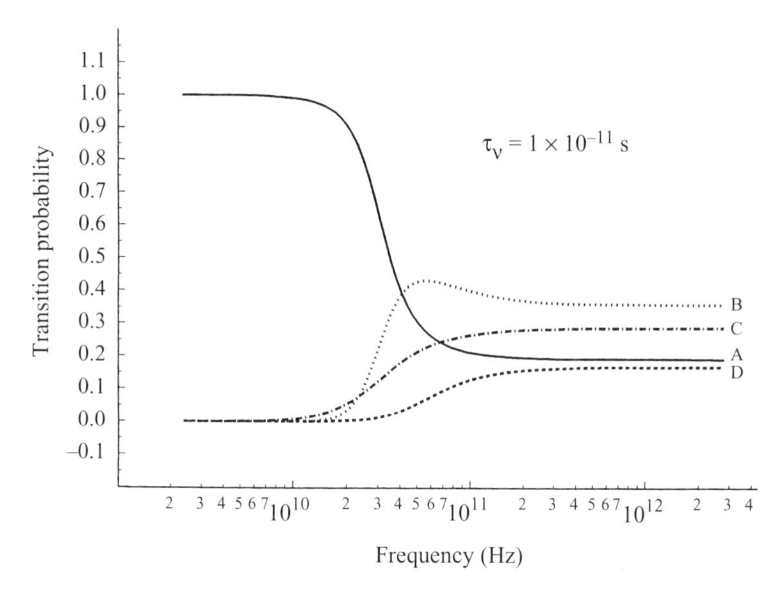

Figure 9.1 (b) Frequency variation of the transition probabilities of the four allowed transitions, calculated from the eigenfunctions of the relaxation matrix, assuming $\tau_v = 1 \times 10^{-11}$ s. Transition probabilities do not correspond to pure ($m_S \rightarrow m_S \pm 1$) transitions.

Lorentzian functions in an analysis to determine approximate values of T_{2e}, a prediction that we found to be borne out by our experiments. Figure 9.1(b) plots the transition probabilities for the allowed transitions as a function of frequency. Below about 10 GHz (depending on the value of τ_v), only one transition has any appreciable intensity, but above that frequency, all four transitions contribute, and a multi-exponential behavior for T_2 is expected. A multi-frequency study of Gd(III) chelates in solution thus also yields information about the ZFS through changes in T_{2e} with frequency.

3 EXPERIMENTAL METHODS

9.3.1 CW EPR SPECTROSCOPY

Observation of electron spin resonance is the most direct way to determine electronic relaxation times of paramagnetic metal ion complexes. However, extremely short electronic relaxation times of Gd(III) in fluid aqueous solutions (T_1 is of the order of 1 ns at 0.3 T) make real-time EPR experiments virtually impossible at the current level of EPR technology, and there are no literature reports of the direct observation of either free induction decay or echo signals from aqueous solutions of Gd(III) chelates except at cryogenic temperatures.

Thus, to date the most frequently employed way to study fluid aqueous solutions of Gd(III) chelates by EPR has been by means of the continuous-wave (CW) method. However, even application of CW EPR is not without difficulties because short relaxation times of Gd(III) chelates and large ZFS values (of the order of 1 GHz) result in broad–several hundred gauss in width–EPR lines at conventional EPR frequencies (9–10 GHz, X-b and, 0.3 T resonance field). Such wide lines demand relatively high concentrations (ca. 1–10 mM) for the experiments. Another problem of X-band CW EPR of Gd(III) complexes is that the spectra at these frequencies can not be well-approximated by Lorentzian functions, which complicates the estimation of spin–spin relaxation times. These problems can be overcome by moving EPR experiments to higher resonance frequencies/magnetic fields. Up until very recently, 35 GHz (Q-band, 1.2 T resonance field for $g = 2$) was the only other higher-than-X-band EPR frequency at which commercial EPR instrumentation was available. Therefore, most research on multi-frequency EPR of Gd(III) chelates has been carried out with home-built EPR instruments.

To date, EPR spectra of liquid aqueous solutions of Gd(III) chelates have been measured at EPR frequencies up to 330 GHz by using various EPR spectrometers of very different designs. Typically, from 10 to 140 GHz, EPR spectrometers utilize waveguide mm-wave technology, which makes the instruments quite narrow-banded. Above 140 GHz or so, waveguides becomes increasingly lossy and quasi-optical technology provides an attractive alternative. Although quasi-optical elements are inherently broad-banded, the mm-wave sources employed are not (the latter are usually either Gunn oscillators with multipliers or far-infrared lasers). Thus, the EPR spectra of Gd(III) chelates are measured only at discrete frequencies, usually with relatively large steps between the frequencies of observation.[1]

High-field EPR spectrometers are generally of either homodyne or heterodyne design. Although in some cases heterodyne design might be advantageous, it also adds some extra cost and complexity. Recently, we have shown that exceptionally good sensitivity in 95 GHz (W-band) EPR experiments can be achieved with a simpler homodyne scheme which uses optimized/matched mm-wave components [28].

The first Illinois 95 GHz (W-band) EPR spectrometer (Mark I), designed and built as a technological research and development project under the NIH Research Resources program, became operational a decade ago. At that time it

[1] Although broad-banded EPR spectrometers have been developed, applications of these instruments remains limited to solid-state samples with high spin concentrations (e.g. see Schwenk H., Konig, D., Sielig M., Schmidt S., Palme W., Luthi B., Zvyagin S., Eccleston, R. S., Azuma M., and Takano M., Physica B, 1997, **237**, 115–116.). Liquid aqueous samples are very lossy at mm-wave frequencies and this calls for highly optimized spectrometer design. In order to maintain high sensitivity, components must be tuned or matched, thus reducing the bandwidth of the spectrometer.

was one of only two 95 GHz spectrometers in the world. Briefly, the Mark I spectrometer has a homodyne-bridge design with a reference arm and is equipped with a narrow-bore Varian XL-200 superconducting magnet operating in persistent mode at 3.3608 T, and an air-cooled scanning coil of 300 gauss range. The Mark I instrument is suitable for radical species with $g = 2$ at a temperature above 150 K, and is described in detail elsewhere [29, 30].

In order to provide wider sweeps, highest sensitivity, cryostatic capability, and greater versatility, we have built a newer W-band instrument (Mark II) that has been frequently upgraded since it became fully operational in 1995. Briefly, the Mark II instrument features the following: an Oxford custom-built sweepable (at up to 0.5 T/min) 7 T superconducting magnet, a water-cooled coaxial solenoid, designed and built at the University of Illinois, with up to $\pm$ 550 G scans under digital (15 bits resolution) computer control, a custom-engineered precision feedback circuit based on an Ultrastab 860R sensor that has a linearity better than 5 ppm and a resolution of 0.05 ppm for control of the magnet-sweep linearity, and an Oxford CF 1200 cryostat modified at the University of Illinois to accommodate a new W-band probehead with remote tuning and a one-axis goniometer for the vertical-bore solenoid. Figure 9.2 shows a generalized scheme for the spectrometer. The microwave bridge incorporates a low-noise Gunn (ZAX, phase noise ca. 75 dB/Hz at 100 kHz) that has an output power of 60 mW and can deliver greater than 30 mW to the cavity. The 3dB loss between the source and the cavity is due primarily to the isolator, circulator, and dielectric waveguide, each of which contributes about 1 dB of insertion loss. The resonator is a frequency-tunable TE_{01n} cavity ($n = 2$ or 3), fabricated from machinable ceramic with either a fired-on gold or thin-wall metallic gold surface (0.025 mm thick gold foil, purchased from Alfa Aesar, Ward Hill, MA). A unique block design for the cavity provides very good mechanical stability and permits the use of a very thin iris (0.004″). The quality factor of the unloaded cavity is 4000. A locally built low-noise video preamplifier is coupled to either a tuned point-contact whisker diode detector or a beam-lead Schottky barrier diode detector. The video preamplifier is integrated with a regulated current bias and uses discrete components for the first-stage amplification. The microwave frequency is measured by an EIP578 frequency counter, with the magnetic field being measured by a Metrolab PT2025 NMR teslameter. A detailed description of this instrument can be found elsewhere [28].

With this spectrometer we were able to achieve exceptional concentration sensitivity even with lossy aqueous samples. Figure 9-3 shows the results of a recent test of our Mark II W-band spectrometer with a 7 μM perdeuterated Tempone (2, 2′, 6, 6′-tetramethyl-4-piperidone-1-nitroxide) aqueous solution at room temperature (The experimental conditions and spectral parameters are given in the figure caption). Converting conversion time to a standard 1 s per data point and accounting for 0.8 G modulation amplitude, 0.8 G linewidth, and three nitrogen hyperfine lines, we derive the concentration sensitivity as 7 nM/G (i.e. 7×10^{-9} M/G). This result does not account for an extrapolation to

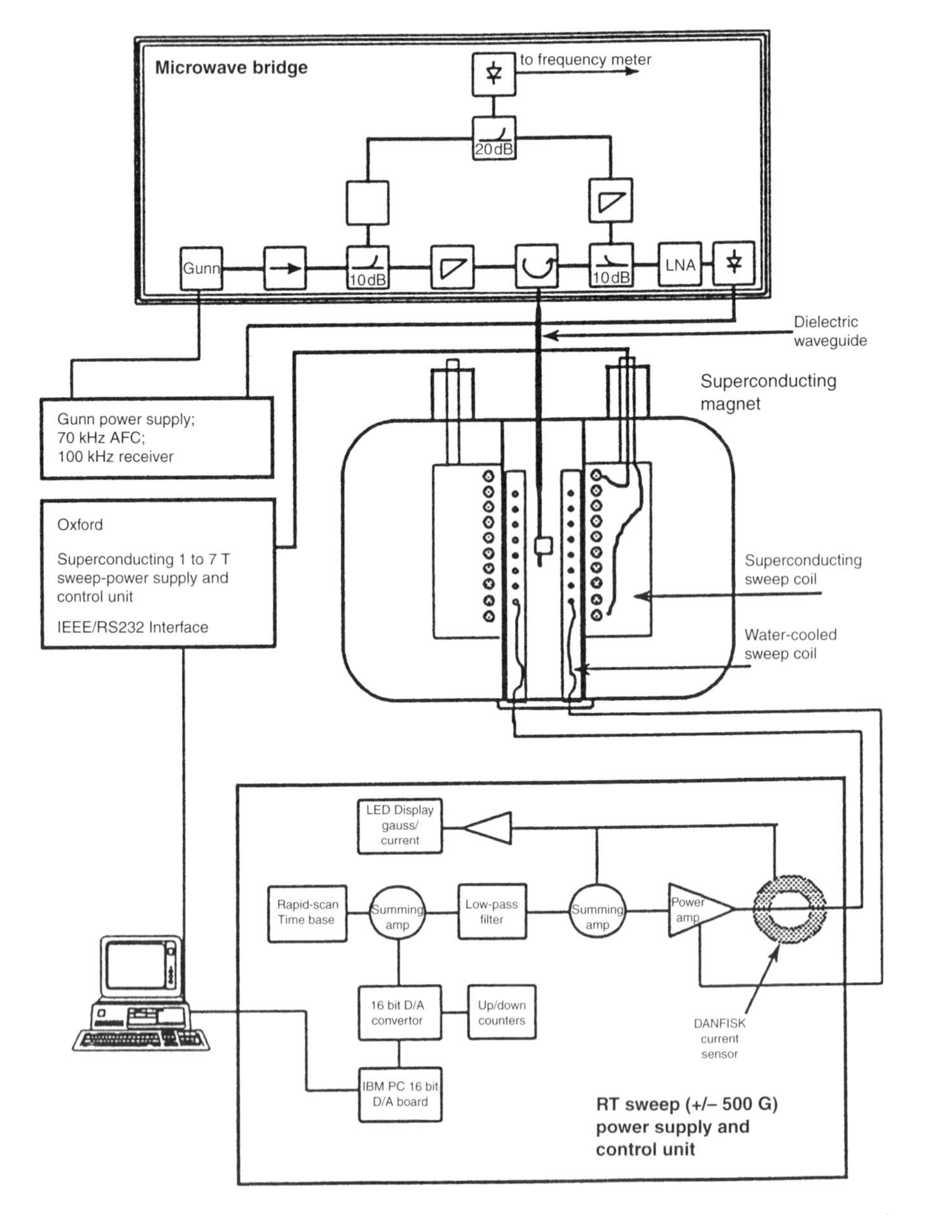

Figure 9.2 Block diagram of the Illinois Mark II W-band EPR spectrometer (95 GHz) at the Illinois EPR Research Center, University of Illinois at Urbana-Champaign. Reprinted from [28]. Copyright (1999) with permission from Springer Verlag.

maximum available power, usually carried out by instrument manufacturers when quoting the results of their test(s). The only well-documented sensitivity considerations for lossy samples in high-frequency (HF) EPR were summarized

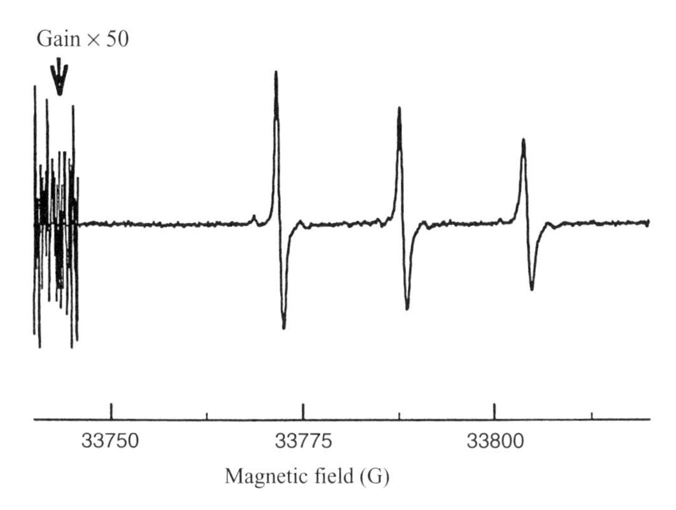

Figure 9.3 Experimental 94.5 GHz EPR spectrum obtained from 7 μM Tempone aqueous solution at room temperature: modulation amplitude 0.8 G at 100 kHz; conversion time 0.12 s per data point: 3.5 mW power from the source. The low-field portion of the spectrum shows the noise level at a 50–fold amplification. (signal-to-noise ratio for the spectrum is 115:1). Reprinted from [28]. Copyright (1999) with permission from Springer Verlag.

by Lebedev [31]. The results predict that, for water, the single-mode resonator concentration sensitivity at 140 GHz should be 1.6 $\times 10^{-7}$ M/G. The signal-to-noise ratio for a Fabry–Perot resonator was calculated to be about two- to threefold better. It is clear that the results demonstrated here exceed the estimates reparted for either a Fabry–Perot resonator or a single-mode cavity given in [32].

At W-band (3.3 T resonance field), the widths of EPR spectra from many Gd(III) complexes are about 15–30 G. Thus, good quality spectra can be obtained even for samples at 20–100 μM concentrations. The concentration sensitivity becomes important when HF EPR is used to monitor changes in Gd(III) EPR spectra, and thus in ZFS upon binding to phospholipids and proteins. Some of these experiments are further discussed later in this chapter.

Between the Q- and W-band EPR frequencies (35 and 95 GHz, respectively), the Gd(III) spectrum undergoes significant changes in both linewidth and apparent g-factor. Both effects are now used to measure the ZFS from aqueous-solution spectra and both are approximately proportional to $1/\nu^2$ where ν is the resonance frequency. After some initial experiments, we realized that an additional EPR frequency between the Q- and W-band might be very useful to rigorously test the existing theory and to study the ZFS effects in high-spin S-state ions (such as Gd(III) or Mn(II)) more thoroughly and precisely. In addition, for optimal measurements the new frequency should be positioned approximately in between the Q- and W-bands on the $1/\nu^2$ scale and be close to the typical fields used in MRI experiments (about 1.5 T). Thus, we have chosen 48 GHz (U-band) as the EPR frequency.

The new U-band mm-wave bridge has been economically constructed by using the basic design of our W-band homodyne bridge (but without a low-noise mm-wave amplifier). A specially designed low-noise Gunn oscillator (Millimeter-Wave Oscillator Co., Longmont, CO) allows for both mechanical and electrical tuning, with the latter being used for automatic frequency control (AFC). For initial testing, the bridge was interfaced with the W-band control electronics and a part of our Varian E-115 Q-band console with a Varian 18″ Q-band electromagnet to generate and control the magnetic field. A Varian cylindrical TE_{011} cavity is retuned to the 48 GHz resonance frequency. This inexpensive U-band set-up provides a sensitivity comparable to that of a Varian E-line Q-band spectrometer.

For experiments at Q-, U- and W-band frequencies, samples were drawn into quartz capillaries (VitroCom, Inc.) with internal diameters of .30, 0.20, or 15 mm (depending on the EPR frequency) and sealed with Critoseal (Fisher Scientific, Pittsburgh, PA). Figure 9.4 shows how the three frequencies (Q, U, and W) span the 35–95 GHz range in a $1/\nu^2$ plot.

G-band (249 GHz) experiments were carried out in the laboratory of Professar Jack Freed (Baker Laboratory of Chemistry, Cornell University, Ithaca,

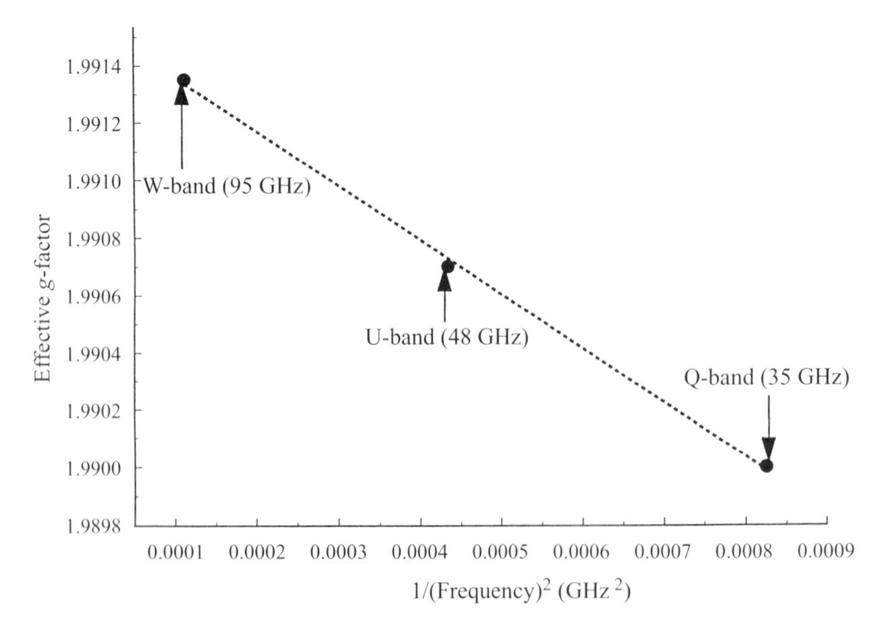

Figure 9.4 Plot of the effective *g*-factor as a fucntion of $1/\nu^2$ for GdDOTA in water at 22 °C, illustrating how the Q, U, and W-band frequencies span the range from 35 to 95 GHz. The slope of the straight line through these data points gives $\Delta^2 = 1.24 \times 10^{19} \mathrm{rad}^2/\mathrm{s}^2$ when analyzed by using Equation (9.8).

NY) with an instrument built in that laboratory and described elsewhere [32, 33]. This spectrometer utilizes a Fabry–Perot cavity and quasi-optical components. The 330 GHz experiments were carried out in the laboratory of Dr Brunel at the National High Magnetic Field Laboratory (Tallahassee, FL) with the expert help of Dr Hans van Tol.

9.3.2 LODEPR

Recently, we have been successful in making T_{1e} measurements on Gd(III) chelates in aqueous solutions at room temperature by means of longitudinally detected EPR (LODEPR) at the X-band, in collaboration with Professor V. A. Atsarkin and co-workers at the Institute of Radio Engineering and Electronics, Russian Academy of Sciences, Moscow, Russia. Prof. Atsarkin has developed an X-band instrument which allows the direct measurement of T_{1e} of aqueous solutions of GdDOTA derivitives [34]. Values of T_1 in the range from 1.7 to 4 ns were recorded; the shortest T_1 that can be observed with this instrument is projected to be in the range of 0.1 ns. This approach should be very helpful in studying relaxation in such systems, and currently is being implemented at 35 GHz.

4 COMPARISON OF THEORY AND EXPERIMENT

Figure 9.5 illustrates the EPR spectra of Gd(III) DTPA aqueous solutions at experimental frequencies of 9.5, 35, 94, and 249 GHz. Table 9.1 summarizes the peak-to-peak linewidths and *effective* g-factors for this complex and for Gd(III) DOTA. Figures 9.6 (a) and 9.6 (b) show the variation of g_{eff} as a function of the inverse square of the experimental frequency for the two complexes. The good linearity of the plots allows us to determine Δ^2 in each system (Equation (9.8)), i.e. $7.0 \times 10^{19} \text{rad}^2/\text{sec}^2$ for Gd DTPA and $1.2(4) \times 10^{19}\ \text{rad}^2/^2$ for Gd-DOTA. Similar values for the ZFS have been obtained by other methods for other aqueous Gd(III) complexes [7, 35–38].

When unchelated Gd(III) was present in a solution together with chelated Gd(III) (Gd-DOTA), the 94 GHz EPR spectrum (Figure 9.7) showed a partial splitting. This splitting disappeared at 35 or 249 GHz. Similar splittings in HF EPR spectra were observed when a lipophilic derivative of DOTA containing an *n*-pentane side-chain, (DOTA-P), was partitioned between aqueous and lipid phases of multilamellar DMPC liposomes [12]. The EPR spectra of this system at 94 and 249 GHz exhibit two resonance lines, corresponding to signals from the aqueous and lipid environments. Similar but less well-resolved spectra are observed when lipophilic chelates, such as DOTA-P and the ethoxybenzoate derivative of DTPA, (DTPA-EOB), interact with human serum albumin and with hepatocytes [39]. We attribute the splitting at 94 GHz from the

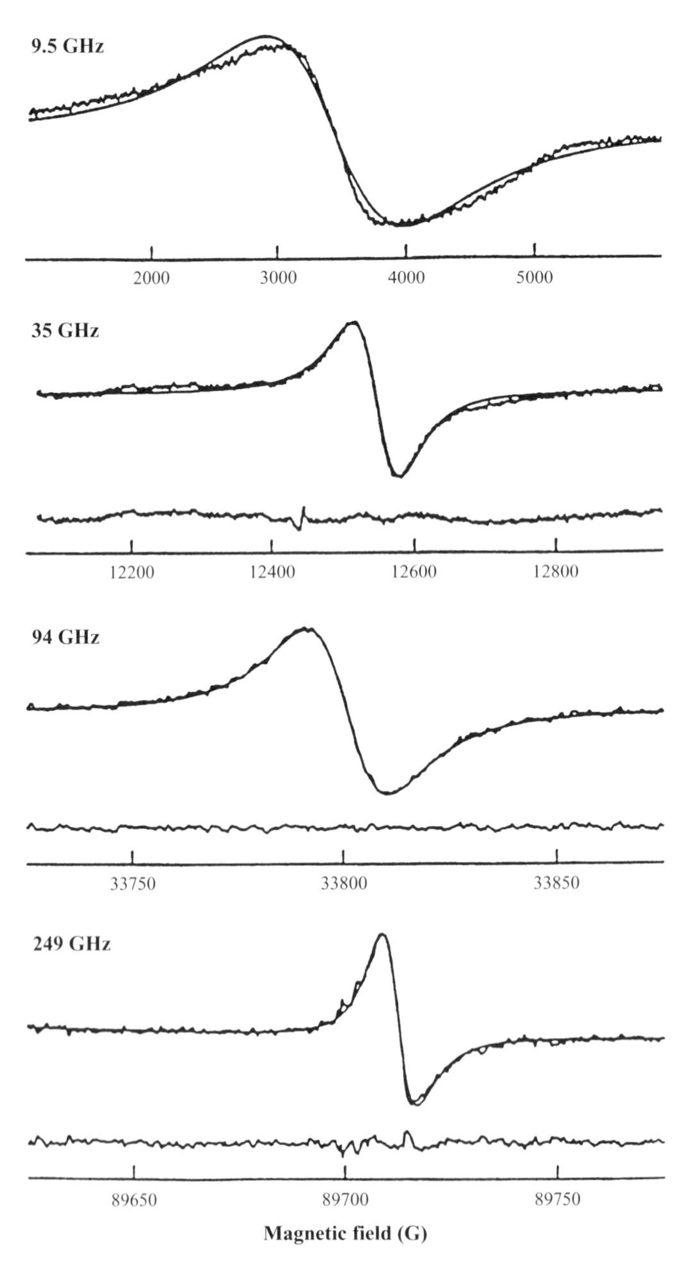

Figure 9.5 EPR spectra of a 1.0 mM aqueous solution of Gd-DTPA, measured at 9.5, 35, 94 and 249 GHz; the solid lines are best-Lorentzian fits to the data. Reproduced with permission from [11].

Table 9.1 Multi-frequency EPR parameters for Gd(III) chelates in water (1.0 mM) at 22 °C.

Experimental frequency (GHz)	Gd-DTPA		Gd-DOTA	
	g_{eff}	ΔB_{pp} (G)	g_{eff}	$\Delta B_{pp}(G)$
9.5	—	620	1.9706(5)	81
34.8	1.9828(5)	63.8	1.9900(0)	25.9
94.1	1.9905(7)	19.1	1.9913(5)	13.6
249	1.9914(3)	7.8	1.9916(1)	5.8

$GdCl_3$/Gd-DOTA solution to the effect of chelation on the static ZFS. The *effective* g-factors for signals arising from the unchelated Gd(III), which has a broader line, and Gd-DOTA, plotted as a function of the inverse square of the experimental frequency, are shown in Figure 9.7(b). The slopes of the plots of g_{eff} as a function of v^{-2} in the two environments are observed to be different, thus leading to the conclusion that Δ_s^2 is different for chelated and unchelated Gd(III). For $GdCl_3$, Δ^2 is found to be 2.2×10^{19} and for Gd-DOTA, $1.2 \times 10^{19} \text{rad}^2/\text{sec}^2$. Figure 9.7(a) also shows that when this system is observed at two different spectrometer frequencies, i.e. 94 and 249 GHz, the absolute magnetic field separation of the two resonances is greater at 94 GHz than it is at 249 GHz, again leading to the conclusion that it is the field-independent ZFS term that is the major cause of difference between g_{eff} in these two resonances.

The spectra of Gd-DTPA at different microwave frequencies (shown in Figure 9.5) are accompanied in each case by the best-fit Lorentzian lineshapes. While the spectrum at the X-band is only approximately Lorentzian, the spectra at higher frequencies are well fitted by the Lorentzian function. This fact can be used to calculate the approximate values of T_{2e} for each frequency, according to the relationship $1/T_{2e} = \Delta B_{pp}\gamma(\sqrt{3})/2$, where ΔB_{pp} is the peak-to-peak linewidth of the first-derivative spectrum. However, in order not to overlook real contributions from the other three transitions (and to avoid forcing Lorentzian behavior on non-Lorentzian lines), we calculated the values of T_{2ei} and the transition probabilities for all transitions in the system, as a function of the magnetic field. We then calculated the time-dependent magnetization, $M(\omega, \Delta^2, \tau_v, t)$, as a sum of all transitions, performed a Fourier transformation of the magnetization function, and calculated the peak-to-peak linewidth of the simulated first-derivative spectrum. The experimental data were fitted by allowing the values of Δ^2 and τ_v to vary, making use of a non-linear least-squares procedure, and then calculating the eigenvalues of the $\mathcal{R}$-matrix at each τ_v value. A field-independent relaxation rate, R_o, also was added to the fitting equation, which then became:

$$\boldsymbol{M}(\omega, \Delta^2, \tau_v, t) = \sum_{i=1}^{7} I(\omega, \tau_v, i)\left[\exp\left(\frac{-t}{T_{2e}(\omega, \tau_v, \Delta^2, i) + \frac{1}{R_0}}\right)\right] \tag{9.10}$$

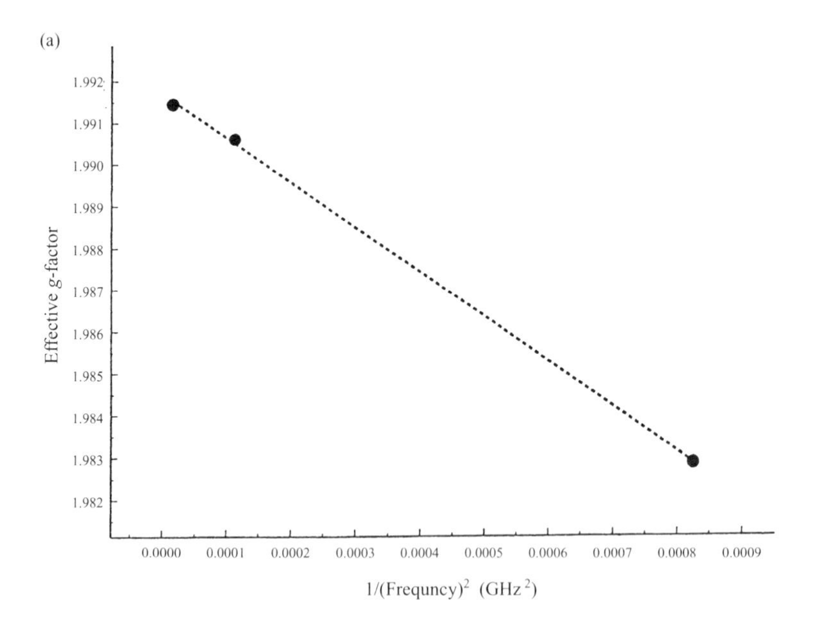

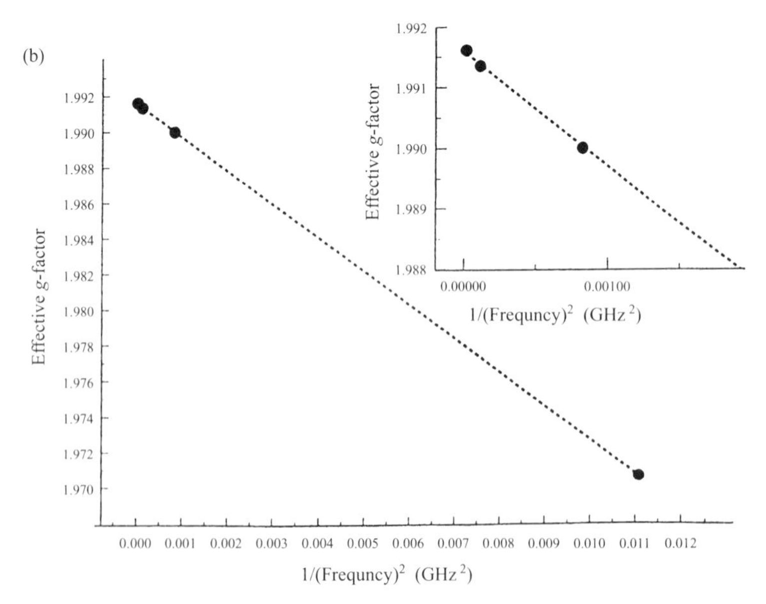

Figure 9.6 The variation of g_{eff} as a function of the inverse square of the experimental frequency for aqueous solutions of Gd-DTPA and Gd-DOTA. The slopes of the best straight lines through the data are used to calculate Δ^2 according to Equation (9.8): Gd-DTPA, 1.0 mM aqueous solution, $\Delta^2 = 7 \times 10^{19}\ \mathrm{rad}^2/\mathrm{s}^2$; (b) Gd-DOTA, 1.0 mM aqueous solution, $\Delta^2 = 1.24 \times 10^{19}\ \mathrm{rad}^2/\mathrm{s}^2$ The inset shows a magnified section of the region at low values of $1/\nu^2$. Reproduced with permission from [11].

Here, $M(\omega, \Delta^2, \tau_v, t)$ is the magnetization in the x–y plane, $I(\omega, \tau_v, i)$ is the transition probability of a particular ($i = 1, \ldots, 7$) spin transition at the experimental frequency ω, and $T_{2e}(\omega, \tau_v, \Delta^2, i) = T_{2ei}$ of Equation (9.9). The value of R_o was held constant for all calculations of a particular Gd^{3+} chelate. A field-independent contribution to the transverse relaxation rate has been observed by other workers [40, 41], although the exact mechanism(s) responsible for R_o is still the subject of investigation. Figure 9.8 shows that this model accounts very well for the frequency dependence of T_{2e} in the Gd-DTPA system. A similar agreement between experiment and calculation was found for data obtained from Gd-DOTA. Since the T_{2ei} values can be expressed as functions of Δ^2 and τ_v, as shown in Equation (9.9), the fitting of transverse relaxation dispersion data also provides another route to estimate the transient ZFS interaction parameters. Table 9.2 summarizes the results for ZFS parameters obtained by fitting the frequency dependencies of g_{eff} and T_{2e}, as well as the results from a Solomon–Bloembergen–Morgan analysis of proton nuclear magnetic relaxation dispersion (NMRD, T_{1p}, data. [7, 35–38].

The approach of Hudson and Lewis can also be used to obtain values for T_{1e} (see [7]). Comparing the T_{1e} values obtained for the Gd DOTA derivitives measured by longitudinally detected EPR (LODEPR) at the X-band with the

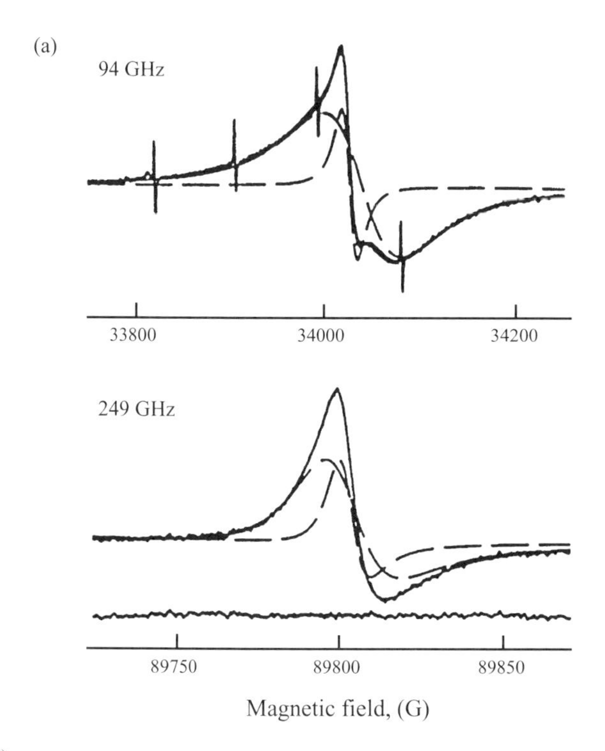

Figure 9.7 (a)

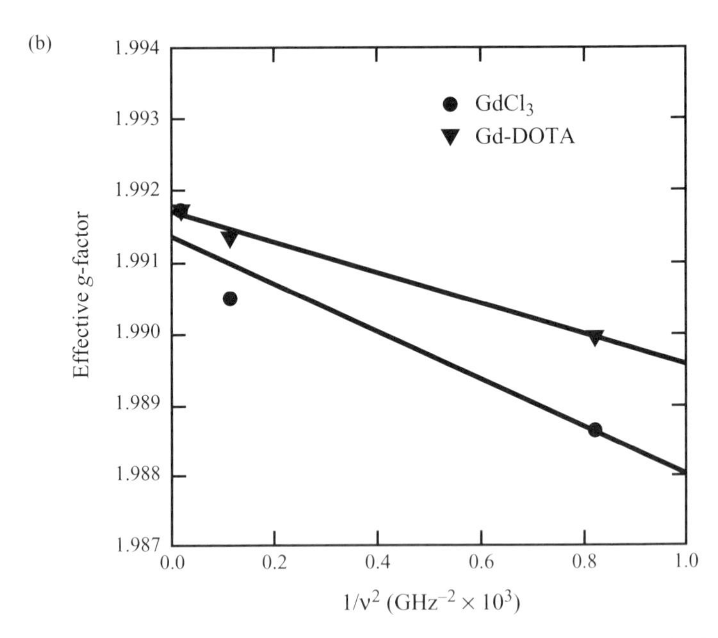

Figure 9.7 (a) 94 and 249 GHz EPR spectra of an aqueous solution of $GdCl_3/Gd-DOTA$. Beneath each spectrum is a least-squares fit of the data, obtained by using a simple two-Lorentzian model: an adjustable phase shift was included in the simulation. The magnetic field separation between the line centers is 14 G at 94 GHz, and 5.8 G at 249 GHz; the sharp lines superimposed on the 94 GHz EPR spectrum are from Mn^{2+}, used as a field marker. (b) Variation of g_{eff} with the inverse square of the experimental frequency for the $GdCl_3$ and Gd-DOTA signals. The best straight lines through the data points allowed the calculation of Δ^2: $GdCl_3$: 2.2×10^{19} rad^2/s^2; Gd-DOTA, 1.2×10^{19} rad^2/s^2. Reproduced with permission from [11].

calculated values (making use of the same Δ^2 and τ_v values found from multi-frequency T_{2e} measurements), gives a theoretical prediction of 1.3 and 2 ns for T_{1e} of the two compounds, while LODEPR measures values of 1.7 and 3.2 ns, respectively. The experimental values are in good agreement with the theoretical predictions, and suggest that when more data are available from this experimental method, both T_{1e} and T_{2e} measurements can be combined to obtain best-fit values for Δ^2 and τ_v.

5 APPLICATIONS TO CONTRAST AGENT RESEARCH

5.1 INTERACTIONS OF Gd(III) CHELATES WITH PHOSPHOLIPID MEMBRANES AND PROTEINS [12].

In the course of rational design of contrast agents, a balance between lipophilic groups and more polar moieties and ionized groups is essential. Polar and

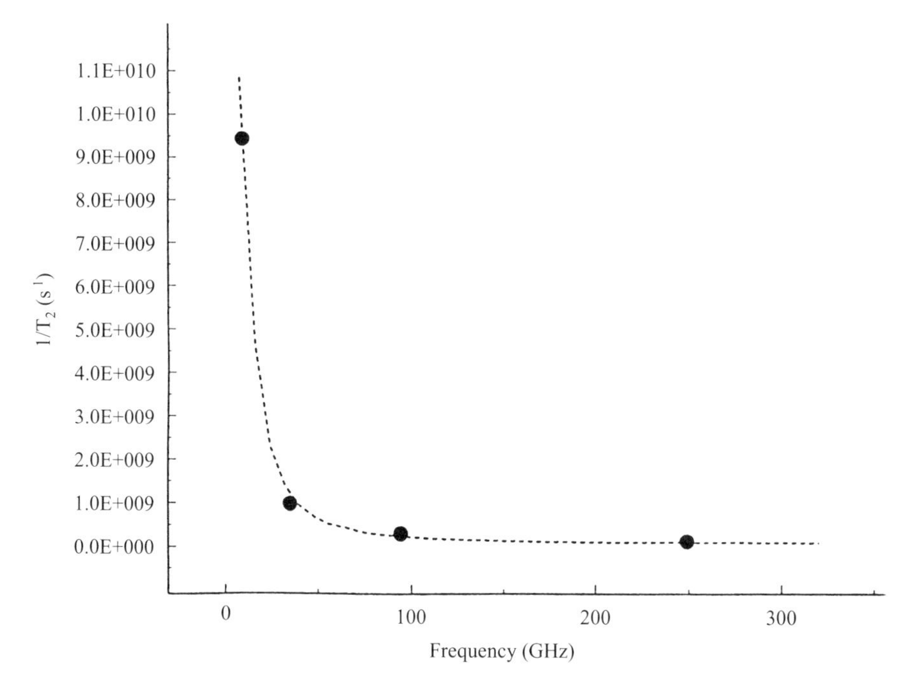

Figure 9.8 Variation of the calculated values of $1/T_{2e}$ with experimental frequency for Gd-DTPA in aqueous solution. Superimposed on the data is the best fit obtained by using the relaxation matrix given by Lewis and Hudson [18], which was used to calculate all allowed values of T_{2ei} and their transition probabilities; from this, the time-dependent x–y magnetization was calculated. A Fourier transformation of $\boldsymbol{M}_{x,y}(t)$ yielded the EPR lineshape. Least-squares optimization of the fit between calculated and observed lineshapes at the four frequencies, allowing Δ^2, τ_v, and R_0 to vary, provided the final parameters. The best-fit values are as follows: $\Delta^2 = 8 \times 10^{19}$ rad^2/s^2, $\tau_v = 2.4 \times 10^{-11}$ s, and $R_0 = 7.6 \times 10^7$ s^{-1}. Reproduced with permisson from [11].

ionized groups are required for water solubility and important for non-covalent interactions with macromolecular binding sites, while lipophilic groups are largely responsible for contrast agent–macromolecule binding. Correlations have been established between lipophilicity and albumin binding [42] and are expected for other macromolecules such as cytosolic proteins or hepatocyte membrane receptors. The overall effect of binding controls pharmacokinetics and, as a result, the biodistribution properties [43]. Lipophilic groups, however, can interact with cell membranes, thus contributing to contrast agent toxicity. Alteration of membrane potentials is considered as a possible mechanism of toxicity, together with non-specific protein conformational effects or enzyme inhibition [44]. Obtaining specific information on contrast agent–membrane interaction is important in attempts to control and balance the numerous consequences of these interactions. Information about the effects of lipophilic Gd^{3+} compounds on the structure of the membranes can be obtained by

Table 9.2 ZFS analysis of Gd(III) chelates, measured for 1.0 mM aqueous solutions at 22 °C.

Complex	Δ^2 (rad^2/s^2)	τ_v (s)
GdDTPA		
NMRD analysis [36]	5.3×10^{19}	1.9×10^{-11}
g_{eff} vs B_0	7.0×10^{19}	—
$1/T_{2e}$ vs B_0[a]	8×10^{19}	2.4×10^{-11}
Powell *et al.* [38]	4.6×10^{19}	2.5×10^{-11}
GdDOTA		
NMRD analysis [35]	7×10^{18}	2.6×10^{-11}
g_{eff} vs B_0	1.24×10^{19}	—
$1/T_{2e}$ vs B_0[a]	1.1×10^{19}	1.26×10^{-11}
Powell *et al.* [38]	1.6×10^{19}	1.1×10^{-11}

[a] Fitted by Equation (9.10), in which $R_0 = 7.6 \times 10^7\ s^{-1}$ for both Gd-DTPA and Gd-DOTA.

monitoring changes in the relaxation times of a spin label located at a specific site within the membrane. Several spin-labeled doxyl stearic (DS) acids were incorporated in a model phospholipid membrane so that the position of the nitroxide moiety across the phospholipid bilayer was well defined. Each nitroxide label probes a very specific area of the membrane. By combining data obtained with different probes, one can measure the distribution of a property (for example, polarity, oxygen permeability and rotational dynamics) across the membrane. One set of experiments recently completed employs molecular oxygen to probe the structural organization of the membranes exposed to lipophilic contrast agents. Because of its small size and high solubility in the hydrocarbon phase, the oxygen molecule can enter pockets transiently formed in the membrane and thus report on even small changes and distortions in the membrane organization. This method is based on spin–spin interactions of the spin label and oxygen, where the latter shortens the electronic T_1 and T_2 relaxation times of the spin label. Thus, changes in T_1 and T_2 of the spin label reflect variations in local oxygen permeability, and can be related to structural changes in the system. Use of oxygen as a molecular probe in combination with site-directed spin labeling has found a wide range of applications in studies of the structure and organization of membranes and membrane–protein systems [45, 46]. We have used this method to study the contrast agent/DMPC (1, 2–dimyristoyl-*sn*-glycero-3-phosphocholine) system in the presence of 20 mM Gd-DOTAP (an *n*-pentane derivative of DOTA) at 30.8 °C. We found that a model of additional Lorentzian broadening (see [46]), describes the oxygen effect extremely well. The analysis shows that in presence of 20 mM Gd-DOTAP, the oxygen permeability of the DMPC bilayer at the position of 12–DS measurably increases (by $6.5 \pm 0.3\,\%$). The oxygen-permeability experiment supports our suggestion that Gd-DOTAP is partitioning within the DMPC membrane.

Partitioning of Gd-DOTAP into a phospholipid membrane should result in a broadening of the EPR spectra of labeled DMPC samples when contrast agent is added. We analyzed spectra collected for deoxygenated samples for the broadening effects, caused by spin–spin interactions between the contrast agent molecules and the nitroxide labels. We have shown that the model of additional Lorentzian broadening describes EPR line shape changes of a nitroxide label by Gd^{3+} complex reasonably well, and can be accurately used to measure the magnitude of the effect. Data obtained for several doxyl stearic acid spin labels at temperatures below and above the main phase transition are summarized in Figure 9.9, which shows the Lorentzian peak-to-peak broadening induced by 10 mM of Gd-DOTAP versus the distance from the center of

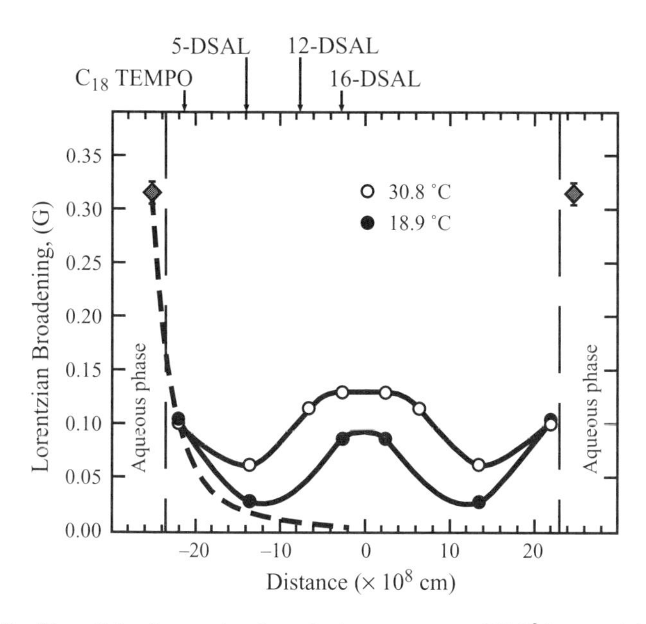

Figure 9.9 Profiles of the Lorentzian broadening parameter $\delta(\Delta B^L_{pp})$ caused by 10 mM of Gd-DOTAP across the DMPC membrane (pH 9.5) measured for two temperatures, i.e. below (18.9 °C) and above (30.8 °C) the main phase transition; the approximate locations of the spin labels in the membrane are shown by arrows. The abbreviations used are as follows: C_{18} TEMPO (tempo stearate), 4-(octadecanoyloxy)-2, 2, 6, 6-tetramethylpiperidine-1-oxyl; 5-DSAL, 5-doxyl-stearic acid spin label; 12-DSAL, 12-doxyl-stearic acid spin label; 16-DSAL, 16-doxyl-stearic acid spin label. The parameter $\delta(\Delta B^L_{pp})$ for the aqueous phase measured with the nitroxide, tempo-choline (4-[*N*, *N*-dimethyl-*N*-(2-hydroxyethyl)]-ammonium-2, 2, 6, 6, -tetramethylpiperidine-1-oxyl, chloride), is represented by the diamonds, while the variations of this parameter for the two locations are within the actual 'size' of the symbols as shown. The approximate locations of the aqueous phases are indicated by the vertical dashed lines, while the dotted line shows the theoretical $1/r^3$ dependence of the line broadening caused by dipole–dipole interactions, assuming the Gd-DOTAP to be localized only at the lipid–water interface. Reprinted from [12]. Copyright (1999), with permission from Elsevier Science.

the bilayer. The broadening is maximal in the bilayer center. This result is in agreement with oxygen permeability experiments and supports our hypothesis that Gd-DOTAP partitions into the phospholipid bilayer with a preferred location at its center. In contrast, we observed no measurable broadening effects upon the addition of 10 mM Gd-EOB-DTPA, as well as no effect of this complex on the oxygen permeability profile. This demonstrates that partitioning of Gd-EOB-DTPA is negligible and that this contrast agent does not distort the local order in the phospholipid bilayer.

We have also shown that electron paramagnetic resonance (EPR) at high magnetic fields, in combination with spin-labeling methods, can provide site-specific information on the binding of Gd(III) complexes to serum albumin and other proteins that bind lipophilically. Experiments also show that HF EPR of Gd(III) complexes allows one to measure binding constants, while multi-frequency EPR data provide an estimate of electronic zero field splitting–an important parameter of paramagnetic contrast agent electronic relaxation. Because of the nature of the ligand, we have suggested that Gd-DOTAP might bind to bovine serum albumin (BSA) in a way similar to fatty acids (FAs) at the binding site II in the sub-domain III. We have labeled BSA at this site with a series of stearic acids (SAs) containing the doxyl nitroxide at the 5-, 7- 12- and 16-carbon positions. As FAs bind to albumin, these probes are located along the FA binding channels at well-defined positions. If the Gd(III) complex binds in the vicinity of the spin probe it will broaden the EPR spectrum of this probe. Because the EPR spectra of nitroxides are also sensitive to polarity, electrostatic potential, free volume, microviscosity, and the elements of the tertiary structure of the labeled protein site, spin-labeling methods allow one to monitor how all of these parameters are affected by contrast agent binding. Figure 9.10 shows the dependence of Lorentzian peak-to-peak broadening as a function of the label position in the FA binding channel. The broadening progressively decreases from 5- to 16-DS acid. This supports our hypothesis that this complex binds to albumin in the vicinity of the main FA binding site. The results also show that at a contrast agent/BSA ratio of 1:1, Gd-DOTAP binds preferentially at the opening of the FA binding channel. A substantial difference in the spin-label–metal interactions along the FA binding channel for Gd-EOB-DTPA demonstrates that this complex binds at the main FA binding site in a non-specific manner. Thus, its binding is very different from that found for Gd-DOTAP.

Significant changes in 95 GHz CW EPR spectra were observed upon binding of the Gd-EOB-DTPA to human serum albumin (HSA). In HSA solution, as the fraction of bound Gd-EOB-DTPA increases, the central component of its 95 GHz EPR spectrum becomes broader and wings in the low- and high-magnetic field regions become apparent. These features are similar to those observed for frozen aqueous solutions, indicating that the motion of the bound contrast agent is significantly restricted. Figure 9.11 shows a 95 GHz EPR spectrum obtained from 1 mM Gd-EOB DTPA in 20% HSA solution (A)

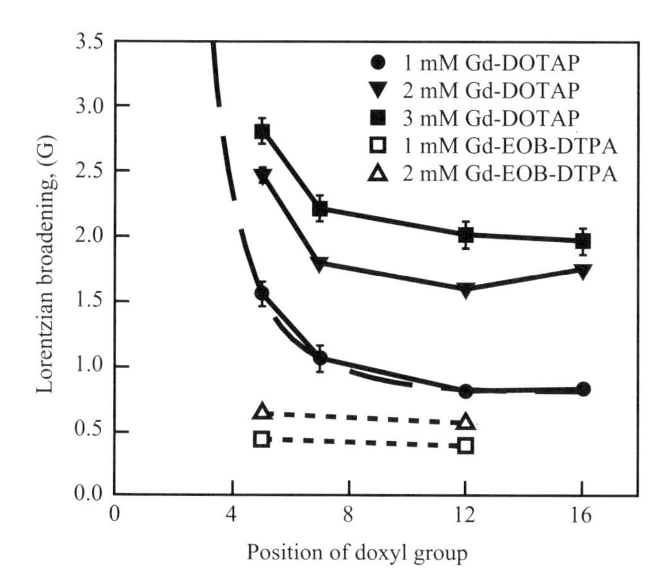

Figure 9.10 Profiles of Lorentzian broadening as a function of the doxyl-stearic acid (DSA) label positions along the fatty acid (FA) binding channel caused by the addition of Gd-DOTAP and Gd-EOB-DTPA to 1 mM solutions of bovine serum albumin (BSA), at DSA/BSA ratios of 1/1. The dashed line represents the expected broadening if the 1 mM Gd-DOTAP is bound at the mouth of the FA binding channel.

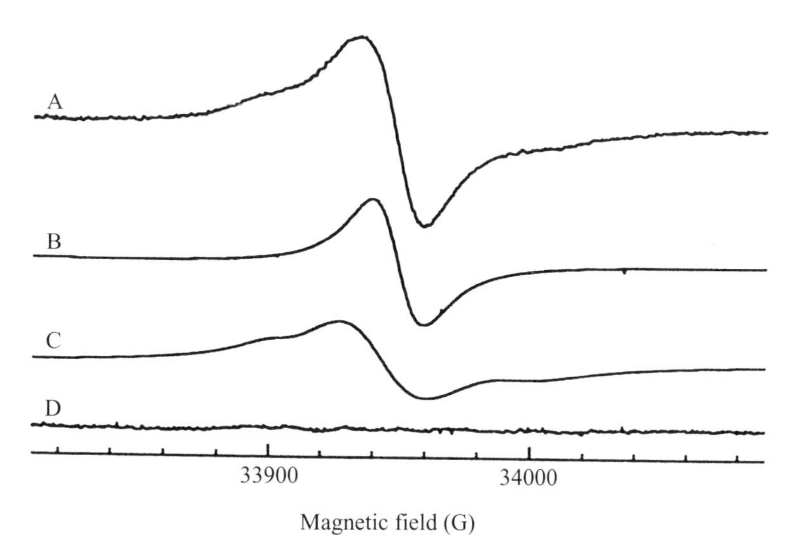

Figure 9.11 95 GHz EPR spectrum obtained from 1 mM Gd-EOB DTPA in 20% HSA solution (A), and least-squares simulations which allow the decomposition of the experimental spectrum into free Gd-EOB-DTPA (B) and bound Gd-EOB-DTPA (C) components. The residual (D), i.e. the difference between the experimental and simulated spectra, is nearly flat, showing excellent agreement with the model.

and least-squares simulations carried out to decompose the experimental spectrum into free Gd-EOB-DTPA (B) and bound Gd-EOB-DTPA (C) components. Residual (D), the difference between experimental and simulated spectra, is nearly flat, thus showing excellent agreement with the model. Spectral simulations of HF EPR spectra of Gd-EOB-DTPA at various HSA concentrations allowed us to calculate the partitioning coefficient for Gd-EOB-DTPA in HSA. This partitioning, measured as a function of HSA concentration, was modeled as a non-covalent binding to either a single or to equivalent multiple sites. The binding constant K_a (or K_{a*n}, where n is a number of equivalent independent binding sites), for Gd-EOB-DTPA was 1×10^3 M^{-1}. Compared to NMRD and proton relaxation enhancement (PRE) titration experiments, in the case of multiple binding sites, HF EPR directly measures the bound/unbound ratio for Gd(III) complexes without any assumptions concerning Gd(III)–HSA adduct relaxivity.

5.2 SENSITIVITY OF ZFS PARAMETERS TO BIOLOGICAL ENVIRONMENTS

Compared to other lanthanides, Gd^{3+} exhibits a high sensitivity of its magnetic parameters to the crystal field. However, low spectral sensitivity, broad lineshapes that are poorly understood, and low spectral resolution of X-band EPR for Gd^{3+} spectra, complicated uses of this characteristic. Many of these difficulties can be overcome by employing EPR at microwave frequencies higher than the X-band. EPR spectra of 10 mM aqueous solutions of Gd-DOTAP (unbuffered, pH $\approx$ 5.6) and Gd-EOB-DTPA at multiple frequencies from 9.5 to 249 GHz, shows a progressive decrease in the EPR linewidth with increasing microwave frequency. For Gd-DOTAP, the peak-to-peak line width decreases from $\Delta B_{pp} \approx 400$ G at 9.5 GHz to 24.6 ± 0.1 G at 94.3 GHz and 9.3 ± 0.1 G at 249 GHz. Similar decreases were observed for Gd-EOB-DTPA, i.e. from $\Delta B_{pp} \approx$ 560 G at 9.5 GHz to about 63 G at 34 GHz to 19.3 ± 0.1 G at 94.3 GHz. This observation is in qualitative agreement with data reported for other Gd^{3+} compounds studied at multiple EPR frequencies (3, 9.5, 35, and 150 GHz) [7, 38]. The shape of the X-band (9.5 GHz) EPR spectrum of Gd-DOTAP can not be described as Lorentzian, while the EPR spectra at 35 GHz and higher are well approximated by Lorentzian functions, thus simplifying the data analysis.

The sensitivity of Gd-DOTAP 94.3 GHz EPR spectra to the microenvironment was utilized to observe the interaction of this complex with the DMPC bilayer. Figure 9.12 shows an experimental EPR spectrum obtained from the system where 10 mM Gd-DOTAP was added to a 5.5% multilamellar DMPC aqueous dispersion (pH = 9.5, $T = 31$°C) at 35 (A), 94.2 (B) and 249 (C) GHz . At all three frequencies, the spectra can be very well fitted by a superposition of two Lorentzian lines with different g-factors and linewidths; the phase shift for the two signals was assumed to be the same. At 94.2 GHz, comparison of the

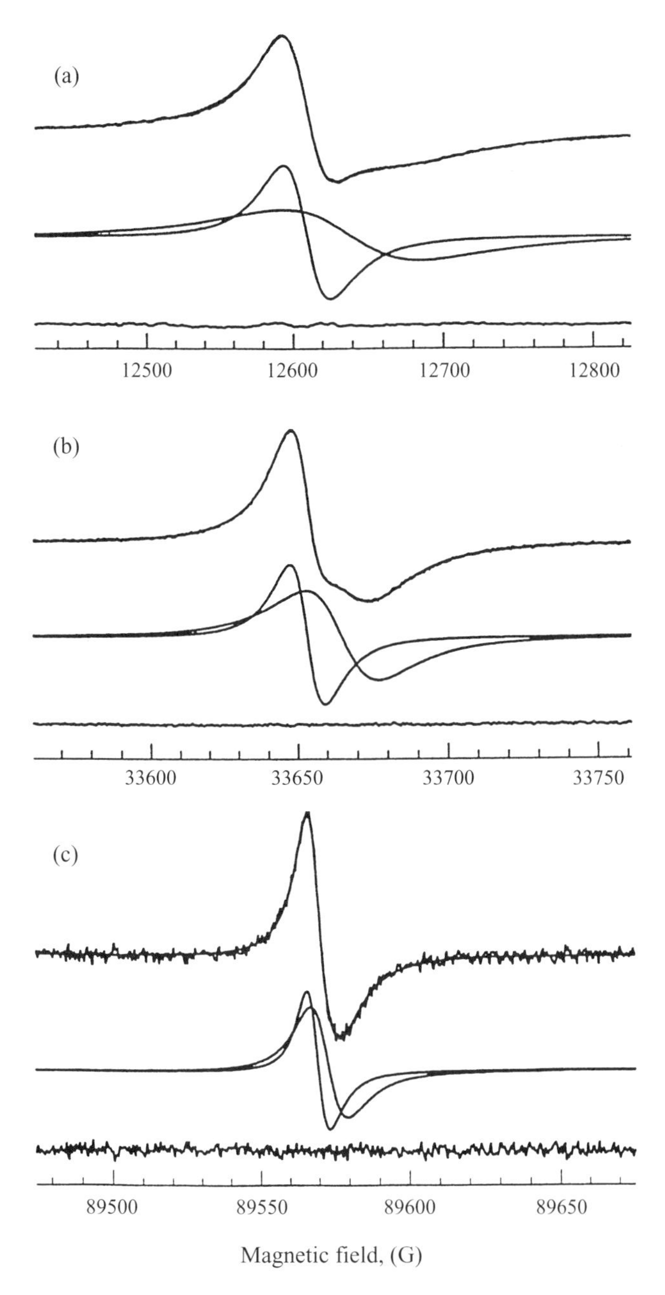

Figure 9.12 Experimental room-temperature EPR spectra obtained from 10 mM Gd-DOTAP added to a 5.5% DMPC aqueous dispersion (pH 9.5), measured at 35 (a), 94.2 (b), and 249 (c) GHz. Reprinted from [12]. Copyright (1999), with permission from Elsevier Science.

width of the narrow line ($\Delta B_{pp}(1) = 12.0$ G) and its g-factor ($g_{eff} = 1.991\,02$) with the signal observed in solution at pH = 9.5, shows that the narrow signal originates from Gd-DOTAP in the aqueous phase. The broader line ($\Delta B_{pp}(2) = 24.3$ G) with a smaller g-factor ($g_{eff} = 1.99030$) is assigned to the Gd^{3+} complex partitioned or interacting with the lipid phase of the phospholipid bilayer. At all three EPR frequencies, the signal assigned to the lipid-associated complex is shifted to higher magnetic fields. The splitting between the lipid and the aqueous components decreased with the EPR frequency; however, this fact can not be explained by the polarity effect on the g-factor that is typically observed for free radicals [47, 48]. For free radicals, the sign of the g-shift with polarity is opposite to that observed in our experiments. The field separation of these two signals decreases with an increase in frequency, thus indicating a field-dependence of the effective g-factors of the signals.

Figure 9.13 shows plots of the apparent g-factors versus $1/\nu^2$ for the aqueous and the lipid signals of Gd-DOTAP in a 5.5 % DMPC dispersion, as measured

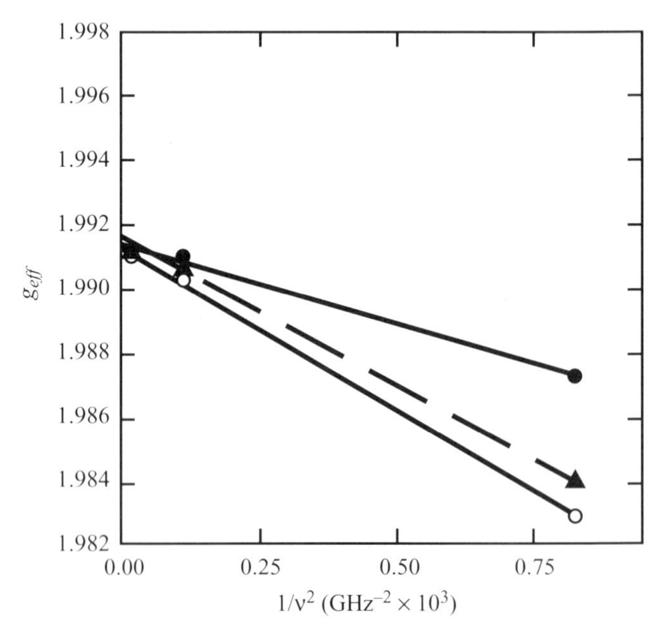

Figure 9.13 Apparent g-factors (g_{eff}) versus $1/\nu^2$ as measured at 35, 95 and 249 GHz and the results of linear regression (Equation (9.8)) for the aqueous (filled circles) and the lipid (open circles) EPR signals observed in a 5.5 % DMPC aqueous dispersion (pH 9.5) after addition of Gd-DOTAP to give a 10 mM concentration. The filled triangles correspond to the apparent g-factors, as measured at 35 and 95 GHz for unbuffered 4 mM aqueous solutions of Gd DTPA-EOB. The estimated ZFS parameter, Δ^2, is 6.6×10^{19} rad^2/s^2 for Gd-DOTAP in the lipid phase, and 3.2×10^{19} rad^2/s^2 in the aqueous phase; this parameter is 6.1×10^{19} rad^2/s^2 for Gd-DTPA-EOB in aqueous solution. Reprinted from [12]. Copyright (1999), with permission from Elsevier Science.

at 35, 94 and 249 GHz, together with the results of linear regression (Equation (9.8)). The estimated ZFS parameter, Δ^2, is 6.6×10^{19} rad^2/s^2 for Gd-DOTAP in the lipid phase, and 3.2×10^{19} rad^2/s^2 in the aqueous phase (pH 9.5); this parameter is 6.1×10^{19} rad^2/s^2 for Gd-DTPA-EOB in aqueous solution. Multi-frequency EPR experiments clearly demonstrate that Gd-DOTAP partitions into phospholipid bilayers and that the ZFS parameter is higher for the Gd-DOTAP which is partitioned in the bilayer. This strengthens an existing opinion that interactions of Gd contrast agents with a biological environment may affect the ZFS parameter, and thus modify the proton-relaxation enhancement.

6 DISCUSSION

Multi-frequency EPR spectroscopy can provide useful information on the ZFS in Gd(III) chelates. Previous calculations of the Δ^2 and τ_v terms for these molecules in aqueous solution relied on an analysis of proton NMRD data, making use of inner-sphere point-dipole models (Solomon–Bloembergen–Morgan theory), which incorporate frequency-dependent terms to calculate the electron spin relaxation rates as they are reflected in proton relaxation enhancement by the paramagnetic species [35, 36, 38, 49, 50]. Such analyses necessarily require the optimization of multi-parameter models, and the uniqueness of the best fits of theory with experiment is always a matter of concern. Recently, ^{17}O NMR relaxation in aqueous solutions of Gd(III) chelates has been studied in variable temperature, pressure, and magnetic-field experiments, and ZFS values were extracted by analysis of the data [40]. Similar variable-temperature and variable-magnetic-field EPR experiments have also been performed, from which ZFS parameters were obtained [7, 51]. In a recent summary of their work incorporating ^{17}O NMR, variable field, temperature, and pressure EPR, and NMRD techniques, Merbach and co-workers give values for the important parameters that characterize several Gd(III) monomer and dimer chelates [9, 38]. Their values for Δ^2 and τ_v in Gd-DTPA and Gd-DOTA (aqueous solutions) are listed in Table 9.2. The EPR experiments reviewed in this chapter make direct use of the magnetic field dependence of the electronic spin Hamiltonian and the relaxation matrix, thus avoiding the complexity of analyzing electron relaxation from its effects on nuclear relaxation. The EPR approach may also avoid the necessity of performing variable-temperature measurements and of considering the effects of temperature on spin relaxation and water exchange rates, which can be particularly complex in biological systems involving living cells or tissues, although the temperature dependence of EPR results is still not well understood. The most desirable approach would make use of both EPR and NMR results in a global optimization, but such an analysis is as yet not always satisfactory with the available theory. Results from previous work agree well with the values of Δ^2 and τ_v obtained here from the analysis of the field dependence of g_{eff}, and transverse relaxation (T_{2e}) dispersion (electron magnetic relaxation dispersion, (EMRD)). The T_{1e} values calculated from the same parameters (Δ^2, τ_v) agree with direct LODEPR measurements, thus further supporting the general theoretical approach.

The approximations used in calculating ZFS values from the analysis of transverse relaxation dispersion data should be mentioned as potential sources of error. It is important to keep in mind that the relaxation model we discussed makes use of the general Bloch–Wangsness–Redfield approach [19]. This approach assumes that $\tau_v \ll T_{2e}$, an inequality that is not strongly satisfied for Gd(III) chelates observed at frequencies of 9.5 GHz and below. Therefore, in order to make better use of EMRD as a method for examining ZFS interactions in these systems, it would be preferable if experimental frequencies were chosen in the range of 15 GHz and higher, although the Δ^2 values calculated from EMRD data, including the 9.5 GHz observation, are almost identical to those values derived from g_{eff} dispersion measurements. It also should be noted that the EPR lineshapes of Gd(III) complexes in solution at 9.5 GHz are not strictly Lorentzian functions, and these deviations from Lorentzian may not arise entirely from contributions to multiple T_{2ei} pathways. The origin of the complex lineshapes at lower EPR frequencies, especially for the Gd-DTPA chelate, warrants further study, including the possibility of contributions from forbidden spin transitions, particularly at lower experimental frequencies. Finally, it is currently unclear if any of the Gd complexes exhibit a *g*-anisotropy sufficient to contribute to the linewidths observed in solution at high EPR frequencies.

In order to describe experimentally observed EPR linewidths, we found it necessary to include an empirical field-independent term R_0 in the rate equation (Equation (9.10)). The value of R_0 was the same for both complexes studied, i.e. $7.6 \times 10^7\, s^{-1}$. Concentrations of Gd in our experiments were low enough to exclude spin–spin broadening as an effective relaxation mechanism (see [38] for a discussion of concentration-dependent linewidths at 5.0 T). In an ^{17}O NMR relaxation study, González *et al.* also found a field-independent process of the same magnitude [40]. While several suggestions have been forwarded as mechanisms for this additional relaxation (e.g. spin-rotation and non-dispersive contribution from ZFS modulation [41]), the exact nature of the process is still under investigation.

The resolution of spectra obtained from chelated and unchelated Gd(III) shown in Figure 9.8 or Gd-DOTAP partitioned between aqueous and lipid phases reported in [10] is seen to be due largely to differences in the ZFS parameters for the complex in the two microenvironments. The best EPR resolution of signals from Gd(III) in two environments is obtained at a spectrometer frequency which depends upon Δ^2 at each site and the relationship between B_0 and the linewidth. In this study, spectra obtained from chelated and unchelated Gd(III) show two signals which are better resolved at 94 GHz than at 249 GHz (Figure 9.12(a)). In order to construct an approximate model for the frequency dependence of spectral resolution, we adopt a simple definition of resolution, as follows:

$$Resolution = \frac{[B_0(1) - B_0(2)]}{< \Delta B_{pp} >_{av}} \quad (9.11)$$

Here, $B_0(1, 2)$ are the line centers of the two signals, calculated from Equation (9.8), and $< \Delta B_{pp} >_{av}$ is the average linewidth of the individual signals. If the line shapes are Lorentzian, and if Reuben's relationship [5] between T_{2e} and B_0 is used to calculate ΔB_{pp} as a function of spectrometer frequency/field, then Figure 9.14 illustrates how resolution varies with experimental frequency for a system in which $\Delta^2(1) = 3 \times 10^{19}$ (rad/s)^2, $\Delta^2(2) = 2 \times 10^{20}$(rad/s)^2, and τ_v for both components is 2×10^{-11} s. The frequency at which maximum resolution is achieved will depend upon the difference in Δ^2 between the two sites and the exact form of lineshape sensitivity to experimental frequency, but the general shape of the curve given in Figure 9.14 agrees with the experimental data.

Environmental effects on Δ^2 and τ_v may reflect a sensitivity of the ZFS to: (i) differences in the interactions with water in the aqueous and lipid phases, and (ii) differences in chelate dynamics in phases with different viscosities and dielectric constants. Koenig has speculated on the differences of water dynamics

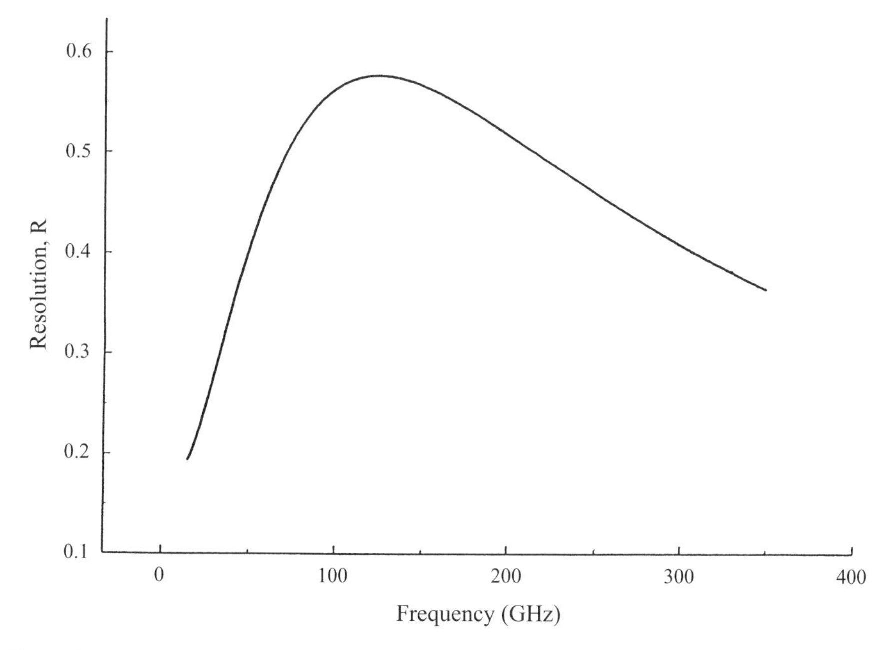

Figure 9.14 Variation of the spectral resolution R with the spectrometer frequency; in the simulation, $\Delta^2(1) = 3 \times 10^{19}$ $(rad/s)^2$, $\Delta^2(2) = 2 \times 10^{20}$ (rad/s)^2 and τ_v for both components is 2×10^{-11} s. Reproduced with permission from [11].

in lipid and aqueous phases [52], pointing out how much more 'dilute' water is in a lipid environment owing to low solubility, and Odelius *et al.* [14] and Kowall *et al.* [53] have carried out molecular dynamics calculations to show how water in the inner hydration sphere of $S > 1/2$ paramagnetic metal ions leads to observed ZFS effects. Thus, it will be important to further investigate the sensitivity of the ZFS to water and chelate dynamics. Since Gd(III) chelates are being developed and applied as paramagnetic contrast agents in magnetic resonance imaging (MRI), this variation in the ZFS parameters with environment may be important in understanding the proton-relaxation-enhancement effects of these compounds in various biological systems, which often contain aqueous and lipid compartments.

Finally, the ZFS parameters in Gd(III) systems can also be studied by simulating the EPR spectra of frozen solutions of the chelates. In work currently being carried out in our laboratory, frozen glassy solutions of Gd(III) chelates are being observed by X-band CW and pulsed EPR at liquid helium temperatures [54]. The powder pattern spectra are then simulated to obtain values for the ZFS parameters D and E. For GdDTPA, this procedure yields a value of $(2/3D^2 + 2E^2) = \Delta^2 = 7.6 \times 10^{19}(\text{rad/s})^2$, in excellent agreement with the values for the magnitude of the ZFS found by multi-frequency measurements of frequency shifts and linewidths in aqueous solutions at room temperature. What is interesting is that the apparent ZFS is relatively constant from 4 to 300 K, in solids, glassy frozen solutions, and in liquids. While more work is needed to better understand the true nature of the ZFS (i.e. Δ_s, Δ_t, Δ^2, D, E, and τ_v), it seems reasonable to conclude from all of the measurements that the same ZFS, or effects that scale as the ZFS, are being observed in all cases that we have so far studied.

An analysis of the magnetic-field dependence of the spin Hamiltonian and the relaxation matrix for several Gd(III) chelates in aqueous solutions has demonstrated that the ZFS parameters Δ^2 and τ_v can be obtained from these data. The multi-frequency EPR method avoids the difficulties posed by other methods that have previously been used to analyze the ZFS effects for metal ions in solution. In spite of the approximations used in this investigation, the values of the ZFS parameters for the Gd(III) chelates in aqueous solution reported here are in good agreement with other results in the literature. The simple, direct multi-frequency/field EPR methods reviewed in this chapter thus constitutes a useful new approach, and further development of this method is warranted.

ACKNOWLEDGMENTS

Partial support for this work was provided through grants from the National Institutes of Health (R01–GM42208, R.B.C; P41–RR01811 and RO1–RR01811, R.L.B; R01–GM25862 and R01–RR07216, J.H.F; F32–CA73156,

T.I.S.). The contents of this review are solely the responsibility of the authors and do not necessarily represent the official views of the National Cancer Institute, the National Institue of General Medical Sciences, or the National Center for Research Resources. Some facilities were provided by the Illinois EPR Research Center, an NIH-funded Resource Center. The Gd(III) complexes were provided by Dr B. Radüchel and colleagues, Schering, AG, Berlin, Germany, and Professor Peter Petillo, Professor Mark J. Nilges, and Mr Hoon Kang, of the University of Illinois at Urbana/Champaign, are thanked for valuable discussions.

REFERENCES

1. Hagen, W. R., *Coord. Chem. Rev.*, 1999, **192**, 209.
2. Liang, Z. C. and Freed, J. H., *J. Phys. Chem. B*, 1999, **103**, 6384.
3. Belford, R. L. and Clarkson, R. B. in *Techniques in Magnetic Resonance for Carbonaceous Solids*, R. Botto and Y. Sanada, (Eds), ACS Advances in Chemistry Series, Vol. 229, American Chemical Society, Washington, DC, 1993, pp. 107–138.
4. Hassan, A. K. Maniero, A. L. van Tol, H. Saylor, C. and Brunel, L. C., *Appl. Magn. Reson.*, 1999, **16**, 299.
5. Reuben, J., *J. Phys. Chem*, 1971, **75**, 3164.
6. Grinberg, O. Y., Dubinskii, A. A. and Lebedev, Ya. S., *Russ. Chem. Rev.* (*Engl.* Transl.), 1983, **52**, 850.
7. Powell, D. H., Merbach, A. E., González, G., Brücher, E., Micskei, K., Ottaviani, M. F., Köhler, K., von Zelewsky, A., Grinberg, O. Ya. and Lebedev, Ya. S., *Helv. Chim. Acta*, 1993, **76**, 2129.
8. Caravan, P., Tóth, E., Rockenbauer, A. and Merbach, A. E., *J. Am. Chem. Soc.*, 1999, **121**, 10403.
9. Borel, A., Tóth, E., Helm, L., Jánossy, A. and Merbach, A. E., 2000, *Phys Chem. Chem. Phys.*, **2**, 1311.
10. Smirnova, T. I., Smirnov, A. I., Belford, R. L. and Clarkson, R. B., *J. Am. Chem. Soc.*, 1998, **120**, 5060.
11. Clarkson, R. B., Smirnov, A. I., Smirnova, T. I., Kang, H., Belford, R. L., Earle, K. and Freed, J., *Mol. Phys.*, 1998, **95**, 1325.
12. Smirnova, T. I., Smirnov, A. I., Belford, R. L. and Clarkson, R. B., *Magn. Reson. Mater. Phys. Biol. Med.*, 1999, **8**, 214.
13. Abragam, A. and Bleaney, B., *Electron Paramagnetic Resonance of Transition Ions*, Oxford University Press, New York, 1970, pp. 335–341.
14. Odelius, M., Ribbing, C. and Kowalewski J., *J. Chem. Phys.*, 1996, **104**, 3181.
15. Bloembergen, N. and Morgan, L. O., *J. Chem. Phys.*, 1961, **34**, 842.
16. McLachlan, A. D., *Proc. R. Soc. London*, **A**, 1964, **280**, 271.
17. Hudson, A. and Luckhurst, G. R., *Mol. Phys.*, 1969, **16**, 395.
18. Hudson, A. and Lewis, J. W. E., *Trans. Faraday Soc.*, 1970, **66**, 1297.
19. Slichter, C. P., *Principles of Magnetic Resonance*; Springer, Berlin, 1980, pp. 167–174.
20. Sharp, R. R., *J. Chem. Phys.*, 1993, **98**, 2507.
21. Bertini, I., Galas, O., Luchinat, C. and Parigi, G., *J. Magn. Reson., Sera. A*, 1995, **113**, 151.
22. Bertini, I., Kowalewski, J., Luchinat, C., Nilsson, T. and Parigi, G., *J. Chem. Phys.*, 1999, **111**, 5795.

23. Abernathy, S. M. and Sharp, R. R., *J. Chem. Phys.*, 1997, **106**, 9032.
24. Fraenkel, G. K., *J. Chem. Phys.* 1965, **42**, 4275.
25. Poupko, R., Baram, A. and Luz, Z., *Mol. Phys.*, 1974, **27**, 1345.
26. Baram, A., Luz, Z. and Alexander, S., *J. Chem. Phys.*, 1973, **58**, 4558.
27. Freed, J. H., Bruno, G. V. and Polnaszek, C. F., *J. Phys. Chem.*, 1971, **75**, 3386.
28. Nilges, M. J., Smirnov, A. I., Clarkson, R. B. and Belford, R. L., *Appl. Magn. Reson.*, 1999, **16**, 167.
29. Wang, W., Belford, R. L., Clarkson, R. B., Davis, P. H., Forrer, J., Nilges, M. J., Timken, M. D., Walczak, T., Thurnauer, M. C., Norris, J. R., Morris, A. L. and Zhang, Y., *Appl. Magn. Res.*, 1994, **6**, 195.
30. Clarkson, R. B., Wang, W., Nilges, M. J., and Belford, R. L. In *Processing and Utilization of High-Sulfur Coal*, R. Markuszewski and T. D. Wheelock, (Eds)., Elsevier, Amsterdam, pp. 1990, 67–77.
31. Lebedev, Ya. S., in *Modern Pulsed and Continuous-Wave Electron Spin Resonance*, Kevan, L. and Bowman, M., (Eds)., Wiley, New York, 1990, pp. 365–403.
32. Lynch, W. B., Earle, K. A., and Freed, J. H. *Rev. Sci. Instrum.*, 1988, **59**, 1345.
33. Barnes, J. P., and Freed, J. H. *Rev. Sci. Instrum.*, 1997, **68**, 2838.
34. Atsarkin, V. A., Demidov, V. V., and Vasneva, G. A., *Phys. Rev. B*, 1995, **52**, 1290.
35. Aime, S., Anelli, P. L., Botta, M., Fedeli, F., Grandi, M., Paoli, P. and Uggeri, F., *Inorg. Chem.*, 1992, **31**, 2422.
36. Uggeri, F., Aime, S., Anelli, P. L., Botta, M., Brocchetta, M., de Haën, C., Ermondi, G., Grandi, M. and Paoli, P., *Inorg. Chem.*, 1995, **34**, 633.
37. Geraldes, C. F. G. C., Urbano, A. M., Alpoim, M. C., Sherry, A. D., Kuan, K.-T., Rajagopalan, R., Maton, F. and Muller, R. N., *Magn. Reson. Imagning*, 1995, **13**, 401.
38. Powell, D. H., Ni Dhubhghaill, O. M., Pubanz, D., Helm, L., Lebedev, Ya. S., Schlaepfer, W. and Merbach, A. E., *J. Am. Chem. Soc.*, 1996, **118**, 333.
39. Smirnova, T. I., Smirnov, A. I., Belford, R. L. and Clarkson, R. B., *Abstracts of 4th Scientific Meeting of the International Society of Magnetic Resonance in Medicine*, New York, April 27–May 3 1996, Society of Magnetic Resonance, Berkeley, CA, 1996, Vol. 3, Abstract 1378, p. 1377.
40. González, G., Powell, D. H., Tissières, V. and Merbach, A. E., *J. Phys. Chem.*, 1994, **98**, 53.
41. Banci, L., Bertini, I. and Luchinat, C., *Nuclear and Electron Relaxation: The Magnetic Nucleus–Unpaired Electron Coupling in Solution*, VCH, New York, 1991.
42. Lauffer, R. B., Vincent, A. C., Padmanabhan, S., Villringer, A., Saini, S., Elmaleh, D. and Brady, T. J., *Magn. Reson. Med.*, 1987, **4**, 582.
43. Caravan, P., Ellison, J., McMurry, T. J and Lauffer, R. B., *Chem. Rev.*, 1999, **99**, 2293.
44. Lauffer, R. B., *Chem. Rev.*, 1987, **87**, 901.
45. Subczynski, W. K., Renk, G. E., Crouch, R. K., Hyde, J. S. and Kusumi, A., *Biophys. J.*, 1992, **63**, 573.
46. Smirnov, A. I., Clarkson, R. B. and Belford R. L., *J. Magn. Res. B*, 1996, **111**, 149.
47. Kawamura, T., Matsunami, S. and Yonezawa, T., *Bull. Chem. Soc. Jpn*, 1967, **40**, 1111.
48. Earle, K. A., Moscicki, J. K., Ge, M., Budil, D. E. and Freed J. H., *Biophys. J.*, 1994, **66**, 1213.
49. Chen, J. W., Auteri, F. P., Budil, D. E., Belford, R. L. and Clarkson, R. B., *J. Phys. Chem.*, 1994, **98**, 13452.
50. Koenig, S. H. and Epstein, M., *J. Chem. Phys.*, 1975, **63**, 2279.
51. Powell, D. H., González, G., Tissières, V., Micskei, K., Brücher, E., Helm, L. and Merbach, A. E., *J. Alloys Compd.*, 1994, **207/208**, 20.

52. Koenig, S. H., in *Encyclopedia of Nuclear Magnetic Resonance*, Grant, D. M. snd Harris, R. K. (Eds), vol. 3, John Wiley &. Sons, Chichester, 1995, pp. 1819–1830.
53. Kowall, Th, Foglia, F., Helm, L. and Merbach, A. E., *J. Phys. Chem.*, 1995, **99**, 13078.
54. Kang, H., Belford, R. L., and Clarkson, R. B., *J. Phys. Chem.*, to be published.

10 Particulate Magnetic Contrast Agents

ROBERT N. MULLER, ALAIN ROCH, JEAN-MARIE COLET, ASSIA OUAKSSIM and PIERRE GILLIS
University of Mons-Hainaut, Mons, Belgium

Although paramagnetic liposomes and micelles belong to the general category of particulate materials and therefore present some common features with contrast agents made of crystalline cores of ferrite, for instance, regarding their biodistribution, they will not be covered in this present chapter which will be restricted to the latter because of their genuine magnetic properties.

1 INTRODUCTION

From the foregoing chapters, we know that among other parameters, the relaxivity of an ion is related to the square of its electronic moment, and hence to the square of its electronic spin, S. If a molecule contains N paramagnetic ions of the same kind, the relaxivity, expressed as usual with respect to the concentration of the ion, will not be very different from what could be expected for isolated spins located on the same carrier. If however, thanks to a *cooperativity*, the individual spins S build up a *superspin* $\boldsymbol{S} = NS$, then the relaxivity will increase as a function of N. This kind of cooperativity has not yet been achieved with multimetallic paramagnetic complexes under physiological conditions of solvent and temperature but is commonly observed in inorganic ferrites such as iron oxides (Fe_3O_4). Very small crystals of such materials are fully magnetized and exhibit large magnetic moments. A dispersion of these grains, smaller than a magnetic domain, does not show magnetic remanence. Such nanocrystals (with diameters of around 5–10 nm), known as superparamagnetic (SPM), are therefore ideal *cores* for the design of new and efficient contrast agents.

The field of applications of superparamagnetic colloids is wide, and encompasses mechanics as well as medicine where they can be used as magnetic resonance (MR) contrast media, [1], but also as therapeutic agents in cancer

The Chemistry of Contrast Agents in Medical Magnetic Resonance Imaging
Edited by A. E. Merbach and É. Tóth.

treatment. These nanosystems can indeed act as high-frequency electromagnetic wave absorbers, able to induce a localized hyperthermia leading to a selective destruction of tumorous cells [2]. The static magnetic properties of nanoparticles are also exploited in the production of magnetically suspended seals and magnetic inks [3].

2 SUPERPARAMAGNETIC MR CONTRAST AGENTS

2.1 STRUCTURE

A common misconception about superparamagnetic materials concerns the dimensions of the system. A *particle* is made of a core of one or more magnetic *grains* or *crystals* embedded in a coating which prevents agglomeration and possibly governs the targeting. We therefore have to distinguish, on the one hand, the size of the magnetic *core*, the *grains* or the *crystals*, and on the other hand, the global size of the *particle*. As discussed later in this chapter, the first dimension plays a major role with regard to the relaxivities, while the second one is of importance for the pharmacokinetics.

We will consider three kinds of particulate magnetic contrast agents, namely the ultrasmall particles (Ultra Small Particles of Iron Oxide (USPIO)), the small particles (Small Particles of Iron Oxide (SPIO)) and the large particles. While the overall size of the former two categories (Figure 10.1) remains well below the micrometer and allows intravenous administration, the large particles can have a diameter of several microns, thus limiting their use in exploration of the gastrointestinal track.

2.2 SYNTHESIS

Although various ferrites containing endogeneous metals can be synthesized, most of the work performed in the field of MR contrast agents has been devoted to iron oxide nanosystems. Their synthesis is rather simple and is based on the oxidative hydrolysis of an iron (II) salt in alkaline medium or on the precipitation from a mixture of iron (II)/iron (III) ions under basic conditions.

A more difficult question which is often based on the trial-and-error approach, is the choice of the experimental conditions leading to an homodisperse population of magnetic grains of suitable size.

The coating of the grains, necessary to prevent their agglomeration by sterical or electrical repulsion, can be achieved during the *one-pot* synthesis where the proper molecules are present (for example, dextran) or be performed by dispersion of the isolated *naked* grains in a medium containing the coating

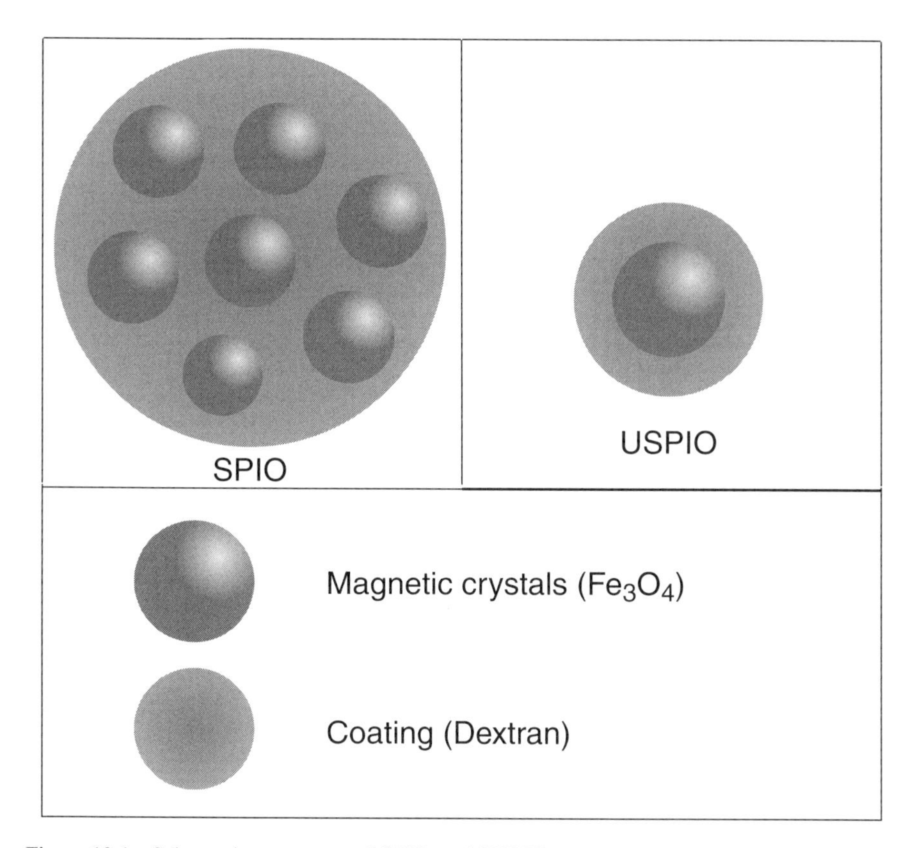

Figure 10.1 Schematic structures of SPIO and USPIO.

molecules. This can also be carried out by adequately substituting the steric barrier used in the initial step of the synthesis.

Finally, a purification phase is usually carried out in order to narrow the distribution size. If the usual strategies can quite easily address the separation according to the overall size of the particles (gel chromatography, centrifugation, etc.), a selection according to the size of the superparamagnetic grains is more difficult.

2.3 CHARACTERIZATION

The relaxivity and the biological behavior of superparamagnetic colloids depend on the size of the crystals, their chemical composition, their magnetic characteristics and the type of coating. Unlike paramagnetic chelates, which can be prepared at a very high level of purity and unambiguously characterized, superparamagnetic systems are rather complex, mainly because of their particulate nature and the intricate relationships linking their chemical, geometrical and magnetic characteristics to their properties as MR contrast agents. Consequently, a whole set of analytical techniques have to be applied to fully apprehend the systems and their efficacies:

- Transmission Electronic Microscopy (TEM), which is a method for direct measurement of the size of the magnetic core. This technique does not give information about the overall size of the particle. On the other hand, it can overestimate the extent of clustering of the grains in the colloidal state due to a possible aggregation taking place during the preparation of the sample for submitting for examination. To be statistically significant, the size determination requires measurements on a large number of crystals. TEM examination can also be coupled with an electron diffraction analysis which, like X-ray diffraction, gives useful information about the chemical composition of the material.
- Small-Angle X-ray Scattering (SAXS), which is used to measure the mean size of crystals in the aqueous suspension by irradiating the sample with an X-ray source [4]. The measurement is based on the analysis of the diffracted beam as a function of the diffusion angle.
- Photo-Correlation Spectroscopy (PCS), in which the brownian motion of the particles modulates the intensity of the light diffused by the suspension when irradiated by a laser beam. The mathematical analysis of this modulation gives a mean value of the hydrodynamic volume, and hence the global diameter of the particles [5].
- Magnetometry, where here the curves representing the magnetization of a sample as a function of the applied magnetic field can confirm the absence of magnetic hysteresis and therefore the superparamagnetic character of the material in suspension. Their interpretation according to the Langevin function provides valuable information about the specific magnetization and the mean diameter of the crystals.
- X-ray diffraction, which is a good way to determine the crystal composition and structure [6], in spite of some difficulties in distinguishing magnetite and maghemite.
- Mössbauer spectroscopy [7], which is an alternative technique for assessing crystal composition (for example, Fe_3O_4 vs γ-Fe_2O_3). This analytical method also gives unique information about the order of magnitude of the Néel relaxation time (τ_N), which is an important characteristic of superparamagnetic particles (see theoretical discussion below (Section 3.1)).
- As shown in previous chapters of this book and illustrated hereafter in the case of superparamagnetic systems, relaxivity profiles recorded over a wide range of magnetic fields (Nuclear Magnetic Relaxation Dispersion (NMRD)) contain a great deal of information about the magnetic and geometrical properties of the systems. Their fitting according to the proper and newly

developed theories (see below) indeed allows the determination of the mean crystal size, the specific magnetization and the Néel relaxation time [8].

It is important to point out that the presence of a size distribution complicates the analysis of the results and can for instance lead to discrepancies between the crystal diameter obtained by the various techniques previously mentioned. The values obtained under these conditions depend on weighting factors which are specific to the respective methodologies, and therefore their comparison and interpretation can become hazardous.

2.4 TYPICAL SUPERPARAMAGNETIC MR CONTRAST AGENTS

As previously mentioned, one can distinguish three different kinds of particulate contrast agents, i.e. the SPIO, the USPIO and the large particles or aggregates. We will give here a brief description of a few examples of those materials.

2.4.1 Ultrasmall Particles of Iron Oxide (USPIO):

- Sinerem® (Guerbet, Aulnay-sous-Bois, France), also called ferumoxtran or AMI-227 (Advanced Magnetics, Cambridge, MA, USA), is obtained by chromatographic size fractionation of AMI-25 (see below) [9]. TEM gives a crystal diameter of 4.3–4.9 nm and PCS measurements indicate a global diameter of ca 50 nm. The magnetization of ferumoxtran at 25 °C and 5 T is about 94.8 emu/gFe, while its relaxivities in 0.5 % agar at 20 MHz are 22.7 and 53.1 $(\mathrm{mM\ s})^{-1}$, respectively, for r_1 and r_2 [10].
- Other preparations of USPIO, as Clariscan® (Nycomed-Amersham, Oslo, Norway), are currently under development [11].

2.4.2 Small Particles of Iron Oxide (SPIO)

- Endorem® (Guerbet, Aulnay-sous-Bois, France), also called ferumoxide or AMI-25 (Advanced Magnetics, Cambridge, MA, USA), is prepared by the co-precipitation of ferric chloride and ferrous chloride in ammonia in the presence of dextran [12]. This contrast agent is composed of non-stoichiometric magnetite (Fe_3O_4) crystals with a radius of 4.3–4.8 nm (TEM). PCS measurements give an intensity-weighted particle diameter of ca 200 nm with a wide size distribution. Its magnetization at 25 °C and 5 T is about 93.6 emu/gFe, while its r_1 and r_2 relaxivities at 37 °C and 20 MHz are 24 and 107 $(\mathrm{mM\ s})^{-1}$, respectively [10].

- Resovist® (Schering AG, Berlin, Germany), also called SH U 555A, is a suspension of nanoparticles of magnetite and maghemite coated with carboxy-dextran. The magnetic part of the particle contains several single crystals, with each about 4.2 nm in diameter as measured by TEM. The overall particle diameter is around 62 nm as measured by PCS. The r_1 and r_2 relaxivities measured in plasma at 20 MHz and 37 °C are, respectively, 20 and 190 $(\mathrm{mM\ s})^{-1}$ [13, 14]. It is worth mentioning here that as long as no structure modification occurs (see below), the reported relaxivities are largely independent of the medium (water, saline, agar gel, plasma, etc.).

2.4.3 Large Particles

- Abdoscan®, also called Oral Magnetic Particles or OMP (Nycomed-Amersham, Oslo, Norway) [15, 16]. This contrast agent is composed of monodisperse polymer particles of a diameter of 3 μm coated with crystals of iron oxide (thickness of about 50 nm). As expected, these particles do not form a stable dispersion and can not be injected. They are used for MRI studies of the abdomen and the pelvis. Such sophisticated material seems, however, superfluous for those applications where only a bulk susceptibility effect is seeked.
- Lumirem® (Guerbet, Aulnay-sous-Bois, France) is another commercial formulation of 'drinkable' iron oxide crystals.

3 RELAXATION INDUCED BY SUPERPARAMAGNETIC PARTICLES

Relaxation induced by superparamagnetic agents can not be described by a unique theoretical approach because of the huge importance of the size and morphological properties of the particles. We will first describe the basic theory which is valid for USPIO, i.e. particles containing only one superparamagnetic crystal (diameter smaller than 15 nm). The basic assumption of this model is a homogeneous dispersion of identical and spherical ferrite crystals.

We will then consider the relaxation induced by SPIO samples, for which several superparamagnetic cores are distributed in the same particle (with a permeable coating). The basic theory describing USPIO must then be adapted to account for the agglomerated structure of the SPIO particles.

Finally, we will investigate the case of large magnetized spheres (compact aggregates of superparamagnetic crystals or diamagnetic microspheres covered with a superparamagnetic layer), where transverse relaxation can only be studied by computer simulations.

3.1 BASIC THEORY (USPIO)

Evaluating and understanding the performance of magnetic colloids used as contrast agents for MRI requires a theory describing the magnetic interactions of superparamagnetic compounds with water protons. As already mentioned, NMRD profiles (i.e. the field dependence of the proton relaxation rate) provide a very powerful tool for testing the pertinence of the theory accounting for those interactions [17]. In addition to the light brought on the relaxation mechanisms, NMRD profiles, which are *fingerprints* of the samples, are also extremely valuable for quality control of the preparations.

Applying the classical outer-sphere relaxation theory to this problem is very appealing. This theory provides longitudinal (R_1) and transverse (R_2) relaxation rates of the water protons which are diffusing nearby the unpaired electrons responsible for the particle magnetization [18]. The magnetic moments typical of superparamagnetic (SPM) particles are much larger than the electron moments, but, at first glance, are not qualitatively different. Curie relaxation [19] plays, however, a central role in the reformulation of this basic theory and leads us to consider two different time-scales in the problem. We have indeed to examine separately, on one hand, the effect of water diffusion through the field inhomogeneities created by the time-averaged value of the particle magnetic moment (whose dependence on the external static field is governed by a Langevin function), and on the other hand, the effect of the fluctuations of the magnetic moment itself [20]. This model accounts quite well for the high-field part of the NMRD profiles ($B_0 > 0.02$ T), but fails to explain a low-field feature which is specific of the USPIO. The NMRD profiles of USPIO show indeed a slight dispersion at low field (below 1 MHz), a feature which completely disappears for SPIO (Figures 10.2 and 10.3). Simulations according to the original outer-sphere theory largely overestimate this low-field dispersion.

This unexpected feature can be understood by considering the crystal anisotropy energy, which reflects the qualitative difference between super paramagnetic and paramagnetic compounds. Indeed, outer-sphere theory assumes a perfectly isotropic environment for the unpaired electrons, a highly questionable assumption for SPM particles since an anisotropy field exists within such particles and forces the magnetic moment of the particle to align along the so-called 'axes of easy magnetization'. When the anisotropy energy is large compared to the thermal energy, it can even lock the moment on one of the 'easy directions', thus possibly preventing any precession around the external field. In this case, the magnetic fluctuations then arise from the jumps of the moment between the easy directions, a process characterized by the Néel correlation time, τ_N.

This high anisotropy limit is met by this seminal and rather phenomenological relaxation model [20]: the contributions responsible for the low-field dispersion, arising from the matching of the electron precession and the electron moment fluctuations created by the Néel relaxation, vanish because of the

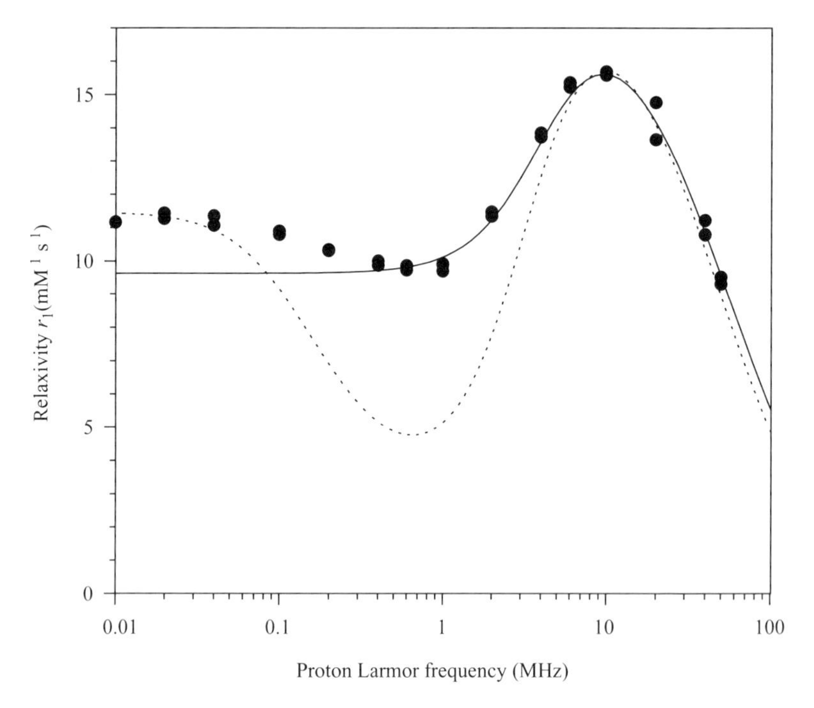

Figure 10.2 Longitudinal NMRD profile of an aqueous suspension of USPIO. The two fits of the experimental data (•) are respectively obtained through successive reformulations of the classical outer-sphere theory: (a) by assuming a high anisotropy energy of the crystal [20] (solid line); (b) by neglecting this energy [24] (dotted profile). Reproduced by permission of the American Institute of Physics from [8].

immobilization of the magnetization along the easy magnetization axes. Within this high-anisotropy limit, the relaxation rates are given by the following general expressions:

$$R_1 = C\left[3J_A\left(\sqrt{2\omega_I\tau_D}\right) < \mu_z >^2 +(10/3)J_F(\omega_I, \tau_D, \tau_N)\Delta\mu_z^2\right] \qquad (10.1)$$

$$R_2 = C\left\{\left[(3/2)J_A\left(\sqrt{2\omega_I\tau_D}\right) + 2J_A(0)\right] < \mu_z >^2 + \left[(4/3)J_F(\omega_I, \tau_D, \tau_N) + 2J_F(0, \tau_D, \tau_N)\right]\Delta\mu_z^2\right\} \qquad (10.2)$$

$$\text{where } C = (32\pi/135\ 000)\gamma_I^2 N_A[M]\tau_D/r^3$$

In these expressions, ω_I is the proton Larmor pulsation (s^{-1}), τ_N is the Néel relaxation time, $\tau_D(r^2/D)$ is the translational correlation time (s), r is the radius

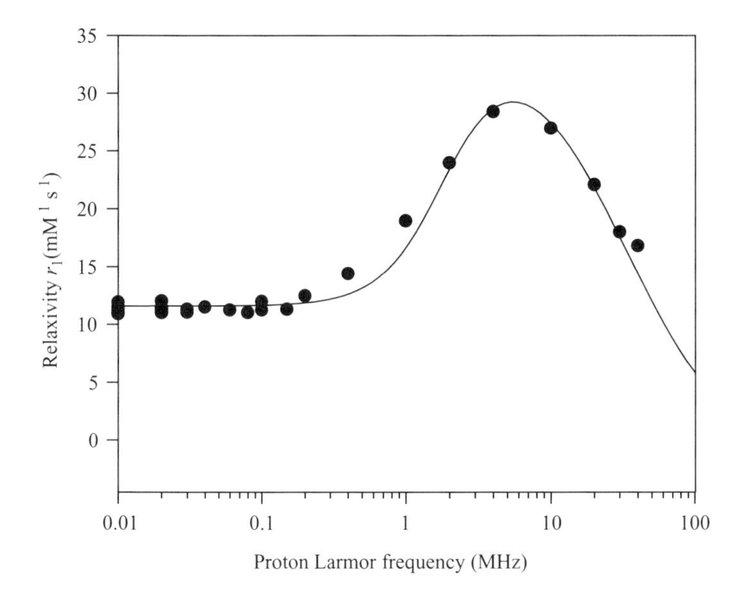

Figure 10.3 Longitudinal NMRD profile of an aqueous suspension of SPIO. The fit of the experimental data is obtained through the reformulation of the classical outer-sphere theory by assuming a high anisotropy energy of the crystal. Reproduced by permission of the American Institute of Physics from [8].

of the crystal, D is the water self-diffusion coefficient, N_A is the Avogadro number, [M] is the concentration in moles per liter of the *crystals*, μ is the magnetic moment of the particles and γ_I is the proton gyromagnetic ratio.
NB: the relaxivities can be calculated from Equations (10.1) and (10.2) by scaling R_1 and R_2 to the proper concentration units (for instance, in mmol of iron per liter).

Equations (10.1) and (10.2) contain two spectral density functions. One arises from the proton diffusion in the non-uniform magnetic field created by the mean value of the SPM particle moment [21], as follows:

$$J_A(z) = \frac{1 + \frac{5z}{8} + \frac{z^2}{8}}{1 + z + \frac{z^2}{2} + \frac{z^3}{6} + \frac{4z^4}{81} + \frac{z^5}{81} + \frac{z^6}{648}} \tag{10.3}$$

while the other, namely:

$$J_F(\omega_I, \tau_D, \tau_N) = Re\left(\frac{1 + \frac{1}{4}(\Omega)^{1/2}}{1 + (\Omega)^{1/2} + \frac{4}{9}(\Omega) + \frac{1}{9}(\Omega)^{3/2}}\right) \tag{10.4}$$

where $\Omega = i\omega_I\tau_D + \tau_D/\tau_N$, is the spectral density function accounting both for proton diffusion and for the fluctuation of the magnetic moment around its mean value [22]: J_A is the limit of J_F when the correlation time characterizing the magnetization fluctuation becomes infinitely long. Distinguishing the magnetization mean value (independent of the electronic fluctuations) from its fluctuating part, i.e. the principle of Curie relaxation as established by Gueron [19], leads to the splitting up of $< \mu_z^2 >$ into $(< \mu_z >^2 + \Delta\mu_z^2)$ each time it appears in the equations, and to the association of J_A to $< \mu_z >^2$ and J_F to $\Delta\mu_z^2$. The terms proportional to $< \mu_z >^2$ define the Curie relaxation, and dominate the other ones at high field, while the fluctuating terms dominate at low field.

In Equations (10.1) and (10.2), $< \mu_z >$ is given by the Langevin function, as follows:

$$< \mu_z >= \mu L(x) = \mu[\coth(x) - 1/x] \tag{10.5}$$

with $x = \mu B_0/(kT)$, while the fluctuating part of the moment was shown to be given by the following [23]:

$$\Delta\mu_z^2 = \mu^2[1 - 2L(x)/x - L^2(x)] \tag{10.6}$$

Equations (10.1) and (10.2), which predict no dispersion at low field, i.e. a field-independent relaxivity until the contribution brought by the Langevin function (first term of both equations) increases with B_0, adequately match the relaxivities for SPIO, but, as mentioned above, fail to describe the low-field part for USPIO (see Figures 10.2 and 10.3), where experiments show a weak low-field dispersion.

Anisotropy energy is known to be particle-size-dependent, more precisely proportional to the particle volume. Now, Equations (10.1) and (10.2) arise from an assumption of rigorous locking of the magnetization along the axes of easy magnetization, a situation corresponding to an infinite anisotropy energy. This assumption becomes less and less valid when the particle size decreases, thus releasing somewhat the magnetization locking, and allowing for precession. Accounting for such a reduced coupling with the anisotropy field requires a new theory [8], aimed at introducing the anisotropy energy as a quantitative parameter of the problem, in order to go beyond the two limits considered so far: Equations (10.1) and (10.2), where this energy has been assumed to be infinite, and the classical outer-sphere theory, adapted to high-susceptibility materials [24], where it is neglected.

Taking into account the anisotropy energy implies that we need to substitute for the unique electron transition frequency (characteristic of the Zeeman interaction) a wide distribution of transition frequencies, generally peaking at values close to the highest frequency of the distribution.

The exchange energy within superparamagnetic crystals is large, so one can consider the ensemble of all individual electronic spins as co-operating to create one large superspin $\boldsymbol{S}$, submitted to $\boldsymbol{B}_0$, the external magnetic field, and to the anisotropy field. The Hamiltonian may therefore be written as follows:

$$H = -h\nu_A S/2\ (\mathbf{1}_S \mathbf{1}_A)^2 - \gamma_S \hbar \mathbf{B}_0 \mathbf{S} \tag{10.7}$$

where $\mathbf{S}$ is the crystal superspin, $\mathbf{1}_S$ is an unitary vector aligned along $\mathbf{S}$, $\mathbf{1}_A$ is an unitary vector pointing in the direction of the anisotropy axis (unique in this simplified picture), and γ_S is the electron gyromagnetic ratio; ν_a is the electron precession frequency at the anisotropy field ($h\nu_A = 2E_A/S$), and $h = 2\pi\hbar$ is the Planck constant.

A quantitative evaluation of the relaxivities as functions of the magnetic field, $\mathbf{B}_0$, requires extensive numerical calculations due to the presence of two different axes (the anisotropy and the external field axes), resulting in non-null off-diagonal elements in the Hamiltonian matrix; furthermore, the anisotropy energy has to be included in the thermal equilibrium density matrix. Figures 10.4 and 10.5 show the attenuation of the low-field dispersion of the calculated NMRD profiles when either the crystal size or the anisotropy field increases.

Theory clearly predicts a difference between the low-field relaxation profiles of small particles (weak dispersion) and larger particles (no dispersion) (see Figures 10.2 and 10.3). Experimental confirmation of this theoretical approach was provided by the NMRD curves of suspensions of colloidal magnetite doped with cobalt, an element which considerably enhances the energy of anisotropy [25]. For these small particles, the low-field dispersion disappears even at low cobalt content.

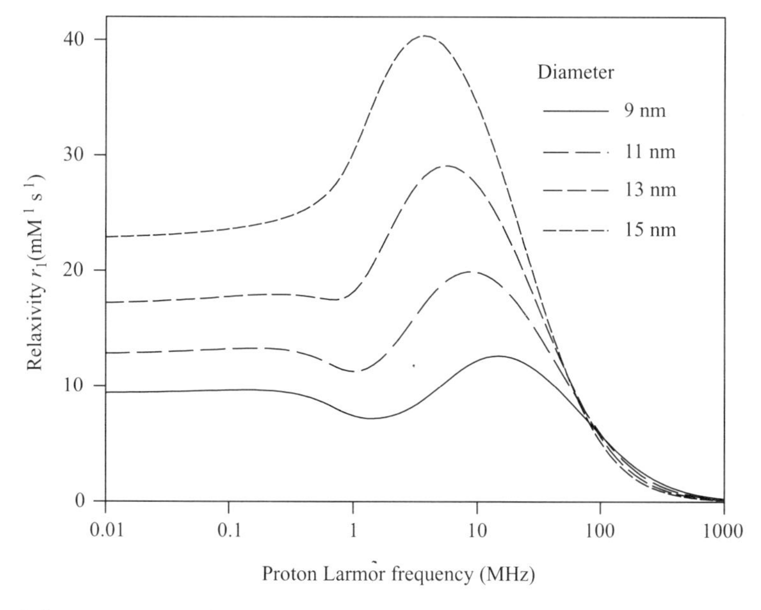

Figure 10.4 Dependence of the longitudinal NMRD profile on the magnetic crystal diameter; $M_{sat} = 2.07 \times 10^5$ A/m; $D = 3.4 \times 10^{-9}$ m^2/s; $\nu_A = 1$ GHz at $T = 310$ K. Reproduced by permission of the American Institute of Physics from [8].

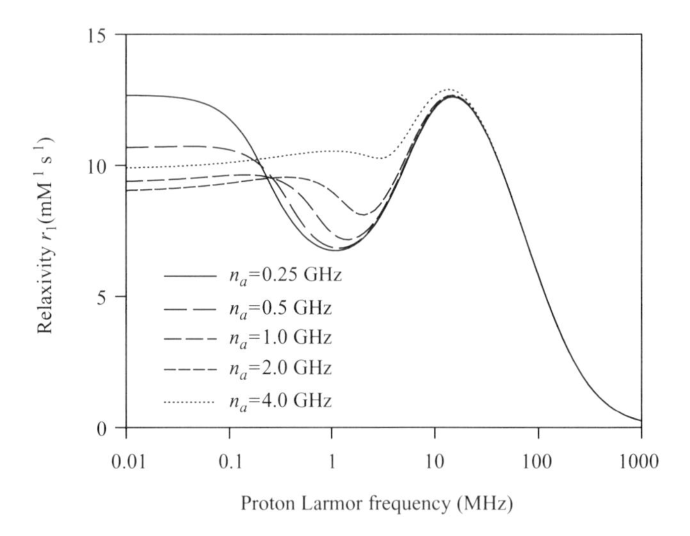

Figure 10.5 Dependence of the longitudinal NMRD profile on the anisotropy energy: diameter = 4.5nm; $M_{sat} = 2.07 \times 10^5$ A/m, $D = 3.4 \times 10^{-9}$ m^2/s; $T = 310$ K. Reproduced by permission of the American Institute of Physics from [8].

Theoretical investigation and experimental measurements converge thus to demonstrate that the anisotropy energy of the superparamagnetic colloids determines their low-field relaxivity. It has to be stressed that the low-field dispersion, predicted by the standard outer-sphere theory, is severely attenuated for small crystals, and completely disappears for crystals of larger size or of high anisotropy energy.

This conclusion is very similar to the conclusion from studies of low-field paramagnetic relaxation, where the zero-field splitting dominates the Zeeman interaction, so that the electron spin–lattice relaxation processes do not influence the paramagnetic enhancement of the proton spin relaxation profile. Finally, it has to be mentioned that, in the high-field range, which is relevant for MRI, the transverse relaxivity is significantly higher than the longitudinal one.

3.2 WEAKLY AGGLOMERATED SYSTEM (SPIO)

Theoretical models [8, 20, 25] describing the field dependence of the relaxation of superparamagnetic colloids are based on a simplifying assumption concerning the crystal distribution: elementary grains, namely spherical ferrite crystals, are supposed to be homogeneously distributed (USPIO).

A clustering can, however, occur (SPIO), resulting in important consequences to r_1 and on r_2 [26, 27]. Two types of effects arise from the aggregation of magnetic grains: on the one hand, effects related to the global structure of the cluster and to the magnetic field distribution around it, and on the other hand, the effects attributable to the inner structure of the aggregate. While the latter ones affect r_1, the former ones predominantly affect r_2 and r_2^* (the transverse relaxivity, measured without refocusing the spin dephasing induced by microscopic magnetic field inhomogeneities).

The cluster itself may be considered as being a large magnetized sphere whose total magnetic moment increases according to Langevin's law. Such a picture accounts for the increase of R_2 observed at high field [26]: the transverse relaxivity of water protons diffusing within the inhomogeneous magnetic field created by the cluster is dominated by an important secular contribution.

On the contrary, the global magnetization of the cluster weakly affects the longitudinal relaxivity at high field since the dispersion of the NMRD profile is known to shift towards lower frequencies when the size of the magnetic core increases, as it is the case for a cluster [27]. Complementarily, the global magnetic moment of the aggregate is too weak at low field (Langevin) to significantly contribute to the longitudinal relaxivity r_1. Longitudinal relaxation thus remains essentially described by the theory developed for USPIO, while for SPIO lacking the low-field dispersion (see Figure 10.3), it has to be considered in the limit of very high anisotropy. Indeed, although the particles are made of grains with the same size as USPIO, their anisotropy is enhanced by the dipolar coupling between the nearby crystals.

In the range of magnetic fields used for MRI, a clustering tends to decrease r_1 and to increase r_2, with the latter effect being the more pronounced.

3.3 MICROSPHERES COVERED WITH FERROMAGNETIC MATERIAL AND STRONGLY AGGLOMERATED PARTICLES

As are all of the relaxation theories, the one presented above is valid as long as the Redfield condition is satisfied. In the case of susceptibility-induced relaxation, this condition may be written as $\Delta\omega\tau_D \leq 1$, where $\Delta\omega$ is the equatorial magnetic field at the surface of the particle (expressed in angular frequency units), and $\tau_D = r^2/D$ is the time required for a water molecule to diffuse over a distance r, the radius of the particle. The Redfield condition thus defines the distinction between 'small-size' and 'large-size' domains.

For bulk magnetite, $\Delta\omega = 3.4 \times 10^7\ \mathrm{s}^{-1}$, so that the boundary corresponds to a radius of about 10 nm. As expected, small-dimension particles induce longitudinal as well as transverse relaxation of water protons through monoexponential evolutions. The observed R_2 and R_1 values are in remarkable agreement with the predictions of the previous paragraph [8].

In the case of large magnetized spheres, the transverse and longitudinal magnetizations exhibit an obvious multiexponential behavior [28]. In addition, the transverse magnetization decay strongly depends on the echo time.

So far, no theoretical approach predicting the longitudinal relaxation typical of such large particles exists. The transverse relaxation is dominated by the secular term and several numerical procedures have successfully predicted this contribution, especially for water diffusing through magnetic-field inhomogeneities [29, 30]. These numerical computations are based on simulations of random walks of water molecules diffusing through local magnetic fields induced by the presence of impenetrable magnetized spheres. The secular contribution is calculated from the phase evolution of the isochromats in the x–y plane and the magnetization decay is obtained by averaging over a large number of isochromats; 180° pulses are simulated by an instantaneous change of the phase sign. Figure 10.6 shows the evolution of R_2 with the size of the particles for systems containing a constant amount of magnetized material. Several cases are presented in the figure i.e. without a 180° refocusing pulse, a situation thus simulating the proton magnetization behavior in a gradient-echo sequence, and seven spin-echo conditions, ranging from short (0.2 ms) to long (10 ms) echo times (TEs) [31].

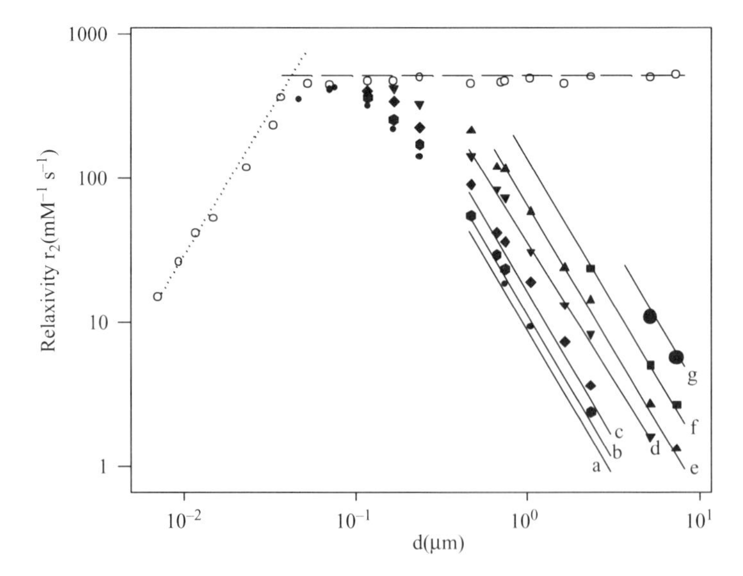

Figure 10.6 Computer-generated data of transverse relaxivity plotted versus the particle radius: volume fraction of spherical Fe_3O_4 particles, 5×10^{-6}; $\Delta\omega = 2.36 \times 10^7$ rad/s; $D = 3.4 \times 10^{-9}$ m^2/s. Gradient echo and CPMG sequences: CPMG TE = 0.2 ms (•, and line (a)); 0.4 ms (⬢ , and line (b)); 1 ms (◆, and line (c)); 4 ms (▼, and line (d)); 10 ms (▲, and line (e)); 20 ms (■, and line (f)); 40 ms (⬤, and line (g)). The dotted line is the rate predicted by outer-sphere theory, while the dashed line represents the static-dephasing-regime (SDR) rate [32]; The empty symbols represent r_2 values [31].

In all of these situations, the relaxation rates are equivalent, provided that the time required for a water molecule to jump from one particle to the next one is much shorter than the TE.

This feature can be easily understood: a 180° refocusing pulse eliminates the reversible decay due to static-field inhomogeneities and will be effective as long as each water molecule has not encountered many magnetized particles between the pulses. If this is not the case, the random spatial distribution of the particles induces an irreversible loss of phase coherence proportional to the defocusing period, i.e. the echo time. At the scale of the molecular diffusion, each water molecule can then be seen to be moving in a quasi-static magnetic field gradient.

Finally, the transverse relaxation rate calculated without spin refocusing (R_2^*), reaches a plateau for diameters above 0.2 μm. The corresponding maximum relaxation rate of about 160 s^{-1} is predicted by the Static Dephasing Regime model [32], which assumes that, as depicted above, transverse relaxation is only attributed to differences in precession frequencies of static spins.

In conclusion, while the dispersed or weakly agglomerated crystals act as R_1 and R_2 enhancers, an extensive aggregation switches their effect towards a major R_2 or R_2^* effect.

Finally, it is worth mentioning that the R_2^* effect produced around a blood vessel containing magnetic particles is governed by the bulk magnetization of the vascular compartment (which is of a large size as compared to the water diffusion) and therefore does not depend on the aggregation status of the grains.

4 BIODISTRIBUTION AND CLEARANCE ASPECTS

The efficacy and the biodistribution of particulate contrast agents are mainly influenced by the administration route, plus the size and the nature of the coating. The size of the nanoparticles seems to be the most critical parameter of the biodistribution [33]. Intravenously administered, the particles with an overall diameter larger than 20 μm are trapped in the pulmonary alveolar capillaries. These kinds of particles are essentially used in nuclear medicine. In order to avoid entrapment in the lungs, and probably also possible side-effects, the size of the magnetic particles designed for MRI examination is kept in the submicron range.

One of the main and original goals of developing superparamagnetic particles was the targeting of the monocyte phagocytic system (MPS) which plays a major role in the clearance of particulate materials circulating in the blood stream. The liver and the spleen together possess more than 80 % of the cells forming the MPS, with the remainder being located in the bone marrow and in the lymph nodes.

The fixed macrophages of the liver (Kupffer cells) and of the spleen efficiently capture large particles with a size ranging from ca 50 to 500 nm [34], while the macrophages circulating in the blood stream or fixed in the bone marrow internalize smaller particles (10–100 nm) [35], which easily cross the fenestrated endothelial wall of sinusoids.

According to these different distributions governed by the size, two classes of nanoparticles can be distinguished. The largest superparamagnetic particles, symbolized by SPIO (Small Particles of Iron Oxide), such as the microspheres of magnetite coated with albumin (1 to 5 μm), silan (1 to 2 μm), starch (300 to 500 nm) or dextran (40 to 150 nm), essentially accumulate in the liver and spleen [36]. These particles are quickly cleared from the blood with a vascular half-life shorter than 15 min. Depending on the coating, the nanoparticles might interact with the blood components. For example, Magnetic Starch Microspheres bind to albumin and then get opsonized by the immunoglobulin IgG [37]. These starch-covered SPIO have been shown to interact with receptors located at the surface of the Kupffer cells [38]. This interaction induces the endocytosis of the nanoparticles and their sequestration in phagolysosomes. The administered dose does not influence the biodistribution to any extent since for doses ranging from 6 to 23 mg (0.1–0.4 mmol) of iron by kilogram of body weight, similar distributions have been reported with 60 to 80 % in the liver and 5 to 7 % in the spleen [39].

The clearance of these particles in the liver varies from 2 to 30 days depending on their size [40]. Pharmacokinetic studies using radioactive nanoparticles have elucidated the metabolism of the particulate iron in the liver and in the spleen and its recycling in the pool of the physiological iron. No sub-acute toxicity related to this iron overload has been reported in animal studies, even at the highest dosage of 250 mg per kilogram of body weight [41]. In humans, the clinically available SPIO are well tolerated after intravenous injection, although a few cases of back pain have been reported with Endorem® [42]. These SPIO (Endorem® and Resovist®) are mainly taken up by the Kupffer cells (> 70 %) and the crystalline core is destroyed in less than seven days by the lysosomes.

The sequestration of the SPIO in sub-cellular vacuoles dramatically modifies their magnetic properties. In the isolated rat liver perfused with magnetic starch microspheres, it has been demonstrated that the internalization of the nanoparticles in Kupffer cells induces a rapid and biphasic evolution of the transverse relaxation rate of the water protons [43]. A very fast decaying component is observed, together with an increase of the relaxation rate of the remaining water tissue. This regime is strongly dependent on the echo time and on the iron concentration, a behavior which is characteristic of the agglomeration of magnetic particles (see Section 3).

For this type of application, the overall size and the coating of the particles which govern the biodistribution are much more important than the geometry of the magnetic cores, since the intracellular uptake and the subsequent agglomeration convert them in T_2^* enhancers regardless of their initial core size.

Due to their strong T_2^* effect, the internalized particles strongly reduce the intensity of the MR signal. Hence, after their incorporation into the MPS, a strong darkening of the liver, the spleen, the bone marrow and the lymph nodes, is observed while very little change is noticed in the pancreas, kidneys, fat and muscles, where the latter do not accumulate the contrast media. Similarly, the detection of liver and spleen pathologies is facilitated because of a reduction or an absence of phagocytic activity. Focal lesions and metastasis of the liver, as well as the spleen, are extremely well visualized after the intravenous administration of SPIO.

We know that when they are not extensively agglomerated, SPM materials are both T_1 and T_2 relaxers. Therefore, when the local concentration of superparamagnetic particles is low enough to minimize the T_2 effect, a brightening of the image due to a T_1 effect can be achieved with the use of T_1-weighted sequences. This can be observed in vessels or in highly vascularized tissues such as hemangiomas [44].

In order to extend the vascular phase, particles smaller than 30 nm (USPIO) have been developed. After intravenous administration, these particles distribute in the vascular space and diffuse across the endothelial wall of the sinusoids, thus eventually reaching the lymph nodes and the bone marrow.

More subtle targeting can be achieved by adequately coating the magnetic cores. This strategy has been nicely illustrated by USPIO coated with arabinogalactan, a polysaccharide bearing terminal galactosyl groups and selectively interacting with the hepatocytes receptors [45]. Due to their small size (ca 10 nm), these particles escape the recognition by fixed macrophages (i.e. Kupffer cells), and reach the space of Disse by crossing the endothelial sinusoidal wall. Then, they interact with receptors for galactosyl residues located on the membrane of the hepatocytes and are subsequently internalized. In contrast to the SPIO which, in the liver, are exclusively taken up by the Kupffer cells, these USPIO are selectively internalized in the hepatocytes. In addition to the anatomical information that they bring, such USPIO targeted to the hepatobiliary system are likely to provide information on the hepatic function.

Intravenous injection is not the only administration route for contrast agents. Oral administration is, of course, obvious for examination of the digestive system [16]. The particles are orally administered as a dispersion in a flavored inert carrier substance which increases the viscosity of the suspension, thus leading to a reduction of material deposition in the intestinal lumen. The particles cross the gastro-intestinal track without any degradation or absorption. Such agents attenuate the signal of the stomach, the duodenum and the small intestine, and allow a better delineation of the surrounding tissues and organs. However, a possible artefact arises from a susceptibility effect which is induced by excessive concentrations of magnetic material.

REFERENCES

1. Muller, R. N. in *Encyclopedia of Nuclear Magnetic Resonance*, Grant, D. M. and Harris, R. K. (Eds), John Wiley & Sons, Chichester, 1995, pp. 1438–44.
2. Hilger, I., Andrä W., Bähring R., Daum A., Hergt R. and Kaiser W. A. *Invest. Radiol.* 1997, **32**, 705.
3. Nakatsuka, K. J. *Magn. Magn. Mater.* 1993, **122**, 387.
4. Ciccariello, S., Benetti, A. and Pollizzi, S. *J. Appl. Phys.* 1991, **69**, 6355.
5. De Jaeger, N. , Demeye, H. , Finsy, R. , Sneyer, R. , Vanderdeelen, J. , van der Meeren, P. and van Laethem, M. *Part. Part. Syst. Charact.* 1991, **8**, 179.
6. Guinier, A. *Théorie et Technique de la Radiocristallographie*, Dunod, Paris, 1964, pp. 462–482.
7. Wertheim, G. K. *Mössbauer Effect: Principles and Applications*, Academic Press, New York, 1964, pp. 239–296.
8. Roch, A. , Muller, R. N. and Gillis, P. *J. Chem. Phys.* 1999, **110**, 5403.
9. Weissleder, R. , Elizondo, G. , Wintterberg, J. , Rabito, C. A. , Bengele, H. H. and Josephson, L. *Radiology* 1990, **175**, 489.
10. Jung, C. W. and Jacobs, P. *Magn. Reson. Imaging* 1995, **13**, 661.
11. Kellar, K. E. , Fuji, D. K. , Guther, W. H. H. , Briley-Saebo, K. , Spiller, M. , Bjornerud, A. and Koenig, S. H. *J. Magn. Reson. Imaging*, 2000, **11**, 488.
12. Groman, E. V. , Josephson, L. and Lewis, J. M. *US Pat. 4827945*, 1989.
13. Reimer, P. , Müller, M. , Marx, C. , Weidermann, D. , Muller, R. N. , Rummeny, E. J. , Ebert, W. , Shamsi, K. and Peters, P. E. *Radiology* 1998, **209**, 831.
14. Bremer, C. , Alkemper, T. , Baermig, J. and Reimer, P. *J. Magn. Reson. Imaging* 1999, **10**, 461.
15. Jacobson, T. and Klaveness, J. PCT Int. Appl. WO 00017, 1985.
16. Rinck, P. A. , Smevik, O. , Nilsen, G. , Klepp, O. , Onsrud, M. , Øksendal A. and Borseth, A. *Radiology* 1991, **178**, 775.
17. Bulte, J. W. M. , Brooks, R. A. , Moskowitz, B. M. , Bryant Jr, L. H. and Frank, J. A. *Magn. Reson. Med.* 1999, **42**, 379.
18. Gillis, P. and Koenig, S. H. *Magn. Reson. Med.* 1987, **5**, 323.
19. Gueron, M. *J. Magn. Reson.* 1975, **19**, 58.
20. Roch, A. and Muller R. N. *Proceedings of the 11th Annual Meeting of the Society of Magnetic Resonance in Medicine*, F. Wehrli (Ed.), Works in Progress 1447, August 8–14, 1992.
21. Ayant, Y. , Belorizky, E. , Alizon J. and Gallice, J. *J. Phys.* 1975, **36**, 991.
22. Freed, J. H. *J. Chem. Phys.* 1978, **68**, 4034.
23. Gillis, P. , Roch, A. and Brooks, R. *J. Magn. Reson.* 1999, **137**, 402.
24. Koenig, S. H. and Kellar, K. *Magn. Reson. Med.* 1995, **34**, 227.
25. Roch, A. , Gillis, P. , Ouakssim, A. and Muller, R. N. *J. Magn. Magn. Mater.* 1999, **201**, 77.
26. Moiny, F. , Gillis, P. , Roch, A. and Muller, R. N. in *Proceedings of the 11th Annual Meeting of the Society of Magnetic Resonance in Medicine* F. Wehrli (Ed.), Works in Progress 1431, August 8–14, 1992.
27. Roch, A. , Gillis, P. and Muller, R. N. in *Proceedings of the 3rd Scientific Meeting of the Society of Magnetic Resonance in Medicine*, F. Wehrli (Ed.), 1992, August 19–25, 1995.
28. Roch, A. , Bach-Gansmo, T. and Muller R. N. *MAGMA* 1993, **1**, 83.
29. Hardy, P. A. and Henkelman, R. M. *Magn. Reson. Imaging* 1989, **7**, 265.
30. Muller, R. N. , Gillis, P. , Moiny, F. and Roch, A. *Magn. Reson. Med.* 1991, **22**, 178.
31. Moiny, F. *Ph. D. Thesis*, University of Mons-Hainaut, Belgium, 1994.
32. Brown, R. J. S. *Phys. Rev.* 1961, **121**, 1379.

33. Weissleder, R. and Reimer, P. *Eur. Radiol.* 1993, **3**, 198.
34. Bach-Gansmo, T., Fahlvik A. K., Ericsson, A. and Hemmingsson, A. *Invest. Radiol.* 1994, **29**, 339.
35. Weissleder, R., Elizondo, G., Wittenberg, J., Lee, A. S., Josephson, L. and Brady, T. J. *Radiology* 1990, **175**, 494.
36. Oksendal, A. N. and Hals, P. A. *J. Magn. Reson. Imaging.* 1993, **3**, 157.
37. Colet, J. M. and Muller; R. N. *MAGMA* 1994, **2**, 303.
38. Colet, J. M., Van Haverbeke, Y. and Muller, R. N. *Invest. Radiol.* 1994, **29**, S223.
39. Majumdar, S., Zoghbi, S. S. and Gore, J. C. *Invest. Radiol.* 1990, **25**, 771.
40. Weissleder, R. and Papisov, M. *Rev. of Magn. Reson. Imaging* 1992, **4**, 1.
41. Stark, D. D. in *Contrast Media in Magnetic Resonance Imaging, International Workshop*, Bydder, G. (Ed.), Bossum Medicom, Berlin, 1990, 281–93.
42. Clément, O., Siauve, N., Lewin, M., de Kerviler, E., Cuénod, C. A. and Frija, G. *Biomed. Pharmacother.* 1998, **52**, 51.
43. Colet, J. M., Piérart, C., Seghi, F., Gabric, I. and Muller, R. N. *J. Magn. Reson.* 1998, **134**, 199.
44. Chambon, C., Clément, O., Le Blanche, A., Schouman-Clayes, E. and Frija, G. *Magn. Reson. Imaging.* 1993, **11**, 509.
45. Reimer, P., Weissleder, R., Lee, A. S., Wittenberg, J. and Brady, T. J. *Radiology* 1990, **177**, 729.

11 Photophysical Aspects of Lanthanide(III) Complexes

JAMES I. BRUCE, MARK P. LOWE and DAVID PARKER
University of Durham, Durham, UK

The luminescent ions Eu^{3+} and Tb^{3+} flank Gd^{3+} in the lanthanide series and studies of their photophysical properties can throw light on the solution behavior of the corresponding Gd species. Measurements of the radiative rate constants for decay of the luminescent Eu or Tb excited states in H_2O and D_2O can be used to assess the complex hydration states, for example, while analysis of high-resolution europium absorption and luminescence spectra aid the determination of complex speciation, structure and symmetry. Therefore, when examining the behavior of a new gadolinium complex, it is both prudent and informative to also prepare the analogous Tb and Eu complexes and to study their photophysical and NMR spectral properties in detail.

1 ELECTRONIC CONFIGURATIONS AND SPECTROSCOPIC TRANSITIONS

For a given lanthanide(III) ion, the energy levels may be calculated by taking into account the interactions between the 4f electrons. The $4f^6$ and $4f^8$ configurations of Eu(III) and Tb(III) species give rise to a large number of states whose relative energy is determined by a combination of interelectronic repulsion, spin–orbit coupling and the ligand field. Interelectronic repulsion splits the conjugate $4f^6$ and $4f^8$ configurations into ${}^{(2S+1)}\Gamma$ spectroscopic terms (119 of them), where Γ = S, P, D and F when the quantum number $L = 0, 1, 2$ or 3. This electrostatic interaction yields terms which are separated by about 10^4 cm^{-1}. Spin–orbit coupling then splits these terms into J states which are typically separated by 10^3 cm^{-1}. There are 295 ${}^{(2S+1)}\Gamma_J$ spectroscopic levels for the f^6 and f^8 configurations whose relative energies are predicted by Hund's rules, as follows:

- the lowest energy term has the highest spin multiplicity $(2S_{max} + 1)$;
- for terms with the same spin multiplicity, that of the higher degeneracy (L_{max}) is the lowest in energy;

The Chemistry of Contrast Agents in Medical Magnetic Resonance Imaging
Edited by A. E. Merbach and É. Tóth.

- the level which is lowest in energy has the smallest *J*-value, if the sub-shell is less than half-full, and the highest *J*-value otherwise.

Thus, for both Eu(III) and Tb(III) ions, the ^7F term is lowest in energy and the ground-state levels are $^7F_6$ for Tb(III) and $^7F_0$ for Eu(III). In principle, the relatively large spin–orbit coupling observed for lanthanides should require application of an intermediate coupling scheme. In practice, however, the simpler Russell–Saunders coupling scheme provides a reasonable approximation [1], and its use is vindicated by the observation that the spectroscopically measured separation between adjacent J levels follows the Landé interval rule quite well, i.e. the separation between levels J and $(J + 1)$ is proportional to J (Figure 11.1).

As a consequence of their relatively small radial extension, the 4f orbitals are effectively shielded from the environment by the $5s^2 5p^6$ arrangement and are only minimally involved in bonding. This has two important implications for lanthanide emission and absorption spectroscopy. First, ligand-field splittings are very small, i.e. typically between 100 and 250 cm^{-1} [2]. Hence the $^7F_1$ level for the Eu^{3+} ion lies about 250 cm^{-1} above $^7F_0$. At ambient temperature it is 13 % thermally populated ($kT \sim 208$ cm^{-1} at 298 K), and this accounts for the magnetic moment of the aqua Eu^{3+} ion (3.61 μ at 298 K). Thus, when an f–f transition occurs between spectroscopic terms with the same $4f^n$ configuration, the spectral bands which arise are very sharp and are similar to the bands observed with the free ions. The second point concerns the probability of these transitions. Electric dipole transitions are forbidden by the parity (Laporte) selection rule, which disallows transitions among orbitals having the same symmetry towards an inversion centre. The interaction with the ligand field, or with its vibrational states, mixes electronic states of different parity into the 4f wavefunction, thus giving rise to forced (or induced) electric dipole transitions [3]. However, the weakness of this interaction means that only small contributions are added to the 4f wavefunction and the oscillator strength remains very low and in the range 10^{-7} to 10^{-5}. Therefore, the absorbance spectra of Ln(III) ions are characterized by very low extinction coefficients, with ε being typically less than 1 dm^3 mol^{-1} cm^{-1}. Although magnetic-dipole-allowed transitions are permitted by the parity rule, their oscillator strengths are extremely low (10^{-9} to 10^{-7}). The intensity of the weak magnetic dipole transitions is rather insensitive to the Ln(III) ionic environment and may conveniently be used as an internal standard when measuring emission (or absorption) intensities. A good example is provided by the $^5D_0 \leftrightarrow {}^7F_1$ transition in Eu(III) complexes.

Selection rules govern the various electronic transitions [4]. In the case of magnetic-dipole-allowed transitions these are as follows: $\Delta S = 0$, $\Delta L = 0$ and $\Delta J = 0, \pm 1$ (with $J = 0 \rightarrow 0$ forbidden). In addition, a transition is only allowed if it conforms to the group theoretical rules appropriate to the site symmetry of the Ln(III) ion. More detailed discussions may be found elsewhere

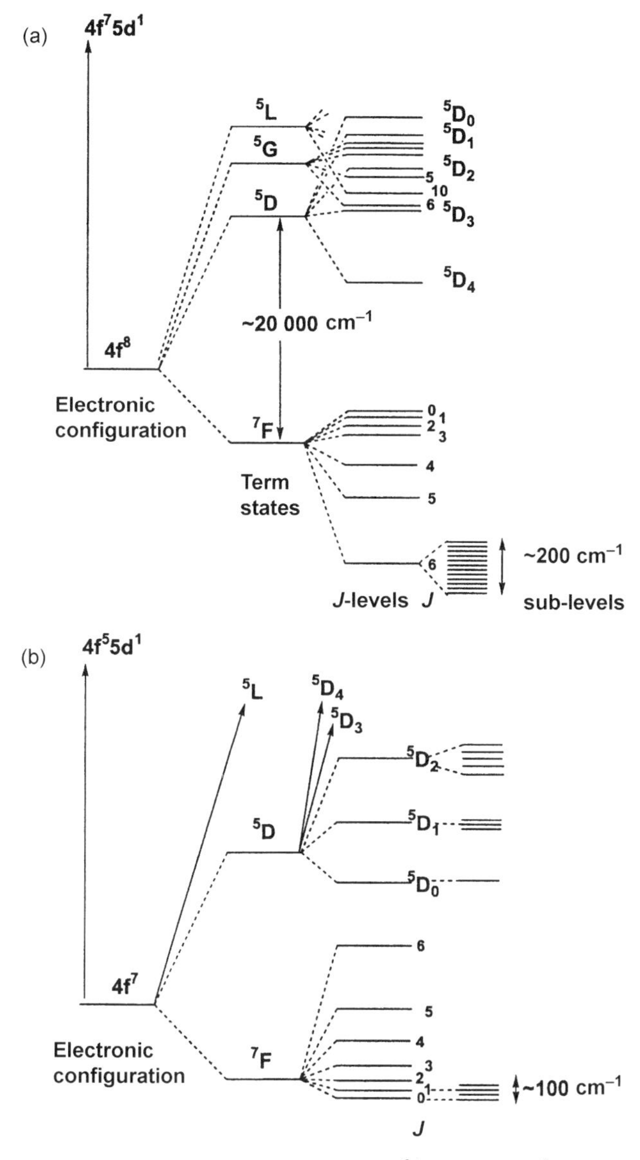

Figure 11.1 Schematic energy level diagrams for Tb^{3+} (a) and Eu^{3+} (b) ions, giving the relative size of interelectronic repulsion, spin–orbit coupling and ligand-field effects. The major (minor) luminescent excited states are $^5D_0$ ($^5D_1$) for Eu^{3+} and $^5D_4$ ($^5D_3$) for Tb^{3+}.

[4]. A simple example illustrates the importance of symmetry: for the $^5D_0$–$^7F_1$ transition in Eu(III) complexes (around 590–600 nm), only two transitions are observed if the complex is axially symmetric and contains a C_3 or C_4 axis. Three transitions are observed for complexes of lower symmetry. Moreover, the

magnitude of the splitting of the two peaks in axially symmetric complexes is directly related to the size of the dipolar NMR shift, as both are determined by the crystal field coefficient A_0^2 [5–7] (see Chapter 8).

1.1 DIRECT AND INDIRECT PHOTOEXCITATION

A direct consequence of the low molar absorption coefficients of the lanthanide ions is that absorption spectra may only be recorded on concentrated solutions, typically > 0.1 M. A further outcome is that excited states are not efficiently populated by continuous sources of radiation, and laser excitation is often required for high-resolution luminescence studies. The high intensity of laser excitation and the intrinsic selectivity – with bandwidths typically < 0.03 nm– coupled with improvements in photon-counting techniques, compensate for this disadvantage. Tunable dye lasers allow continuous selective excitation of the $^5D_0$ (Eu) and $^5D_4$ (Tb) levels: using the dye Rhodamine 6G, the maximum power output is at 580 nm, ideal for excitation of Eu(III), while the Tb(III) ion may be excited to the $^5D_4$ state by emission from the dye Coumarin 102, or alternatively by using the 488 nm line of the argon laser (Figure 11.2). Other transitions can also be excited selectively, for example $^7F_1 \rightarrow {}^5D_2$ and $^7F_0 \rightarrow {}^5D_2$ (Eu) by using the weaker lines of the argon ion laser at 472.7 and 465.8 nm [8, 9]. More recently, the advent of fairly powerful light-emitting diodes (365, 430 and 490 nm) offers scope for further development.

The Eu^{3+} ion is relatively easily reduced in aqueous solution ($E_{1/2} = -0.35$ V for the Eu aqua ion, compared to -1.15 V for $[Eu(DOTA)]^-$), so that in the absorption spectrum of Eu complexes a broad ligand-to-metal charge-transfer band may be observed in the mid ultraviolet region. This is not the case for the corresponding Tb^{3+} complexes, which are much more difficult to reduce and a charge-transfer band (often metal-to-ligand charge transfer (MLCT)) is only observed below 250 nm. Therefore, for certain Eu(III) complexes, (e.g. of DTPA and DOTA) it is possible to excite the europium via the ultraviolet LMCT band ($\varepsilon \sim 50$ at 280 nm). A more efficient method relies upon sensitized emission, requiring the presence of a suitable chromophore (or antenna) which can absorb light at a convenient wavelength and transfer its excitation energy to the proximate metal ion. If the antenna has a relatively high extinction coefficient and the energy transfer process is efficient, then the 'effective' molar absorption coefficient of the bound Ln^{3+} ion is greatly increased and intense luminescence results from excitation with conventional light sources [10]. A simple aromatic chromophore is particularly suitable for Tb(III) complexes, so that in its complexes with BOPTA (**1**) or EOB-DTPA (**2**), indirect excitation at 255 or 280 nm is followed by efficient energy transfer via the aromatic triplet excited state [11] to the terbium $^5D_4$ level. Intramolecular energy transfer is believed to involve the triplet excited state of the chromophore which therefore must possess a triplet energy level above those of the emissive $^5D_0$ (Eu, 17 240

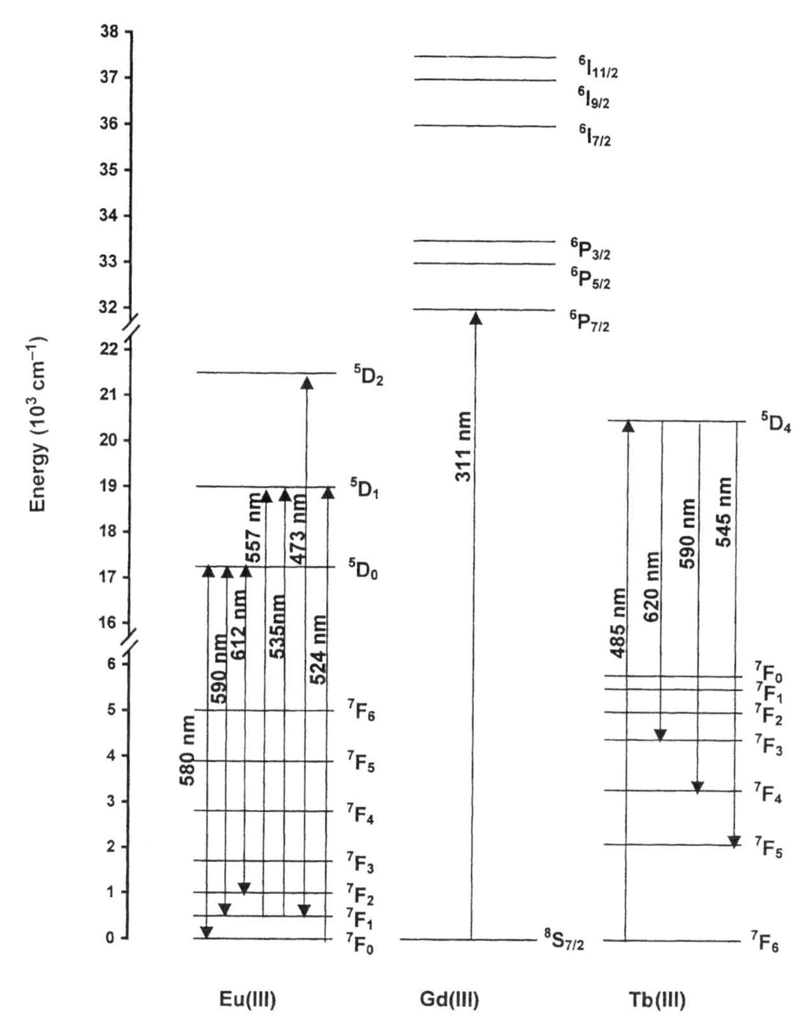

Figure 11.2 Selected absorption and emission transitions for complexes of Eu(III), Gd(III) and Tb(III).

cm^{-1}) and $^5D_4$ (Tb, 20 400 cm^{-1}) levels. Much weaker emission spectra are obtained in the Eu complexes of **1** and **2**, following excitation of the aryl chromophore. In these cases, the intermediate aryl singlet excited state is efficiently quenched by rapid electron transfer to the Eu^{3+} ion [12, 13]; such a process is inhibited for aryl chromophores which are less easily oxidized, e.g. those bearing CO_2R or CF_3 substituents or which are positively charged [14, 15].

Gadolinium complexes of **1** and **2** may not be excited via such a sensitization mechanism. The Gd^{3+} ion possesses a $^8S_{7/2}$ ground state and its first excited state ($^6P_{7/2}$) lies some 32 150 cm^{-1} towards the higher energy levels, well above the aryl triplet excited states. The emission from this excited state occurs at 311 nm, but in complexes with **1** and **2** the aromatic groups *quench* the excited Gd(III) ion, thus leading to formation of the aryl triplet and enhanced ligand phosphorescence [16].

1.2 ABSORPTION SPECTROSCOPY

Despite the very low absorbance of the Eu(III) ion in solution ($\varepsilon < 0.2$ dm^3 mol^{-1} cm^{-1} at 394 and 375 nm), for highly soluble complexes, sufficiently concentrated samples may be obtained (> 0.1 M) to allow f–f absorption spectra to be obtained. An added impetus for such work is that the time-scale for electronic transitions is sub-picosecond, so that discernible species undergoing exchange processes may be separately identified. Particular attention has focused on studying the unique $\Delta J = 0$ transition in Eu(III) complexes, for which each distinct chemical species can give rise to a separate absorption line in the region 579–581 nm. For example, it has been suggested that the two transitions observed with $[Eu(EDTA)]^-$ are due to two complexes differing in the value of q ($q = 2$ and 3) [17, 18]. In related work, the coordination equilibrium between the species of apparent differing coordination number has been studied. The ligand DO3A (**3a**) has seven donors and the observed band positions in the 580 nm region were consistent with a dissociative exchange process, as follows:

$$H_2O + [Eu(DO3A)H_2O] \rightleftharpoons [Eu(DO3A)(H_2O)_2] \qquad (11.1)$$

Variable temperature and pressure measurements allowed the activation parameters to be obtained (e.g. $\Delta V^{\#} = -7.5\ (\pm 1.3)$ cm^3 mol^{-1} at 298 K) [19]. Such intensity measurements assume that the molar absorptivities for the two observed species are equal – which may not always be the case if the exchanging species possess different local symmetry and structures.

1.3 LUMINESCENCE FROM Eu^{3+} COMPLEXES

When the europium(III) ion is excited from the $^7F_0$ ground state to higher states, it undergoes radiationless decay to the long-lived $^5D_0$ state, from which virtually all luminescence to the ^7F_J manifold arises in solution. The strongest transitions occur between states with rather low J-values, so that detailed interpretation is possible without the need for analysis of the crystal-field splitting. Of particular interest are the $\Delta J = 0$ (580 nm), $\Delta J = 1$ ($\sim$ 590) and $\Delta J = 2$ (616 nm) transitions (Table 11.1). The $\Delta J = 2$ transition has an intensity which is particularly sensitive to the chemical environment: it is termed

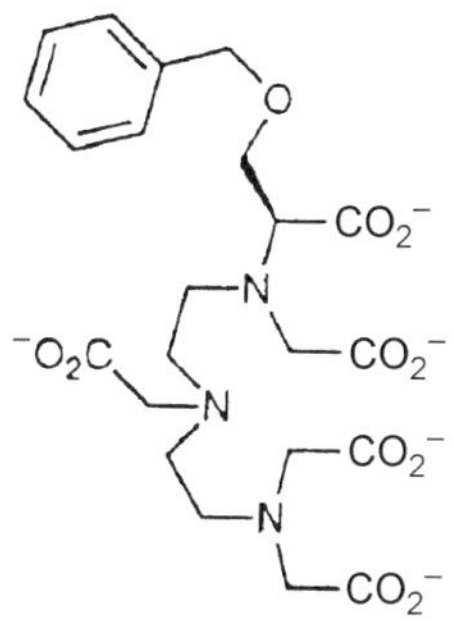

1

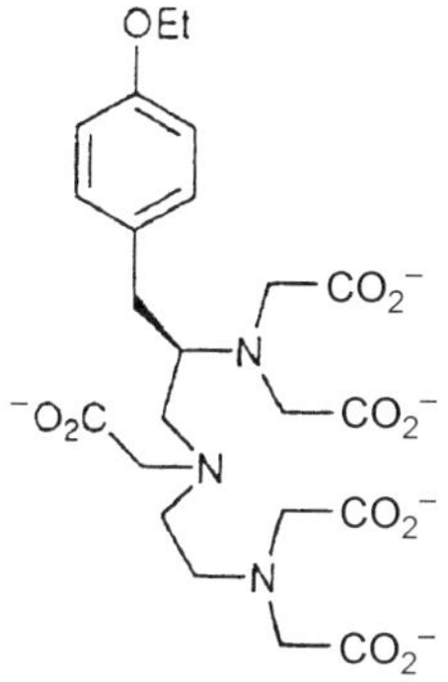

2

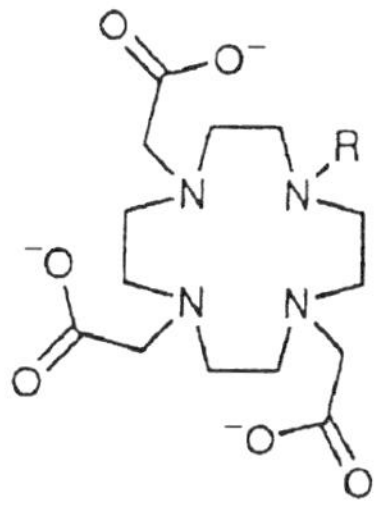

3a R = H
3b R = Me

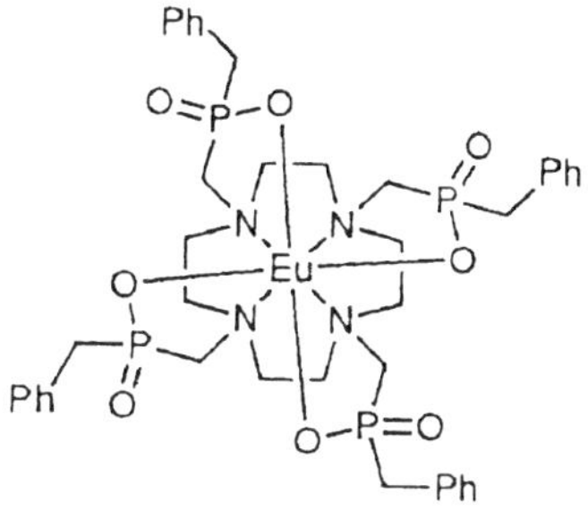

[Eu **4**]$^-$

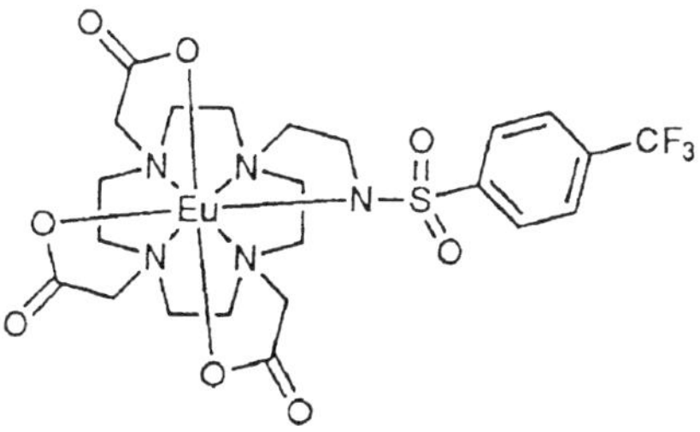

[Eu **5**]$^-$

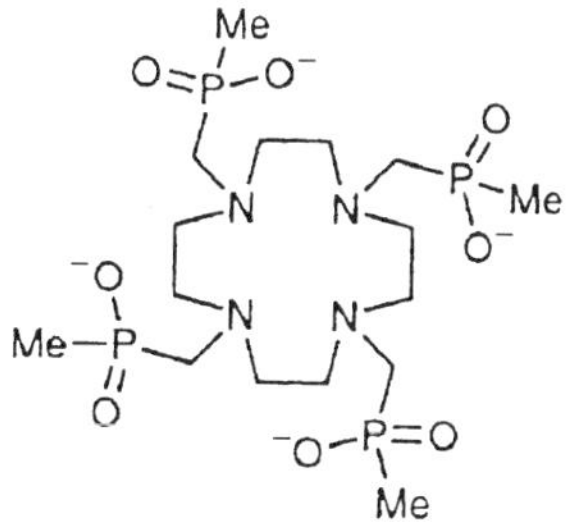

6

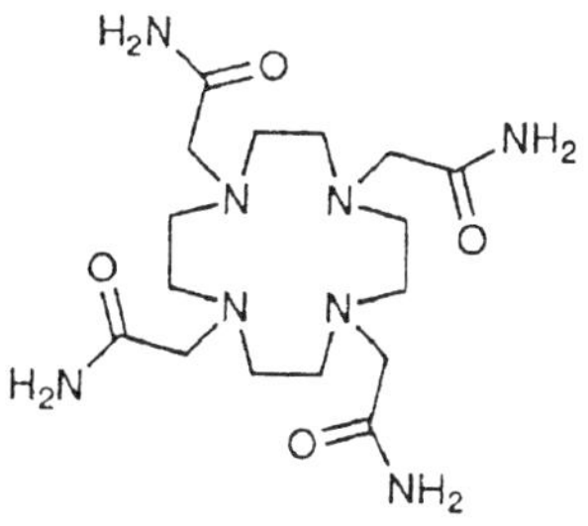

7

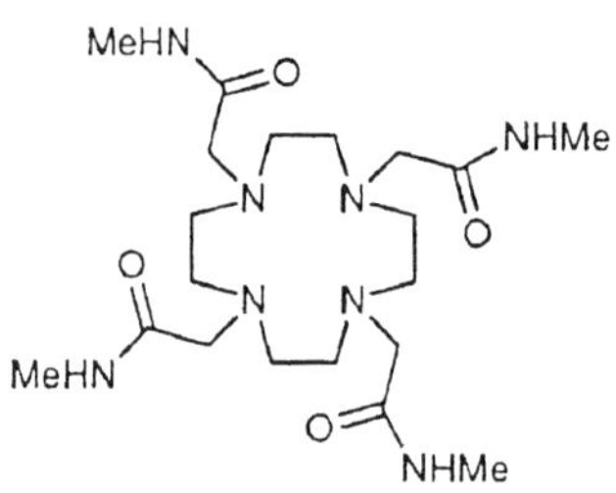

8

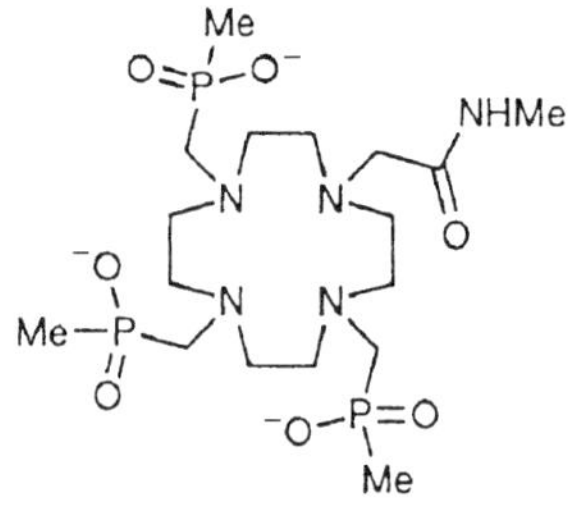

9

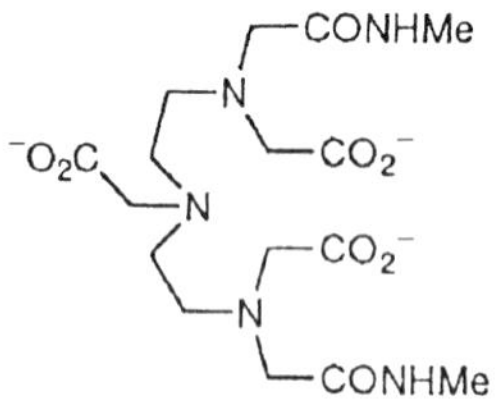

10

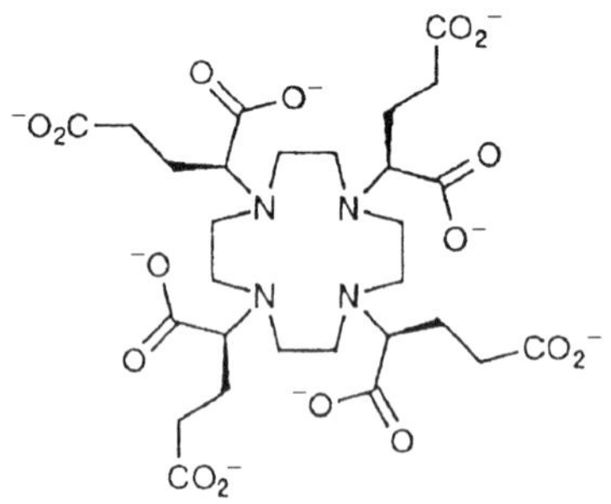

11

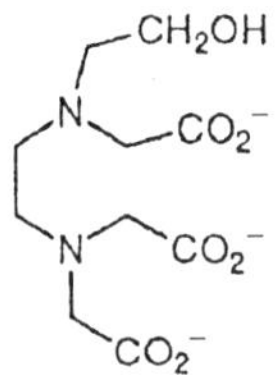

12

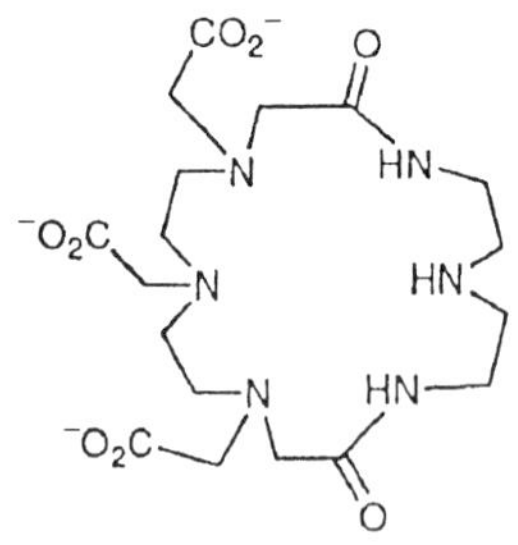

13

[**Chart 11.1**]

'hypersensitive', a property which is shared by the $\Delta J = 4$ transition near 700 nm. The origins of this hypersensitivity remain to be clarified, and were earlier related to the degree of covalency and dynamic coupling [20]. Analysis of more recent examples suggests that the intensity of this band is lower for axially symmetric complexes [21] (e.g. $[Eu(DOTA)]^-$ and the tetraphosphinate $[Eu\ \mathbf{4}]^-$; see Figures 11.3 and 11.4) and increases with the degree of polarizability of the donor atom type. Thus, in comparing the $\Delta J = 2$ transitions for $[Eu(DOTA)]^-$ and $[Eu(DTPA)]^{2-}$ (Figure 11.3), the lower intensity of the two allowed $\Delta J = 2$ bands in the C_4-symmetric former case may be contrasted with the higher intensity observed in the DTPA complex, which possesses lower symmetry.

Table 11.1 Characteristics of the $^5D_0$–^7F_J transitions for europium(III) complexes in solution[a].

λ range (*nm*)	*J*	Intensity	Comments
577–581	0	Weak and variable	Single forbidden transition; intensity increases in C_n and C_{nv} symmetry by *J*-mixing but is reduced for m[b] versus M[b] DOTA-like isomers
585–600	1	Strong	Magnetic-dipole allowed; three transitions in low symmetry, with two for C_3 or C_4; intensity almost independent of environment; strong CPL[c]
611–624	2	Strong/very strong	Hypersensitive; increases in intensity as symmetry lowers and as donor atom type is changed (polarizability?)
640–655	3	Very weak	Forbidden and always weak
680–710	4	Medium/strong	Sensitive to Eu coordination geometry and donor atom type/polarizability

[a] Transitions to the $J = 5$ and $J = 6$ states are very weak; transitions from the higher energy $^5D_1$ level are also very weak, with the $\Delta J = 0$ band at 530–540 nm sensitive to Eu environment.
[b] M, major isomer; m, minor isomer.
[c] CPL, circular polarized luminescence (see [54] and [55] for examples).

The number of bands in the $\Delta J = 2$ manifold is strictly determined by the site symmetry: thus, two bands (A and E) are observed for the twisted square-antiprismatic and C_4-symmetric tetrabenzylphosphinate complex, [Eu **4**]$^-$ [22, 23], whereas at least three bands may be discerned for the less symmetric neutral N_5O_3-bound complex [Eu **5**]$^-$.

The magnetic-dipole-allowed transition, $^5D_0$–$^7F_1$, possesses an intensity which varies relatively little from one coordination environment to another [4], and there are three transitions allowed in low symmetry, but only two if the complex possesses a C_3 or C_4 axis. For [Eu (DOTA)]$^-$ (Figure 11.5), there are two major $\Delta J = 1$ bands, while the third (central) band relates to the presence of the minor (m) isomer (ratio m/M 1:4, where M is the major isomer). In the related C_4-symmetric complex possessing four α-substituents (in an *RRRR* configuration, see Figure 11.5), the m/M ratio is 4:1 and the central transition increases in intensity accordingly. Profound changes in the $\Delta J = 4$ manifold also characterize this difference between the m and M isomers.

The highest energy emission band is for the formally forbidden $\Delta J = 0$ transition, which gains intensity through *J*-mixing in C_n or C_{nv} symmetry. The initial and final states are non-degenerate so that in principle one component is expected for each chemically distinct Eu species [24–26]. Probing this transition requires high-resolution laser spectroscopy. The lower intensity of this transition which is generally seen in twisted square-antiprismatic complexes (e.g. [Eu **4**]$^-$, Figure 11.4) contrasts with the higher intensity seen with the

(a)

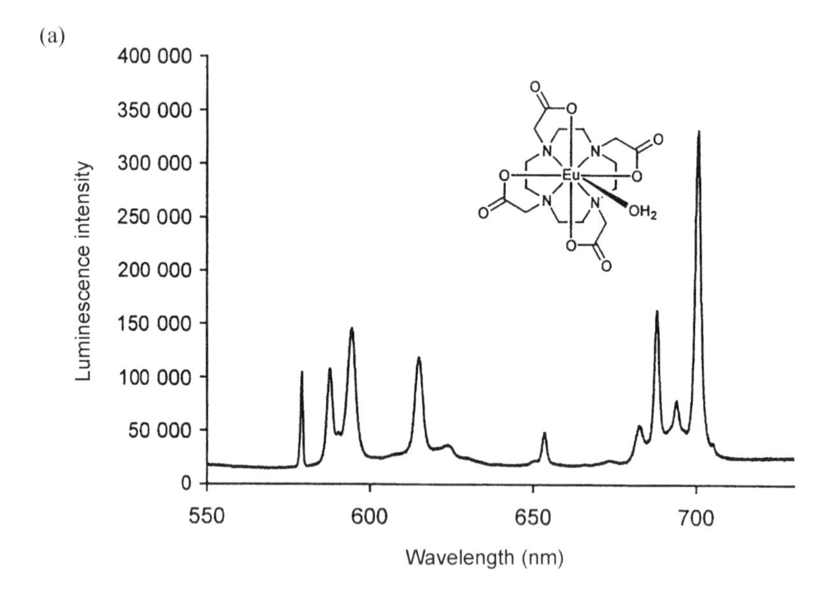

(b)

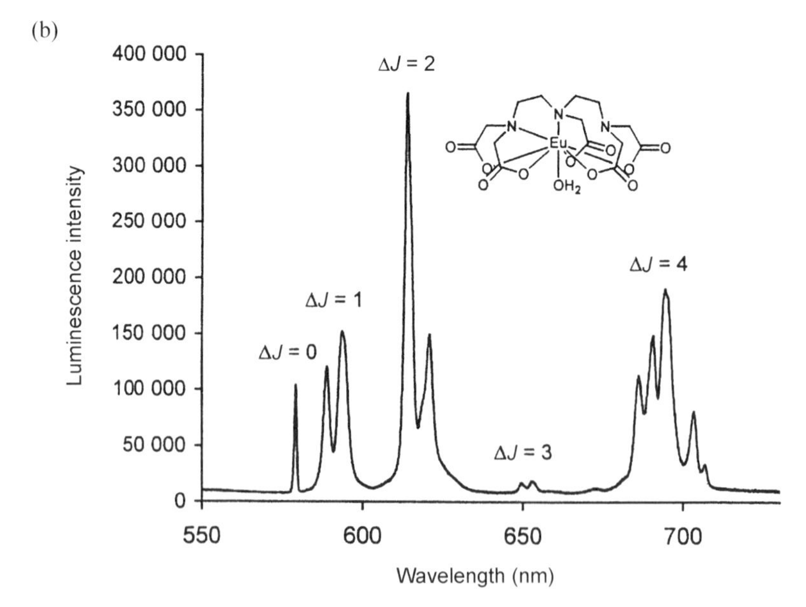

Figure 11.3 Europium luminescence emission spectra for $[Eu(DOTA)]^-$ (a) and $[Eu(DTPA)]^{2-}$ (b) (0.1 mM, 295 K, H_2O, $\lambda_{exc} = 397$ nm), showing the transitions from the $^5D_0$ state to the $^7F_0$ (580 nm), $^7F_1$ (590 nm), $^7F_2$ (618 nm), $^7F_3$ (655 nm) and $^7F_4$ (705 nm) states.

(a)

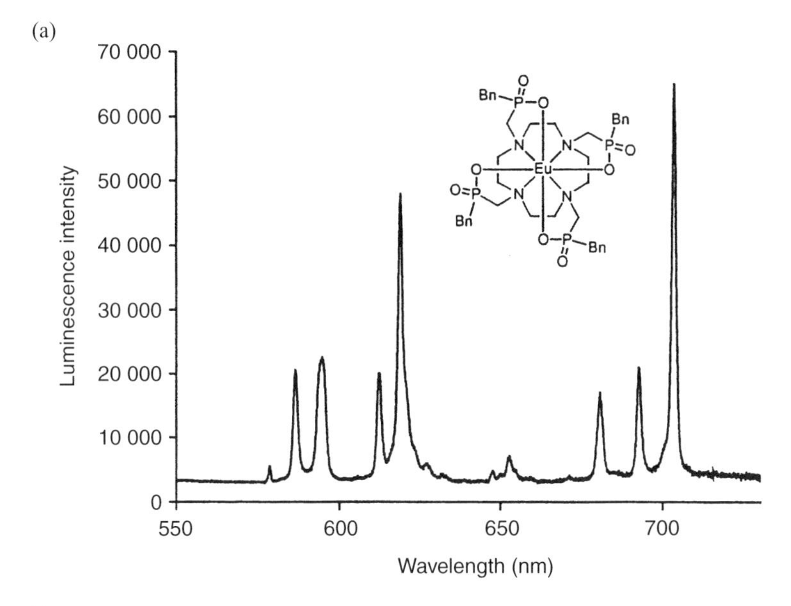

(b)

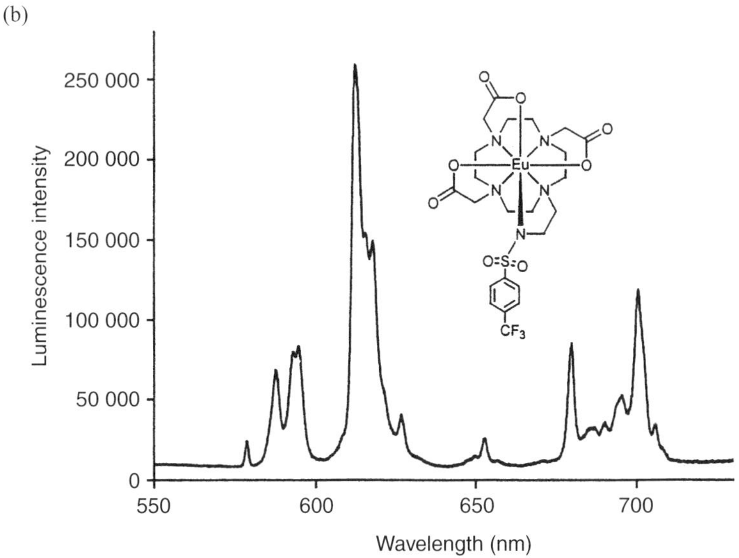

Figure 11.4 Europium luminescence emission spectra for [Eu **4**]$^-$(a) and [Eu **5**]$^-$(b) (293 K, H_2O, $\lambda_{exc} = 250$ and 270 nm respectively). Under higher resolution, the $\Delta J = 0$ band for [Eu **4**]$^-$ shows only one component (bandwidth < 0.3 nm).

(a)

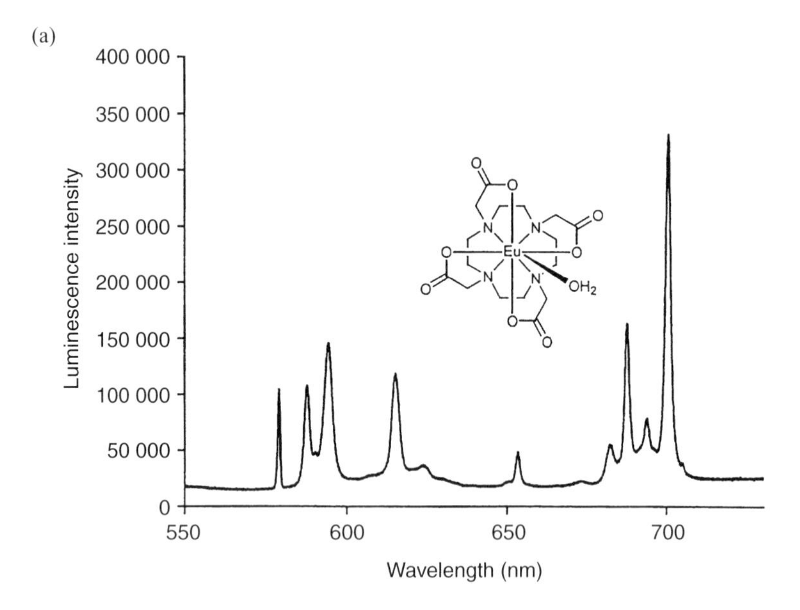

(b)

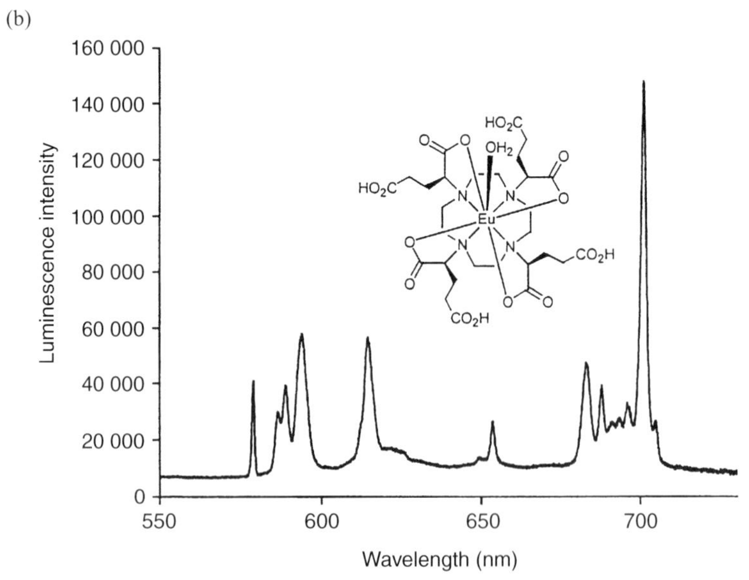

Figure 11.5 Europium luminescence emission spectra for $[Eu(DOTA)]^-$ (4:1 ratio of M/m isomers) (a) and the (*RRRR*)-tetra(carboxyethyl) derivative (1:4 ratio of M/m) (b), highlighting the smaller separation of the A and E transitions in the $\Delta J = 1$ manifold and the profound differences in the $\Delta J = 4$ transitions (295 K, H_2O, $\lambda_{exc} = 397$ nm, pH 5.5).

predominant M isomer for $[Eu(DOTA)]^-$, notwithstanding their similar C_4 symmetry. For this reason, care is needed in comparing relative peak intensities in this region, and even more so if the two species being compared have different numbers of coordinated water molecules, as the quantum yields for emission will differ as a consequence of the different degrees of excited state quenching by energy-matched OH oscillators (see Section 2).

1.4 LUMINESCENCE FROM Tb^{3+}, Gd^{3+} AND OTHER IONS

In emission, a consequence of the low transition probabilities for the Ln^{3+} ions is that they should have long excited-state lifetimes – of the order of milliseconds. Theoretical treatments of luminescence allow radiative lifetimes to be estimated by using the Einstein coefficient to express the rate of relaxation of an excited state to a given ground state [2, 20]. The energies of the emissive levels of selected Ln^{3+} ions are collated together with their *calculated* lifetimes in aqueous media in Table 11.2. In solution, experimental excited-state lifetimes are considerably lower than these values. This is usually a consequence of vibrational deactivation by energy-matched oscillators (especially OH, see Section 2). Thus, the excited Sm^{3+} and Dy^{3+} are efficiently deactivated by OH (and OD) oscillators and Nd^{3+}, Er^{3+} and Ho^{3+} possess many states of intermediate energy and are rapidly quenched by OH and CH oscillators. Only complexes of Eu^{3+}, Gd^{3+} and Tb^{3+} – with much larger energy gaps between the emissive state and the ground state, and without any intermediate energy levels – exhibit long-lived (> 0.1 ms) luminescence in aqueous media. The excited states of Gd^{3+} complexes are not susceptible to OH or OD quenching and might appear ideal as luminescence probes. However, the large energy gap means that emission is in the ultraviolet (311 nm) and excitation must be achieved at shorter wavelengths than this (e.g. 273–275 nm for direct excitation), thus precluding sensitization via triplet energy transfer from aryl chromophores. The absence of spectral fine-structure, coupled with the possibility of energy transfer *from* the excited Gd^{3+} $^6P_{7/2}$ to high-lying aryl triplet states, adds to the disadvantages.

On the other hand, terbium complexes give rise to long-lived and generally intense emission spectra following direct or indirect excitation. The multiplicity $(2J + 1)$ of energy levels in the lower states of the ground state manifold gives

Table 11.2 Calculated radiative lifetimes of the emissive excited states of selected aqueous Ln^{3+} ions [2, 20].

Lanthanide	Nd $^4F_{3/2}$	Sm $^4G_{5/2}$	Eu $^5D_0$	Gd $^6P_{7/2}$	Tb $^5D_4$	Dy $^4F_{9/2}$	Ho $^5S_2$
Energy (cm^{-1})	11 460	17 900	17 277	32 200	20 500	21 100	18 500
Calculated radiative lifetime (ms)	0.42	6.26	9.67	10.9	9.02	1.85	0.37

rise to complex spectra, particularly for the most intense $\Delta J = 2, 1, 0$ and -1 transitions at around 490, 545, 590 and 620 nm, respectively [4, 14]. Spectral analysis is generally not attempted in solution, although the magnetic-dipole allowed $^5D_4$–$^7F_5$ transition is most often used in lifetime (Section 2.2) and chiroptical studies. The remaining transitions to the higher lying $^7F_2$, $^7F_1$ and $^7F_0$ bands (see Figures 11.1 and 11.2) are very weak and have not been properly explored for aiding speciation or structural analyses.

2 LUMINESCENCE QUENCHING BY VIBRATIONAL ENERGY TRANSFER

Quenching of the luminescence from excited Tb^{3+} and Eu^{3+} ions in solution occurs by means of a vibrational energy transfer process involving high-energy vibrations of solvent molecules or of the bound ligand. Originally, evidence for the role of vibrational quenching was provided by studies into the effect of substitution of D_2O for H_2O in aqueous solutions of lanthanide ions [28–31]. The substitution of D_2O for H_2O causes no change in the absorption spectrum of a Ln^{3+} ion, so that changes in intensity or lifetime reflect the effect of isotopic substitution on the emissive state [32]. Kropp and Windsor showed that the luminescence of Eu^{3+} and Tb^{3+} complexes was more intense in D_2O than in H_2O [28a, 32] and that the lifetimes were up to an order of magnitude longer in D_2O. Intensity ratios in H_2O and D_2O bore an inverse relationship to the energy gap between the emissive state and the next lower level [29]. Energy transfer occurs to O–H stretching vibrations and the rate is proportional to the number of OH oscillators associated with each Ln^{3+} ion. The probability of an OH (or O–D) oscillator becoming excited from its ground vibrational level v' to a higher level v'' decreases rapidly as v'' increases, owing to the poorer Franck–Condon overlap of the two wavefunctions. Thus, energy transfer to O–D oscillators in Eu/Tb complexes was found to be about 200 times slower than for OH, as a consequence of the lower vibrational frequency of the OD vibration [33]. Energy transfer also becomes less probable as the gap (ΔE) between the emissive and the next lower level increases, consistent with the differing sensitivity of the excited Ln^{3+} ions to vibrational quenching.

These effects are illustrated for Eu^{3+} and Tb^{3+} in Figure 11.6, which highlights the greater sensitivity of excited Eu^{3+} ions to OH quenching, with each OH oscillator acting independently so that the overall effect is a function of the number of bound and closely diffusing water molecules.

2.1 THEORETICAL BACKGROUND

In the general case, the rate constant for depopulation of the lanthanide excited state in water may be partitioned as the sum of the different quenching contributions, as follows:

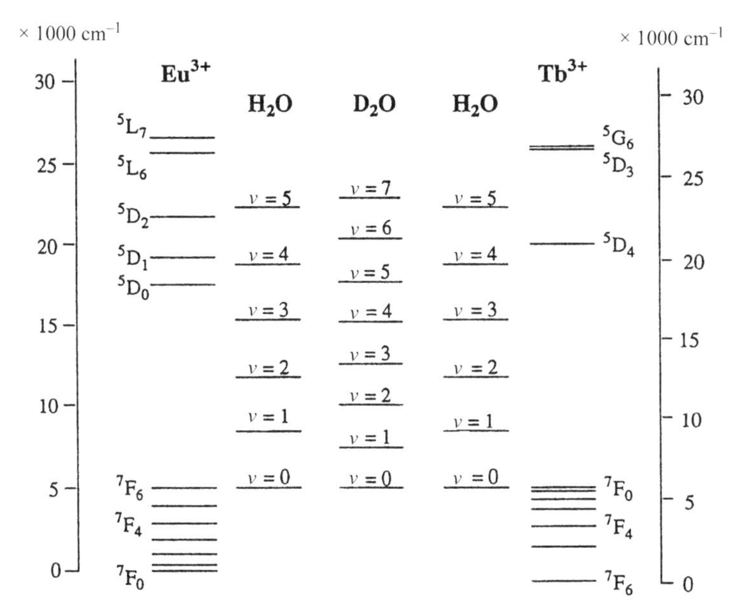

Figure 11.6 Vibrational quenching of Eu^{3+} and Tb^{3+} emissive states by water. An offset has been applied so that the lowest vibrational level of OH/OD is shown at the same energy as the highest level of the ground-state manifold of the Ln^{3+} ion. No anharmonicity is assumed in the vibrational ladder, with $\nu_{OH} = 3405$ and $\nu_{OD} = 2520\ cm^{-1}$.

$$k_{H_2O} = k_{nat} + k_{nr} + \sum k_{XH} + \sum k_{vib'} \tag{11.2}$$

where k_{nat} is the natural radiative rate constant, k_{nr} is the rate constant for non-radiative de-excitation, and $\sum k_{XH}$ and $\sum k_{vib'}$ are the sums of the rate constants for energy transfer to proximate matched XH and other oscillators, respectively. In D_2O, if it is assumed that all exchangeable XH oscillators do not quench, then the equation simplifies by eliminating this term, so that the difference in measured rate constants in H_2O and D_2O will be directly proportional to the quenching effect of the XH oscillators only, as follows:

$$\Delta k = k_{H_2O} - k_{D_2O} = \sum k_{XH} \tag{11.3}$$

Horrocks and Sudnick went on to assume that OH quenching predominated and that only coordinated water molecules contributed to the measured Δk values [26]. By comparison of the Δk values obtained for a range of complexes with the number of water molecules, q, determined by X-ray crystallography, they devised the following relationship:

$$q = A(k_{H_2O} - k_{D_2O}) \quad (11.4)$$

where A is an empirically determined proportionality constant, reflecting the sensitivity of each Ln^{3+} ion to OH_2 quenching. For Tb complexes, $A = 4.2$ ms^{-1} and for Eu, $A = 1.05\ ms^{-1}$ The estimated error was assessed to be ± 0.5 water molecules, a rather high level of uncertainty [26], given the precision of lifetime measurements (± 10 %).

The general energy transfer formula developed by Förster [34] has been proposed to explain the loss of electronic energy from a luminophore to the vibrational energy of a solvent molecules [35]. A multipole–multipole coulombic interaction is postulated wherein dipole–dipole interactions dominate. The rate of energy transfer between the two centers, k_{12}, varies with distance as $1/r^6$ (Equation (11.5)), in which $\bar{v}$ the energy of the resonant transition, $g_1(\bar{v})$ and $g_2(\bar{v})$ are, respectively, the normalized lineshapes of the emission and absorption transitions, and f_1 and f_2 are the oscillator strengths of the emission transition of the lanthanide and the absorption of the vibrational transition of the oscillator, respectively. This relationship highlights the sensitivity of the rate of quenching to the distance between a lanthanide ion and a proximate oscillator [36]:

$$k_{12} = c_{d-d}r^{-6} \quad (11.5a)$$

where:

$$c_{d\text{-}d} = \left[\frac{(3e^4 f_1 f_2)}{8\pi^2 m^2 c^3 \eta^4 \bar{v}^2}\right] \int g_1(\bar{v}) g_2(\bar{v}) d\bar{v} \quad (11.5b)$$

in which m is the electron mass, c is the speed of light, η is the refractive index, and $\bar{v}$ is the energy of the resonant transition.

2.2 QUENCHING BY OH, NH AND RELATED OSCILLATORS: EVALUATION OF q

Other energy-matched, exchangeable XH oscillators may also quench the excited state of Tb^{3+} and Eu^{3+}, including amide NH [37], amine NH [16, 36, 38] and C–H. [36,39, 40]. In the context of complexes used in MRI, only NH groups are significant in this respect, as the other oscillators do not normally undergo deuterium exchange under ambient conditions. These oscillators are often more distant from the Ln^{3+} center in kinetically stable complexes and hence their quenching effect is less marked than the effect of OH oscillators. The effect of distance on the efficiency of vibrational deactivation is illustrated in Figure 11.7 for OH and NH oscillators. It can be seen that at 3.6 Å, vibrational quenching by OH groups is still significant (ca 25 %, a similar effect to that seen in the solid state) [35b], while even at 4.5 Å there is still a discernible effect. The distances correspond closely to typical values for second-sphere and

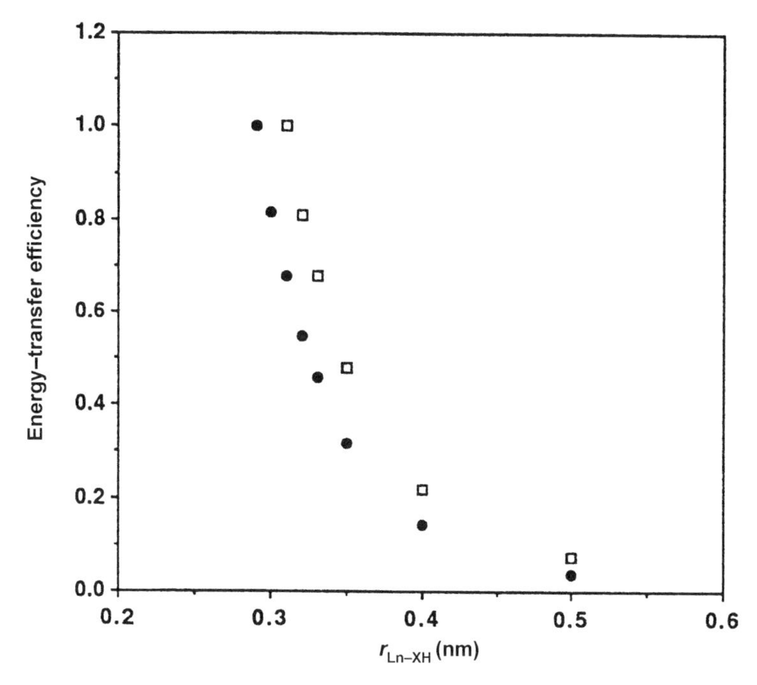

Figure 11.7 Calculated effects of distance on the efficiency of vibrational energy transfer between a Ln^{3+} ion and quenching OH (•) and NH (□) oscillators. The distance of 2.90 Å (0.29 nm) represents the case for a coordinated (inner-sphere) water molecule (i.e. $q = 1$), while r = 3.10 Å (0.31 nm) for maximal Ln–NH quenching; an r^{-6} dependence is assumed (Equation (11.5)).

outer-sphere hydration and highlight the quenching effect of *unbound* water molecules [36].

As a consequence of these effects, a revised protocol for assessing the q-values has been introduced, wherein Equation (11.4) is modified to account for the quenching effect of closely diffusing (unbound) water molecules and, where appropriate, the effect of other exchangeable XH oscillators, notably amide NH groups [36], to give the following:

$$q_{\mathrm{Eu}} = 1.2[(k_{\mathrm{H_2O}} - k_{\mathrm{D_2O}}) - 0.25] \tag{11.6}$$

$$q_{\mathrm{Tb}} = 5[(k_{\mathrm{H_2O}} - k_{\mathrm{D_2O}}) - 0.06] \tag{11.7}$$

These revised equations, include a correction factor for unbound water molecules which was derived by analysis of over 20 related hydrophilic complexes. In addition, and for Eu complexes only, an additional subtraction of 0.075 ms^{-1} should be made, per coordinated amide NH group (Figure 11.7; Ln^{3+} bound-amide NH distance is ca 5 Å). Examples of Eu and Tb complexes for which such measurements have been made are collated in

Table 11.3 Rate constants for depopulation of the excited states of selected europium complexes in H_2O and D_2O (293 K, + 10 %) and derived q-values using Equation (11.6) [36].

Complex	k_{H_2O} (ms^{-1})	k_{D_2O} (ms^{-1})	Δk	Δk_{corr}	q
$[Eu(DOTA)]^-$	1.60	0.53	1.07	0.82	0.98
$[Eu(DTPA)]^{2-}$	1.59	0.44	1.15	0.90	1.08
$[Eu(EDTA)]^-$	2.90	0.48	2.42	2.17	2.60
[Eu **4**]$^-$	0.63	0.48	0.15	0	0
[Eu **6**]$^-$	0.80	0.54	0.26	0.01	0.01
[Eu **7**]$^{3+a}$	1.93	0.46	1.47	0.77	0.92
[Eu **8**]$^{3+}$	1.82	0.47	1.35	0.80	0.97
[Eu **9**]^b	1.32	0.54	0.78	0.45	0.54
[Eu **3a**]	2.27	0.52	1.75	1.50	1.80
[Eu **3b**]	2.38	0.80	1.58	1.33	1.60
[Eu **10**]	1.74	0.47	1.27	0.87	1.04

[a] The primary amide NH oscillators are diastereotopic, and one set of four NH groups are more than 0.35 Å distant from the Eu ion than the other four: an arbitrary reduction of 50 % for their quenching effect has been made (Figure 11.6).
[b] Six other related monoamidetriphosphinate complexes gave q-values in the range 0.4 to 0.7, although NMR and high-resolution emission spectroscopic analysis ($\Delta J = 0$), revealed one predominant (> 90 %) isomer [23, 41].

Table 11.4 Rate constants for depopulation of the excited states of selected terbium complexes in H_2O and D_2O (293 K, ± 10 %) and derived q-values using Equation (11.7).

Complex	k_{H_2O} (ms^{-1})	k_{D_2O} (ms^{-1})	Δk	Δk_{corr}	q
$[Tb(DOTA)]^-$	0.66	0.39	0.27	0.21	1.05
$[Tb(DTPA)]^{2-}$	0.69	0.41	0.28	0.22	1.10
$[Tb(EDTA)]^-$	0.93	0.29	0.64	0.58	2.90
[Tb **4**]$^-$	0.24	0.22	0.02	0	0
[Tb **6**]$^-$	0.34	0.27	0.07	0.01	0.05
[Tb **8**]$^{3+a}$	0.60	0.30	0.30	0.24	1.20
[Tb **9**]$^{a,\ b}$	0.31	0.23	0.08	0.02	0.10
[Tb **10**]^a	0.70	0.43	0.27	0.21	1.05

[a] On the time-scale of the experiment, amide NH/ND exchange does not occur in D_2O [36].
[b] Six other similar mono-amide triphosphinate complexes also gave a q-value of 0.1 [36].

Tables 11.3 (Eu) and 11.4 (Tb). There will still be an error in such analyses, e.g. the presence of local hydrogen-bond donor/acceptor groups and complex hydrophobicity will determine the mean distance of closest approach of diffusing water molecules [36, 41], but it is likely to be less than 20%, particularly for complexes where $q \leq 2$.

2.3 CRITICAL ASSESSMENT OF *q* AND SELECTED EXAMPLES

There are some obvious analogies to be drawn between the quenching interaction of water molecules with an emissive excited state and the relaxivity behavior of gadolinium complexes. In each case, there is a $1/r^6$ dependence on the distance between the lanthanide center and the water proton, and the interactions may be understood in terms of inner-sphere, second-sphere and outer-sphere effects [36]. A consideration of particular examples highlights the importance of this correlation.

Inspection of the data presented in Tables 11.3 and 11.4 (including the footnotes) reveals a marked difference between the correlated q-values of europium ($0.4 < q < 0.7$) and terbium ($q = 0.1$) complexes with a series of monoamidetriphosphinate complexes. Proton and phosphorus-31 NMR studies had revealed the presence of one major (> 90 %) twisted square-antiprismatic diastereoisomeric species, corroborated by high-resolution analysis of the $\Delta J = 0$ transition at 579 nm for the Eu complexes [23, 41]. Related tetraphosphinate complexes take up a similar structure and exhibit q-values of 0.2 (Eu, Tb, Yb), consistent with X-ray structural analyses [22]. The 'partial' hydration of the Eu complexes has been interpreted in terms of a second-sphere of hydration in which the amide carbonyl group may serve as a hydrogen-bond acceptor for a closely diffusing water molecule. The 'break' in hydration at gadolinium is distinct, but is typical of the discontinuities observed in the physical properties of the central Ln^{3+} complexes [42]. Moreover, the Gd complexes of this same class of monoamidetriphosphinates give R_{1p}-values of ca 3 mM^{-1} s^{-1} (298 K, 20 MHz), ^{17}O VT NMR behavior consistent with the absence of a bound water molecule and fitted NMRD profiles that allow the Gd–H distance to be estimated with some confidence. The correlation between the 'NMRD-fitted' distance r and the Δk values ($k_{H_2O} - k_{D_2O}$; *uncorrected* for outer-sphere effects) for the corresponding Eu complexes in quite reasonable (Figure 11.8), suggesting that a 'break' in hydration occurs between Gd and Tb.

When non-integral q-values are measured which fall outside a 20 % error margin, then there may be more than one factor to consider to explain the effect. First, there may be present an equilibrium mixture of complexes of differing hydration states, as is probably the case for $[Eu(EDTA)]^-$ [21], and as defined for certain lanthanide complexes of DOTA [43]. Given that the emissive time-scale for Eu and Tb complexes is of the order of milliseconds and that the rate of water exchange is likely to be faster (although not always the case) [44], then the luminescence-derived q-value represents a weighted time-average figure. Secondly, the sensitivity of q to distance (Figure 11.7 shows that a lengthening of the time-averaged $Ln–OH_2$ bond distance may reduce q significantly, for example, an 0.1 Å lengthening reduces q from 1.0 to 0.82. Crystallographic evidence for $Ln–OH_2$ distances up to 0.13 Å longer than expected in complexes with coordination numbers of nine have been reported [22, 45, 46], and each corresponds to the occurrence of a twisted

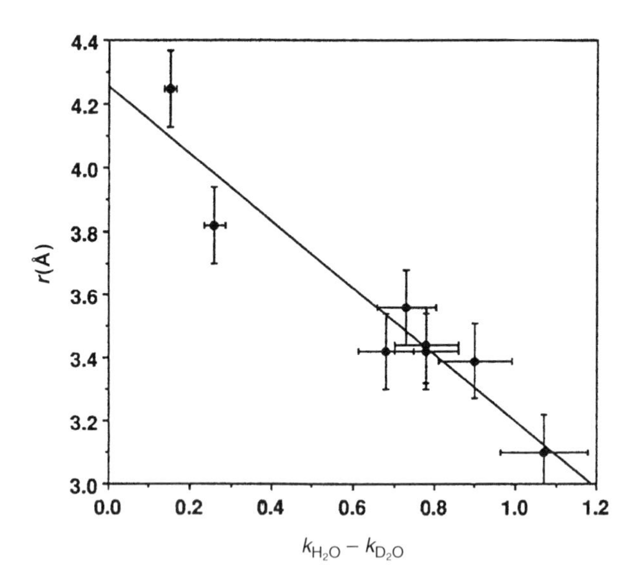

Figure 11.8 Correlation between the NMRD-derived distance r in eight Gd complexes and the luminescence quenching effect of OH oscillators ($k_{H_2O} - k_{D_2O}$) for the corresponding Eu complexes of the tetraphosphinates [Ln **4**]$^-$ and [Ln **6**]$^-$, [Ln(DOTA)$^-$] and five neutral monoamide triphosphinate ligands.

square-antriprismatic structure. Such a situation may occur for the Tb complex of (*RRRR*)-**11** for which $q_{Tb} = 0.60$, and NMR reveals a 2:1 ratio of twisted/regular square antiprismatic isomers [36], while the corresponding Gd complexes exhibit relaxivity values, ^{17}O NMR and NMRD-profile behavior, consistent with a $q = 1$ complex. In addition, the q-values of 1.8 and 1.6 measured for the europium complexes with DO3A (**3a**), and its N–Me analogue (**3b**), could be considered to indicate that one of the two diastereotopic water molecules is , on average, more distant from Eu than the other, in a predominant twisted square anti-prismatic coordination environment [36]. However, the presence of two complexes with $q = 1$ and $q = 2$ may also explain this non-integral q-value [19]. Finally, as discussed above, for cases when $0.3 < q < 0.3 < q < 0.8$, then the presence of a well defined second-sphere of hydration may be indicated – but only after having established the absence of complexes with differing coordination numbers.

3 DETERMINATION OF COMPLEX STABILITY, DYNAMICS AND SPECIATION

The determination of equilibrium stability constants for lanthanide complexes using classical pH metric methods is rendered difficult when the kinetics of formation are slow or the formation constants are very high (see Chapter 6).

Alternative methods have included the use of a chromogenic chelate (arsenazo III) as a competitive spectral indicator [47, 48] and Eu^{3+} luminescence spectroscopy using ligand/ligand or Eu/M^{n+} competition. The $^7F_0$–$^5D_0$ transition in Eu complexes gives rise to a unique spectral band and in the absence of accidental degeneracy, each Eu complex produces a single excitation band. By using pulsed laser excitation, characteristic Eu^{3+} excited-state lifetimes may be measured for a given complex and this technique has been used to measure relative (to Eu^{3+}) binding constants of many different metal ions by competition between Eu^{3+} and other ions for a given ligand [49]. Similarly, differences in the $^7F_0$–$^5D_0$ excitation spectra (Figure 11.9) have been used to determine stability constants for a given ligand, relative to EDTA as a reference ligand in ligand/EDTA competition experiments [50]. The total intensity of the $\Delta J = 0$ band, I_t, represents the sum of the contributions from the two complexes present, according to the following expression:

$$I_t = I_1 + I_2 = k_1[\text{EuL}] + k_2[\text{EuEDTA}]^- \qquad (11.8)$$

where k_1 and k_2 are experimentally determined proportionality constants measured separately on [EuL] and $[EuEDTA]^-$. The $^5D_0$ excited-state lifetime is also a function of each individual complex, e.g. $[EuEDTA]^-$, 315 μs, $[EuDTPA]^{2-}$ 627μs, [Eu-DO3A] 291 μs, [Eu **12**] 265 μs, and [Eu **13**] 591 μs. By using these τ-values, the luminescence decay curve may be fitted to two

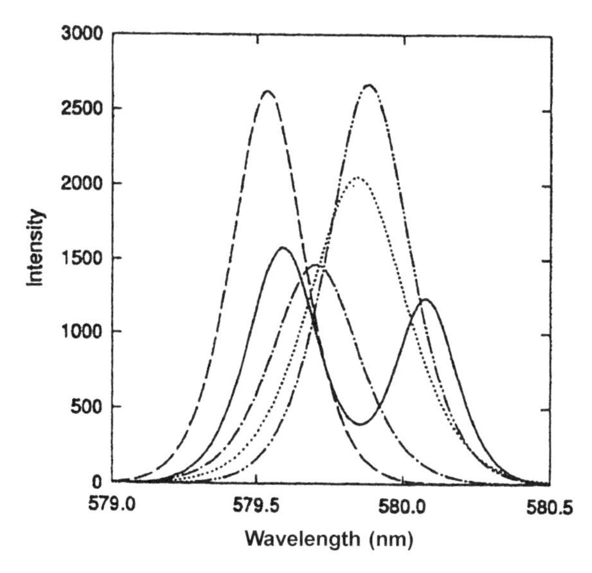

Figure 11.9 Regression analysis fits to the $^7F_0$–$^5D_0$ excitation spectra of various Eu^{3+} complexes: [Eu EDTA]$^-$ (–); [Eu **12**] (–); [Eu DTPA]$^{2-}$ (–···–); [EuDO3A] (–·–·); [Eu **13**] (....) (298 K, [EuL] = 2 μM). Reprinted with permission from [50]. Copyright (1996) American Chemical Society.

exponential functions (provided that τ_1 and τ_2 are sufficiently different) according to Equation (11.9), allowing I_1 and I_2 to be determined and hence the complex concentrations (Equation (11.8)):

$$I_t(t) = I_1 \exp\left(\frac{-t}{\tau_1}\right) + I_2 \exp\left(\frac{-t}{\tau_2}\right) \tag{11.9}$$

The dynamics of the ligand exchange reactions may also be studied by time-resolved luminescence methods [51, 52]. The time dependence of luminescence emission is a function of the rate of chemical exchange, relative to the de-excitation rate. When the chemical exchange rate is of a similar order of magnitude to the luminescent decay rate (Equation (11.10)), the time-dependence of the observed luminescence may be precisely defined mathematically [4, 51, 52].

$$\begin{array}{ccc} \mathrm{LnL} + \mathrm{L'} & \underset{k_b}{\overset{k_a}{\rightleftharpoons}} & \mathrm{LnL'} + \mathrm{L} \\ \downarrow k_{em} & & \downarrow k'_{em} \end{array} \tag{11.10}$$

Such a situation has been observed for water exchange with cationic europium tetraamide complexes e.g. [Eu **8**] in the millisecond regime, and with related emissive ytterbium complexes (λ_{em} 980 nm) on the microsecond time-scale [53].

4 SUMMARY AND OUTLOOK

Measurements of the luminescence decay rate and absorption emission spectra of Eu and Tb complexes afford a great deal of useful information, relating to the structure and dynamics of gadolinium MRI contrast agents. More precise measurements of the complex hydration state, q, have been devised [36], and detailed analysis of the Eu emission spectra in particular, affords unique information on local symmetry and speciation ($^7F_0$–$^5D_0$ transition) [50]. In future, a better understanding of the factors that determine the intensity of hypersensitive $\Delta J = 2$ and $\Delta J = 4$ transitions in Eu emission, should lead to further direct information relating to complex geometry, coordination number and donor atom type. The increased use of circularly polarized emission spectroscopy [54, 55] to simplify spectral interpretation may also be envisaged, particularly as the polarization of emission from racemic complexes may be studied directly by using circularly polarized excitation.

REFERENCES

1. Reisfeld, R. and Jorgensen, C.K., *Lasers and Excited States of Rare Earths*, Springer-Verlag, Berlin, 1977.
2. Gschneinder, K.A. and Eyring, L.R. (Eds), *Handbook on the Physics and Chemistry of the Rare Earths*, Vol. 23, Elsevier, Amsterdam, 1996.

3. Porcher, P. and Caro P., *J. Lumin.*, 1980, **21**, 207.
4. Bunzli, J.-C.G., in *Lanthanide Probes in Life, Chemical and Earth Sciences, Theory and Practice*, Bunzli, J.-C.G. and Choppin, G.R. (Eds), Elsevier, Amsterdam, 1989, Ch. 7, pp. 219–290.
5. Judd, B.R., *Mol. Phys.*, 1959, **2**, 407.
6. Bleaney, B., *J. Magn. Reson.*, 1972, **8**, 91.
7. Golding, R.M. and Pyykko, P., *Mol. Phys.*, 1973, **26**, 1389.
8. Hilmes, G.L. and Riehl, J.P., *Phys. Chem.*, 1983, **87**, 3300.
9. Huskowska, E., Maupin, C.L., Parker, D., Williams, J.A.G. and Riehl, J.P., *Enantiomer*, 1997, **2**, 381.
10. Weissman, S.J., *J. Chem. Phys.*, 1942, **10**, 214.
11. Crosby, G.A., Whan, R.E. and Alire, R.E., *J. Chem. Phys.*, 1961, **34**, 743.
12. Abusaleh, A. and Meares, C.F., *Photochem. Photobiol.*, 1984, **39**, 763.
13. Sabbatini, N., Indelli, M.T., Gandolfin, M.T. and Balzani, U., *J. Phys. Chem.*, 1982, **86**, 3585.
14. Parker, D. and Williams, J.A.G., *J. Chem. Soc., Dalton Trans.*, 1996, 3613.
15. (a) Parker, D., *Coord. Chem. Rev.*, 2000, **205**, 109; (b) Dickins, R.S., Howard, J.A.K., Maupin, C.L., Moloney, J.M., Parker, D., Riehl, J.P., Siligardi, G. and Williams, J.A.G., *Chem. Eur. J.*, 1999, **5**, 1095.
16. Anelli, P.L., Balzani, L., Prodi, L. and Uggeri, F., *Gazz. Chim. Ital.*, 1991, **121**, 359.
17. Geier, G. and Jorgensen, C.K., *Chem. Phys. Lett.*, 1971, **9**, 263.
18. Graeppi, N., Powell, D.H., Laurenzcy, G., Zékány, L. and Merbach, A.E., *Inorg. Chim. Acta*, 1994, **235**, 311.
19. Tóth, E., Ni Dhubhghaill, O.M., Besson, G., Helm, L. and Merbach, A.E., *Magn. Reson. Chem.*, 1999, **37**, 701.
20. Carnall, W.T., in *Handbook on the Physics and Chemistry of the Rare Earths*, Gschneinder, K.A. and Eyring, L.R. (Eds), North Holland, Amsterdam, 1977, Vol. 3, Ch. 24.
21. Bryden, C.C. and Reilley, C.N., *Anal. Chem.*, 1982, **54**, 610.
22. Aime, S., Batsanov, A.S., Botta, M., Dickins, R.S., Faulkner, S., Foster, C.E., Harrison, A., Howard, J.A.K., Moloney, J.M., Norman, T.J., Parker, D., Royle, L. and Williams, J.A.G., *J. Chem. Soc., Dalton Trans.*, 1997, 3623.
23. Aime, S., Botta, M., Parker, D. and Williams, J.A.G., *J. Chem. Soc., Dalton Trans.*, 1995, 2259.
24. (a) Albin, M. and Horrocks, W. de W. Jr, *Inorg. Chem.*, 1985, **24**, 895; (b) Bunzli, J.C.G., Giorgetti, A. and Harrison, D., *J. Chem. Soc., Dalton Trans.*, 1985, 885.
25. Amin, S., Voss, D.A., Horrocks, W. de W. Jr, Lake, C.H., Churchill, M.R. and Morrow, J.R., *Inorg. Chem.*, 1995, **34**, 3294.
26. Horrocks, W. de W. Jr and Sudnick, D.R., *Acc. Chem. Res.*, 1981, **14**, 384.
27. Stcin, G. and Wurzberg, E., *J. Chem. Phys.*, 1975, **62**, 208.
28. (a) Kropp, J.L. and Windsor, M.W., *J. Chem. Phys.*, 1963, **39**, 2769; (b) Heller, A., *J. Am. Chem. Soc.*, 1966, **88**, 2058.
29. (a) Haas, Y. and Stein, G., *Chem. Phys. Lett.*, 1971, **11**, 143; (b) Haas, Y. and Stein, G., *J. Phys. Chem.*, 1971, **75**, 3677.
30. Stein, G. and Wurzberg, E., *J. Chem. Phys.*, 1975, **62**, 208.
31. Horrocks, W. de W. Jr, Schmidt, G.R., Sudnick, D.R., Kittrell, C. and Bernheim, R.A., *J. Am. Chem. Soc.*, 1977, **99**, 2378.
32. Kropp, J.L. and Windsor, M.W., *J. Chem. Phys.*, 1965, **42**, 1599.
33. Haas, Y. and Stein, G., *J. Phys. Chem.*, 1972, **76**, 1093.
34. Förster, Th., *Ann. Phys. (Leipzig)*, 1948, **2**, 55.
35. (a) Ermolaev, V.S. and Sveshnikova, E.B., *Chem. Phys. Lett.*, 1973, **23**, 349; (b) May, P.S. and Richardson, F.S., *Chem. Phys. Lett.*, 1991, **179**, 277.

36. Beeby, A., Clarkson, I.M., Dickins, R.S., Faulkner, S., Parker, D., Royle, L., de Sousa, A.S., Williams, J.A.G. and Woods, M., *J. Chem. Soc., Perkin Trans 2*, 1999, 493.
37. (a) Dickins, R.S., Parker, D., de Sousa, A.S. and Williams, J.A.G., *J. Chem. Soc., Chem. Commun.*, 1996, 697; (b) Parker, D. and Williams, J.A.G., *J. Chem. Soc., Perkin Trans. 2*, 1995, 1305.
38. Wang, Z., Choppin, G.R., Bernardo, P.D., Zanonato, P.L., Portonova, R. and Tolazzi, M., *J. Chem. Soc., Dalton Trans.*, 1993, 2791.
39. (a) Oude-Wolbers, M.P., van Veggel, F.C.J.M., Snellink-Ruel, B.H.M., Hofstraat, J.W., Guerts, F.A.J. and Reinhoudt, D.N., *J. Am. Chem. Soc.*, 1997, **119**, 138; (b) Oude-Wolbers, M. P., van Veggel, F.C.J.M., Hofstraat, J.W., Guerts, F.A.J. and Reinhoudt, D.N., *J. Chem. Soc., Perkin Trans. 2*, 1998, 2141.
40. Hemmila, I., Mukkala, V.-M. and Takalo, H., *J. Fluoresc.*, 1995, **5**, 159.
41. Aime, S., Botta, M., Parker, D. and Williams, J.A.G., *J. Chem. Soc., Dalton Trans.*, 1996, 17.
42. Richens, D.T., *The Chemistry of Aqua Ions*, Wiley, Chichester, 1997.
43. Aime, S., Botta, M., Fasano, M., Marques, M.P.M., Geraldes, C.F.G.C., Pubanz, D. and Merbach, A.E., *Inorg. Chem.*, 1997, **36**, 2059.
44. Aime, S., Barge, A., Botta, M., Howard, J.A.K., Moloney, J.M., Parker, D., de Sousa, A.S. and Woods, M., *J. Am. Chem. Soc.*, 1999, **121**, 5762.
45. Spirlet, M.-R., Rebizant, J., Wang, X., Jin, T., Gilsoul, D., Comblin, V., Maton, F., Muller, R.N. and Desreux, J.F., *J. Chem. Soc., Dalton Trans.*, 1997, 497.
46. Rohovec, J., Vojitsek, P., Hermann, P., Mosinger, J., Zak, Z. and Lukes, I., *J. Chem. Soc., Dalton Trans.*, 1999, 3585.
47. Cacheris, W.P. Nickle, S.K. and Sherry, A.D., *Inorg. Chem.*, 1987, **26**, 958.
48. Kumar, K., Chang, C.A. and Tweedle, M.F., *Inorg. Chem.*, 1993, **32**, 587.
49. Albin, M., Farber, G.K. and Horrocks, W. De W. Jr, *Inorg. Chem.*, 1984, **23**, 1648.
50. Wu, S.L. and Horrocks, W. De W. Jr, *Anal. Chem.*, 1996, **68**, 394.
51. Emolaev, V.L. and Gruzdev, V.P., *Inorg. Chim. Acta*, 1984, **95**, 179.
52. Horrocks, W. de W. Jr, Arkle, V.K., Liotta, F.J. and Sudnick, D.R., *J. Am. Chem. Soc.*, 1983, **105**, 3455.
53. Batsanov, A.S., Beeby, A., Bruce, J.I., Howard, J.A.K., Kenwright, A.M. and Parker, D., *Chem. Commun.*, 1999, 1011.
54. Riehl, J.P. and Richardson, F.S., *Chem. Rev.*, 1986, **86**, 1.
55. (a) Dickins, R.S., Gunnlaugsson, T., Parker, D. and Peacock, R.D., *Chem. Commun.*, 1998, 1643; (b) Aime, S., Botta, M., Howard, J.A.K. and Kataky, R., Lowe, M.P., Moloney, J.M., Parker, D. and de Sousa, A.S., *Chem. Commun.*, 1999, 1047.

INDEX